# Pine Wilt Disease: A Worldwide Threat to Forest Ecosystems

Manuel M. Mota · Paulo Vieira
Editors

# Pine Wilt Disease: A Worldwide Threat to Forest Ecosystems

Springer

*Editors*
Manuel M. Mota
ICAM – Universidade de Évora
Portugal

Paulo Vieira
ICAM – Universidade de Évora
Portugal

ISBN: 978-90-481-7886-5 e-ISBN: 978-1-4020-8455-3



Printed on acid-free paper

9 8 7 6 5 4 3 2 1

springer.com

# Preface

Pine wilt disease (PWD) is unquestionably a major threat to forest ecosystems worldwide. After seriously affecting Eastern Asian countries, the challenge is now in Europe, following its detection in Portugal in 1999 and its subsequent spread.

For foresters, these were really very bad news and, in order for adequate action to be taken, scientists had to teach politicians about the seriousness of the problem. That is never an easy task, but it was successfully done at that time, mainly by the continued effort of Professor Manuel Mota.

The challenge of having political decisions based on good science is fundamental for the success of any program, but especially in difficult situations such as those arising by the introduction of harmful organisms in new ecosystems. The success of the dialogue between science and policy requires intelligent partners from each side, which is not always necessarily the case...

Examples of lack of recognition of problems raised by science are unfortunately abundant throughout the history of science. The recent recognition of the efforts of the Intergovernmental Panel on Climate Change (IPCC) and Al Gore with the Nobel Prize, and the continued failure in taking appropriate actions by major political players is a dramatic modern example of the difficulty of this dialogue...

These are some of the reasons why I think that this book plays a fundamental role in the issue of pine wilt disease: Firstly, the book addresses a very important problem that threatens the ecological and economical balance of many forested areas worldwide. Secondly, it assembles contributions of the best specialists worldwide in the various facets of the problem. Thirdly, it summarises knowledge in an attempt to make it useful for adequate action. Finally, it provides insights for future developments in scientific research.

I had already the privilege of addressing some words of recognition to the participants of the PWD Conference at the Gulbenkian Foundation in Lisbon in July 2006 where I was very much impressed with the importance and the quality of the contributions. As Director of the Portuguese Forest Services (DGRF) at that time and until November 2007, I must stress that this Conference was very instrumental in setting the stage for discussions and for the planning of new strategies in dealing with the issue of the presence of the pinewood nematode in Portugal.

For these new strategies important scientific contributions were given by Edmundo Sousa (another relevant participant to the Conference) in addressing the issues related to the spread of the insect vector.

I would like to take this opportunity to stress my recognition for the tremendous and unique work done by the team of the Forest Services, coordinated by José Manuel Rodrigues (a contributor to this book) that resulted in the establishment in early 2007 of a clearcut belt 430 km long and 3 km wide around the affected zone. This strategy was adopted and financially supported by the European Commission, which sent several missions to Portugal. The mission leaders, in November 2007, expressed satisfaction with the success of this extremely difficult operation.

We do not know, at this moment, what will be the final effectiveness of this new strategy. We do know, however, that without this major effort the hope of success for the eradication program would be minimal. I am sure that, until November 2007, the Forest Services did everything it was possible, by dedicating human and financial resources, by taking the necessary risks and facing lack of understanding, to ensure that appropriate action was taken, making use of the best science available. And I hope that this effort will be continued with the same strength in the future...

I am certain that this book, absolutely necessary for those who want to act in a responsible manner in the very difficult combat against the spread of pine wilt disease, constitutes also a fundamental contribution for the advancement of science and for the stimulus of future research in this field.

For the courageous editors and for the excellent contributors to the conference and the book, I would like to reiterate my sincere recognition and gratitude, that I am sure will be shared by all of those who care for forests around the world.

Thank you!

Lisbon
February 2008

Francisco Castro Rego

# Contents

# Contributors

Pierre Abad
Interactions Biotiques en Santé végétale, UMR1301 INRA-UNSA-CNRS, Sophia Antipolis, France, pierre.abad@sophia.inra.fr

Süleyman Akbulut
Duzce University, Duzce Forest Faculty, 81300 Düzce, Turkey, akbulutsuleyman@yahoo.com

Eric Allen
Natural Resources Canada, Canadian Forest Service, Victoria, V8Z 1M5, British Columbia, Canada, eallen@pfc.cfs.nrcan.gc.ca

Susana C. Arcos
Dept. Agroecologia, Instituto de Ciencias Agrarias, CCMA, Consejo Superior de Investigaciones Científicas Serrano 115 dpdo, Madrid, 28006, scobacho@ccma.csic.es

Maria Arias
Dept. Agroecologia, Instituto de Ciencias Agrarias, CCMA, Consejo Superior de Investigaciones Científicas Serrano 115 dpdo, Madrid, 28006, avelion.garcia@ciemat.es

Ismail Baysal
Abant Izzet Baysal University, Duzce Forest Faculty, 81620, Düzce, Turkey

António Bello
Dept. Agroecologia, Instituto de Ciencias Agrarias, CCMA, Consejo Superior de Investigaciones Científicas Serrano 115 dpdo, Madrid, 28006, antonio.bello@ccma.csic.es

Dale Bergdahl
Department of Forestry, University of Vermont, USA, dbergdah@uvm.edu

Helen Braasch
Kantstraße 5, D-14471 Potsdam, Germany, h.braasch@t-online.de

Helmut Bröther
Department of Customer Protection, Agriculture and Land Consolidation, Brandenburg, Steinplatz 1, D-15806 Zossen, OT Wünsdorf, Germany, ute.schoenfeld@lvlf.brandenburg.de

Wolfgang Burgermeister
Julius Kuehn Institute, Federal Research Centre for Cultivated Plants, Institute for Epidemiology and Pathogen Diagnostics, Messeweg 11-12, D-38104 Braunschweig, Germany

Chantal Castagnone
Interactions Biotiques en Santé Végétale, UMR 1301 INRA-UNSA-CNRS, Sophia Antipolis, France, chantal.castagnone@sophia.inra.fr

Philippe Castagnone-Sereno
Interactions Biotiques en Santé Végétale, UMR 1301 INRA-UNSA-CNRS, Sophia Antipolis, France, philippe.castagnone@sophia.inra.fr

L. Castresana
Dept. Entomologia, Escuela Superior de Ingenieros de Montes, 28040, Madrid, Spain

Xianfeng Chen
Technical Centre, Ningbo Entry-exit Inspection and Quarantine Bureau, 9 Mayuan Road, Ningbo, Zhejiang, China

Youngjin Cheong
Forest Pest and Disease Division, Korea Forest Research Institute, Seoul 130-712, Korea

Miguel A. Dìez-Rojo
Dept. Agroecologia, Instituto de Ciencias Agrarias, CCMA, Consejo Superior de Investigaciones Científicas Serrano 115 dpdo, Madrid, 28006, diez.rojo@ccma.csic.es

Jinyan Dong
Key Laboratory for Conservation and Utilization of Bioresources, Yunnan University, 650091 Kunming, China

Jonathan D. Eisenback
Dept. Plant Pathology, Physiology and Weed Science, Virginia Polytechnic Institute and State University, 103 Price Hall, Blacksburg, VA 24061, USA, jon@vt.edu

Miguel Escuer
Dept. Agroecologia, CCMA, CSIC Serrano, Madrid, Spain

Hugh Evans
Forest Research, Alice Holt Lodge, Farnham, Surrey, GU10 4LH, England, hugh.evans@forestry.gsi.gov.uk

Sam Evans
Forest Research, Alice Holt Lodge, Farnham, Surrey, GU10 4LH, England, sam.evans@forestry.gsi.gov.uk

Cécile François
Interactions Biotiques en Santé végétale, UMR 1301 INRA-UNSA-CNRS, Sophia Antipolis, France

Kenji Fukuda
Institute of Environmental Studies, Graduate School of Frontier Sciences, The University of Tokyo, Kashiwanoha 5-1-5, Kashiwa-shi, Chiba 277-8563, Japan

Kazuyoshi Futai
Laboratory of Environmental Mycoscience, Graduate School of Agriculture, Kyoto University, Sakyo-ku, Kyoto 606-8502 Japan, futai@kais.kyoto-u.ac.jp

A. Garcìa-Àlvarez
Dept. Agroecologia, CCMA, CSIC, Serrano 115 dpdo, 28006 Madrid, Spain

Margaret Green
Centre for Plant Health, Canadian Food and Inspection Agency, Sidney, V8L 1H3, British Columbia, Canada, greenmg@inspection.gc.ca

Jianfeng Gu
Technical Centre, Ningbo Entry-exit Inspection and Quarantine Bureau, 9 Mayuan Road, Ningbo, Zhejiang, China, gujf@nbciq.gov.cn

Koichi Hasegawa
Institute for Biological Function, Chubu University, 1200 Matsumoto, Kasugai 487-8501, Japan, Laboratory of Environmental Mycoscience, Graduate School of Agriculture, Kyoto, Japan hasegawaelegans@hotmail.com

Mike Hodda
CSIRO Entomology, GPO Box 1700, Canberra ACT 2601, Austrália, mike.hodda@csiro.au

Ute Hoyer-Tomiczek
Federal Research & Training Centre for Forests, Natural Hazards & Landscape (BFW), Department for Forest Protection, Seckendorff-Gudent-Weg 8, A-1131 Vienna, Austria, ute.hoyer@bfw.gv.at

Leland Humble
Natural Resources Canada, Canadian Forest Service, Victoria, V8Z 1M5, British Columbia, Canada, lhumble@pfc.cfr.nrcan.gc.ca

Takefumi Ikeda
Department of Forest Science, Kyoto Prefectural University, Shimogamo-hangicho, Sakyo, Kyoto 606-8522, Japan, tikeda@kpu.ac.jp

Makihiko Ikegami
Forest Research, Alice Holt Lodge, Farnham, Surrey, GU10 4LH, England, makihiko.ikegami@forestry.gsi.gov.uk

Ryûtarô Iwata
Department of Forest Science and Resources, College of Bioresource Sciences, Nihon University, Fujisawa, Kanagawa 252-8510, Japan

Witold Karnkowski
Plant Health and Seed Inspection Service Central Laboratory, ul. Żwirki i Wigury 73, 87-100 Toruń, Poland, w.karnkowski@piorin.gov.pl

Miho Kawai
Environmental Molecular Biology Laboratory, RIKEN, Saitama 351-0198, Japan

Hajime Kosaka
Hokkaido Research Center, Forestry and Forest Products Research Institute, Sapporo 062-8516, Japan, hkosaka@ffpri.affrc.go.jp

Keiko Kuroda
Forestry and Forest Products Research Institute, Kansai Research Center, Momoyama, Fushimi, Kyoto 612-0855, Japan, keiko@affrc.go.jp

Yan-xue Lai
Forest and Plant Quarantine Station of Ningbo City, Ningbo, 315000 Zhejiang, China, zhangyf@cnluye.com

Cornelia Lange
Max-Planck Institute for Genetics, Berlin, Germany, lange_c@molgem.mpg.de

Simon A. Lawson
Department of Primary Industries and Fisheries, Forestry Building, Gate 3 80 Meiers Rd, Indooroopilly, QLD 4068, Australia, simon.lawson@dpi.qld.gov.au

Isabel Leal
Natural Resources Canada, Canadian Forest Service, Victoria, V8Z 1M5, British Columbia, Canada, ileal@pfc.cfs.nrcan.gc.ca

Guohong Li
Key Laboratory for Conservation and Utilization of Bioresources, Yunnan University, 650091 Kunming, China

Marc Linit
College of Agriculture, Food & Natural Resources, University of Missouri, Columbia, MO, USA

Tadashi Maehara
Experimental Station at Tanashi, The University Forests, Graduate School of Agricultural and Life Sciences, The University of Tokyo, Tokyo 188–0002, Japan

Noritoshi Maehara
Tohoku Research Center, Forestry and Forest Products Research Institute, 92–95 Nabeyashiki, Shimo-Kuriyagawa, Morioka, Iwate 020–0123, Japan, maehara@ffpri.affrc.go.jp

Yasuharu Mamiya
5-6-8 Kitanodai, Hachioji, 192-0913 Tokyo, Japan (Formerly Tamagawa University) cbl01545@nifty.com

J. Pedro Mansilla
Estación Fitopatológica do Areeiro Provincial de Pontevedra, Subida a la Robreda, Lourizán, 36153 Pontevedra, Spain

C. Martínez
Dept. Agroecologia, CCMA, CSIC Serrano, Madrid, Spain

Kai Metge
Institute for Biosafety of Genetically Modified Plants, Julius Kuehn-Institute (JKI), Federal Research Centre for Cultivated Plants, Quedlinburg, Germany, kai.metge@jki.bind.de

Johji Miwa
Institute for Biological Function, Chubu University, Kasugai, Japan, Laboratory of Developmental Genetics, Graduate School of Bioscience and Biotechnology, Chubu University, 1200 Matsumoto, Kasugai 487-8501, Japan

Manuel M. Mota
NemaLab-ICAM, Departmento de Biologia, Universidade de Évora, 7002-554 Évora, Portugal, mmota@uevora.pt

Katsunori Nakamura
Tohoku Research Center, Forestry and Forest Products Research Institute, Morioka 020-0123, Japan, knakam@affrc.go.jp

Lila Nambiar
Biosciences Division, Victorian Department of primary Industries, Private Bag 15, Ferntree Gully Delivery Centre, Victoria 3156 Australia

A. Notario
Dept. Entomologia, Escuela Superior de Ingenieros de Montes, 28040 Madrid, Spain

Ivan Pascoe
Biosciences Division, Victorian Department of primary Industries, Private Bag 15, Ferntree Gully Delivery Centre, Victoria 3156 Australia

Janina Polomski
Swiss Federal Research Institute WSL, Birmensdorf 8903, Switzerland, Janina.polomski@wsl.ch

Daniel Rigling
Swiss Federal Research Institute WSL, Birmensdorf 8903, Switzerland

Lee Robertson
Dept. Agroecologia, Instituto de Ciencias Agrarias, CCMA, Consejo Superior de Investigaciones Científicas Serrano 115 dpdo, Madrid, 28006, lee.r@ccma.csic.es

José Manuel Rodrigues
Direcção-Geral dos Recursos Florestais, Direcção de Serviços de Desenvolvimento Florestal, Divisão de Protecção e Conservação Florestal, Av. João Crisóstomo, 26-28, 6° andar, 1069-040 Lisboa, Portugal, prolunp@dgrf.min-agricultura.pt

Michael Rott
Centre for Plant Health, Canadian Food and Inspection Agency, Sidney, V8L 1H3, British Columbia, Canada, rottm@inspection.gc.ca

Alexander Ryss
Zoological Institute, RSA, Universitetskaya nab. 1, 199034 St. Petersburg, Russia, nema@zin.ru

Daisuke Sakaue
Experimental Station at Tanashi, The University Forests, The University of Tokyo, Midori-cho 1-1-8, Nishitokyo, Tokyo 188-0002, Japan

R. Sanz
Dept. Agroecologia, CCMA, CSIC, Serrano 115 dpdo, 28006 Madrid, Spain

Shiroma Sathyapala
Biosecurity New Zealand, Ministry of Agriculture and Forestry, PO Box 2526, Wellington, New Zealand, shiroma.sathyapala@maf.govt.nz

Ute Schönfeld
Department of Customer Protection, Agriculture and Land Consolidation, Brandenburg, Steinplatz 1, D-15806 Zossen, OT Wünsdorf, Germany, ute.schoenfeld@lvlf.brandenburg.de

Fritz Schweingruber
Swiss Federal Research Institute WSL, Birmensdorf 8903, Switzerland

Metin Serin
The Ministry of Environment and Forestry, Western Black Sea Forest Research Institute Bolu, Turkey

Sangchul Shin
Forest Pest and Disease Division, Korea Forest Research Institute, Seoul 130-712, Korea

Etsuko Shoda-Kagaya
Department of Forest Entomology, Forestry and Forest Products Research Institute, Matsunosato 1, Tsukuba, Ibaraki 305-8687, Japan, eteshoda@affrc.go.jp

David Smith
School for Ecosystem Science, University of Melbourne, 500 Yarra Bvd, Richmond, Victoria 3121, Australia

Ivan Smith
School for Ecosystem Science, University of Melbourne, 500 Yarra Bvd, Richmond, Victoria 3121, Australia

Rina Sriwati
Plant Protection Department, Agriculture Faculty, Syiah Kuala University, Darussalam, Banda Aceh and Laboratory of Environmental Mycoscience, Graduate School of Agriculture, Kyoto University, Sakyo-ku, Kyoto 606-8502 Japan
Present address: Plant Protection Department, Agriculture Faculty, Syiah Kuala University, Banda Aceh, Indonesia, rin_aceh@yahoo.com

Kazuo Suzuki
College of Bioresource Sciences, Nihon University, Kameino 1866, Fujisawa, Kanagawa 252-8510, Japan

Shuhei Takemoto
Laboratory of Environmental Mycoscience, Graduate School of Agriculture, Kyoto University, Sakyo-ku, Kyoto 606-8502 Japan, ts.kais@hotmail.com.co.jp

Yuko Takeuchi
Laboratory of Environmental Mycoscience, Graduate School of Agriculture, Kyoto University, Sakyo-ku, Kyoto 606-8502 Japan

Christian Tomiczek
Federal Research & Training Centre for Forests, Natural Hazards & Landscape (BFW), Department for Forest Protection, Seckendorff-Gudent-Weg 8, A-1131 Vienna, Austria, christian.tomiczek@bfw.gv.at

Shin Ugawa
Institute of Environmental Studies, Graduate School of Frontier Sciences, The University of Tokyo, Kashiwanoha 5-1-5, Kashiwa-shi, Chiba 277-8563, Japan, shin-u@nenv.k.u-tokyo.ac.jp

Paulo Vieira
NemaLab-ICAM, Departmento de Biologia, Universidade de Évora, 7002-554 Évora, Portugal, pvieira@uevora.pt

YU Wang
Department of Forest Protection, Faculty of Forest Resources and Environment, Nanjing Forestry University, Longpan Road, Nanjing, 210037, China, njwy01@hotmail.com

John Webster
Department of Biological Sciences, Simon Fraser University, Canada

Chunxia Xie
Forestry Academy of Jiangsu Province, Nanjing, 211153, China

Fuyuan Xu
Forestry Academy of Jiangsu Province, Nanjing, 211153, China

Keqin Xu
Forestry Disease & Pest Control Station of Jiangsu Province, Nanjing, 210013, China

Toshihiro Yamada
Experimental Station at Tanashi, The University Forests, The University of Tokyo, Midori-cho 1-1-8, Nishitokyo, Tokyo 188-0002, Japan

Akiomi Yamane
1859-4 Hananoi, Kashiwa, Chiba 277-0812, Japan

Beşir Yüksel
Abant Izzet Baysal University, Duzce Forest Faculty, 81620, Düzce, Turkey

Jiancheng Zhang
Technical Centre, Ningbo Entry-exit Inspection and Quarantine Bureau, 9 Mayuan Road, Ningbo, Zhejiang, China

Keqin Zhang
Key Laboratory for Conservation and Utilization of Bioresources, Yunnan University, 650091 Kunming, China, kqzhanglll@yahoo.com.ch

Pei Zhang
Forestry Academy of Jiangsu Province, Nanjing, 211153, China

# Part I
# Pine Wilt Disease: Global Issues, Trade and Economic Impact

**John Webster and Manuel Mota**

## Summary

Pine wilt disease (PWD) is perhaps the most serious threat to pine forests worldwide. Since it's discovery in the early XXth century by Japanese forest researchers, and the relationship with its causative agent, the pinewood nematode (PWN) *Bursaphelenchus xylophilus*, in the 1970s, PWD has wreaked havoc wherever it appears. Firstly, in the Far East (Japan, China and Korea) and now, more recently in 1999, in the EU (Portugal).

The forest sector in Portugal plays a major role in the Portuguese economy with a 12% contribution to the industrial gross domestic product, 3.2% of the gross domestic product, 10% of foreign trade and 5% of national employment. Maritime pine (*Pinus pinaster*) is one of the most important pine productions, and industrial activity, such as the production of wood and resin, as well as coastal protection associated with sand dunes. Also, stone pine (*Pinus pinea*) plays an important role in the economy with a share derived from the exports of high-quality pineon seed. Thus, the tremendous economical and ecological impact of the introduction of a pest and pathogen such as the PWN, although as far as is known, the only species susceptible to the nematode is maritime pine.

Immediately following detection, the research team involved (Univ. Évora, INIAP) informed the national plant quarantine and forest authorities, which relayed the information to Brussels and the appropriate EU authorities. A task force (GANP), followed by a national program (PROLUNP) was established. Since then, national surveys have been taking place, involving MADRP (Ministry of Agriculture), the University of Évora and several private corporations (e.g. UNAC). Forest growers in the area are particularly interested and involved since the area owned by the growers organizations totals 700 000 ha, and is largely affected by PWD. Detection of the disease has led to serious consequences and restrictions regarding exploration and commercialization of wood. A precautionary phytosanitary strip, 3 km wide, has been recently (2007) established surrounding the affected area. The Portuguese government, through its national program PROLUNP, has been deeply involved since 1999, and in conjunction with the EU (Permanent Phytosanitary Committee, and FVAO) and committed to controlling this nematode and the potential spread to the rest of the country and to the rest of the EU.

The global impact of the presence of *Bursaphelenchus xylophilus* or the threat of its introduction and the resulting pine wilt disease in forested areas in different parts of the world is of increasing concern economically. The concern is exacerbated by the prevailing debate on climate change and the putative impact this could have on the vulnerability of the world's pine forests to this disease. The scientific and regulatory approach taken in different jurisdictions to the threat of pine wilt disease varies from country to country depending on the perceived vulnerability of their pine forests to the disease and/or to the economic cost due to lost trade in wood products.

Much of the research surrounding pine wilt disease has been located in the northern hemisphere, especially in southern Europe and in the warmer, coastal, Asian countries. However, there is an increased focus on this problem also in those countries in the southern hemisphere where plantations of susceptible pine have been established over the years. The forestry sector in Australia and New Zealand are on "high alert" for this disease and are practicing strict quarantine procedures at all ports of entry for wood products. As well, there is heightened awareness, as there is worldwide, for the need to monitor wood packaging materials for all imported goods.

In carrying out the necessary monitoring and assessment of products for *B. xylophilus* and its vectors substantial costs are incurred especially when decisions have to be made rapidly and regardless of whether the outcome is positive or negative. Australia's response recently to the appearance of some dying pines in a plantation illustrated the high sensitivity of some countries to this disease. Some $200 000 was spent on the assessment in order to save a potential loss of millions of dollars to the disease. This rapid, co-ordinated response to the report was for naught, because once identified it was found not to be *B. xylophilus*. This illustrates the particular importance of taking the responsibility at all levels of management to secure the site and the need of a rapid, reliable diagnostic method for small nematode samples for use in the field.

Australia is particularly concerned about the vulnerability of its 1million hectares of planted forests, 80% of which are *Pinus* species, to attack from incursions of one or more species of the insect vector. *Monochamus alternatus* incursions in wood pallets have been reported from Brisbane, Queensland. The climate of this part of Australia is such that the *Pinus* plantations are particularly vulnerable to the potential outcome of such incursions, and the state of Queensland is developing a risk management strategy and a proactive breeding programme in response to this putative threat.

New Zealand has 1.6 million hectares of planted forests, and 89% of the commercial forest is *Pinus radiata*. Although the climate where these forests are located tends to be somewhat cooler than that in Australia the potential for establishment and development of the disease in that country is believed to be high. The passage alone of 200 000 $m^3$/year of wood packaging through New Zealand ports is itself sufficient to require response. The potential incursion of insect vectors of pinewood nematode through the port system is regarded as high and is monitored carefully.

The enormous expansion of global trade and the continued use of unprocessed/inadequately-processed wood for packaging purposes is a challenge for all trading nations as such wood packaging material often harbours disease or pest species. The extent of this problem is readily illustrated by the expanding economies and exports of countries in south-east Asia, China, Japan and Korea have significant areas of forestland infested with *B. xylophilus*. These countries too are among the largest exporting countries of manufactured goods. Despite the attempts of authorities to ensure that only properly treated wood is used in the crating and packaging of goods *B. xylophilus* and/or its insect vector infested materials is being recorded at ports worldwide. This reminds us, therefore, of the ease with which this nematode pest can gain access to forest lands in new geographic locations through inappropriate use, treatment or monitoring of wood products. It especially highlights the necessity to find an alternative to using low-grade lumber for packaging purposes.

Lest we should believe that all wood products are always carriers of *B. xylophilus* and its vectors, it should be remembered that international trade of all kinds has occurred for thousands of years and that lumber-born pests and diseases do not have worldwide distribution. Other physico-biological factors have a significant role in the occurrence, establishment and sustainability of a disease. The question is often raised as to why the whole of southern Europe doesn't already have *B. xylophilus* and pine wilt disease. European countries have traded with countries that are infested with *B. xylophilus* for hundreds of years. Turkey is an example of a country that appears to be highly vulnerable to pine wilt disease due to its extensive forests in the warm, southern region where the vector, *Monochamus galloprovincialis*, occurs. However, there is no record of the presence of *B. xylophilus* occurring there despite the importation of substantial quantities of wood from several countries

In many respects, Portugal illustrates both the challenge and the dilemma. In recent times *B. xylophilus* was discovered there in the warm coastal region. The research, administrative and quarantine authorities responded rapidly and *B. xylophilus* appears to have been confined to the region in which it was found. The rapid response would seem to have "saved the day" for Portugal. Nevertheless, it raises again the long-standing questions, how long had *B. xylophilus* been in Portugal before it was found? If Lisbon was the port of entry, which seems very likely, why had *B. xylophilus* not entered Lisbon many years earlier and established populations and the pine wilt disease? Will the infestation in Portugal be sustainable and will it spread or will it die out within a few years? We still do not have sufficient understanding of the biology of this pest to know the answers to these questions.

# National Eradication Programme for the Pinewood Nematode

José M. Rodrigues

## Introduction

The pinewood nematode (PWN),[1] *Bursaphelenchus xylophilus*, is listed as a harmful organism to plants or plant products by the European Union (EU) (Annex II, Council Directive 2000/29/EC of 8 May 2000). Its introduction into and spread within all Member States must be banned. This organism, the causal agent of pine wilt disease (PWD), is a serious pest and pathogen of forest tree species, in particular among the genus *Pinus*; its presence in the territory of a member state obliges the country to notify the partners and to adopt immediate safeguard measures. The subjects of contamination are plants of the genus *Abies, Cedrus, Larix, Picea, Pinus, Pseudotsuga*, and *Tsuga*, with the exception of their fruits and seeds, and wood of conifers (Coniferales), originating from non-European countries.

In May 1999, the PWN was detected in Portugal, in dead maritime pine stands located in the Setúbal Peninsula. Following Council Directive 2000/29/EC, the Portuguese authorities informed the European Community and implemented a phytosanitary strategy with the purpose of controlling and eradicating the pest, a programme known as the National Eradication Programme for the Pinewood Nematode (PROLUNP). At the EU level, the situation has been discussed at the Permanent Phytosanitary Committee. Since the pest was recorded in Portugal, several inspection missions have been carried out by the Food and Veterinary Office (DG SANCO). The legal basis for the implementation of this Program is the Executive-Law n.° 154/2005 (Sept. 6th), which establishes the general phytosanitary

J.M. Rodrigues
Direcção-Geral dos Recursos Florestais, Direcção de Serviços de Desenvolvimento Florestal, Divisão de Protecção e Conservação Florestal, 1069-040 Lisboa, Portugal
e-mail: prolunp@dgrf.min-agricultura.pt

[1] The PWN has had devastating effects on pines forests in East Asian countries, as in Japan, for instance. The nematode is transported as fourth-stage dispersal juveniles by cerambycid beetles of the genus *Monochamus*; in Portugal the PWN was found associated with the species *M. galloprovincialis*, which can attack and infect healthy trees and colonise weakened trees with its offspring.

M.M. Mota, P. Vieira (eds.), *Pine Wilt Disease: A Worldwide Threat to Forest Ecosystems*, 

rules for Portugal, and Regulation n.º 103/2006 (Feb. 6th) as amended by Regulation n.º 815/2006 (Aug. 16th) and Regulation n.º 321/2007 (March 23rd).

## Phytosanitary Strategy

In general, the pursued phytosanitary strategy, delineated to avoid the dispersion of the disease, has been the elimination of decline symptomatic trees,[2] identified through the execution of surveys (during the autumn-winter period), complemented with the control of the insect vector population (during the spring-summer period) and the control of coniferous wood flows (during all year).

Even though PROLUNP covers all mainland Portugal, the fact that the PWN is confined to a certain region, led to the definition of a Demarcated Area (DA), subdivided into an Affected Zone (AZ),[3] a Buffer Zone (BZ)[4] and the remainder of the territory, the Free Zone (FZ)[5] in which risk areas can be found, i.e. places where conifer wood (raw and processed) is stored, and subject of periodic monitorization (Fig. 1). Two critical locations (CL), i.e. clearly delimited areas in which there is a

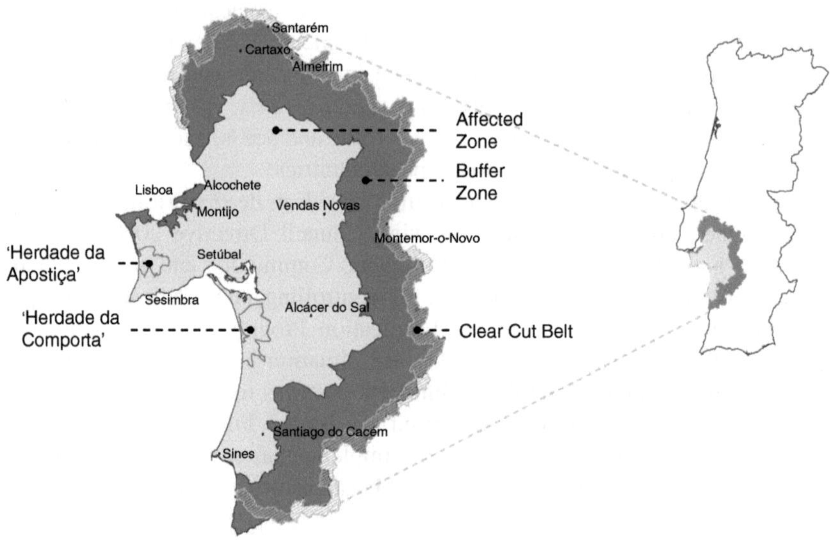


**Fig. 1** PROLUNP territorial coverage. The Demarcated Area (Affected Zone and Buffer Zone), Clear Cut Belt and Critical Locations are depicted

---

[2] Decline symptomatic trees – trees found to be infested by the PWN, showing symptoms of poor health, or located in salvage areas (cf. Annex to Commission Decision 2006/133/EC).

[3] Affected Zone – area in which the pine wood nematode is known to occur.

[4] Buffer Zone – area surrounding the Affected Zone, of no less than 20 km width, where the pine wood nematode is not known to occur.

[5] Free Zone – Area of the territory in which the PWN does not occur.

higher incidence of decline symptomatic trees, are located within the Affected Zone, namely Herdade da Comporta and Herdade da Apostiça.

The Affected Zone covers, currently, 510.000 ha and is surrounded by a Buffer Zone (BZ) of approximately 500.000 ha. The sum of both (1.010.000 ha) constitutes the Demarcated Area (DA), which is subjected to periodic survey, eradication and insect vector control actions and where all forestry activities relating to conifers are subjected to intensive control. The Demarcated Area has changed over the years as result of the evolution of the disease (Fig. 2).

The results of the 2005/2006 surveying and eradication campaign indicated a considerable increase of symptomatic trees throughout the Demarcated Area. This increase was even more evident in the Affected Zone. Several samples collected from the Buffer Zone tested positive for PWN and therefore the Affected Zone and the Demarcated Area limits were redefined. Furthermore, it was decided to create a corridor free from the PWN and its vector host trees, *Picea orientalis, Pinus halepensis, P. nigra, P. nigra laricio, P. pinaster, P. radiata* and *P. sylvestris*, in the periphery of the Demarcated Area, the Clear Cut Belt (CCB), with the purpose of minimizing the possibilities of disease dispersion, as proposed in the 2006 Action

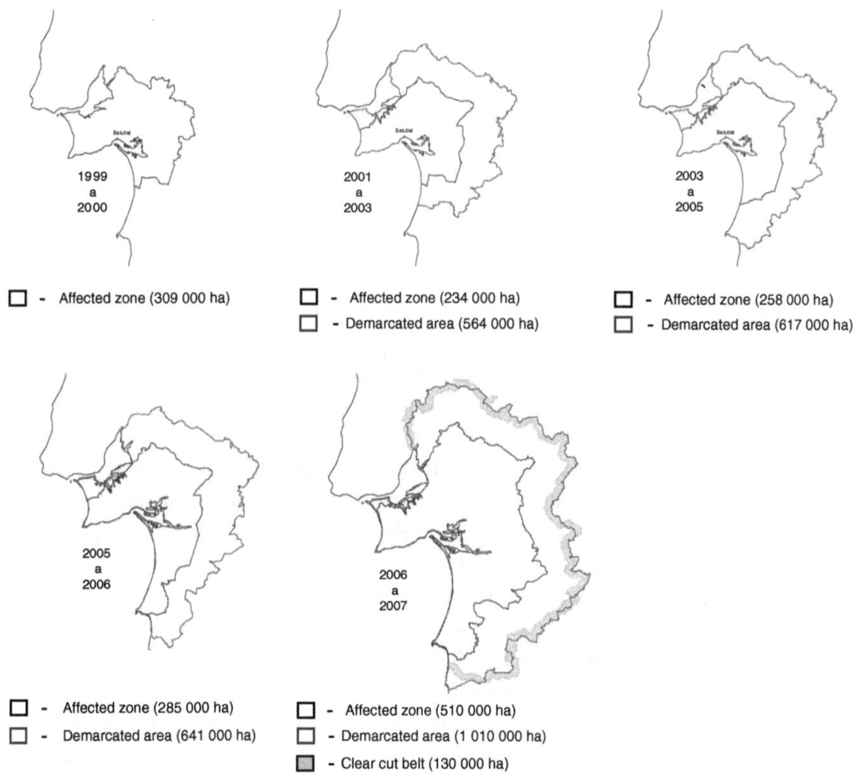


**Fig. 2** Evolution of the Demarcated Area and Affected Zone

Plan, presented to the Standing Committee on Plant Health and in accordance with Article 2 of the Commission Decision 2006/133/EC (revised July 2006).

## *Survey in the Free Zone*

The survey carried out in this area aims to monitor conifer forests, focusing in particular on surrounding areas where there is a lot of wood risk materials, whether wood in its natural state or processed wood (risk areas). It also aims to inspect and assess conifers located in permanent plots in each risk area, test all the collected material for PWN in duly accredited laboratories for this purpose and ensure that the methods and procedures provided for in the EU monitoring protocol for PWN are applied correctly.

The survey activities in the Free Zone will also include areas which may be highly attractive for breeding of *Monochamus galloprovincialis*, especially those surrounding the demarcated area, where the survey of coniferied stands was intensify by establishing 200 extra plots. In these plots, samples are taken also from non-symptomatic trees at different heights, including canopy level, and are incubated in order to screen for the presence of PWN.

## *Survey and Eradication in the Demarcated Area*

The aim of the monitoring and the eradication actions within the Demarcated Area is to detect and eliminate all the trees showing symptoms of decline. This area is divided for survey purposes into 136 units of approximately 7 500 ha each. The symptomatic trees are registered within a specific and appropriate geo-referenced matrix, using specific 150 ha maps (Fig. 3) and screening analysis for the PWN presence carried out, in all the identified trees, for the ones located in the Buffer Zone, and in a sample of randomly chosen trees, for the ones located in the Affected Zone. Surveying has been conducted mainly by Forest Owners' Associations, as their knowledge regarding the local sensibilities is valuable. Samples were collected from symptomatic and non-symptomatic trees, taken at different heights, including canopy level, in order to screen for the presence of PWN; some of those samples were incubated (see Chapter 4. for details). Monitoring work is usually expected to start around November and to be concluded in the beginning of February, starting from the periphery to the interior of the Demarcated Area.

The eradication activity consists in the elimination of all the conifers identified by the surveying action guaranteeing the destruction/processing of all the felled trees, according to the Law (Regulation of the Ministry of Agriculture, Rural Development and Fisheries n. ° 103/2006, Feb. 6th, as amended by Regulation n. ° 815/2006, Aug. 16th, and Regulation n. ° 321/2007, March 23rd). The forest owners, farmers or usufructuaries are accountable for the eradication and are informed about the basic lines of action via public notices sent to the local administration, published

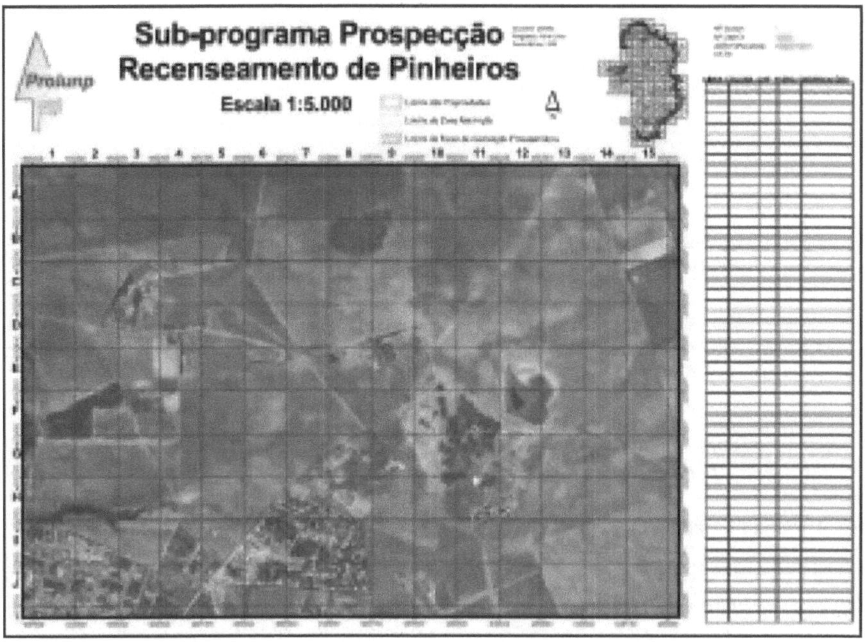


**Fig. 3** Example of a surveying 150 ha map

in newspapers and available online; when the owners do not perform the actions themselves, the government must replace them. To do so, it is necessary to sub-contract the eradication services from private companies.

The eradication procedure is expected to guarantee the felling and elimination of all trees identified with decline symptoms, during the period of December 1st to April 1st, the non-flying period of the insect vector's life cycle.

## *Implementation of a Barrier Free from PWN Vector Hosts (Clear Cut Belt)*

The establishment of a Clear Cut Belt intended to set up a corridor (3 km wide) free from PWN vector hosts, roughly following the limits of the most recently defined Demarcated Area, mostly in the Buffer Zone. In this corridor, with an area of about 130 000 ha, all conifers regarded as hosts of *M. galloprovincialis* must be detected, located and eliminated, both declining and healthy ones. To do so, it was necessary to sub-contract the eradication services from private companies.

The Commission Decision 923/2006/CE (Dec. 13th), created a financial contribution for 2006 and 2007 to cover expenditure incurred by Portugal for the purpose of controlling the PWN, considered compensation payment of a compensation for the value of the wood to tree owners or beneficial owners.

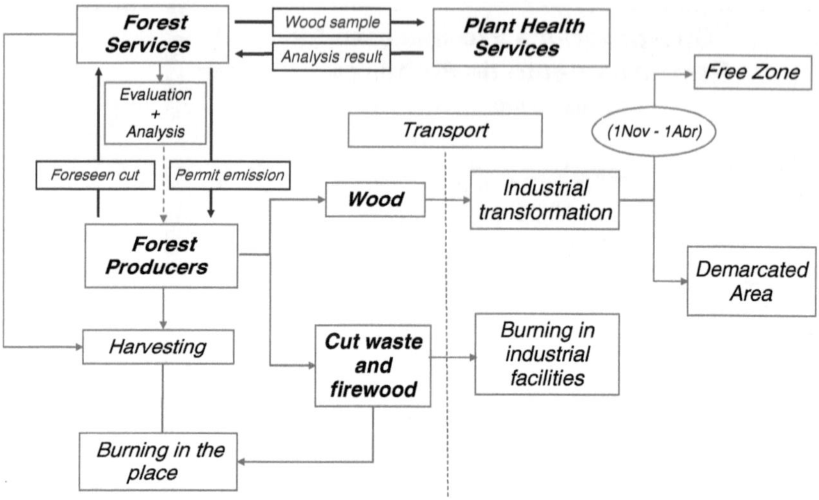


**Fig. 4** Wood flow control system

## *Insect Vector Control*

Besides de identification and elimination of decline symptomatic trees, the control of the insect vector populations can be used as an additional strategy to control PWN dispersion. This has been done through the use of a network of traps set along the outer limit of the Affected Zone, which capture the insect during its flight period (spring and summer). Research is in progress regarding the development of a more appropriate and more effective trap.

## *Inspection and Control of Coniferous Wood*

PROLUNP set up a wood trace back system, which compels owners to apply for a conifers' felling and transport permit in the Demarcated Area. After inspection, the phytosanitary inspectors authorize the cuts (permit emission) and the wood destination (Fig. 4), granted that the notification is in accordance with the legal disposaitions. The inspectors also control the authorized destinations in order to assure the fulfillment of phytosanitary measures.

## The Disease in Portugal

The number of decline symptomatic trees, potentially infested with PWN, has been increasing since the disease was detected, a trend which is not confirmed by the 2006/2007 data. In what concerns the Buffer Zone, the number of eradicated trees has been generally the same along the years (Fig. 5). However, it is important to note

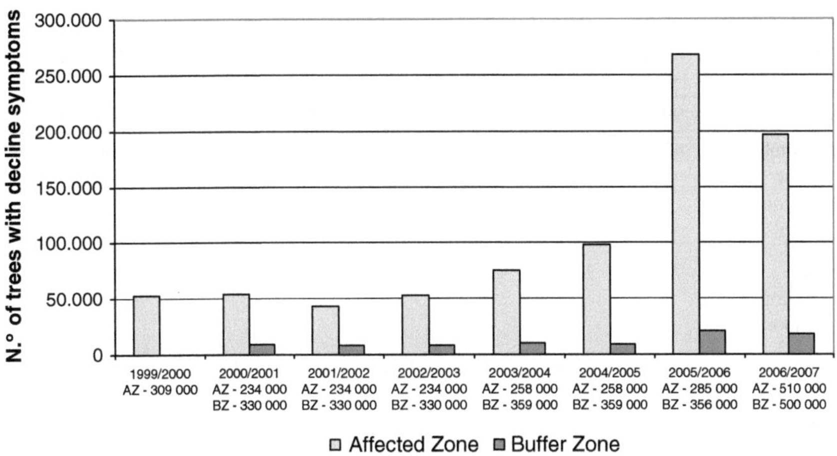


**Fig. 5** Decline symptomatic trees evolution in the Demarcated Area. Also depicted is the area covered by the Affected Zone and by the Demarcated Area, in ha

that the Affected Area has been changing along the years and also that the decline symptomatic trees are not necessarily infested with the PWN, research showing that the conifers' decline causal agents, in the region, are rather diverse, biotic and abiotic. This can be deduced by the analysis of the graphic presented on Fig. 6 that shows a distinct trend in both indexes [Number of decline symptom trees/DA pine stand area] and [Estimated number of positive trees/Number of decline symptomatic trees], along the different campaigns. This suggests that other decline causal agents, rather than PWN, are present and might be responsible for the decline increase.

In the Affected Zone, decline symptomatic trees are concentrated in some important production areas, the "critical locations". Table 1 indicates the evolution of total number of decline symptomatic trees identified in the Affected Zone, as well

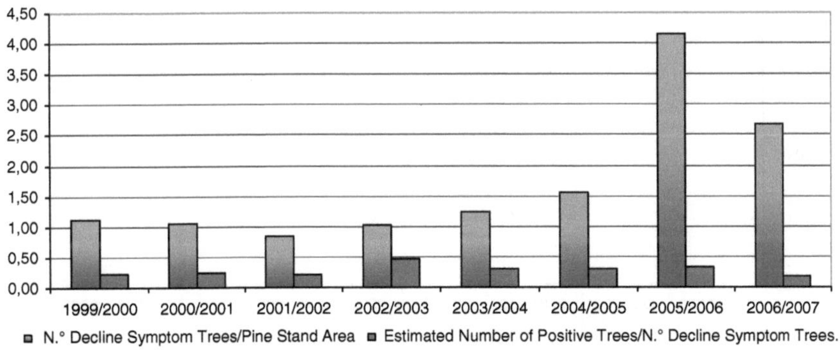


**Fig. 6** Evolution of the indexes [Decline symptomatic trees/DA Pine stand area] and [Estimated number of positive trees/Decline symptomatic trees]

**Table 1** Evolution of the number of decline symptomatic trees in the Affected Zone and in the Critical Locations, Comporta e Apostiça. For reference the estimated number of maritime pines, in the Affected Zone, is 7 millions

| CAMPAIGN | 99/00 | 00/01 | 01/02 | 02/03 | 03/04 | 04/05 | 05/06 | 06/07 |
|---|---|---|---|---|---|---|---|---|
| Total number of decline symptomatic trees in AZ[(1)] | 45.531 | 57.402 | 46.068 | 57.061 | 71.107 | 95.302 | 240.097 | 163.892 |
| Critical Location of Comporta | 7,42% | 21,88% | 49,55% | 54,55% | 59,82% | 64,57% | 60,00% | 58,33% |
| Critical Location of Apostiça | 14,34% | 9,76% | 3,15% | 7,51% | 3,12% | 4,66% | 5,14% | 1,09% |
| Critical Locations (Total) | 21,75% | 31,64% | 52,70% | 62,07% | 62,93% | 69,23% | 65,14% | 59,42% |
| Remaining Affected Zone | 78,25% | 68,36% | 47,30% | 37,93% | 37,07% | 30,77% | 34,86% | 40,58% |

[(1)] For the sake of comparison, the percentages shown refer to the Affected Zone limits stated on the Regulation n.º 1572/2003, from December, 27th, 258.000 ha.

as the percentual evolution of these numbers considering the critical locations and the reminder Affected Zone.

A general analysis of the Demarcated Area, shows that although the absolute number of symptomatic trees in the Affected Zone had increased, there has been a percentual reduction of this number in the Critical Locations, where 59, 42% of the DA decline symptomatic trees are located, as shown in Fig. 7.

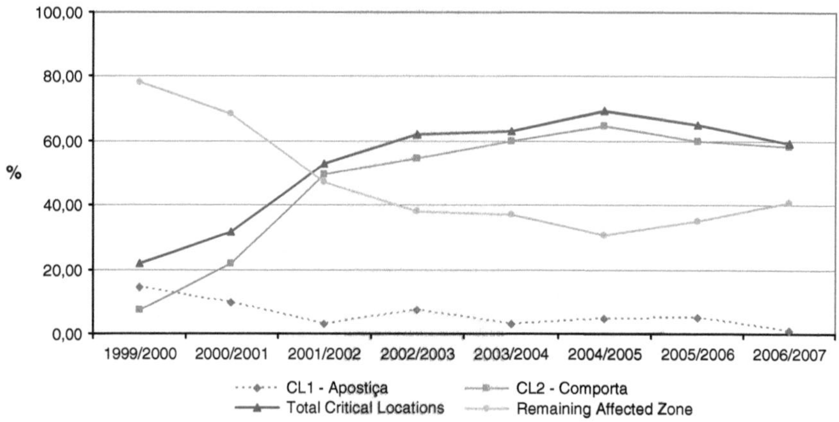


**Fig. 7** Decline symptomatic trees evolution in the Demarcated Area

## The 2006/2007 Survey/Eradication Campaign

A total number of 218 895 trees was identified as symptomatic, occupying 28 667 ha, from which 4 595 were located in the Clear Cut Belt. In the Demarcated Area, 214 300 decline symptomatic trees have been detected, from which 17 770 were located in the Buffer Zone and 196 530 in the Affected Zone. A total number of 5 797 samples were collected and analysed in order to screen for the presence of PWN, from which 1 232 were located in the Affected Zone, 3 703 in the Buffer Zone and 862 in the Clear Cut Belt. Details regarding the number of samples collected from symptomatic and non-symptomatic trees, at DBH height or at the canopy level and the number of samples incubated, are provided in Table 2.

In 2006/2007, 249 samples tested positive for the PWN in the Affected Zone. No positive samples were found in the Clear Cut Belt and in the Free Zone.

The total number of eradicated trees was 1 202 601, from which 218 895 were located at the Demarcated Area and 983 706 at the Clear Cut Belt; in this corridor, a large number of trees with DBH < 10 cm have been detected and eradicated (3758 054). Table 3 summarize the 2006/2007 survey and eradication campaign.

The number of eradicated trees was far beyond the number initially estimated, in what regards non-symptomatic trees. In the Clear Cut Belt there was 4 741 760 trees cut (4 041 760 trees more than initially estimated), including 3 758 054 specimens with DBH < 10.

**Table 2** Number of samples collected in the different PROLUNP set regions, at Diameter at Breast Height (DBH) and at canopy level; it is also presented the number of samples incubated

| *N.º* of samples collected | Free Zone | | | Demarcated Area | | Clear Cut Belt |
|---|---|---|---|---|---|---|
| | Risk Areas (1193 plots) | Lisbon (78 plots) | 200 plots | Affected Zone | Buffer Zone | |
| From symptomatic trees at DBH level | 167 | 8 | 412 | 809 | 3 505 | 473 |
| From symptomatic trees at canopy level | | 0 | 68 | 85 | 102 | 264 |
| From non-symptomatic trees at DBH level | 60[1] | 37 | 467 | 297 | 51 | 113 |
| From non-symptomatic trees at canopy level | 0 | 0 | 129 | 41 | 45 | 12 |
| Incubated | 0 | 0 | 1.010 | 126 | 147 | 276 |
| **TOTAL** | **227** | **45** | **1 076** | **1 232** | **3 703** | **862** |

[1] Include 58 samples collected from material stored at Risk Areas.

**Table 3** 2006/2007 survey and eradication campaign results

| 2006/2007 campaign | Demarcated Area | | Clear Cut Belt |
|---|---|---|---|
| | Affected Zone | Buffer Zone | |
| N.° of symptomatic trees identified | 196 530 | 17 770 | 4 595 |
| N.° of decline symptomatic trees eradicated | 196 530 | 17 770 | 4 595 |
| N.° of non-symptomatic trees eradicated | Not applicable | Not applicable | 983 706 + 3 758 054 |

## Actions Planned

Actions to be implemented will vary according with the area of intervention (Free Zone and Demarcated Area), as follows:

Free Zone Survey
Demarcated Area
Survey
Eradication
Insect Vector and Scolitids' Control
Inspection and Control
Forest Reconversion
Public Awareness
Research and Development

# Incursion Management in the Face of Multiple Uncertainties: A Case Study of an Unidentified Nematode Associated with Dying Pines Near Melbourne, Australia

**Mike Hodda, David Smith, Ian Smith, Lila Nambiar and Ian Pascoe**

**Abstract** In late November 1999, dying pine trees were observed near the docks in Melbourne. The cause was initially identified as *Bursaphelenchus xylophilus*, the pinewood nematode. However, it was soon discovered that it was another nematode, *Bursaphelenchus hunanensis*, which was associated with the dying pine trees. Very little was known about the biology or pathogenicity of this species, except that it had never before been recorded in Australia. Other dying trees were soon discovered with the nematode, and deciding on an appropriate response became a critical issue. This paper describes the subsequent events in the face of the uncertainty regarding the pathogenicity of the nematode, its origin, its vector, and its biology, particularly dispersal. More general principles can be drawn from this experience regarding the management of incursions of pinewood and other nematodes. There may also be important lessons regarding spread of nematodes associated with wood and insects.

## Introduction

This paper is about the response to an apparent incursion of an exotic nematode of uncertain pathological effect. The nematode is from the genus *Bursaphelenchus*, which includes the pinewood nematode (*B. xylophilus*), a major quarantine pest. The nematode was originally thought to be *B. xylophilus*, but was instead the little known species *B. hunanensis*. This paper describes how the situation was handled, particularly in the light of uncertainty about the biological characteristics of the nematode, its pathogenicity, and its relationship to the local fauna. It is important to document such events so that successful and unsuccessful incursions can be compared, the processes of invasion and of becoming a pest can be better understood, and improved responses to future incursions can be planned and implemented. Some

M. Hodda
CSIRO Entomology, GPO Box 1700, Canberra ACT 2601, Australia
e-mail: mike.hodda@csiro.au

M.M. Mota, P. Vieira (eds.), *Pine Wilt Disease: A Worldwide Threat to Forest Ecosystems*, 

general issues in the management of incursions are discussed. The benefit-cost ratio of action is calculated as indicating that erring on the cautious side was justified in this case, so that taking appropriate action to contain and eradicate the potential pest was the correct response. The eradication campaign shows that with an appropriate, adaptive response to incursions in place, eradication of nematodes can be achieved. This, too, is an important result in considering the most appropriate responses to the incursion of exotic nematodes.

## The Situation and Initial Response

In January 2000, government officers from a local Council reported the rapid decline of a mature pine tree (*Pinus halepensis*) in the botanic gardens at Williamstown, near the main port of the city of Melbourne, Australia (37°51′S 144°53′E). The tree was reported as having declined rapidly, with the needles turning yellow to brown and the twigs becoming dry and brittle. Symptoms first appeared in early summer and developed over a 4–6 week period. The dead pines retained their needles. These symptoms are very similar to those of Pine Wilt Disease (PWD) (Evans et al., 1996).

In February, wood samples were taken from the tree and submitted to the local diagnostic service. Initially the cause was thought to be a fungal disease, but the only fungi extracted after extensive sampling were *Sphaeropsis* spp. isolated from the branches and trunk. *Sphaeropsis* spp. are known as shoot blight, and were formerly in the genus *Diplodia. Sphaeropsis* spp. are known to cause severe damage only in trees under stress from unfavourable environmental conditions, and often kill only current-season buds, shoots, and 2nd-year cones. Cankers and resin exudations on infected shoots, branches and main stems are also characteristic of *Sphaeropsis* infections. However, there were none of these symptoms, and no other fungal pathogens could be found. Foliage chloride was tested because the tree was near the coast, but this was below the threshold considered harmful for pines.

It was decided that further investigations were warranted for two reasons. First, because the tree was located near a port handling a large volume of international cargo, and there was concern about incursions of exotic pests and pathogens. Second, there are large areas of plantation pine forests located in the state. Pines are the basis of a large plantation softwood industry producing about 10 million cubic metres of timber per year. Furthermore, the only species of the genus *Pinus* in Australia are introduced, and subject to few diseases: this is one of the reasons for the large plantation area. There is thus a strong interest in any disease—local or exotic—which may affect this resource.

Following the failure to find any fungi likely to have killed the tree, the local diagnostic service searched for nematodes. Samples of wood were cut into chips and the nematodes were extracted over several days using Whitehead-Hemming trays (Hooper, 1986). Large numbers of nematodes were found of several different types. One of the types was identified as possibly being *B. xylophilus*, the cause of PWD.

This species had never been found in Australasia before (MacLeod et al., 1994), and is a major quarantine pest (Evans et al., 1996; EPPO, 2003). The tentative diagnosis was forwarded to the federal quarantine authorities, who convened a meeting of the advisory committee on plant health. The committee requested that more nematodes be extracted and forwarded to the main nematode diagnostic laboratory in Australia, at CSIRO Entomology in the city of Canberra.

At the CSIRO laboratory, three types of nematodes were found in the wood samples, but the absence of *B. xylophilus* could be confirmed definitely based on the morphology of the nematodes, within the statistical restrictions of sampling a large volume of material. All of the Aphelenchida present differed from *B. xylophilus* in basic characters, such as stylet length, tail shape and the development of the anus. Two of the nematodes were identified as belonging to the genera *Aphelenchoides* and *Ektaphelenchus*. No species in these genera are pathogenic to trees (Hodda, 2003), so identification to species was not attempted at this stage. The third type of nematode was identified morphologically as probably belonging to the species *B. hunanensis*. There were no adult males present, which made positive identification problematic. Attempts to culture the nematode using the methods devised for *B. xylophilus* (e.g. Bolla and Jordan, 1982; Braasch et al., 1995, 1999a; Hoyer et al., 1998) were unsuccessful, including using both local fungi and fungal cultures used for several *Bursaphelenchus* species, which were obtained from Dr Thomas Schroeder (BBA, Braunschweig, Germany). The diagnosis of *B. hunanensis* raised several issues about the differential diagnosis of the genus *Bursaphelenchus* from *Aphelenchoides* and *Laimaphelenchus*. These issues have also arisen in other studies of aphelenchid nematodes (Braasch, 2004; **?**, **?**). There have been no molecular studies of this species.

*B. hunanensis* has been recorded in the literature only from Hunan Province, China, where it was associated with dead *Pinus massoniana* Lamb (Yin et al., 1988). The only known records of any species of the genus *Bursaphelenchus* in Australia were as follow.

1. An unidentified species was found on *Hyleops glabratus*, the hoop-pine stitch beetle (Coleoptera, Curculionidae) on 7 July 1972 about a thousand kilometres away in another state (north-eastern NSW). This record was from subtropical wet forest and not associated with dying trees. The record was from the specimen database of the Queensland Museum, Brisbane, but the specimen was no longer in good condition, and the identification was doubtful.
2. A species described as close to *B. sexdentati* was found on *Ips grandicollis* which were attacking *Pinus taeda* in a pine plantation in similar areas to the first report (north-eastern NSW) (Stone, 1990; Stone and Simpson, 1990, 1991).

There were few other records of sampling for nematodes in trees of any sort in Australia (Hodda, 2003), and sampling of other, healthy trees nearby did not find *B. hunanensis*.

The conclusion from the data available at this stage was that *B. hunanensis* was probably exotic, and possibly associated with pathology of the trees in some way. However, neither of these conclusions could be definitive given the data available. This sort of situation may be more common in quarantine situations than is often thought, given the increasing recognition of cryptic species, races, biotypes, pathotypes and other hitherto unknown structure within what were previously considered uniform "species", as well as the vast number of unknown taxa. Nematodes are not the only organisms where this may happen.

## Response Stage Two

The advisory committee on plant health decided that further action was justified. Several factors were taken into account in deciding this: that *B. hunanensis* was apparently exotic, the serious quarantine pest status of one species in the genus *Bursaphelenchus*, plus the uncertain pest status of some others, the high monetary value of nearby pine forests, and the value of an immediate response. A detailed evaluation of these factors in the decision is presented in the section "evaluation of response".

The infested tree was isolated, then the above-ground parts were cut down and carefully removed in a sealed truck to the local garbage tip, where it was burnt. The roots were excavated for a radius of about a metre around the stump and to a similar depth, then removed in a sealed truck for deep burial at the same tip.

The next steps were to ascertain if there was any spread of the nematode, and identify any insect vectors involved. Light traps were set up around the area where possible, and checked regularly.

In deciding where to sample for the nematode, data on the flight radius for vectors of *B. hunanensis* were sought. Unfortunately, there is no data available for this species, as is the case for many species of the genus *Bursaphelenchus*, so the data for *B. xylophilus* was used. This raises another issue where the best-known species in a genus is almost invariably the most severe pest, not necessarily the most representative. This issue is discussed further in the section "evaluation of response".

Known and potential insect vectors of *B. xylophilus* include at least 40 species of beetles (Coleoptera): 19 species of the genus *Monochamus* (pine sawyers), 10 species in 9 other genera of the family Cerambycidae, 4 species in genera of Curculionidae, and a species of Buprestidae (Hodda, 2006). The flight range of most of these is, again, unknown, so again the only information available comes from the species that are the most severe economic pests. *Monochamus alternatus* are the main vector in Japan, and generally disperse less than 100 m (Ido and Kobayashi, 1977). In Japan, about 75% of beetles are recaptured within 100 m of release, but dispersal of 2.5 Km has been recorded (Ido and Kobayashi, 1977; Fujioka, 1993; Yoshimura et al., 1999; Takasu et al., 2000). The annual expansion of range for *B. xylophilus* on *M. alternatus* was estimated at 2–15 Km (Togashi et al., 2004). The issue of dispersal is also discussed further in the section "evaluation of response".

Sampling of all trees within a 1 Km radius, with sampling of dead or dying coniferous trees only within a further radius of 4 Km, was decided upon as combining a

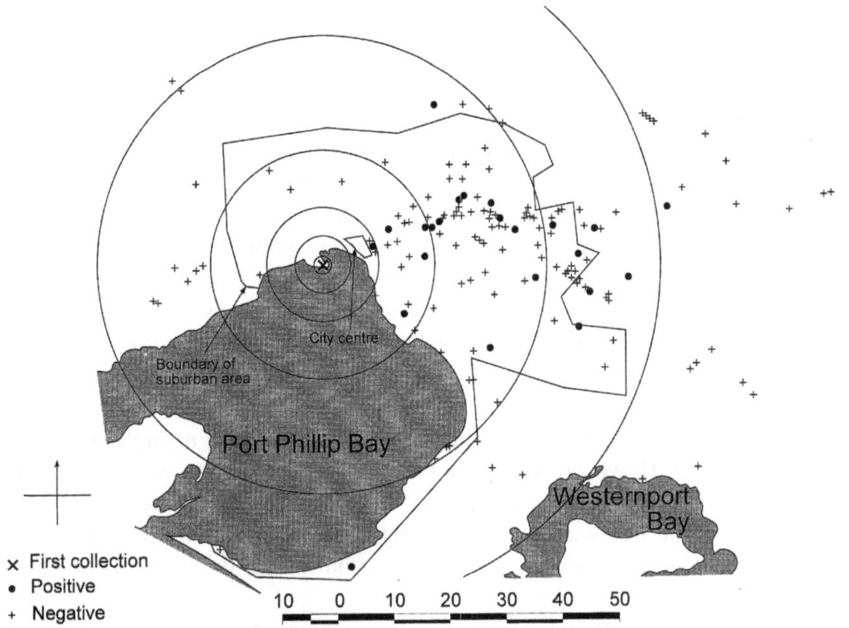


**Fig. 1** Trees sampled for nematodes around Melbourne. Concentric circles or arcs correspond to distances of 1, 5, 10, 20, 40 and 60 Km from the first tree

high probability of containing all or most nematodes if vectored by *Monochamus* (Fig. 1). In deciding this, the assumption was that this was a recent introduction. It was assumed that if the introduction caused significant pathology and occurred some time in the past, symptoms would have been noticed before they were. The known pathology of *B. xylophilus* makes this assumption somewhat tenuous. Disease symptoms only occur above a certain temperature, and are more likely if trees are under stress, particularly from drought, high temperatures, chemical compounds or shading (Tanaka, 1975; Rutherford and Webster, 1986; Kaneko, 1989; Rutherford et al., 1990; Evans et al., 1996; Kawaguchi et al., 1999; Mamiya, 1999; Braasch, 2000). Melbourne winters and springs are cool and damp, so few symptoms would be expected until early summer, even if introduction was up to 6 months previously.

The other factor in choosing this sampling scheme was feasibility. In a basically urban environment, sampling of all trees within the 1 Km radius was considered likely to be feasible, but sampling every tree within a radius of 5 Km was thought unrealistic. There were no estimates of the number of coniferous trees in the area on which to base this assessment.

To complete the sampling, trees in the 1 Km core area were located visually by driving all local streets. All were then sampled for nematodes by cutting discs from any branches which appeared unhealthy, as well as taking small cores from the trunk at various heights. Whitehead-Hemming trays were used for extraction of nematodes. Using these methods, nematodes were detected in none of the 21 trees located within the 1 Km radius study area, nor in any of the trees sampled in the 1–5 Km radius area. (Cypress trees on the foreshore were dying back due to high chloride in foliage, the result of salt spray.)

Samples from pine forests from other states of Australia were taken by local forestry agencies to confirm that *B. hunanensis* really was exotic. Pines in poor condition were preferentially sampled in Western Australia, South Australia, Tasmania, NSW and Queensland using similar methods to those used in Victoria, and submitted to the CSIRO laboratory in Canberra. None of the 10 samples had any of the species isolated from the original dying tree. This was consistent with the nematodes being exotic, although it was hardly a sufficient number of samples to prove this was so with any degree of certainty. This issue is discussed further in the section "evaluation of response".

At this stage (March 2000), it seemed that there was only a single tree which had been infested. However, in May, a single dying *Pinus* spp. was reported approximately 10 Km from the original tree, and on investigation found to contain *B. hunanensis*. The tree was immediately removed using similar protocols to the tree at Williamstown. No single cause of death other than the nematode could be found, despite extensive testing.

A survey of all dead or dying coniferous trees within a 5 Km radius of the second tree with *B. hunanensis* was commenced immediately, with trees located as before from the roads. In this survey, a further tree containing *B. hunanensis* was located near the edge of the 5 Km radius, and 15 Km from the original tree. It was becoming obvious that the assumptions regarding dispersal were inconsistent with observations, so a new strategy was adopted. Aerial surveys were conducted by helicopter for any coniferous tree with symptoms within a radius of 50 Km of the original tree, and the public of the entire state were invited in print, radio and television to report coniferous trees which had died rapidly. Large areas of largely native (*Eucalyptus* spp.) forest east of Melbourne were also surveyed visually for obviously diseased trees from the main roads. Light and pheromone traps were deployed throughout Melbourne in an attempt to catch any vector, particularly if it too was exotic. The measures continued through June and July.

All 110 trees identified in the aerial surveys and 36 trees reported by the public were investigated for the presence of nematodes. Trees reported by the public were located as far as 500 Km away. A total of 33 trees contained *B. hunanensis*, the furthest 60 Km from the original tree and presumed source at the docks (Fig. 1). All were removed as soon as practicable after large samples of wood from the trunk and branches were removed. These large wooden billets were placed in drums to recover potential vectors.

The exotic cerambycid *Arhopalus rusticus* and the globally widespread *Ips grandicollis*, both potential vectors for PWD (Lieutier and Vallet, 1982; Linit et al., 1983), were found in the light and pheromone traps. A few *A. rusticus* were found in one of the infested trees. No *Monochamus* were found in the light or pheromone traps. No nematodes were found in extensive searches of the exterior and interior of the beetles. *B. hunanensis* was common, but not universal in suspect trees, and no *B. xylophilus* were found.

## Response Stage Three

Surveillance was maintained for the following 18 months, particularly for the following two summers (December 2000–March 2001, and December 2001–March 2002). A further 40 trees were tested because of rapid death during the summer of 2000–2001, with only two having *B. hunanensis*. These trees were removed. During the summer of 2001–2002, only 5 trees were tested and none had *B. hunanensis*. *B. hunanensis* was not isolated from any healthy tree, nor was it found in all dying trees.

## *Pathogenicity of* **B. hunanensis**

Attempts to evaluate the pathogenicity of *B. hunanensis* were conducted in parallel to the eradication campaign. Nematodes extracted from wood were directly inoculated into 3-year-old *Pinus radiata* (the predominant species in local plantations) and *P. halepensis* (the species of the first tree affected). Trees of this age were the only ones available at short notice. Other studies have often found that pathological responses are more likely in young trees (McNamara, 2004). After 3 months, no nematodes could be reisolated from the trees and there were no symptoms.

Direct inoculation of healthy trees with wood plugs from trees known to have the nematode also failed to reproduce the symptoms. No nematodes or fungi could be re-isolated from trees so treated. Attempts to grow the nematode in culture were unsuccessful (see above: stage 1). Other potential causes of pathogenicity were considered but rejected as being the main cause. The state was in the fourth year of drought, but this was unlikely to be a main cause because affected trees were often within groups of otherwise healthy trees. Several of the trees were within parks where the trees received some degree of care and watering. Likewise, although *Diplodia* was isolated from many dying trees, it was not found in all trees. Nor were there symptoms consistent with the usual manifestation of the disease: whole trees were affected rather than single branches, and only one individual in a group of trees was dying rather than all. Other fungi and insects—*Armillaria* spp., *Ophiostoma* spp., *Phytophthora* spp. and *Ips grandicollis*—were rejected as being primary causes of mortality on similar grounds. Salt levels were tested and found outside the generally accepted pathogenic range in most affected trees. Physical wounding, earthworks etc. may have affected some of the trees that died, but were certainly not involved in most, including the first two trees affected.

The tests for pathogenicity were therefore inconclusive. *B. hunanensis* could not be confirmed as associated with symptoms, except statistically, and Koch's Postulates were not satisfied. However, no other single potential cause, biological, chemical or physical could be unambiguously identified either. Because of the nature of the disease, doubts have been expressed as to whether Koch's Postulates can ever be used to demonstrate the pathogenicity of *B. xylophilus* (McNamara, 2004).

In the present case, the most likely explanation may be that a complex of factors was involved. This is discussed further below.

## *Distribution of Trees Containing* B. hunanensis

A summary of the trees from which *B. hunanensis* was isolated is presented in Fig. 1. The pattern resembles a plume from the first tree and presumed origin at Williamstown. Most trees containing the nematode were in an arc between bearings of 60° and 120° from the first tree, with the mean direction about 70–90° (East-Nor-East to East). This is the general direction of prevailing winds.

Another feature of the figure is that there are few infested trees within 10–15 Km of the presumed origin. In the direction of the plume, shorter distances correspond with the city centre (where there are few trees) and the open waters of the bay. The furthest tree was about 62 Km away, but most were within 20–50 Km. None of the trees outside the immediate environs of Melbourne contained *B. hunanensis*.

The main exception to the general pattern was a tree located about 52 Km away at bearing 170°. In this general direction this is the closest landfall.

The pattern is generally consistent with a single dispersal flight by wind-borne or wind-aided vectors. However, the distances involved are considerably greater than are normally associated with vectors of *B. xylophilus*. Within the context of preparedness for, and control measures following, possible incursions by nematodes from the genus *Bursaphelenchus*, the fact that at least some species of the genus can be transported these distances is of particular note. The identity of the vector would be most interesting, but as *B. hunanensis* now appears to have been eradicated, the vector will probably never be known. The rapid decline in trees containing the nematode over the years subsequent to its first isolation, and the inability to find any insects with nematodes, indicated that whatever the vector was, it was inefficient in terms of the percentage of vectors carrying nematodes and possibly also the number of nematodes carried on the vector.

However inefficient in terms of numbers, the distances an organism can be carried is more significant in terms of quarantine. This is especially so if there are other vectors which may be more efficient, but only travel short distances. The short-distance efficient vector could then transmit nematodes from foci created by the inefficient, but long-distance vector some distance away from initial infestations.

Most trees had relatively few *B. hunanensis* (less than 10 individuals per 10 g of wood), but some had high numbers of nematodes within them (up to 7000 individuals per 10 g of wood). The vast majority of nematodes were juveniles. These are population characteristics consistent with non-breeding, inviable populations. The nematodes may have found few appropriate hosts, habitats, food sources, or appropriate vectors beyond their initial ones.

# Discussion

## *Initial Analysis of Risk*

The likely impact of an exotic organism relative to the cost of the options of eradication, containment, control or doing nothing is important information in considering the best response to an incursion. The likely impact can be estimated by considering each part of the invasion process separately: the likelihood of success in invasion, the consequences of a successful invasion, and the success and cost of mitigation strategies if the organism becomes established. The likelihood of success in invasion can be further divided into processes of arrival, establishment, spread and persistence. Consequences can be divided into direct effects on humans, crops or beneficial organisms, indirect effects such as loss of ecosystem services or trade restrictions, and the cost of control or management strategies, including the costs of what is precluded by the pest or management strategy.

The problem with estimating the risk this way is that a lot of data is required: not all may be available, and quality of different parts may vary widely (Stohlgren and Schnase, 2006). This situation is common with nematodes (Hockland et al., 2006). Thus there is an estimation process with some degree of uncertainty. The role of uncertainty is discussed in the following section.

In the initial consideration of the risk posed by *B. hunanensis*, the basic information was evaluated as follows. This was the situation after *B. hunanensis* had been found in a single dying tree, with no other information except the report of *B. hunanensis* from the type locality.

- ***Cost of successful establishment of B. hunanensis***. The worst case scenario is similar to that for *B. xylophilus*. In Japan, mortality (without mitigation measures) of *P. radiata* (the most common species in Victoria) was about 80% (Mamiya, 2004). Spread has been about 2–15 Km yr$^{-1}$ (Togashi et al., 2004), and the value of *P. radiata* plantations is about AUD15 000 to AUD 30 000 Ha$^{-1}$ (Southern Tablelands Farm Forestry Network, pers comm.). With these figures, the total value of potential losses ranges from about AUD 0.2–12 million in the first year, and 0.35–19 billion over 20 years.

  Another way of estimating is using the national turnover of forest products (AUD 18 billion: Parsons et al., 2006) or returns for exports (AUD 2 billion, Australian Bureau of Statistics, 2005), and the proportion of forests potentially affected (30%: Mamiya, 2004). Using these figures, the estimate of potential effects are AUD 6 billion in annual turnover, and 0.7 billion annually in exports. At the initial stage, the only record of *B. hunanensis* on *P. halepensis* was the one we had, so pathogenicity in our situation was uncertain, discussed further below in section using other species data. The costs of trade restrictions, amenity value etc. would have to be added to these figures. Damage to ecosystems may also occur, though quantification is lacking (Batabyal and Beladi, 2006; Perrings et al., 2000).

- ***Containment***. This was not considered a feasible option, given the urban setting with large, unregulated movement of plant and other material. It was also considered an unlikely because of the spread in most countries in which it had been introduced (Webster, 1999, 2004; Yoshimura et al., 1999).
- ***Eradication***. Cost unknown but thought relatively small because of the sparse distribution of potential hosts in the urban area. Note that relative costs and feasibility of containment and eradication may invert in a less-inhabited, heavily-forested environment, or one where access was restricted.
- ***Control costs if established***. Based on the experiences of China and Japan, complete control appeared unlikely, and mitigation the best likely to be achieved (Evans et al., 1996; Mamiya, 1988; Yang, 2004). Furthermore, in the local context, mitigation measures were likely to be difficult because of large areas and low workforce numbers. The costs of this option were thought likely to be high.

The conclusion had to be that a response was probably warranted because of the economic importance of pine forests, and further, that to attempt eradication was probably the best response. This was what was implemented.

The main factor in the decision was the large potential cost of a nematode with effects similar to *B. xylophilus*. Even though the probability that *B. hunanensis* would have such effects was very low, the high cost meant that action was warranted. This is a general feature of risk management: if the unmitigated negative effect is very large, then action is required even if the likelihood is very low (Gigerenzer, 2002).

## *Uncertainty*

Uncertainty is a common feature of many quarantine situations: an incursion by definition involves an organism in a place that it does not normally occur. Where the organism is a relatively well-known exotic, the outcome can be predicted with reasonable certainty in some situations.

- There may have been previous incursions of the organism.
- The new landscape is highly modified so that it is very similar in many ways to the landscape which is the origin of the exotic. The commonest examples are many agricultural landscapes, which have been deliberately modified to at least partially mimic the origin of a particular crop (and many of its pests).
- The organism has been involved in numerous documented incursions into other new places, and so there are numerous precedents.

In many other situations, like that described for *B. hunanensis* in Australia, the organism and its response in a new environment are inadequately known, so the outcome of an incursion cannot be predicted with certainty. Whether it is exotic or not may even be uncertain. In these cases there needs to be a prediction as to whether the organism will become a pest, with a measure of the uncertainty of the prediction.

There are several ways to estimate the likelihood that an exotic nematode will become a pest. The simplest calculates the percentage of known species that are pests. For animals in general this is about 1% (Williamson, 1996; Williamson and Fitter, 1996). For nematodes, the figure may be about 2% (about 200 species are recognized as pests (Nickle, 1992), out of 11 050 species described at a similar time (Andrassy, 1992)). More sophisticated estimates might involve the percentage of exotics that are pests (Dark, 2004; Rejmanek and Randall, 2004), but this is not known for nematodes.

Of most relevance is whether uncertainty should increase or decrease the reason to respond. There are reasons to do both.

Reasons that a high degree of uncertainty should increase the risk—and consequently the likelihood of a response—include the simple statistical observation that a high degree of uncertainty generally means that there will be a wider range of possibilities. This means that there is a greater chance of severe negative effects. The possibility of severe negative effects (disasters) generally means a high level of concern, almost irrespective of the probability (Vitousek et al., 1996; Mack et al., 2000; Gigerenzer, 2002).

There is also the possibility that there will be strong selective pressures for the evolution of pathotypes, races or biotypes in pests, and that these will be unknown. Recent evidence suggests that evolution may occur more rapidly than previously thought (Rodriguez-Trelles and Rodriguez, 1998; Bradshaw and Holzapfel, 2006; Umina et al., 2005; Franks et al., 2007). The genus *Bursaphelenchus* has long been considered likely to evolve rapidly (Giblin-Davis, 1993; Evans et al., 1996). The importance for quarantine of such characteristics of many pest species is becoming increasingly recognised (Brasier, 2001; Kohn, 2004).

Reasons that a high degree of uncertainty should lessen the risk include the observation that, in nematodes at least, knowledge about a particular species may be related to pest status. Pest species are studied more than those that are benign precisely because of their economic effects, so a higher proportion of pest species may be known than benign. Conversely, pests may be a smaller proportion of the species about which we know little—and are consequently highly uncertain—than they are among the well-known species. Hence, there may be a lower probability that a poorly known or unknown nematode will be a pest than there is for known species. If this is the case, there should be less justification for action for poorly-known or unknown species.

There are also reasons for the observation of pests frequently being better-studied than benign nematodes meaning that action is justified. This is because many quarantine pests are not major pathogens in their native range. The best examples of this in nematodes are some of the biggest quarantine threats: *B. xylophilus* and the potato cyst nematodes (PCN) *Globodera rostochiensis* and *G. pallida* (Mamiya, 1984; Wingfield, 1987; Lehman, 2004; Hockland et al., 2006). *B. xylophilus* was not studied in its native range until after it was recognized as the cause of PWD elsewhere (Mamiya, 1984). Likewise the PCN (Franco et al., 1998). Thus species which become pests out of their native range may actually be less likely to be well-studied because they are unlikely to be major problems in their native range.

It is also possible that a species being poorly known may have no effect on the probability that it will become a pest in a new environment. This is because characteristics in the native range are often poor at predicting whether an animal will become an invasive pest in another place (Williamson, 1996; Kolar and Lodge, 2001; Marchetti et al., 2004; Arim et al., 2006; Inderjit and Drake, 2006). Whether this applies to nematodes requires research: many of the worst nematode pests are apparently the same everywhere (e.g. Mota et al., 2006).

The fact that the host is also out of its native range may also increase the probability of pathological interactions. It is has been frequently observed that *B. xylophilus* only affects many species out of their native range (Furuno et al., 1993; Mamiya, 2004). All species of *Pinus* are exotic to Australia.

## *Estimation of Impact Using Known Relatives*

One way to predict the possible effects of an organism about which little is known is to use the nearest relative about which the information exists. This may lead to an over estimate of the adverse effects of the poorly-known species because that species is likely to be a pest, because, as suggested above, pests are more likely to be studied than their benign relatives due to their economic impact. Unless there are more studies of free-living nematodes, this will remain a factor in decision making, but it should be acknowledged more than currently.

Whatever the bias, estimation of impact using known relatives may give an excellent estimate of the "worst-case scenario". For many risk analysis methodologies, the worst case is the most significant (Gigerenzer, 2002).

## *Eradication*

There have been few examinations of the options available for nematode quarantine, and few documented cases of eradication. Emphasis in nematode quarantine is mostly on prevention of incursions through hygiene and disinfestation (Hockland et al., 2006). When incursion management is discussed, it is generally with regard to containment (e.g. McNamara and Smith, 1998; Whitehead and Turner, 1998; Watson, 2004; Hockland et al., 2006), only very occasionally with regard to eradication (e.g. Marshall, 1998).

It is very noteworthy that eradication of nematodes can work, as reported here. It should be considered as an option in at least some nematode incursions. The circumstances under which eradication will be the best option require further investigation.

## *Cost of Delay*

There is a cost in delaying action in the face of uncertainty. In either of the options of eradication or containment, delay inevitably increases the radius of the area which needs to be treated. Cost is, of course, generally related to area, and area is related to the square of the radius, so cost increases rapidly. The chance that a suitable habitat may be located for establishment increases in a similar way.

In the context of the incursion of *B. hunanensis*, a second flight of a vector similar to the presumed original flight was estimated to cost AUD 1 million. This is a substantial cost of delay, and justified at least commencing responses before all data were available.

The correct identification of the nematodes to at least genus by the local identification service also facilitated a quick response. Maintaining a degree of awareness of nematodes and some expertise in nematode identification at a local level proved valuable in the present case.

## *Value of Systematics*

Having expertise or diagnostic tools for both local and exotic species is an important part of being able to respond rapidly (discussed above). There are two issues: one is the general difficulty of all but experts in differentiating nematode species (Eyualem and Blaxter, 2003; Powers, 2004; Sturhan, 1996). The other is that only a small percentage of all nematode species have been described, and an even smaller percentage of species have had intra-specific variability recorded. This means that it is often difficult to diagnose species, either on the basis of evidence or by judgement. This difficulty is relevant in quarantine situations when the organism found is nearly, but not exactly, like a known species, either local or exotic. Is it then the same or different? This is a frequent occurrence in parts of the world like Australia where knowledge of the fauna remains poor, and undescribed species are common. This is a reason to maintain general taxonomists with general knowledge of important groups of organisms.

Apart from being able to provide the most rapid response, having expertise within the country avoids release of partial information: the first report of this incursion was in fact a report from NZ, based on material sent for identification of potential beetle vectors (Ridley et al., 2001; Smith et al., 2007). However, in the absence of tools or expertise for exotic species, a personal network with overseas experts can be a surrogate, so that an unfamiliar species can be referred to someone familiar with it.

Whatever the expertise available, a robust phylogenetic systematic framework in which to place both known and unknown species, is of considerable value in reducing uncertainty and improving predictions in quarantine situations such as that described here. An unknown species needs to be placed into an existing taxonomic framework to determine the closest relatives, where they are found, and how distant is the relationship. In the case of all of the three species found initially at Williamstown, there was uncertainty caused by the desperate need for revision of the entire phylum Aphelenchida (Hunt, 1993).

Having a robust phylogenetic classification is particularly important for unknown or poorly-known species. It allows the nearest relative to be identified, and used to predict the otherwise unknown risks associated with the species (as discussed above). If material is insufficient to allow identification to species, then a phylogenetic classification enables the maximum information to be gleaned from identification to species group, genus, sub-family, family or superfamily.

In the events described herein, this issue arose in the identification of *B. hunanensis* and the genera *Aphelenchoides* and *Bursaphelenchus*. There are considerable morphological overlaps between the females of these genera, and nematodes with a long stylet, long dorsally overlapping oesophageal glands, a short post-uterine sac (PUS), and without a vulval flap could be in either genus (Baujard, 1980; Hunt, 1993). There are other characters to separate males, but males are not known in all species of *Aphelenchoides*, nor are they present at all times in *Bursaphelenchus* spp. (Hunt, 1993; Ryss et al., 2005).

It is better still if there have been attempts to map traits—such as parasitism, hosts, or vectors—onto phylogenies. This allows estimation of the likelihood that a new species will have a greater or lesser potential for pathogenicity. For example, pathogenicity may have evolved several times within a genus, so that a new species in the genus has a heightened chance of being pathogenic. Similarly, a large number of species within a genus may be pathogenic, but from only one evolutionary event. If pathogenicity is absent within the larger Family or Ordinal grouping, then the chances that a new species of uncertain placement within the genus, Family or Order may be less likely to be pathogenic.

There is some debate about the origins of pathogenicity within the genus *Bursaphelenchus*. Some hypothesise that pathogenicity has evolved in an all-or-nothing fashion at most twice in the genus of about 75 species (including *B. cocophilus*: Giblin-Davis et al., 2003; McNamara, 2004; Ryss et al., 2005). Others hypothesise that there is a gradualistic transition between non-pathogenic and pathogenic species, with some species partly pathogenic or only pathogenic under certain circumstances (Braasch et al., 1999b; Kanzaki and Futai, 2006; Michalopoulos-Skarmoutsos et al., 2004). The evidence is currently equivocal, but may become clearer if there is a bacterial complex involved in causing disease (Han et al., 2003, 2006; Zhao et al., 2003).

The genome of *B. xylophilus* is highly plastic, with consequent potential for changes in pathogenicity (Evans et al., 1996; Jones et al., 2005; Kikuchi et al. 2004, 2005, 2006, 2007). Two forms of *B. xylophilus* within North America, are recognised by some: one has a round tail, usually occurs in *Pinus* spp. and is mostly associated with disease; the other has a more pointed or mucronate tail, occurs mostly on fir or spruce (occasionally on pine and other conifers) and is usually benign (Bolla et al., 1986, 1987). Populations maintained as laboratory cultures can change in pathogenicity and biochemical composition over time (Bolla et al., 1986, Kiyohara and Bolla, 1990).

How many other species in the genus *Bursaphelenchus* share this plasticity is unknown. However, it is possible that plasticity is a characteristic of the genus. *Bursaphelenchus* has had a spectacular evolutionary radiation and is the second-largest genus within the Order Aphelenchida: only *Aphelenchoides* with over 200 species is larger (Hodda, 2003).

It is important to note that systematic resources as discussed in this section cannot be generated instantly in the case of an urgent need when there is an incursion.

## *Knowing the Local Fauna*

It is important in quarantine situations such as with *B. hunanensis* that the local fauna be reasonably well described and sampled. Without descriptions of many local species at all, it may be uncertain whether an interception is a species occurring locally or not. In the absence of formal descriptions, voucher material and samples which can be checked rapidly are useful. It may be impossible to obtain material to tell whether an organism occurs locally or not within the time frame needed for quick decisions on action (as discussed below: sampling problems). It is worth knowing the local fauna. This is a worldwide problem for a wide range of nematodes (e.g. Bello et al., 2005 in Spain; Bongers et al., 2003 in Costa Rica; Braasch and Enzian, 2004 in Europe; Queneherve and van den Berg, 2005 in French West Indies).

Before the incursion of *B. hunanensis*, the Australian fauna of Aphelenchida associated with pines and insects was almost totally unknown, with two records only (Queensland Museum, Stone 1990; Stone and Simpson, 1990, 1991). Considerable efforts have now gone into rectifying this situation (e.g. **?** **?**, Zhao et al. 2006a, b, 2007). Decisions as to the best action after *B. hunanensis* was first found would have been much easier if the information and collections available now had been available then.

## *Sampling*

Three sampling issues arose in responding to the incursion of *B. hunanensis*: the area necessary to search, the way to search, and how to tell when *B. hunanensis* is absent. It was necessary to sample for the absence of *B. hunanensis* in two circumstances: to verify that it really was absent from other areas and states of Australia, and to verify eradication.

The area to search for *B. hunanensis* was partly dependent on estimates of the dispersal of possible vectors (discussed in the next section), and partly on practical considerations (what resources were available and the ways that the nematode could be found).

As the initial search area was fairly small (1 Km radius), and potential hosts for the nematode (coniferous trees) were sparse, comprehensive manual searching was feasible, both in terms of logistics (one team in a vehicle was easy to obtain), and in terms of manually sampling every tree. In the outer (5 Km radius) area, there were no estimates of the number of trees, unlike in a forest, but locating and sampling every potential host could have been a very large task. There is no estimate of the percentage of trees sampled, but the targeting of diseased trees only made the task feasible. When the area was expanded enormously after *B. hunanensis* was found outside the initial search area, targeting of diseased trees only made what could have been an enormous task feasible. Targeted diseased trees were also easy to identify from the air, and so made the use of a helicopter possible. Aerial searching is often overlooked as a possibility for sampling large areas because of difficulties in identification from the air (Mullerova et al., 2005).

Throughout the searches for *B. hunanensis*, completely covering the area of the incursion was more important than improving the detection rate within the area. This was because of the high level of uncertainty surrounding the vector(s), the importance of restricting spread outside the urban area, and the need to minimise the area from which eradication would be required. Missing nematodes that subsequently dispersed further increases the cost of eradication enormously (as discussed above). The fact that initial estimates of the radius of dispersal were very much too low is discussed further below.

In considering the lessons from the experience with *B. hunanensis*, one of the best lessons was that relying on reports from the public was relatively effective and efficient. Within the suburban environment, a high proportion of the dying trees were reported, and even in rural areas, trees were reported from hundreds of kilometres away.

The other major issue in searching for *B. hunanensis* was sampling to verify that it was in fact exotic, and that it had been eradicated. Sampling for the absence of an organism is often very costly because sampling effort for a given reliability increases inversely with abundance. One lesson here was that ensuring total absence is not necessary if a minimum viable population density can be estimated (Anderson, 2005). The sampling effort to ensure that abundance is below that which is viable may be considerably less than that required to ensure that the population is essentially zero (with the same degree of reliability). Estimates of minimum viable populations are, of course, subject to considerable uncertainty, as discussed above. Minimum viable populations for *B. xylophilus* have been estimated (Togashi, 1985; Togashi and Shigesada, 2006).

## *Dispersal*

The dispersal distance is key information for many parts of dealing with an incursion such as that of *B. hunanensis*. From the feasibility of eradication or the area that needed searching to estimating the minimum viable population, dispersal distance was among the most important pieces of information.

Hence considerable effort was justified in trying to locate the vector. The fact that a vector was not found increased the uncertainty involved in all actions, but perhaps not by much. Potential vectors may behave very differently in different environments. For example, a species of *Dendroctonus* spp. disperses approximately 16 Km in North America (Smith, 1971), but up to 35 Km in China (Zhang et al., 2002), and attacks trees in a different way (Yan et al., 2005). Vector behaviour may also be affected by the phoretic nematodes (Aikawa et al., 2003).

Whatever the vector, a main conclusion from the results reported here is that the nematode dispersed considerably farther than originally estimated. Part of this may have been due to assistance of the vector by wind: the distribution of trees with the nematode is a classic plume in the direction of prevailing winds. Part of the reason for the great distance of dispersal may have been due to the plume being over water or urban areas for shorter distances from the presumed origin (Fig. 1). Part of the reason for the greater distances may be simply that some vectors can travel these distances. Most studies on this topic have concentrated on dispersal by vectors of *B. xylophilus* which are efficient in causing PWD (e.g. Linit et al., 1983; Wingfield and Blanchette, 1983; Kobayashi et al., 1984; Sato et al., 1987). However, *B. xylophilus* also has a non-disease cycle (Wingfield, 1987), and may also have other vectors. These vectors may feed in different ways or be attracted to dead trees only, so that they are not efficient in transmitting PWD, but can nevertheless disperse the nematode (e.g. Arakawa and Togashi, 2002). There may also be vectors which carry few nematodes—and are therefore unlikely to be detected—but which can travel considerable distances. The rapid spread of *B. xylophilus* after the incursion in Portugal in 1999, and in other countries also suggests that dispersal distances may be greater than current estimates. Likewise, studies from Australia on other aphelenchids associated with dead and dying trees (*Aphelenchoides, Ptychaphelenchus* and *Laimaphelenchus*), have shown that these nematodes may disperse over large areas (Hodda and Falez, 2008). These studies were made subsequent to the incursion of *B. hunanensis*.

## *Interactions and Disease Complexes*

The pathology of the tree deaths observed is worthy of discussion. As outlined above, the studies on pathology of *B. hunanensis* were inconclusive, and a disease complex was suggested as a possible explanation.

It has been suggested recently that a bacterium (*Pseudomonas fluorescens*) is associated with pathology by nematodes in pines (Han et al., 2003; Zhao et al., 2003). This bacterium may cause symptoms whether associated with *B. xylophilus* or *B. mucronatus* (Han et al., 2006). The ability of other species of *Bursaphelenchus* to carry the bacterium requires further testing, but may be possible. A toxin may also be associated with some strains of the bacterium (Oku, 1988, 1990).

There is also the possibility that *B. hunanensis* is only pathogenic under some circumstances. This may be the case with several species of the genus *Bursaphelenchus* other than *B. xylophilus*. These may include *B. leoni, B. sexdentati* and *B. hellenicus* (Braasch, 2000; Caroppo et al., 2000; Skarmoutsos and Michalopoulos-Skarmoutsos, 2000; Michalopoulos-Skarmoutsos et al., 2004; Kanzaki and Futai, 2006). The pathogenicity of *B. mucronatus* in some circumstances is disputed (Kruglik, 2001; Giblin-Davis et al., 2003; Kulinich, 2004; McNamara, 2004).

It is also possible that the pathogenicity only occurred in trees where several factors weakened their resistance to disease. Such factors may have included genetic susceptibility and environmental stress (drought, high temperatures, chemicals or shading), which can increase susceptibility to PWD (Tanaka, 1975; Rutherford and Webster, 1986; Kaneko 1989; Rutherford et al., 1990; Evans et al., 1996; Kawaguchi et al., 1999; Mamiya, 1999; Braasch, 2000), as well as a range of diseases (Manion, 1981; Nilsen and Orcutt, 1996; Orcutt and Nilsen, 2000). Trees carrying *B. xylophilus* can be asymptomatic for long periods before the disease manifestation is triggered (Bergdahl and Halik, 1999).

How the risk and consequences of a disease complex are evaluated in quarantine situations requires further investigations.

## *Cost-Benefit Analysis*

In the worst case scenario, the ultimate cost of an incursion of a nematode with a similar effect on local pine plantations as *B. xylophilus* in Japan can be estimated at about \$3 billion per year. Softwood forestry is worth about AUD18 billion $yr^{-1}$ (Parsons et al., 2006), and losses average 30% $yr^{-1}$ in Japan (Mamiya, 2004). The total cost of the eradication campaign was approximately AUD 0.2 million. If *B. hunanensis* was associated with timber losses of the same magnitude as *B. xylophilus*, then the benefit to cost ratio of the eradication campaign over 30 years (the rotation cycle of the trees planted at the time of the incursion), would be about 1 million to 1, a very good ratio. Put another way, it is worth conducting 1 million eradication campaigns where the benefits are uncertain if this results in stopping one pest which will cause damage of the magnitude of that estimated for *B. xylophilus*. The figure of 1 million is a worst-case scenario, but even if this is over-estimated by several orders of magnitude, many eradications where there is uncertainty over the level of threat are justified. The conclusion that action was justified is robust to considerable uncertainty if there is the possibility of a major effect, however remote. We believe that there was the possibility of a major effect in this case, as discussed above.

Many estimates of the probability that an exotic animal will become a pest are about 1% (Williamson, 1996). Much research is directed at being able to predict the phylogenetic, ecological or molecular characteristics of the 1% of species that will become pests, but at present predictions are still unreliable (Williamson, 1996; Kolar and Lodge, 2001; Marchetti et al., 2004; Arim et al., 2006; Inderjit and Drake, 2006). If the chance of an exotic organism such as *B. hunanensis* becoming a pest are 1 in 100, then the eradication campaign was even more justified.

## Conclusions

There are many questions about this incursion that will never be answered because of the successful eradication campaign. However, the experiences and suggestions that follow from the events surrounding the incursion of *B. hunanensis* may prove useful in other quarantine incursions, particularly regarding nematodes, and most particularly involving nematodes associated with trees and insects. The main summary points are as follow.

1. More research is needed on which nematodes and other organisms become pests, the conditions under which it is most likely to occur, and the ecosystems most subject to invasion.
2. There are many research issues around predicting the characteristics of an unknown or poorly-known species, and how representative are the species about which most is known.
3. A robust phylogenetic systematic framework within which to place nematode species has considerable value in terms of telling new species from variants of existing species. It is also useful in allowing inadequate material to be identified to the best taxonomic level possible. This sort of information cannot be generated instantly, but relies on existing expertise.
4. Studies mapping the origins of particular traits, particularly the origins of parasitism, may be very useful.
5. A good knowledge of the local fauna, and access to knowledge about the faunas of other areas are important in deciding whether a species is exotic or not. This information is lacking for many groups of nematodes for many parts of the world.
6. The means, speed and distance of dispersal are critical things to know for this group of nematodes.
7. Dispersal of Aphelenchida generally, and *Bursaphelenchus* in particular, may be further than currently assumed on basis of current studies of efficient vectors in pine wilt disease.
8. There is considerable value in an immediate response to incursions by nematodes with the ability to disperse considerable distances. The value is likely to be greater than the extra costs associated with uncertainty about the organism's identity or characteristics. A low cost of eradication because of a rapid response and sparse potential new habitat can have major effect on the economics of eradication (see point 14).
9. Training of staff likely to be involved in detection of exotic organisms in the early stages of incursions proved very useful in this case.
10. The lowest viable populations of exotic invaders is very useful information, if available.
11. Just because an organism is not a major pest in its native environment, doesn't mean that it will not be a pest in a new environment.
12. The presence of exotics that may be part of disease complexes complicates the options and management of incursions.
13. Eradication of exotic nematodes can be successful, and a valid response to incursions.
14. Cost-benefit ratios of eradication programmes for nematodes can be very favourable under almost any assumptions of their cost or the probability of the exotic nematode becoming a pest.

## References

AIKAWA, T., TOGASHI, K. AND KOSAKA, H. (2003). Different developmental responses of virulent and avirulent isolates of the pinewood nematode, *Bursaphelenchus xylophilus* (Nematoda: Aphelenchoididae), to the insect vector, *Monochamus alternatus* (Coleoptera: Cerambycidae). *Environmental Entomology* 32: 96–102.

ANDERSON, M.C. (2005). Potential applications of population viability analysis to risk assessment for invasive species. *Human and Ecological Risk Assessment* 11: 1083–1095.

ANDRASSY, I. (1992). A short census of free-living nematodes. *Fundamental and Applied Nematology* 15: 187–188.

ARAKAWA, Y AND TOGASHI, K. (2002). Newly discovered transmission pathway of *Bursaphelenchus xylophilus* from males of the beetle *Monochamus alternatus* to *Pinus densiflora* trees via oviposition wounds. *Journal of Nematology* 34: 396–404.

ARIM, M., ABADES, S.R., NEILL, P.E., LIMA, M. AND MARQUET, P.A. (2006). Spread dynamics of invasive species. *Proceedings of the National Academy of Sciences of the United States of America* 103: 374–378.

AUSTRALIAN BUREAU OF STATISTICS (2005). *Year Book 2005.* Australian Bureau of Statistics, Canberra.

BATABYAL, A.A. AND BELADI, H. (2006). International trade and biological invasions: A queuing theoretic analysis of the prevention problem. *European Journal of Operational Research* 170: 758–770.

BAUJARD, P. (1980). Trois nouvelles especes de *Bursaphelenchus* (Nematoda: Tylenchida) et remarques sur le genre. *Revue de Nematologie* 3: 167–177.

BELLO, A., ROBERTSON, L., DIEZ-ROJO, M.A. AND ARIAS, M. (2005). A re-evaluation of the geographical distribution of quarantine nematodes reported in Spain. *Nematologia Mediterranea* 33: 209–216.

BERGDAHL, D.R. AND HALIK, S. (1999). Inoculated *Pinus sylvestris* serve as long-term hosts for *Bursaphelenchus xylophilus*. In *Sustainability of Pine Forests in Relation to Pine Wilt and Decline. Proceedings of an International Symposium, Tokyo, Japan, 26–30 October 1998* (eds. K. Futai, K. Togashi and T. Ikeda), pp. 73–78. Shokado Shote, Kyoto, Japan.

BOLLA, R.I. AND JORDAN, W. (1982). Cultivation of pine wilt nematode, *Bursaphelenchus xylophilus*, in axenic culture media. *Journal of Nematology* 14: 377–381.

BOLLA, R.I., WEAVER, C., KOSLOWSKI, P., FITZSIMMONS, K. AND WINTER, R.E.K. (1987). Characterization of a nonparasitic isolate of *Bursaphelenchus xylophilus. Journal of Nematology* 19: 304–310.

BOLLA, R.I., WINTER, R.E.K., FITZSIMMONS, K. AND LINIT, M.J. (1986). Pathotypes of the pinewood nematode *Bursaphelenchus xylophilus. Journal of Nematology* 18: 230–238.

BONGERS, T., ESQUIVEL, A. AND ARIAS, H. (2003). Preliminary results of the Costa Rican nematode inventory. *Journal of Nematode Morphology and Systematics* 6: 91–94.

BRAASCH, H. (2000). Influence of temperature and water supply on mortality of 3-year-old pines inoculated with *Bursaphelenchus xylophilu*s and *B. mucronatus. Nachrichtenblatt des Deutschen Pflanzenschutzdienstes* 52: 244–249.

BRAASCH, H. (2004). Morphology of *Bursaphelenchus xylophilus* compared with other Bursaphelenchus species. In *The pine wood nematode*, Bursaphelenchus xylophilus. *Nematology Monographs and Perspectives, vol. 1* (eds. M. Mota and P. Vieira), pp. 127–143. E. J. Brill, Leiden, The Netherlands.

BRAASCH, H., BURGERMEISTER, W. AND PASTRIK, K.H. (1995). Differentiation of three *Bursaphelenchus* species by means of RAPD-PCR. *Nachrichtenblatt des Deutschen Pflanzenschutzdienstes* 47: 310–314.

BRAASCH, H. AND ENZIAN, S. (2004). The pine wood nematode problem in Europe – present situation and outlook. In *The pine wood nematode*, Bursaphelenchus xylophilus. *Nematology Monographs and Perspectives, vol. 1* (eds. M. Mota and P. Vieira), pp. 77–91. E. J. Brill, Leiden, The Netherlands.

BRAASCH, H., METGE, K. AND BURGERMEISTER, W. (1999a). *Bursaphelenchus* species (Nematoda, Parasitaphelenchidae) found in coniferous trees in Germany and their ITS-RFLP patterns. *Nachrichtenblatt des Deutschen Pflanzenschutzdienstes* 51: 312–320.

BRAASCH, H., CAROPPO, S., AMBROGIONI, L., MICHALOPOULOS, H., SKARMOUTSOS, G. AND TOMICZEK, C. (1999b). Pathogenicity of various *Bursaphelenchus* species to pines and implications to European forests. In *The pine wood nematode,* Bursaphelenchus xylophilus. *Nematology Monographs and Perspectives, vol. 1* (eds. M. Mota and P. Vieira), pp. 57–64. E. J. Brill, Leiden, The Netherlands.

BRADSHAW, W.E. AND HOLZAPFEL, C.M. (2006). Climate change – Evolutionary response to rapid climate change. *Science* 312: 1477–1478.

BRASIER, C.M. (2001). Rapid evolution of introduced plant pathogens via interspecific hybridization. *Bioscience* 51: 123–133.

CAROPPO, S., CAVALLI, M., CONIGLIO, D. AND AMBROGIONI, L. (2000). Pathogenicity studies with various *Bursaphelenchus* populations on conifer seedlings under controlled and open air conditions. *Redia* 83: 61–75.

DARK, S.J. (2004). The biogeography of invasive alien plants in California: an application of GIS and spatial regression analysis. *Diversity and Distributions* 10: 1–9.

DRAKE, J.A. (2006). The ecology of non-native invasive plant species: are there consistent patterns? *CAB Reviews: Perspectives in Agriculture, Veterinary Science, Nutrition and Natural Resources* 1: 1–11.

EPPO (2003). National regulatory and control systems. *Bursaphelenchus xylophilus* and its vectors: procedures for official control. *Bulletin OEPP/EPPO Bulletin* 33: 301–304.

EVANS, H.F., MCNAMARA, D.G., BRAASCH, H., CHADOEUF, J. AND MAGNUSSON, C. (1996). Pest risk analysis (PRA) for the territories of the European Union (as PRA area) on *Bursaphelenchus xylophilus* and its vectors in the genus *Monochamus. Bulletin OEPP/EPPO Bulletin* 26: 199–249.

EYUALEM, A. AND BLAXTER, M. (2003). Comparison of biological, molecular, and morphological methods of species identification in a set of cultured *Panagrolaimus* isolates. *Journal of Nematology* 35: 119–128.

FRANCO, J., OROS, R., MAIN, G. AND ORTUNO, N. (1998). Potato cyst nematodes (*Globodera* species) in South America. In *Potato Cyst Nematodes, Biology, Distribution and Control* (eds. R.J. Marks and B.B. Brodie), pp. 239–269. CAB, Wallingford UK.

FRANKS, S.J., SIM, S. AND WEIS, A.E. (2007). Rapid evolution of flowering time by an annual plant in response to a climate fluctuation. *Proceeding of the National Academy of Science* 104: 1278–1282.

FUJIOKA, H. (1993). A report on the habitat of *Monochamus alternatus* Hope in Akita Prefecture. *Bulletin of the Akita Forest Technology Centre* 2: 40–56.

FURUNO, T., NAKAI, I., UENAKA, K. AND KAYA, H. (1993). Pine Wilt on exotic species introduced in Kamaigamo and Shirahama Experimental Stations of Kyoto University: variation in resistance within the genus *Pinus* to Pine Wilt Nematode, *Bursaphelenchus xylophilus* (Steiner & Buhrer). *Reports of the Kyoto University Forest* 51: 23–36.

GIBLIN-DAVIS, R.M. (1993). Interactions of nematodes with insects. In *Nematode Interactions* (ed. W.M. Khan), pp. 302–344. Chapman & Hall, London.

GIBLIN-DAVIS, R.M., DAVIES, K.A., MORRIS, K. AND THOMAS, W.K. (2003). Evolution of parasitism in insect-transmitted plant nematodes. *Journal of Nematology* 35: 133–141.

GIGERENZER, G. (2002). *Reckoning with risk.* Penguin, London.

HAN, Z., HONG, Y., CAO, Y. AND ZHAO, B. (2006). Pathogenicity of bacteria carried by the pine wood nematode. In *Proceedings of the symposium Pine Wilt Disease: a worldwide threat to forest ecosystems* (eds. W. Burgermeister, K. Futai, M. Linit, M. Mota, P.,Vieira and J. Webster). Lisbon, Portugal (http://www.nemalab.uevora.pt/).

HAN, Z.M., HONG, Y.D. AND ZHAO, B.G. (2003). A study on pathogenicity of bacteria carried by pine wood nematodes. *Journal of Phytopathology* 151: 683–689.

HOCKLAND, S., INSERRA, R.N., MILLAR, L. AND LEHMAN, P.S. (2006). International plant health – putting legislation into practice. In *Plant Nematology* (eds. R.N. Perry and M. Moens), pp. 327–345. CABI, Wallingford UK.

HODDA, M. (2003). A comprehensive checklist of the nematode Order Aphelenchida. http://www.ento.csiro.au/science/nematodes/checklist_dec2003.rtf.

HODDA, M. (2006). *Diagnostic Protocol: Pine Wilt Nematode (Bursaphelenchus xylophilus)*. Department of Agriculture, Fisheries and Forestry, Canberra, 38 pp.

HODDA, M. (2009). *Ptychaphelenchus eucalypticola* gen. nov., sp. nov. (Nematoda: Aphelenchida: Aphelenchoidoididae) from wood and bark of *Eucalyptus macrorhyncha* in Australia, with a key to genera of Aphelenchoididae and Parasitaphelenchidae. *Transactions of the Royal Society of South Australia* 132: in press.

HODDA, M. AND FALEZ, E. (2008). Effects of fire on bark and wood dwelling nematodes and their insect associates in native forests of the Australian Capital Territory. *Journal of Wildland Fire* (submitted).

HOOPER, D.J. (1986). Extraction of nematodes from plant material. In *Laboratory Methods for Work with Plant and Soil Nematodes* (ed. J.F. Southey), pp. 51–58. HMSO, London.

HOYER, U., BURGERMEISTER, W. AND BRAASCH, H. (1998). Identification of *Bursaphelenchus* species (Nematoda, Aphelenchoididae) on the basis of amplified ribosomal DNA (ITS-RFLP). *Nachrichtenblatt des Deutschen Pflanzenschutzdienstes* 50: 273–277.

HUNT, D.J. (1993). *Aphelenchida, Longidoridae and Trichodoridae: Their Systematics and Bionomics*. CAB International, Wallingford UK.

IDO, N. AND KOBAYASHI, K. (1977). Dispersal of *Monochamus alternatus*. In *Studies on the control of pine wilt disease*, pp. 87–88. Secretariat of Agriculture, Forestry and Fisheries Research Council, Ministry of Agriculture, Forestry and Fisheries, Tokyo.

JONES, J.T., FURLANETTO, C. AND KIKUCHI, T. (2005). Horizontal gene transfer from bacteria and fungi as a driving force in the evolution of plant parasitism in nematodes. *Nematology* 7: 641–646.

KANEKO, S. (1989). Effect of light-intensity on the development of pine wilt disease. *Canadian Journal of Botany* 67: 1861–1864.

KANZAKI, N. AND FUTAI, K. (2006). Is *Bursaphelenchus mucronatus* a weak pathogen to the Japanese red pine? *Nematology* 8: 485–489.

KAWAGUCHI, E., GYOKUSEN, K. AND SAITO, A. (1999). Behaviour of *Bursaphelenchus xylophilus* and the development of pine wilt disease under shaded condition. In *Sustainability of Pine Forests in Relation to Pine Wilt and Decline. Proceedings of an International Symposium, Tokyo, Japan, 26–30 October 1998* (eds. K. Futai, K. Togashi and T. Ikeda), pp. 39–41. Shokado Shote, Kyoto, Japan.

KIKUCHI, T., AIKAWA, T., KOSAKA, H., PRITCHARD, L., OGURA, N. AND JONES, J.T. (2007). EST analysis of the pine wood nematode *Bursaphelenchus xylophilus* and *B. mucronatus*. *Molecular and Biochemical Parasitology* 155: 9–17.

KIKUCHI, T., JONES, J.T., AIKAWA, T., KOSAKA, H. AND OGURA, N. (2004). A family of GHF45 cellulases from the pine wood nematode *Bursaphelenchus xylophilus*. *FEBS Letters* 572: 201–205.

KIKUCHI, T., SHIBUYA, H., AIKAWA, T. AND JONES, J.T. (2006). Cloning and characterization of pectate lyases secreted by the pine wood nematode *Bursaphelenchus xylophilus*. *Molecular Plant-Microbe Interactions* 19: 280–287.

KIKUCHI, T., SHIBUYA, H. AND JONES, J.T. (2005). Molecular and biochemical characterization of an Endo-ß-1,3-glucanase from the pine wood nematode *Bursaphelenchus xylophilus*. *Biochemical Journal* 389: 117–125.

KIYOHARA, T. AND BOLLA, R.I. (1990). Pathogenic variability among populations of the pinewood nematode, *Bursaphelenchus xylophilus*. *Forest Science* 36: 1061–1076.

KOBAYASHI, F., YAMANE, A. AND IKEDA, T. (1984). The Japanese pine sawyer beetle as the vector of pine wilt disease. *Annual Review of Entomology* 29: 115–135.

KOHN, L.M. (2004). Applying comparative genomics to plant disease epidemiology. *Phytoprotection* 85: 45–48.

KOLAR, C.S. AND LODGE, D.M. (2001). Progress in invasion biology: predicting invaders. *Trends in Ecology & Evolution* 16: 199–204.

KRUGLIK, I.A. (2001). The pathogenicity of Far Eastern isolate of *Bursaphelenchus mucronatus. Vestnik Dal'nevostochnogo Otdeleniya Rossiiskoi Akademii Nauk* 1 95: 72–76.

KULINICH, O. (2004). Surveys for the pine wood nematode in Russia. In *The Pine Wood Nematode*, Bursaphelenchus xylophilus. *Nematology Monographs and Perspectives, vol. 1* (eds. M. Mota and P. Vieira), pp. 65–75. E. J. Brill, Leiden, The Netherlands.

LEHMAN, P.F. (2004). Cost-benefits of nematode management through regulatory programs. In *Nematology: Advances and Perspectives, vol. 2: Nematode Management and Utilization* (eds. Z.X. Chen, S.Y. Chen and D.W. Dickson), pp. 1133–1177. CABI, Wallingford UK.

LIEUTIER, F. AND VALLET, E. (1982). Observations sur les nematodes parasites et associes aux principaux Scolytidae ravageurs du pin sylvestre en forets d'Orleans et de Sologne. *Acta Oecologica, Oecologica Applicata* 3: 131–148.

LINIT, M.J., KONDO, E. AND SMITH, M.T. (1983). Insects associated with the pinewood nematode, *Bursaphelenchus xylophilus* (Nematoda: Aphelenchoididae), in Missouri. *Environmental Entomology* 12: 467–470.

MACK, R.N., SIMBERLOFF, D., LONSDALE, W.M., EVANS, H., CLOUT, M. AND BAZZAZ, F.A. (2000). Biotic invasions: causes, epidemiology, global consequences, and control. *Ecological Applications* 10: 689–710.

MACLEOD, R., REAY, F. AND SMYTH, J. (1994). *Plant Nematodes of Australia Listed by Plant and by Genus*. NSW Agriculture, Sydney.

MAMIYA, Y. (1984). The Pine Wood Nematode. In *Plant & Insect Nematodes* (ed. W.R. Nickle), pp. 589–626. Marcel Dekker, New York.

MAMIYA, Y. (1988). History of pine wilt disease in Japan. *Journal of Nematology* 20: 219–226.

MAMIYA, Y. (1999). Review on the pathogenicity of *Bursaphelenchus mucronatus*. In *Sustainability of Pine Forests in Relation to Pine Wilt and Decline. Proceedings of an International Symposium, Tokyo, Japan, 26–30 October 1998* (eds. K. Futai, K. Togashi and T. Ikeda), pp. 57–64. Shokado Shote, Kyoto, Japan.

MAMIYA, Y. (2004). Pine wilt disease in Japan. In *The Pine Wood Nematode*, Bursaphelenchus xylophilus. *Nematology Monographs and Perspectives, vol. 1* (eds. M. Mota and P. Vieira), pp. 9–20. E. J. Brill, Leiden, The Netherlands.

MANION, P.D. (1981). *Tree disease concepts*. Prentice-Hall, Englewood Cliffs, NJ USA.

MARCHETTI, M.P., MOYLE, P.B. AND LEVINE, R. (2004). Invasive species profiling? Exploring the characteristics of non-native fishes across invasion stages in California. *Freshwater Biology* 49: 646–661.

MARSHALL, J.W. (1998). Potato cyst nematodes (*Globodera* species) in New Zealand and Australia. In *Potato Cyst Nematodes, Biology, Distribution and Control* (eds. R.J. Marks and B.B. Brodie), pp. 353–394. CAB, Wallingford UK.

MCNAMARA, D.G. (2004). Quarantine concerns about the methods used to demonstrate pathogenicity of *Bursaphelenchus* spp. In *The pine wood nematode*, Bursaphelenchus xylophilus. *Nematology Monographs and Perspectives, vol. 1* (eds. M. Mota and P. Vieira), pp. 187–189. E. J. Brill, Leiden, The Netherlands.

MCNAMARA, D.G. AND SMITH, I.M. (1998). National control measures for *Globodera* spp. *Bulletin OEPP/EPPO Bulletin* 28: 503–506.

MICHALOPOULOS-SKARMOUTSOS, H., SKARMOUTSOS, G., KALAPANIDA, M. AND KARAGEORGOS, A. (2004). Surveying and recording of nematodes of the genus *Bursaphelenchus* in conifer forests in Greece and pathogenicity of the most important species. In *The Pine Wood Nematode*, Bursaphelenchus xylophilus. *Nematology Monographs and Perspectives, vol. 1* (eds. M. Mota and P. Vieira), pp. 113–126. E. J. Brill, Leiden, The Netherlands.

MOTA, M., TAKEMOTO, S., TAKEUCHI, Y., HARA, N. AND FUTAI, K. (2006). Comparative studies between Portuguese and Japanese isolates of the pinewood nematode. *Journal of Nematology* 38: 429–433.

MULLEROVA, J., PYSEK, P. AND JAROSIK, V. (2005). Aerial photographs as a tool for assessing the regional dynamics of the invasive plant species *Heracleum mantegazzianum. Journal of Applied Ecology* 42: 1042–1053.
NICKLE, W.R. (1970). A taxonomic review of the genera of the Aphelenchoidea (Fuchs, 1937) Thorne, 1949 (Nematoda: Tylenchida). *Journal of Nematology* 4: 375–392.
NICKLE, W.R. (1992). *Manual of Agricultural Nematology*. Marcel Dekker, New York.
NILSEN, E.T. AND ORCUTT, D.M. (1996). *Physiology of plants under stress. Abiotic factors.* John Wiley and Sons, New York.
OKU, H. (1988). Role of phytotoxins in pine wilt disease. *Journal of Nematology* 20: 245–251.
OKU, H. (1990). Phytotoxins in pine wilt disease. *Nippon Nogeikagaku Kaishi-Journal of the Japan Society for Bioscience Biotechnology and Agrochemistry* 64: 1254–1257.
ORCUTT, D.M. AND NILSEN, E.T. (2000). *The Physiology of Plants Under Stress: Soil and Biotic Factors.* John Wiley and Sons, New York.
PARSONS, M., GAVRAN, M. AND DAVIDSON, J. (2006). *Australia's plantations 2006.* Department of Agriculture, Fisheries & Forestry, Canberra.
PERRINGS, C., WILLIAMSON, M. AND DALMAZZONE, S. (2000). Introduction. In *The Economics of Biological Invasions* (eds. C. Perrings, M. Williamson and S. Dalmazzone), pp. 1–13. Edward Elgar, Cheltenham & Northampton.
POWERS, T.O. (2004). Nematode molecular diagnostics: from bands to barcodes. *Annual Review of Phytopathology* 42: 367–383.
QUENEHERVE, P. AND VAN DEN BERG, E. (2005). List of the phytoparasitic nematodes (Tylenchida and Dorylaimida) of the French overseas departments (Guadeloupe, Martinique and Guyana) and regulatory considerations. *Bulletin OEPP/EPPO Bulletin* 35: 519–530.
REJMANEK, M. AND RANDALL, J.M. (2004). The total number of naturalized species can be a reliable predictor of the number of alien pest species. *Diversity and Distributions* 10: 367–369.
RIDLEY, G., BAIN, J. AND DICK, M. (2001). Exotic nematode found in pine trees in Melbourne, Victoria. *New Zealand Journal of Forestry* 46: 41–42.
RODRIGUEZ-TRELLES, F. AND RODRIGUEZ, M.A. (1998). Rapid micro-evolution and loss of chromosomal diversity in *Drosophila* in response to climate warming. *Evolutionary Ecology* 12: 829–838.
RUTHERFORD, T.A., MAMIYA, Y. AND WEBSTER, J.M. (1990). Nematode-induced pine wilt disease: factors influencing its occurrence and distribution. *Forest Science* 36: 145–155.
RUTHERFORD, T.A. AND WEBSTER, J. (1986). Distribution of pine wilt disease with respect to temperature in North America, Japan, and Europe. *Canadian Journal of Forest Research* 17: 1050–1059.
RYSS, A., VIEIRA, P., MOTA, M. AND KULINICH, O. (2005). A synopsis of the genus *Bursaphelenchus* Fuchs, 1937 (Aphelenchida: Parasitaphelenchidae) with keys to species. *Nematology* 7: 393–458.
SATO, H., SAKUYAMA, T. AND KOBAYASHI, M. (1987). Transmission of *Bursaphelenchus xylophilus* (Steiner et Buhrer) Nickle (Nematoda: Aphelenchoididae) by *Monochamus saltuarius* (Gebler) (Coleoptera: Cerambycidae). *Journal of the Japanese Forestry Society* 69: 492–496.
SKARMOUTSOS, G. AND MICHALOPOULOS-SKARMOUTSOS, H. (2000). Pathogenicity of *Bursaphelenchus sexdentati, Bursaphelenchus leoni* and *Bursaphelenchus hellenicus* on European pine seedlings. *Forest Pathology* 30: 149–156.
SMITH, D.I., HODDA M., SMITH I.W., NAMBIAR L. AND PASCOE I.G. (2007). *Bursaphelenchus hunanensis* on *Pinus* species in Victoria, Australia. *Australasian Plant Pathology* 37: 308–317.
SMITH, R.H. (1971). Red turpentine beetle. *US Department of Agriculture Forest Pest Leaflet* 55: 1–8.
STOHLGREN, T.J. AND SCHNASE, J.L. (2006). Risk analysis for biological hazards: what we need to know about invasive species. *Risk Analysis* 26: 163–173.
STONE, C. (1990). Parasitic and phoretic nematodes associated with *Ips grandicollis* (Coleoptera: Scolytidae) in New South Wales, Australia. *Nematologica* 36: 478–480.

STONE, C. AND SIMPSON, J.A. (1990). Species associations in *Ips grandicollis* galleries in *Pinus taeda*. *New Zealand Journal of Forestry Science* 20: 75–96.

STONE, C. AND SIMPSON, J.A. (1991). Effect of six chemicals on the insects, mites, nematodes and fungi associated with *Ips grandicollis* (Eichhoff) (Coleoptera: Scolytidae) in northeastern New South Wales. *Journal of the Australian Entomological Society* 30: 21–28.

STURHAN, D. (1996). On the significance of traditional taxonomy and the importance of the "German Nematode Collection". *Mitteilungen aus der Biologischen Bundesanstalt fur Land- und Forstwirtschaft Berlin-Dahlem* 317: 66–74.

TAKASU, F., YAMAMOTO, N., KAWASAKI, K., TOGASHI, K. AND SHIGESADA, N. (2000). Modelling the expansion of an introduced tree disease. *Biological Invasions* 2: 141–150.

TANAKA, K. (1975). The effect of sulphur dioxide on the development of pine wilt disease caused by *Bursaphelenchus lignicolus*. *Transactions for the Meeting of Japanese Forestry Society* 86: 287–289.

TOGASHI, K. (1985). Transmission curves of *Bursaphelenchus xylophilus* (Nematoda: Aphelenchoididae) from its vector, *Monochamus alternatus* (Coleoptera: Cerambycidae), to pine trees with reference to population performance. *Applied Entomology and Zoology* 20: 246–251.

TOGASHI, K., CHUNG, Y. AND SHIBATA, J. (2004). Spread of an introduced tree pest organism—the pinewood nematode. In *Ecological Issues in a Changing World-Status, Response and Strategy* (eds. S.K. Hong, J.A. Lee, B.S. Ihm, A. Farina, Y. Son, E.S. Kim and J.C. Choe), pp. 173–188. Kluwer, Dordrecht.

TOGASHI, K. AND SHIGESADA, N. (2006). Spread of the pinewood nematode vectored by the Japanese pine sawyer: modelling and analytical approaches. *Population Ecology* 48: 271–283.

UMINA, P.A., WEEKS, A.R., KEARNEY, M.R., MCKECHNIE, S.W. AND HOFFMANN, A.A. (2005). A rapid shift in a classic clinal pattern in *Drosophila* reflecting climate change. *Science* 308: 691–693.

VITOUSEK, P.M., D'ANTONIO, C.M., LOOPE, L.L. AND WESTBROOKS, R. (1996). Biological invasions as global environmental change. *American Scientist* 84: 468–478.

WATSON, R.N. (2004). Internal biosecurity – a realistic objective for plant nematodes? *New Zealand Plant Protection* 57: 151–155.

WEBSTER, J.M. (1999). Pine Wilt Disease: a worldwide survey. In *Sustainability of Pine Forests in Relation to Pine Wilt and Decline. Proceedings of an International Symposium, Tokyo, Japan, 26–30 October 1998* (eds. K. Futai, K. Togashi and T. Ikeda), pp. 254–260. Shokado Shote, Kyoto, Japan.

WEBSTER, J.M. (2004). The pine wood nematode: implications of factors past and present for pine wilt disease. In *The Pine Wood Nematode*, Bursaphelenchus xylophilus. *Nematology Monographs and Perspectives, vol. 1* (eds. M. Mota and P. Vieira), pp. 55–64. E. J. Brill, Leiden, The Netherlands.

WHITEHEAD, A.G. AND TURNER, S.J. (1998). Management and regulatory control strategies for potato cyst nematodes (*Globodera rostochiensis* and *Globodera pallida*). In *Potato Cyst Nematodes, Biology, Distribution and Control* (eds. R.J. Marks and B.B. Brodie), pp. 135–152. CAB, Wallingford UK.

WILLIAMSON, M. (1996). *Biological Invasions*. Chapman & Hall, London.

WILLIAMSON, M. AND FITTER, A. (1996). The varying success of invaders. *Ecology* 77: 1661–1666.

WINGFIELD, M.J. (1987). *Pathogenicity of the pine wood nematode*. APS Press, St Paul MN.

WINGFIELD, M.J. AND BLANCHETTE, R.A. (1983). The pine-wood nematode, *Bursaphelenchus xylophilus*, in Minnesota and Wisconsin: insect associates and transmission studies. *Canadian Journal of Forest Research* 13: 1068–1076.

YAN, Z.L., SUN, J.H., DON, O. AND ZHANG, Z. (2005). The red turpentine beetle, *Dendroctonus valens* LeConte (Scolytidae): an exotic invasive pest of pine in China. *Biodiversity and Conservation* 14: 1735–1760.

YANG, B. (2004). The history, dispersal and potential threat of pine wilt nematode in China. In *The Pine Wood Nematode*, Bursaphelenchus xylophilus. *Nematology Monographs and Perspectives, vol. 1* (eds. M. Mota and P. Vieira), pp. 21–24. E. J. Brill, Leiden, The Netherlands.

YIN, K., FANG, Y. AND TARJAN, A.C. (1988). A key to species in the genus *Bursaphelenchus* with a description of *Bursaphelenchus hunanensis* sp. n. (Nematoda: Aphelenchoididae) found in pinewood in Hunan Province, China. *Proceedings of the Helminthological Society of Washington* 55: 1–11.

YOSHIMURA, A., KAWASAKI, K., TAKASU, F., TOGASHI, K., FUTAI, K. AND SHIGESADA, N. (1999). Modelling the spread of pine wilt disease caused by nematodes with pine sawyers as vector. *Ecology* 80: 1691–1702.

ZHANG, L.Y., CHEN, Q.C. AND ZHANG, X.B. (2002). Studies on the morphological characters and bionomics of *Dendroctonus valens* LeConte. *Scientia Silvae Sinicae* 38: 95–99.

ZHAO, B.G., WANG, H.L. AND HAN, S.F. (2003). Distribution and pathogenicity of bacteria species carried by *Bursaphelenchus xylophilus* in China. *Nematology* 5: 899–906.

ZHAO, Z.Q., DAVIES, K.A., RILEY, I.T. AND NOBBS, J.M. (2006a). *Laimaphelenchus australis* sp nov (Nematoda: Aphelenchina) from exotic pines, *Pinus radiata* and *P. pinaster*, in Australia. *Zootaxa* 1248: 35–44.

ZHAO, Z.Q., DAVIES, K.A., RILEY, I.T. AND NOBBS, J.M. (2006b). *Laimaphelenchus preissii* sp nov (Nematoda: Aphelenchina) from native pine *Callitris preissii* in South Australia. *Transactions of the Royal Society of South Australia* 130: 10–16.

ZHAO, Z.Q., DAVIES, K.A., RILEY, I.T. AND NOBBS, J.M. (2007). *Laimaphelenchus heidelbergi* sp. nov. (Nematoda: Aphelenchina) from Victoria, Australia, and emendment of the diagnosis of the genus. *Transactions of the Royal Society of South Australia* 131: 182–191.

# The Risk of Pine Wilt Disease to Australia and New Zealand

**Simon A. Lawson and Shiroma Sathyapala**

**Abstract** Currently Australia and New Zealand are free of *Bursaphelenchus xylophilus*, the pinewood nematode (PWN), which causes the pine wilt disease, and its primary vectors *Monochamus* spp. Both countries have increasing interest in this pest due to the significant area of exotic pine plantations, predominantly *Pinus radiata*, and native conifer species that could be susceptible to PWN. Although the current phytosanitary measures and surveillance programmes in New Zealand and Australia minimise the risk of entry of PWN, both countries have favourable climatic conditions for its establishment. In addition, there are significant uncertainties in the susceptibility of *Pinus* species to PWN and the potential vector status of native cerambycid beetles. A screening study was conducted in 2003 using 12-month old seedlings from Queensland of *P. elliottii, P. caribaea.* var *hondurensis* and $F_1$ and $F_2$ hybrid clones of these parents to determine their potential susceptibility to PWN and *Bursaphelenchus mucronatus*. Results illustrated the limitations of screening young plants in determining the susceptibility of older trees. The pest risk assessment conducted in 2004 and the screening study confirmed that maintenance of strong quarantine and surveillance programs and development of molecular or physiological markers for identification of PWN resistance in planting stocks of *Pinus* species and native species is indispensable to prevent the introduction of pine wilt disease in New Zealand and Australia.

## Introduction

### *History,* Biology *and* Ecology

Pine wilt disease (PWD), caused by the pinewood nematode (PWN), *Bursaphelenchus xylophilus*, is one of the most significant and devastating diseases affecting *Pinus* spp. worldwide. A native of North America, where it causes mortality

S.A. Lawson
Department of Primary Industries and Fisheries, Forestry Building, Gate 3 80 Meiers Rd, Indooroopilly, QLD 4068, Australia
e-mail: simon.lawson@dpi.qld.gov.au

M.M. Mota, P. Vieira (eds.), *Pine Wilt Disease: A Worldwide Threat to Forest Ecosystems*, 

mainly to exotic *Pinus* species (Dropkin et al., 1981; Wingfield et al., 1984; Linit and Tamura, 1987), it has since spread to Japan (Kobayashi et al., 1984), China in 1982 (Cheng et al., 1983; Li, 1983), Taiwan in 1983 (Tzean and Jan, 1985) and South Korea in 1988 (Choi and Moon, 1989; Enda, 1989). In 1999, the nematode was found in Portugal (Mota et al., 1999) where an eradication campaign has been ongoing (Sousa et al., 2002). Severe mortality to endemic *Pinus* spp. has occurred in all areas where the nematode has become established and where an efficient vector has been present. These vectors have so far all proven to be cerambycid beetles in the genus *Monochamus* (Coleoptera: Cerambycidae). In northeast Asia, the primary vector is the pine sawyer beetle *Monochamus alternatus*, while in Portugal this role is taken by *M. galloprovincialis*. Worldwide there has yet to be a successful introduction of the nematode in the absence of *Monochamus* vectors or a successful establishment of the beetle with or without the nematode. Australia and New Zealand have no native *Monochamus* species and none has thus far been introduced. There are a number of other potential insect vectors of the nematode present in both countries, but similar taxa overseas in either the native range of the nematode or where introductions have occurred have not been demonstrated to be efficient vectors.

## Forestry in Australia and New Zealand

### *Forestry in New Zealand*

New Zealand's forest resource covers over 8.2 million hectares, or nearly 30% of New Zealand's total land area. Indigenous forests make up the majority of this with 6.4 million hectares and planted production forests accounts for the remaining 1.8 million hectares. The forestry sector plays an increasingly important role in New Zealand's economy. In 2006, forestry contributed 3.4% of New Zealand's national income. New Zealand's forest industry supplied 8.8% of Asia Pacific's forest products trade and 1.1% share of the world trade. The dominant species in plantation forest is *Pinus radiata*, which makes up 89% of the total planted forests. Douglas fir (*Pseudotsuga menziesii*) makes up 6.4%, eucalyptus 1.8% and the remainder is a variety of hardwood and softwood species. The major tree species in the indigenous forests are kauri (Araucariaceae), beech (Nothofagaceae), rimu (Podocarpaceae) and tawa (Cupressaceae). Native conifers consist of mainly Podocarpaceae, and Araucariaceae. There are also some exotic conifer species such as spruce, fir and larch which are mainly grown in urban gardens and have limited distribution in New Zealand.

### *Forestry in Australia*

Australia's 164 million hectares of forests are dominated mainly by eucalypts. Australia has no native *Pinus* species, but there are a number of native conifer species in the Araucariaceae, Cupressaceae, Podocarpaceae, and Taxodiaceae, including economically important species such as hoop pine (*Araucaria cunninghamii*) and cypress pine (*Callitris glauca*). Over 40,000 ha of *A. cunninghamii* plantations have been established in Queensland, while *C. glauca* is harvested on a sustainable basis from native stands in New South Wales and Queensland. Other conifers, such as Huon Pine (*Dacrydium franklinii*) and King Billy Pine (*Athrotaxis selaginoides*) in Tasmania are of considerable conservation importance. The susceptibility of these native conifers to pine wilt disease is currently unknown.

In Australia, almost 1 million ha of exotic *Pinus* spp. plantations have been established since the 1950s (Parsons et al., 2006), with this area remaining largely static over the last 10 years. Seventy five percent of the total plantation area is comprised of *Pinus radiata. P. pinaster* is grown to a limited extent in plantations in Western Australia, while *P. caribaea* var. *hondurensis* and *P. elliottii* are used widely in plantations in subtropical and tropical New South Wales and Queensland. *P. caribaea* var. *hondurensis* × *P. elliottii* hybrid clones have now replaced *P. elliottii* as the major taxa used in plantations in southeast Queensland, with about 53,000 ha of these hybrids now established (Forestry Plantations Queensland, 2006). The forest plantation industry in Australia in 2006 was worth approximately AUD$18 billion and employed around 30,000 people directly (Parsons et al., 2006), with approximately 60% of this value attributable to softwood plantations.

A focus for the softwood industry in both Australia and New Zealand has been the prevention of introductions of pests and diseases. This strategy has been largely successful; for example only a handful of significantly impacting insect pests have been introduced into Australia over the last 50 years, the most important and damaging of which has been the wood wasp *Sirex noctilio* (Hymenoptera: Siricidae) which became established in Australia in 1953 (Morgan and Griffith, 1989). In the most recent outbreak between 1989 and 1992 *Sirex* caused AUD$12 million of damage.

## Interceptions

### *New Zealand*

To date no species of *Bursaphelenchus* have been detected in New Zealand or intercepted on goods imported into New Zealand (Sathyapala, 2004). However, *Monochamus* spp., the insect vectors of *B. xylophilus*, have been intercepted on imported goods in New Zealand on numerous occasions.

Although nematodes are not routinely tested for at the New Zealand border, there were some targeted examinations of intercepted insect vectors for PWN in 2004 and 2005. These examinations revealed the absence of PWN larvae in insect vectors. The interceptions records demonstrated that the insect vectors could enter into New Zealand as hitch hikers. In June 2004, a post border interception of male and female *M. alternatus* associated with cardboard packaging was reported. However, no dauer larvae of PWN were found in the beetles.

## *Australia*

There have been 12 separate interceptions of *Monochamus* spp. at Australian ports since 2000 (B. Crowe, *personal communication*). Of these, five were interceptions of *M. alternatus* and one interception of *M. impluviatus*, from regions where pine wilt disease is known to occur. The other six interceptions were of *Monochamus* species from countries known to be free of PWD. Three recent interceptions in Australia associated with possible introductions of this nematode or its vectors are worthy of further comment. In 2000, an incursion of what was later determined to be *B. hunanensis* was recorded in Melbourne (Hodda et al., this volume), while in Brisbane in May 2001 and June 2005 post-border interceptions of *M. alternatus* adults emerging from pallets sourced from China were recorded. The discovery of *Bursaphelenchus* nematodes in a dying tree at the Melbourne Botanic gardens in 2000 prompted a successful 2-year eradication campaign in Victoria (Hodda et al., this volume) and wider scale surveys for *Pinus* mortality potentially associated with nematodes in other states. Responses to the beetle interceptions in Brisbane centred on locating infested pallets, which were then fumigated and deep buried. Surveys for pine mortality in areas surrounding the interceptions were carried out and traps with attractant baits used around the interception sites to detect any remaining beetles for six to twelve months following the first discoveries. None of these interceptions has led to the known establishment of the nematode or vector.

# The Threat

## *Potential Economic and Environmental Impacts*

### New Zealand

The spread of PWD has the potential to have a major impact on New Zealand's forestry based industries if *B. xylophilus* became established in New Zealand. Ninety five percent of commercial forestry in New Zealand is based on plantations of *P. radiata*. This industry accounted for approximately three billion New Zealand dollars of exports in 2006, as well as a domestic market of equivalent size. If biotic and abiotic factors are as favourable to *B. xylophilus* in New Zealand as they are in countries like Japan, where there was vast devastation to *Pinus* trees, the mortality rate could be as high as 60%. The economic impact in New Zealand under those circumstances would cause a direct loss in productivity of plantation forests, and would create difficulties in exporting logs and timber, especially to countries which already have quarantine barriers against *B. xylophilus*.

The devastating impact PWD has had on native conifer forests in Japan, China, Korea, and Taiwan has led to untreated wood from PWN affected areas being banned in the European Union and China. Extra heat-treatment for untreated lumber is required by these importing regions to ensure the wood products are free of living nematodes and their beetle vectors. Likewise, if PWN became established in

New Zealand, countries importing unprocessed *P. radiata* (e.g. logs, timber, wood chips) from New Zealand would require the wood to be heat-treated before export as a phytosanitary measure. The countries most likely to require this extra heat treatment would be those that do not have, or have restricted distributions of *B. xylophilus*, and have significant areas of natural and/or plantation forests of susceptible *Pinus* species. Currently, this would include New Zealand's major forestry export markets (including Australia, China, Korea and Japan). The extra cost of heat-treating unprocessed wood products such as logs would substantially reduce or eliminate the profitability of those products. It is unlikely that any added value would be gained from such a treatment of logs.

Rapid spread of pine wilt in countries like Japan, China and Korea has destroyed large areas of pine forests, causing significant changes in their ecosystems. Extensive tree losses due to PWN have created severe impact on erosion control, sand stabilisation, wind protection, and maintenance of aesthetic value (e.g. Abe and Tani, 1985). *Pinus* species within forests impacted by PWD have largely been replaced with evergreen broad-leaved trees. This has reduced the species richness and resulted in changes to the associated flora and fauna in the forest (Fujihara, 1996).

The most aesthetically and culturally significant native conifer species in New Zealand belong to the Araucariaceae and Podocapaceae families. There are 17 species of these families in New Zealand. It is not known if New Zealand's endemic conifer species are susceptible to PWN infection, or if *B. xylophilus* would have any impact on the native nematode populations in New Zealand. As these conifer species have a very high aesthetic value in New Zealand, large losses of trees due to PWD will have major social, cultural and environmental impacts.

Pine wilt disease could have a direct impact on susceptible exotic species such as fir *(Abies)*, spruce *(Picea)*, and larch *(Larix)* grown in urban areas for recreational and amenity value. High tree mortality of pine and Douglas-fir (*Pseudotsuga menziesii*) plantations and the large scale removal of diseased trees as control measures would increase erosion and result in micro–climate modification of affected areas. Ultimately, adverse impacts on native flora and fauna habitats, indigenous ecosystems and biodiversity could ensue.

## Australia

*Pinus* spp. plantations contribute significantly to the Australian economy (see Forestry in Australia section). These plantations are particularly important to regional and rural economies, which would be severely impacted on if tree mortality similar to that caused by PWD in Japan since the 1970s were replicated in Australia.

Aside from local economic impacts, Australia is a net importer of softwood products and local production is almost exclusively used domestically. Thus, a significant reduction in output of forest products from *Pinus* plantations due to PWD would mean that more of these products would need to be imported, further eroding Australia's balance of trade position in wood products. To illustrate this, Australia ran a trade deficit in forest products of $2.0 billion in 2004–05, and produced just under 90% of its sawn timber needs, of which 73% was derived from softwood plantations (Australia Yearbook, 2007).

Less is known regarding possible environmental effects of establishment of *B. xylophilus* in Australia. The susceptibility of native conifers to infection by the nematode or to maturation feeding by *Monochamus* spp. vectors is untested. However, based on the experience of those countries where PWD has been introduced, it has been almost entirely *Pinus* species that have been affected and it could be argued that this would most likely be the case in Australia should an introduction occur. *Pinus* spp. are common urban and amenity trees, especially in temperate Australia, and many species are widely used in agriculture as windbreaks. Wider environmental effects would therefore be likely to be more minor than the severe economic effects that could be expected. As outlined above for New Zealand, environmental side effects of widespread tree mortality in plantations could lead to increased soil erosion, run-off, and decreased water quality in streams and reservoirs near these plantations.

## *Likelihood of Entry on Pathways*

The pest risk analysis conducted by MAF Biosecurity Authority, New Zealand in 2004 identified that the most likely pathway for the introduction of PWN into New Zealand would be through an importation of untreated forest products produced from host tree species harvested from PWN infested forests. This includes untreated solid wood packaging material, logs, poles, piles, rounds and sleepers and sawn wood.

Around 95% of the *Monochamus* spp. (vectors of PWN) intercepted on material imported into New Zealand was on wood packaging material. The likelihood of entry of adult beetles carrying dauer larvae of PWN is very high when imports are from countries where PWN is present and widespread. Currently, the majority of imported wood packaging material arrives in New Zealand in containers from Australia and countries in South East Asia. Projected future trade patterns indicate an increase in the importation of wooden packaging material from China, Hong Kong and Taiwan. It is possible, however, that much of this wood packaging material could originate from North America.

Poles, piles, rounds, and sleepers (logs) include any wood pieces larger that 300 mm in minimum thickness (cross-section). *Monochamus* spp. preferentially oviposit in the bark of freshly cut trees and trees damaged by other biotic and abiotic factors. The removal of bark from harvested trees reduces survival of beetles but, if the larvae enter the wood before the debarking process, they could develop to the adult stage. Logs imported into New Zealand are required to be debarked, but none of the chemical/heat or fumigation treatments are mandatory for many of the PWN host species. A high temperature heat treatment is required for logs from *Pinus* species and Douglas fir (70 °C for 4 hours). Upon arrival to New Zealand, only 10% of untreated logs are visually inspected for pest and diseases. Visual inspection of logs is very unlikely to detect microscopic PWN. However, larvae of *Monochamus* spp. may be found from evidence of tunnelling. Thus, logs may act as a pathway for entry of PWN, but are less likely to be an entry pathway for *Monochamus* spp.

The probability of *Monochamus* spp. and PWN being present in sawn wood (wood sawn longitudinally, with or without its natural rounded surface, without bark and no larger than 300 mm in thickness) is similar to that in untreated logs. Although the processing method of sawing could reduce the *Monochamus* spp. survival, some larvae could survive the process and continue to live in the sawn wood.

The majority of the wood from conifer tree species imported into New Zealand or Australia originates in the USA and Canada, where *B. xylophilus* and *Monochamus* spp. are endemic. However, *Thuja plicata*, the main species of sawn wood (comprising 95% of imported softwood) is not known to be a host of either *B. xylophilus* or *Monochamus* spp. (Evans et al., 1996).

New Zealand's current import requirements for sawn timber of *Pinus* species originating from areas considered by the MAF not to be free of *Fusarium circinatum* (Pitch Canker disease) require mandatory heat treatment to 70 °C (core temperature) for 4 hours. Taking into account the current phytosanitary measures, the overall likelihood of entry of *Monochamus* spp. and PWN by sawn wood or logs is considered negligible.

## *Likelihood of Establishment*

It is known that transmission of *B. xylophilus* to new host trees is primarily mediated by the insect vector *Monochamus* spp. during their maturation feeding or oviposition. Generally, adult *Monochamus* spp. have a short dispersal range of a few hundred metres, but they can fly up to 3 km depending on the availability of trees on which they feed and oviposit (Kobayashi et al., 1984). If *B. xylophilus* enters Australia or New Zealand with *Monochamus* spp. in wood packaging for instance, and is transferred to an area close to a forest or a saw mill, there is a high likelihood the nematode will be transferred to a host tree.

Several beetle species have the potential to provide a phoretic vehicle for *B. xylophilus* (Mamiya, 1984). None of the reported insect vectors of the Cerambycidae genera have been recorded as established in Australia or New Zealand. Absence of the *Monochamus* spp. and other likely insect vectors of PWN in these countries reduce the probability of direct transfer of *B. xylophilus* to the host trees. Hosking et al. (1989) reported that New Zealand does not have any suitable native insect vector species. However, possible transmission of *B. xylophilus* by native New Zealand insect species has not been studied extensively. Hence, it is not known to what extent local native and exotic insects could act as a vector of PWN in New Zealand. The most likely candidates are the cerambycids e.g. *Hexatricha pulverulenta* and *Arhopalus ferus*, which are both abundant in New Zealand. The risk of PWN becoming established in Australia or New Zealand appears to be dependent therefore upon the establishment of *Monochamus* spp. or other suitable insect vectors, either prior to, or in conjunction with, the introduction of *B. xylophilus*.

If *B. xylophilus* were to establish in Australia or New Zealand, it could be expected to spread rapidly throughout the country where host trees are existing, but only if a suitable insect vector is available. Other non-vector means of spread would be isolated localised occurrences as insect vectors are essential for the dispersal of the nematode. The rapid spread of the Sirex wood wasp (*Sirex noctilio*) within Australia since the 1950s is a model for the potential spread of PWN through plantation and other host trees in the landscape via an efficient insect vector.

## *Suitability of Environmental Conditions*

*Bursaphelenchus xylophilus* is endemic in the southern states of Canada, possibly up to 60° latitude, but PWD is rarely expressed north of 40° latitude. Temperature appears to be more limiting to the pathogenicity of the disease than the survival of the nematode. The geographic range of vector beetles will also affect the survival of *B. xylophilus* because without a vector the nematode would not survive. According to Braasch and Enzian (2004), the potential development of PWD is predicted to be extremely high when long term mean summer temperature is more than 25 °C and long term annual rainfall is less than 600 mm. Conversely, the expected development of PWD would be low at temperatures less then 18 °C and where annual rainfall is more than 600 mm.

The mean summer temperature across the country does not exceed 19 °C in New Zealand. This includes areas with annual rainfall of less than 600 mm. Only in the most northern areas of the North Island do mean February (the warmest month) temperatures exceed 20 °C (20.3 °C). However, this area is unlikely to show extensive pine wilt symptoms in the event of establishment of PWN since the annual rainfall is greater than 600 mm. In Australia, the mean summer temperature exceeds 18 °C over much of the continent, excluding Tasmania, southern Victoria, the lower southeast of South Australia and at higher altitudes along the east coast. Most major pine growing areas in Australia have mean annual rainfall that exceeds 600 mm; however, rainfall in Australia is highly variable and droughts are frequent. Trends to decreased rainfall over the past 50 years have been observed in south-eastern and south-western Australia as well as south-eastern Queensland and current climate models predict that these trends will worsen by 2030 (CSIRO, 2007). Conditions are thus likely to become more favourable for PWN development over much of the existing *Pinus* plantation estate in Australia.

## *Likelihood of Spread*

When evaluating the potential threat that the pinewood nematode poses in Australia and New Zealand it is essential that we know, firstly, the likelihood of the vector prospering if introduced, and, secondly, the susceptibility of the *Pinus* taxa to the nematode.

To evaluate the first of these factors, two simple indices were developed in Japan to assess the risk of severe infestations of the beetle-nematode association. These are the flight index (FI), which is the number of days per year where the maximum temperature is greater than 18 °C and rainfall is less than 10 mm, and the *Monochamus/Bursaphelenchus* (MB) index, which is the annual summation of residues of mean monthly temperature greater than 15 °C (Taketani et al., 1975; Kishi, 1995). If the beetle and vector are both present, severe infestations can be expected when the flight index exceeds 80 days or where the MB index exceeds 40. When representative temperature and rainfall data (1999–2000) for Brisbane in southeast Queensland were used in these indices they both far exceeded the critical values (mean FI = 330, mean MB = 61). This would indicate that, if vector and nematode became established in the presence of susceptible hosts, severe outbreaks could be expected in southeast Queensland (Lawson unpublished data). Calculating the MB index for other major locations in Australia using long-term (1961–1990) temperature data (Australia Bureau of Meteorology) indicated that Sydney (MB = 43) and Perth (MB = 45) both exceeded the index value that would indicate that severe outbreaks could be expected. Melbourne and Hobart (MB = 22 and 5, respectively) had much lower index values, as did Mount Gambier (MB = 11) at the heart of plantations in the Green Triangle region. In the tropics, Melville Island (Northern Territory) and Cardwell (Queensland) had expectedly high MB index values of 148 and 111, respectively. Conditions are therefore more favourable for the spread of PWD from central New South Wales northwards and in south-western Western Australia.

## Management

The establishment of PWN in New Zealand and Australia is mainly dependent on the availability of a suitable insect vector. Therefore, the primary focus of any further measures for possible pathways of entry is to mitigate the risk of entry of insect vectors of PWN in to each country to an appropriate level. As *Monochamus* spp. are poor colonisers of new territories, the likelihood of *Monochamus* spp. becoming established in New Zealand and Australia is considered low with the current phytosanitary measures in place for imported host material. If established by other non-vector methods however, *B. xylophilus* would only cause isolated occurrences of PWD.

Surveillance for *Monochamus* spp. in high risk areas such as ports, sites processing containers, timber yards, and parks and nurseries, is considered essential for the early identification of the presence of pathogenic *Bursaphelenchus* spp. in New Zealand and Australia. The surveillance for *Monochamus* spp. could be included in a wood boring bark beetle surveillance system for high risk sites. Early detection of *Monochamus* spp. is essential in preventing the establishment of PWN.

## Susceptibility of Exotic *Pinus* spp. in Australia and New Zealand

### *Current Knowledge*

Since the first outbreaks of pine wilt disease in Japan in the early 1900s and the significant expansion of damage to pines that has occurred there from the 1970s onward, much data has been accumulated on the relative susceptibility of *Pinus* spp. and other conifers to *B. xylophilus* (e.g. Kishi, 1995; Evans et al., 1996). *Pinus* spp. that have co-evolved with the nematode in its endemic range in North America have generally shown high levels of resistance under normal climatic conditions and in the absence of other predisposing pests or diseases. Introductions of the nematode outside its endemic range (i.e. Japan, Korea, Taiwan, China, and Portugal) have shown that *Pinus* spp. in these areas have not evolved tolerance or resistance to the nematode despite the presence of other *Bursaphelenchus* spp. (such as *B. mucronatus*) in these regions.

Of the exotic pine species introduced to Australia and New Zealand, two of the dominant species planted, *P. radiata* and *P. elliottii*, both originated from North America within the known range of *B. xylophilus: P. radiata* from a limited natural distribution on the west coast of the U.S.A and *P. elliottii* from the south-eastern U.S.A. The other dominant species of interest in Australia, *P. caribaea* var. *hondurensis*, occurs in the eastern half of Central America south-east from the Yucatán peninsula and in published maps of the distribution of *B. xylophilus* (CABI/EPPO, 1999) its range lies somewhat outside of the natural range of the nematode. Evans et al. (1996) in their Pest Risk Analysis of *B. xylophilus* for the European Union categorised 56 *Pinus* taxa according to their susceptibility to the nematode: they listed *P. elliottii* as resistant and *P. radiata* and *P. caribaea* var. *hondurensis* being of intermediate resistance, with the caveat that such lists should be treated with caution because of conflicting records of resistance from different studies. An example is Furuno et al. (1993) who assessed species susceptibility by examining nematode-induced mortality of a range of exotic (to Japan) *Pinus* species planted in and around Kyoto. In contrast to Evans et al. (1996), they described *P. radiata* as being highly susceptible, with most species from western North America showing some degree of susceptibility while those from the east were almost all resistant. In most cases, disagreement over the susceptibility status of a species arise from the differing ages at which trees have been assessed, with mortality of older, mature trees considered a more accurate measure of susceptibility than seedlings or younger trees. Illustrating this point is a study by Dwinell (1985) that showed very high susceptibility of 3-year old seedlings of *P. elliottii* to inoculation with *B. xylophilus*, while other studies such as those used by Evans et al. (1996) and Furuno et al. (1993) indicated that this species is resistant after testing more mature trees.

The abovementioned studies indicate there is sufficient variability and uncertainty in the susceptibility assessments of the major pine plantation species grown in Australia and New Zealand to warrant further research to better define the risk that *B. xylophilus* poses to these species in Australia and New Zealand.

## *The Need for Screening*

Apart from uncertainty in the susceptibility rating of the major plantation *Pinus* species in New Zealand and Australia, there are other factors that suggest that more intensive screening of these species against *B. xylophilus* should be carried out to further define the potential risk to plantations that establishment of the nematode would pose. In addition, planting of resistant genotypes should be proactively promoted as an insurance against possible nematode establishment in the future. For example, plantation *Pinus* spp. in Australia and New Zealand have now had over 50 years of intensive genetic improvement that has maximised growth rates and improved adaptation of germplasm to local environmental and plantation management conditions. It is known for other tree species, such as eucalypts, that genetic selection for long periods in the absence of pests and diseases can lead to a break down in tolerance or resistance mechanisms e.g. Dungey et al. (1997). At present, risk assessment for these key *Pinus* taxa has been based solely on assessments overseas on genotypes that are now many generations of breeding removed from the germplasm that is currently being used in commercial operational plantings in Australia and New Zealand. Until this currently utilised germplasm is screened for resistance/tolerance to PWN, the level of uncertainty about PWD risk will remain high.

## *Preliminary Screening of Queensland Plantation Taxa 2003*

As outlined in the Current Knowledge section, considerable variation exists in results of studies that have assessed the resistance of various *Pinus* taxa to PWN around the world. Of the two natural species grown in Queensland, *P. elliottii* is considered resistant while *P. c.* var. *hondurensis* is thought to be of intermediate resistance, although because the latter species natural range lies outside that of *B. xylophilus* it may not have had the opportunity to co-evolve resistance to the nematode. A complicating factor is that considerable progress has been made in Queensland in the development and use in plantations of $F_1$ and $F_2$ hybrids of *P. elliottii* and *P. c.* var. *hondurensis* since 1996. The literature on the effect of hybridisation of tree species on pest and disease resistance is inconclusive, with the full range of effects from increased susceptibility to enhanced resistance being observed. Fritz et al. (1999), in a review of resistance of hybrids to pests, diseases and parasites found a trend towards an increased susceptibility of hybrid plants. In the case of the four *Pinus* hybrids reviewed, three inherited susceptibility to an insect pest through a parent, and one was susceptible in its own right. This indicates that even if we could be confident of the susceptibility of the parent species, we cannot be confident of the susceptibility of the hybrid progeny.

To address this gap in our knowledge and the difficulty it poses for formulating eradication plans for potential incursions and management, a research program was developed with the Forestry & Forest Products Research Institute of Japan to test the susceptibility to PWN of 12-month old seedlings/cuttings of *P. elliottii* (PEE), *P. c.*

var. *hondurensis* (PCH), and five $F_1$ & one $F_2$ hybrid clones of these two species. The six clones tested were those that were commonly being used in operational plantings in southeast Queensland at that time. These taxa were also tested for their potential susceptibility to the less pathogenic Eurasian species *B. mucronatus* for comparative purposes.

## *Planting Material*

One-year old seedlings and cuttings of the eight taxa to be tested were shipped (bare-rooted, in compliance with phytosanitary requirements) to Japan in May and June 2003. Numbers of each taxon shipped varied according to availability in nurseries at the time, but ranged from 30 (for hybrid clones) to approximately 100 (PCH). More plants of the two distinct species were available than any of the hybrid clones and were at a more advanced stage of growth, especially in the case of PCH cuttings. Survival of plants during transit was very high.

## *Methods*

After arrival in Japan and clearance through quarantine, plants were transplanted into a field nursery at the Forestry and Forest Products Research Institute, Tsukuba, watered regularly and their health condition assessed weekly. Plants were transplanted as blocks of single taxa, three rows wide with spacing of approx. 0.5 m between rows and 30 cm within rows. Blocks of these taxa were isolated from each other by 0.5 m. Approximately 1 month after transplanting, each plant was randomly assigned to one of three inoculum treatments: 10,000 dispersal stage nematodes of *B. xylophilus* (strain Ka-4, one of the most virulent strains) or *B. mucronatus* in aqueous suspension (0.05 ml), or the same volume of distilled water as controls. Plants were inoculated on 25 July 2003 using the incision method (Kosaka et al., 2001), whereby a small cut was made into the phloem and the nematode or control solution injected immediately with a micropipette. Condition of plants was assessed three times post-inoculation, at 37, 82 and 110 days to obtain progression of mortality.

## Results and Discussion

Results of inoculations are shown in Table 1. None of the eight taxa assessed showed significant susceptibility to *B. mucronatus*. This nematode is of Eurasian origin and so PEE and PCH could not have directly co-evolved natural resistance to it. It is thus suggested that co-evolution of PEE and PCH with PWN may also confer resistance or tolerance to the less virulent *B. mucronatus*. Of the two parent species tested, PEE showed the highest susceptibility to PWN, with PCH showing only a small, non-significant elevation in mortality of PWN-inoculated plants above the controls.

**Table 1** Percent mortality of 12 month old *Pinus* taxa 112 days post-inoculation with 10,000 dispersal stage *B. xylophilus* (Ka-4 strain), *B. mucronatus* or with distilled water control. Mortalities followed by different letters across rows differ significantly at $P < 0.05$ level (Chi-squared pairwise comparison)

| Taxa/Clone | Control | *B. mucronatus* | *B. xylophilus* | Susceptibility |
|---|---|---|---|---|
| PCH | 24 a | 20 a | 35 a | – |
| PEE | 6a | 9 a | 94 b | ++ |
| Clone A ($F_2$) | 0 a | 6 a | 18 a | – |
| Clone B ($F_1$) | 33 a | 33 a | 50 a | – |
| Clone C ($F_1$) | 19 a | 19 a | 56 b | + |
| Clone D ($F_1$) | 25 a | 0 b | 57 a | – |
| Clone E ($F_1$) | 40 a | 44 a | 60 a | – |
| Clone F ($F_1$) | 14 a | 14 a | 86 b | ++ |
| Pooled Clones | 19 a | 18 a | 53 b | + |

This result is in agreement with other studies cited above that indicate PEE seedlings (if not more mature trees) are susceptible to PWN. The present study did not test susceptibility of older material.

Hybrids between PEE and PCH showed wide variation in their susceptibility to PWN, with two of the six taxa tested (clones C and F) showing significantly higher mortality than the control treatment. Clone F was the more susceptible of the two, exhibiting 86% mortality of plants inoculated, almost as high as that exhibited by PEE. Clone A, the only $F_2$ hybrid tested, showed the lowest mortality in relation to inoculation with PWN. The remaining clones inoculated with PWN exhibited elevated mortality compared to controls, but, due to their generally poor condition when inoculated (due to transportation and replanting), control mortality was also higher than expected. As a group, the clones displayed significantly higher mortality when inoculated with PWN than the controls. The results may therefore underestimate the susceptibility of clones to PWN.

Parentage of clones is shown in Table 2. It can be seen that the two clones that showed significantly higher mortality due to PWN inoculation, C and F, shared the same PEE parent (designated EA here) but had different PCH parents (designated CD and CE, respectively). Clones B and E, which did not show significantly higher mortality due to PWN also shared the same PEE parent (EA) but again had different PCH parents (CB and CA, respectively). Clone D did not share a parent with any

**Table 2** Parentage of clones use in inoculation trials. Letter combinations refer to individual parents. See the text for further explanation

| Clone | *P. elliottii* parent | *P. caribaea* parent |
|---|---|---|
| A | (EE × CA) | (ED × CF) |
| B | EA | CB |
| C | EA | CDa |
| D | EB | CC |
| E | EA | CA |
| F | EA | CEa |

other clone. These results suggest that variability in susceptibility to PWN may be more strongly inherited through the PCH parent than through the PEE parent. The parentage of the $F_2$ hybrid (clone A) is more complicated, but it does share one of its PCH parents with clone E, which also did not show susceptibility to PWN.

These results indicate that future proactive clonal breeding programs seeking to maximise tolerance of the PEE × PCH hybrid to PWN should concentrate on identifying the PCH parent material that is associated with increased levels of tolerance/resistance, rather than on the PEE parents which were shown here to have less influence. Results from this preliminary study also illustrate the limitations in screening young (1–2 year-old) plants when the ultimate aim is to determine the susceptibility of older trees. Very high susceptibility of young PEE seedlings confirmed the results of earlier studies with this species, although it has been established that healthy older trees are highly resistant (Dwinell, 1985).

## *Further Taxa Screening*

Screening studies of young material such as this can therefore produce false positives (i.e. plants appear to be susceptible when they are not), but it has not been established if false negatives (i.e. plants seem resistant when they are not) also occur in studies of this kind. Preferably, field inoculations of older material should be used to corroborate results from screening of young plants. Even so, there are limitations to this approach: for Australia and New Zealand, within-country inoculations of older trees are not possible due to quarantine considerations. Even if this were possible, such a screening program would necessarily lag behind material being used in breeding programs.

There is therefore a demonstrated need to develop molecular, physiological or anatomical markers that can be used to identify PWN resistance early in tree breeding cycles and to characterise the current plantation resource in terms of its susceptibility and therefore the risk that PWD poses. Kuroda (2006, this volume) has shown that lower numbers of nematodes occur in the xylem of resistant *P. densiflora* Siebold & Zucc. cultivars and that there are consistent differences between resistant and susceptible cultivars in the size and number of resin canals in the xylem, with resistant cultivars having a lower number of smaller sized resin canals than susceptible cultivars. These anatomical characters may be a promising avenue to attempt to screen the resistance status of the current *Pinus* plantation resource in Australia and New Zealand.

## *Quarantine and Surveillance*

### Pathways

Strong quarantine and surveillance pre-border, border and post-border remains the most important means of excluding PWD from Australia and New Zealand. Both countries maintain strong quarantine inspection regimes, but it is becoming more difficult to maintain effective inspection rates of all at-risk material as the volume of world trade is increasing. With recent and ongoing liberalisation of world trade and the subsequent increase in trade volume, there has been a shift to identifying high risk pathways and regulating these so as to minimise the risk of invasive species spreading. This risk minimisation is being carried out in association with the design and implementation of early detection/rapid response systems to detect those incursions that elude pathway controls. This latter strategy also addresses the fact that exotic incursions are difficult to detect in the wider environment (such as plantations) before their numbers and distribution have increased enough to make an eradication campaign either very expensive and/or with a low probability of success.

The highest risk pathway for the introduction of PWD is via timber packaging and dunnage, and a new set of international rules covering treatment and certification of timber packaging, ISPM-15 (FAO, 2006), has now been implemented since 2006. If compliance with ISPM-15 can be maintained, risk of introduction of nematode-carrying insect vectors (i.e. *Monochamus* spp.) from infested areas should be significantly reduced. However, there has been at least one example in Australia where *Monochamus* beetles carrying *Bursaphelenchus* sp. have emerged from pallet material that was stamped as ISPM-15 compliant, so there are still manifest risks associated with this system due to poor or non-compliance.

## Early Detection/Rapid Response

Extensive forest health surveys are still an effective means of delimiting and scoping the extent of damage by forest pests and diseases, but they are of more limited usefulness in detecting low level, incipient populations of exotic introductions (see Wardlaw et al., in press). By the time that forest health surveys detect exotic incursions, the pest is often well-established and the chances of an eradication program being successful are significantly reduced. Early Detection and Rapid Response (EDRR) systems are increasingly being seen as a more effective method of preventing establishment of invasive pests of forestry and enabling effective eradication responses when pests are detected. Early detection is usually targeted on hazard sites where the risk of entry of quarantinable organisms is high, such as sea and air ports, and associated cargo unloading facilities that are often located in suburban areas.

In the case of PWD, kairomone attractant lures (based on ethanol and α-pinene) and intercept traps are available that are effective in trapping *Monochamus* spp. vectors with a relatively high degree of sensitivity, and these are currently being used in a pilot study in some ports and hazard sites in Australia. Similar systems have been trialled in New Zealand (Brockerhoff et al., 2006) and are being used operationally in the U.S.A and Canada (Sweeney et al., 2007; Duerr, 2005). Real-time molecular diagnostic tools for PWN are now also being developed that will enable rapid identification of *B. xylophilus* for quarantine and pest risk assessment purposes (see Francois et al., 2007).

## Conclusions

1. Potential pathways (via timber packaging and lumber) exist for the entry of pine wilt nematode via its vector, *Monochamus* spp. cerambycid beetles, into Australia and New Zealand and a number of interceptions of potential vectors have been made. Determination of the likely vector status of candidate insects established in or native to New Zealand and Australia is needed.
2. While establishment is considered unlikely (especially since both countries have no known efficient vectors), favourable environmental conditions exist for the establishment and spread of *Monochamus* spp. beetles and PWN in both countries, with the thermal requirements more favourable in Australia than in New Zealand.
3. Should an introduction occur, considerable uncertainty exists in regard to the susceptibility of the *Pinus* taxa currently being grown in these two countries. Therefore, there is a need for extensive screening of current and future *Pinus* planting stock to test for susceptibility to *B. xylophilus*. Testing of native conifer species is also highly desirable. Results from preliminary screening of taxa used in plantations in Queensland has underlined the need for the development of robust molecular or physiological markers that would assist in enabling rapid within-country screening for resistance.
4. Strong pre-border, border, and post-border quarantine and surveillance needs to be maintained to prevent introductions of vector and nematode. In particular, further development and refinement of hazard site surveillance systems is seen as an effective and sensitive method for early detection of incursions of the vector, increasing the likelihood that should an introduction occur, it could be successfully eradicated.

**Acknowledgments** SAL gratefully acknowledges the Japan Society for the Promotion of Science and the Australian Academy of Science for supporting the screening work in Japan through a JSPS Invitation short-term fellowship (2002 & 2003), and the Forestry and Forest Products Research Institute of Japan for their help and cooperation in providing the facilities and expertise required to carry out the research. Special thanks to Drs Isamu Okochi, Hajime Kosaka and Takuya Aikawa of FFPRI for their advice and assistance, to Dr Garth Nikles for his input and advice on pine genetics and Rebeccah Aigner and Janet McDonald for help in preparation of plants for shipment. Grateful thanks also to Forestry Plantations Queensland for providing the pine material for screening and for funding SAL's attendance at the pine wilt disease: a worldwide threat to forest ecosystems international symposium Lisbon, July 2006.

## References

Abe, T. and Tani, M. (1985). Streamflow changes after killing of pine trees by the pine-wood nematode. *Journal of the Japanese Forestry Society* 67: 261–270.

Braasch, H. and Enzian, S. (2004). The pinewood nematode problem in Europe – present situation and outlook. In Mota, M. and Vieira, P. (Eds). *The pinewood nematode, Bursaphelenchus xylophilus. Nematology monographs and perspectives, Volume 1*. Leiden, The Netherlands, E.J. Brill, pp. 77–91.

BROCKERHOFF, E.G., JONES, C.D., KIMBERLEY, M.O., SUCKLING, D.M. AND DONALDSON, T. (2006). Nationwide survey for invasive wood-boring and bark beetles (Coleoptera) using traps baited with pheromones and kairomones. *Forest Ecology and Management* 228: 234–240.

BUREAU OF METEOROLOGY (2007). Climate Data Online. http://www.bom.gov.au/climate/averages/index.shtml

CABI/EPPO (1999). *Distribution maps of plant diseases.*

CHENG, H.R., LIN, M., LI, W. AND FANG, Z. (1983). The occurrence of a pine wilting disease caused by a nematode found in Nanjing. *Forest Pest and Disease* 4: 1–5.

CHOI, Y.E. AND MOON, Y.S. (1989). Survey on distribution of pine wood nematode (*Bursaphelenchus xylophilus*) and its pathogenicity to pine trees in Korea. *Korean Journal of Plant Pathology* 5: 277–286.

CSIRO (2007). Climate change in Australia – Technical report 2007. http://www.climatechangeinaustralia.gov.au/ resources.php

DROPKIN, V.H., FOUDIN, A., KONDO, E., LINIT, M., SMITH, M. AND ROBBINS, K. (1981). Pinewood nematode: a threat to U.S. forests? *Plant Disease* 65: 1022–1027.

DUNGEY, H.S., POTTS, B.M., CARNEGIE, A.J. AND ADES, P.K. (1997). *Mycosphaerella* leaf disease: genetic variation in damage to *Eucalyptus nitens, Eucalyptus globulus*, and their F1 hybrid. *Canadian Journal of Forest Research* 27: 750–759.

DWINELL, L.D. (1985). Relative susceptibilities of five pine species to three populations of the pinewood nematode. *Plant Disease* 69: 440–442.

ENDA, N. (1989). The status of pine wilting disease caused by *Bursaphelenchus xylophilus* (Steiner et Buhrer) Nickle and its control in Korea. *Journal of Korean Forestry Society* 78: 248–253.

EVANS, H.F., MCNAMARA, D.G., BRAASCH, H., CHADOEUF, J. AND MAGNUSSON, C. (1996). Pest risk analysis (PRA) for the territories of the European Union (as PRA area) on *Bursaphelenchus xylophilus* and its vectors in the genus *Monochamus. Bulletin OEPP* 26: 199–249.

FAO (2006). *International standards for phytosanitary measures.* ISPM No. 15 Guidelines for regulating wood packaging material in international trade (2002) with modifications to Annex I (2006). Secretariat of the International Plant Protection Convention, 11 pp.

FORESTRY PLANTATIONS QUEENSLAND (2006). *Pocket Facts 2006.* http://www.fpq.qld.gov.au/data/portal/00000005/content/00751001161069120411.pdf

FRANCOIS, C., CASTAGNONE, C., BOONHAM, N., TOMLINSON, J., LAWSON, R., HOCKLAND, S., QUILL, J., VIEIRA, P., MOTA, M. AND CASTAGNONE-SERENO, P. (2007). Satellite DNA as a target for TaqMan real-time PCR detection of the pinewood nematode, *Bursaphelenchus xylophilus. Molecular Plant Pathology* 8: 803–809.

FUJIHARA, M. (1996). Development of secondary pine forests after pine wilt disease in western Japan. *Journal of Vegetation Science* 75: 729–738.

FURUNO, T., NAKAI, I., UENAKA, K. AND HAYA, K. (1993). [The pine wilt upon the exotic pine species introduced in Kamigamo and Shirahama experiment stations of Kyoto University – various resistances among the genus *Pinus* to pine wood nematode, *Bursaphelenchus xylophilus* Steiner and Buhrer.] *The Reports of the Kyoto University Forests* 25: 20–34.

HOSKING, G.P. (1989). Pine wilt nematode: an example of active risk assessment. *New Zealand Journal of Forestry Science* 19: 2–3, 335–337.

KISHI, Y. (1995). *The pinewood nematode and the Japanese pine sawyer.* Thomas Company Ltd., Tokyo, Japan, 302 pp.

KOBAYASHI, F., YAMANE, A. AND IKEDA, T. (1984). The Japanese pine sawyer beetle as the vector of pine wilt disease. *Annual Review of Entomology* 29: 115–135.

KOSAKA, H., AIKAWA, T., OGURA, N., TABATA, K. AND KIYOHARA, T. (2001). Pine wilt disease caused by the pine wood nematode: the induced resistance of pine trees by the avirulent isolates of nematode. *European Journal of Plant Pathology* 107: 667–675.

KURODA, K. (2006). Defense systems of *Pinus densiflora* cultivars selected as resistant to pine wilt disease. Pp. 47 (Abst.) *in* Pine wilt disease: A worldwide threat to forest ecosystems. International symposium. 10–14 July, Lisbon.

LI, G.W. (1983). Discovery of and preliminary investigation on pine wood nematodes in China. *Forest Science and Technology* 7: 25–28.

LINIT, M.J. AND TAMURA, H. (1987). Relative susceptibility of four pine species to infection by pinewood nematode. *Journal of Nematology* 19: 44–50.

MAF FORESTRY TRADE STATISTICS (2006). http://www.maf.govt.nz/statistics/primaryindustries/forestry/forestry-production-and-exports/2006-releases/jun06-forestry-production-and-trade.htm

MAMIYA, Y. (1984). The pinewood nematode. In *Plant and Insect Nematodes* (ed. W.R. Nickle) pp. 589–626. New York and Basel. 925pp.

MORGAN, F.D. AND GRIFFITH, J.A. (1989). Forty years of *Sirex noctilio* and *Ips grandicollis* in Australia. *New Zealand Journal of Forestry Science* 19: 198–209.

MOTA, M.M., BRAASCH, H., BRAVO, M.A., PENAS, A.C., BURGERMEISTER, W., METGE, K. AND SOUSA, E. (1999). First report of *Bursaphelenchus xylophilus* in Portugal and in Europe. *Nematology* 1: 727–734.

PARSONS, M., GAVRAN, M. AND DAVIDSON, J. (2006). *Australia's Plantations 2006*. Bureau of Rural Sciences, Canberra. http://affashop.gov.au/PdfFiles/plantations_100dpi.pdf

SATHYAPALA, S. (2004). *Pest Risk Analysis. Biosecurity Risk to New Zealand of Pinewood Nematode (Bursaphelenchus xylophilus)*. MAF Biosecurity Authority Forest Biosecurity group, 37 pp.

SOUSA, E., NAVES, P., BONIFÁCIO, L., BRAVO, M.A., PENAS, A.C., PIRES, J. AND SERRÃO, M. (2002). Preliminary survey for insects associated with *Bursaphelenchus xylophilus* in Portugal. *Bulletin OEPP* 32: 499–502.

SWEENEY, S., DE GROOT, P., HUMBLE, L., MACDONALD, L., PRICE, J., MOKRZYCKI, T. AND GUTOWSKI, J.M. (2007). Detection of wood-boring species in semiochemical-baited traps. In *Alien Invasive Species and International Trade* (eds. H. Evans and T. Oszako), pp. 139–144. IUFRO Unit 7.03.12 Monograph. Forest Research Institute, Warsaw. (Extended Abst.). http://www.forestresearch.gov.uk/pdf/IUFRO_Jedlnia_2006.pdf/$FILE/IUFRO_Jedlnia_2006.pdf

TAKETANI, A., OKUDA, M. AND HOSODA, R. (1975). The meteorological analysis on the epidemic mortality of Pine trees, with special reference to the accumulated temperature. *Journal of the Japanese Forestry Society* 57: 169–175.

THE YEAR BOOK AUSTRALIA (2007). *Australian Bureau of Statistics*. http://www.abs.gov.au/ausstats/abs@.nsf/productsbytitle/894BC04F84CABB3FCA2572360001B746?opendocument

TZEAN, S. AND JAN, S. (1985). The occurrence of pinewood nematode, *Bursaphelenchus xylophilus* in Taiwan. *Proceedings of the 6th ROC symposium of electron microscopy*, 38–39.

WINGFIELD, M.J., BLANCHETTE, R.A. AND NICHOLLS, T.H. (1984). Is the pine wood nematode an important pathogen in the United States? *Journal of Forestry* 82: 232–237.

# Pine Wilt Disease: A Threat to Pine Forests in Turkey?

**Süleyman Akbulut, Beşir Yüksel, Ismail Baysal, Paulo Vieira and Manuel Mota**

**Abstract** The pinewood nematode (PWN), is the causal agent of pine wilt disease, and constitute one of the most important pathogens of conifer forests. In 1999, this nematode was found and identified for the first time in Portugal and in Europe. The detection of this quarantine pest in Portugal has prompted the need to know more about the distribution of *Bursaphelenchus* spp. in coniferous trees in Europe in order to describe the geographic range of the species and to act quickly in case of the nematode's unwanted introduction into other European regions. Pine forest has a wide distribution in Turkey, which increases the number of susceptible host trees for the PWN. Because of these reasons, some regions of Turkey were surveyed for the presence of the nematode. Three different species of *Bursaphelenchus* were found, however, *B. xylophilus* was not detected. The detection of *B. mucronatus*, very similar to *B. xylophilus* biologically and morphologically, is very important. The presence of this species indicates that *B. xylophilus* could spread easily into the conifer forests of Turkey. Biological characteristics of *M. galloprovincialis* were compared with *M. carolinensis*, the North American insect vector, and some of them were found to be similar.

## Introduction

The pinewood nematode (PWN), *Bursaphelenchus xylophilus*, causes serious damage to pine forests in Japan and China (Mamiya, 1988). It has spread through Taiwan and Korea as well (Mamiya, 1998). Recently, the nematode was isolated from *Pinus pinaster* (maritime pine) wood in Portugal for the first time in Europe (Mota et al., 1999). The detection of this A1 quarantine organism (EPPO/European Plant Protection Organization) in one of the member states of the European Union (EU)

S. Akbulut
Duzce University, Duzce Forest Faculty, 81300 Düzce, Turkey
e-mail: akbulutsuleyman@yahoo.com

M.M. Mota, P. Vieira (eds.), *Pine Wilt Disease: A Worldwide Threat to Forest Ecosystems*, 

forced the implementation of specific measures to control and eradicate this nematode and its insect vector from the affected area and to conduct surveys to confirm the absence of the PWN in pine forests of all EU member countries. Turkey, as a member of EPPO, follows the phytosanitary regulations for quarantine organisms in Europe.

Turkey is in an important transitional geographic area between Europe and Asia, which may potentially increase the possibility of inadvertent introduction of the PWN from infested regions such as East Asia. Because of increasing global trade, exportation and importation of wood products among countries increases the threat of pine wilt disease to uninfested regions of the world. Turkey imports wood products from different countries every year. Consequently, the possible introduction of the PWN and other exotic pests is a constant threat to the country.

The presence and distribution of *Bursaphelenchus* species in Turkey is poorly known, and limited to the first report of the genus by Vieira et al. (2003), and by the occurrence of *B. mucronatus* by Akbulut et al. (2006). There are several reasons to conduct and develop studies related to the potential of pine wilt disease establishment in Turkey. One is geographic location, since most of the trade routes between Asia and Europe pass through Turkey which may increase a possible introduction of non indigenous pests.

Secondly, the total forest area covers about 27% of the country's total land (21 million ha) (Fig. 1) (Anonymous, 2006a). Conifer species (pure stands) cover 54% of this total land (Fig. 2). In Turkey, there are three widely distributed native pine species; *Pinus brutia, P. nigra* and *P. sylvestris*. They cover an area of almost 10.9 million ha of the country's territory. This wide distribution of pine species in Turkey increases the number of susceptible host trees for the PWN. Despite the

**Fig. 1** Forest sites of Turkey (www.ogm.gov.tr)

**Fig. 2** Distribution of forest types in Turkey

richness of the forest area, Turkey has to import industrial wood from different countries every year to cover the gap between supply and demand (Fig. 3).

The third reason for survey is the presence of at least one *Monochamus* species, *M. galloprovincialis*, in Turkey. It has been previously reported from a number of different sites within the country (Çanakçıoğlu and Mol, 1998; Özdikmen et al., 2005). These studies provide only information on general distribution and some morphological characters of the beetle. The pine sawyer beetle, *M. galloprovincialis* was found to be the vector of the PWN in Portugal (Sousa et al., 2001). According to Hellrigl (1971), five species of *Monochamus* occur in Europe; *M. galloprovincialis, M. sartor, M. sutor, M. urussovi* and *M. saltuarius*. Other *Monochamus* species reported in Europe (Hellrigl, 1971) have not been found in Turkey yet.

Detailed biology and ecology of *M. galloprovincialis* in Turkey were unknown. Recently, several studies on the biology and ecology of *M. galloprovincialis* have been conducted. These studies suggested that both *P. sylvestris* and *P. nigra* are suitable hosts for the development of the beetle (unpublished data). This result indicates that *P. sylvestris* and *P. nigra* stands, located in different regions of Turkey, may increase the chance of rapid growth of the PWN's populations in case of introduction.

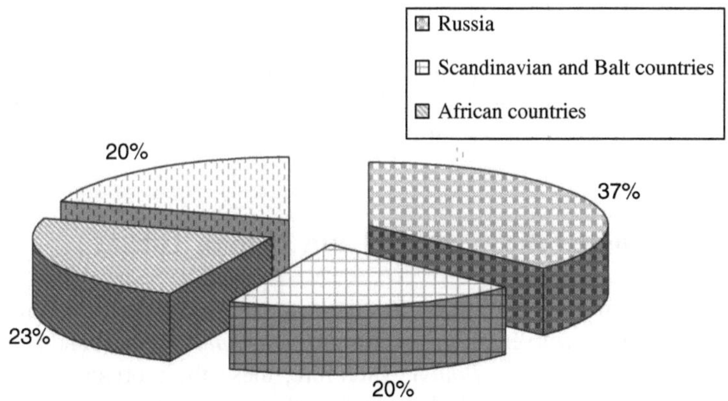


**Fig. 3** Wood imported countries (DPT, 2001)

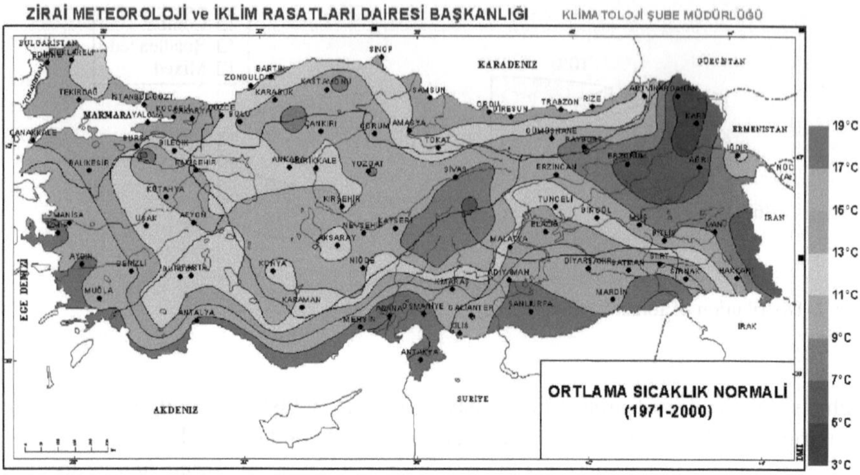


**Fig. 4** Annual mean temperature distribution in different parts of Turkey (www.meteor.gov.tr, 2006b)

The other reason to study pine wilt disease is the presence of highly suitable climatic conditions for the development of the PWN in Turkey. Distribution of annual mean temperature values is presented in Fig. 4. Annual mean temperature was over 12°C between 1928 and 2005. The temperature is over 20 °C between June and September in most parts of Turkey particularly in south and west Anatolia (Anonymous, 2006b).

High temperature (over 20 °C) is important for the development of pine wilt disease. The absence of disease in Europe (except for the recent introduction in Portugal) may be related to the relative absence of large forest areas located in hot regions (Webster, 2004). Most of the *P. sylvestris* forests are distributed throughout the cooler northern part of Europe. Braasch and Enzian (2004) stated that temperatures greater than 20 °C for at least an 8-week period increases the number of vulnerable pine trees to the PWN. These climatic conditions are always present in most parts of Turkey.

A survey of *B. xylophilus* was initiated in 2003 in Turkey, in view of the considerations presented. The survey has not been completed yet. In addition to survey, the potential insect vector species of the nematode have also been investigated.

## Studies Conducted Between 2003 and 2006 in Turkey

The pine wilt disease complex involves three organisms: the nematode (*B. xylophilus*); the insect vector, in general a *Monochamus* species; and a tree host, mostly a pine species (genus *Pinus*). Therefore, these three organisms and their interactions must be studied. In Turkey, the first step was to conduct a survey for the presence of the PWN in pine forests. The second step was to find potential insect vectors and to investigate their biological and ecological characteristics. The third step was to carry out different experiments based on the results of the nematode survey and insect vector studies.

## *Survey of Bursaphelenchus Species*

### Materials and Methods

A survey was conducted between 2003 and 2006 in 11 different Regional Forestry Directorates (Ankara, Artvin, Bolu, Istanbul, Amasya, Trabzon, Çanakkale, Bursa, Balıkesir, Izmir and Mersin) between 2003 and 2006. Selection of these sites was decided according to the following criteria: 1. forests adjacent to harbors; 2. forests adjacent to wood industry centers; and 3. forests in which wilting of trees was observed. Wood samples (40–80 g each) were collected from pine trees displaying declining symptoms, at 1.5 m of the trunk level (DBH), using a Pressler borer, from both sides of each tree, and stored in polythene bags. Nematodes were extracted using a modified Baermann funnel technique, and processed within 48 h. The collected nematodes were inoculated on *Botrytis cinerea*, growing in malt agar, and incubated for 2 weeks at 25 °C.

For nematode identification, special attention was given to the group of species closely similar to *B. xylophilus* (*xylophilus*-group *sensu* Ryss et al., 2005). Identification was made to species level, using morphological and molecular methodologies. For optical microscopic studies (Olympus BX50), nematodes were fixed with hot formalin (4%), processed to anhydrous glycerin and mounted on permanent slides according to the "express technique" described by Ryss (2003). The molecular analysis was performed following the methodology described in Cenis (1993) for DNA extraction, and the ITS-RFLP profiles were obtained following the methodology previously described (Hoyer et al., 1998; Mota et al., 1999).

### Results and Discussions

A total of 1254 samples were collected from the study sites. Over 400 samples have been processed and the nematodes identified. The remaining samples have been checked for the presence of *Bursaphelenchus* species. Some of them revealed several *Bursaphelenchus* species (Table 1). Eighteen of 400 samples contained *Bursaphelenchus* species. Three species of *Bursaphelenchus* were identified: *B. mucronatus* Mamiya and Enda, 1979 from *Pinus nigra, P. sylvestris, B. pinophilus* Brzeski and Baujard, 1997 from *P. nigra* and *B. sexdentati* Rühm, 1960 from *P. pinaster. B. mucronatus* was found in the areas of Artvin and Düzce, *B. sexdentati* was recorded from Düzce and Ankara, and *B. pinophilus* distributed only in Ankara. The identification of *Bursaphelenchus* species found in other cities has not yet been completed.

This survey is important to provide information on *Bursaphelenchus* species from Turkey. The presence of *B. mucronatus*, a closely related species to

**Table 1** *Bursaphelenchus* survey results between 2003 and 2006 in Turkey

| Name of Regional Forestry Directorate | Tree Species | No. of samples collected | No. of samples with *Bursaphelenchus species* |
|---|---|---|---|
| Artvin | *P. sylvestris* | 66 | 6 |
| Trabzon | *P. sylvestris, Picea orientalis* | 80 | 1 |
| Amasya (Samsun) | *P. sylvestris, P. nigra, P. brutia* | 71 | – |
| Ankara | *P. nigra* | 30 | 3 |
| Istanbul | *P. nigra, P. sylvestris* | 46 | – |
| Bolu (Düzce) | *P. sylvestris, P. nigra, P. pinaster* | 74 | 8 |
| Bursa | *P. brutia, P. nigra, P. pinaster* | 200 | – |
| Balikesir | *P. brutia, P. nigra* | 213 | ? |
| Çanakkale | *P. brutia, P. nigra, P. pinaster* | 166 | – |
| Izmir | *P. brutia, P. nigra, P. pinea* | 177 | ? |
| Mersin | *P. brutia* | 131 | ? |

*B. xylophilus*, is very important, since it indicates that Turkey has suitable conditions for the establishment and development of *B. xylophilus* populations. The survey of the PWN has not been completed yet. The remaining regions, located in the south and western parts of Turkey will be surveyed in the coming years.

## *Comparison of Some Biological Characteristics of* M. carolinensis *and* M. galloprovincialis

### Materials and Methods

In this study, colony data of *M. carolinensis* reared on *P. banksiana* Lamb. and colony data of *M. galloprovincialis* reared on *P. sylvestris* L. were used. Data collections were made at different time periods. Comparisons were made between the number of eggs laid, the number of larval entry holes, the number of adults emerged, generation survivorship, survivorship from egg to larva, and survivorship from larva to adult.

### Results and Discussions

A total of 27 trees from *P. banksiana* and 23 from *P. sylvestris* were used to compare the beetles. Log surface area and log volume averaged approximately 2000 $cm^2$ and 7100 $cm^3$ respectively for *P. sylvestris*. *P. banksiana* had a lower values for

**Table 2** Summary statistics of holding time, log metrics, numbers of eggs laid, laval entry holes, adult emerged and survivorship percentages for *M. galloprovincialis* and *M. carolinensis*

| Variable | N | | Mean | |
|---|---|---|---|---|
| | *M.g.* | *M.c.* | *M. galloprovincialis* | *M. carolinensis* (Akbulut et al., 2004) |
| Holding time (d) | 23 | 27 | $10 \pm 5$ | $11 \pm 5$ |
| Log area ($cm^2$) | 23 | 27 | $2046 \pm 618.8$ | $1504.0 \pm 152.2$ |
| Log volume ($cm^3$) | 23 | 27 | $7109.7 \pm 4476.4$ | $4548.0 \pm 905.5$ |
| No. eggs laid | 23 | 27 | $74 \pm 38$ | $127 \pm 61$ |
| No. larval entry holes | 23 | 27 | $13 \pm 3$ | $19 \pm 6$ |
| No. adult emerged | 23 | 27 | $7 \pm 2$ | $8 \pm 4$ |
| Generation survivorship % | 23 | 27 | $13 \pm 5$ | $12 \pm 15$ |
| Apparent survivorship (Larvae to adult %) | 23 | 27 | $22 \pm 7$ | $21 \pm 6$ |
| Apparent survivorship (Larvae to adult %) | 23 | 27 | $56 \pm 14$ | $41 \pm 19$ |
| Adult density (Adults/$dm^3$) | 23 | 27 | $1 \pm 1$ | $2 \pm 1$ |

both variables (Table 2). *M. carolinensis* laid significantly more eggs and produced significantly more larval entry holes than *M. galloprovincialis* ($p = 0.0011$, $p = 0.0001$). On the other hand, apparent survivorship from larva to adult was significantly greater for *M. galloprovincialis* than for *M. carolinensis* ($p = 0.0026$), and the number of adults emerged and generation survivorship did not differ significantly between the two beetle species ($p = 0.4558$, $p = 0.8264$).

In this study, we wanted to examine and compare the life histories of the two vector beetles of the PWN. Although the two species were reared on different host trees, there are remarkable similarities in some biological characteristics of both vectors that suggest that *M. galloprovincialis* would have a similar vector pattern as *M. carolinensis*. This is important to note, considering the potential threat of pine wilt disease to native European pine forests, and the rich body of knowledge already gathered on *M. carolinensis*.

Since the first report of the PWN from Portugal (Mota et al., 1999) studies on biological and ecological characteristics of *M. galloprovincialis* have been undertaking more intensely in several European countries, such as Portugal, Spain and recently Turkey (Naves et al., 2006a, b, Naves et al., 2007).

## Conclusions

Pine wilt disease is an important threat to susceptible pine forests of the world. In this study, several *Bursaphelenchus* species were found in Turkey. Fortunately, *B. xylophilus* has not been detected yet. The presence of *B. mucronatus* indicates that the conditions are suitable for possible introduction and establishment of *B. xylophilus* in pine forests of Turkey.

Potential vector beetle species of *B. xylophilus* and other *Bursaphelenchus* species are not well known in Turkey. The pine sawyer beetle, *M. galloprovincialis* (Olivier), was found to be the vector of the PWN in Portugal (Sousa et al., 2001). The presence of *M. galloprovincialis* has been previously reported from a number of different sites in Turkey (Özdikmen et al., 2005). The biology and ecology of *M. galloprovincialis* have been studied under laboratory conditions since 2002 but the data have not been published yet. Investigation of the biology and ecology of *M. galloprovincialis* is critical to the potential management of pine wilt in Turkey.

In conclusion, Turkey is located at a very important transitional area between Europe and Asia. The PWN is present in both Europe (Portugal) and Asia, which increases the possible introduction of the nematode into Turkey. Two necessary components for the spread of pine wilt disease, a vector beetle and suitable host trees, are already present. Climatic conditions for development of the PWN are highly suitable in a large area of Turkey. Therefore, it is important to study the pine wilt disease complex for prevention of possible introduction of the PWN and to control and eradicate both the nematode and insect vector.

## References

AKBULUT, S., VIEIRA, P., RYSS, A., YUKSEL, B., KETEN, A., MOTA, M. AND VALADAS, V. (2006). Preliminary survey of the PWN in Turkey. *Bulletin OEPP/EPPO Bulletin* 36: 538–542.

ANONYMOUS (2006a). *Forest of Turkey*. The Ministry of Environment and Forest of Turkey, General Forestry Directorate, 160 pp.

ANONYMOUS (2006b). www.meteor.gov.tr

BRAASCH, H. AND ENZIAN, S. (2004). The PWN problem in Europe-present situation and outlook. In *The pine wood nematode,* Bursaphelenchus xylophilus. *Nematology Monographs and Perspectives, vol. 1* (eds. M. Mota and P. Vieira), pp. 77–91. E. J. Brill, Leiden, The Netherlands.

BRZESKI, M.W. AND BAUJARD, P. (1997). Morphology and morphometrics of *Bursaphelenchus* (Nematoda: Aphelenchoididae) species from pine wood of Poland. *Annales Zoologici* 47: 305–319.

ÇANAKÇIOĞLU, H. AND MOL, T. (1998). [*Forest Entomology: Harmful and Beneficial Insects.*] Istanbul University, Pub. No: 4063, 541 pp.

CENIS, J.L. (1993). Identification of four major *Meloidogyne* spp. by random amplified polymorphic DNA (RAPD-PCR). *Phytopathology* 83: 76–80.

DPT (2001). T.R. Prime Ministry State Planning Organization. *Long-term strategy and Eight Five-year development Plan (2001–2005)*. Ankara, Turkey.

HELLRIGL, K.G. (1971). Die bionomie der Europaischen Monochamus-arten (Coleopt., Cerambycidae) und ihre bedeutung fur die forst-und holzwirtschaft. *Redia* LII: 367–509.

HOYER, U., BURGERMEISTER, W. AND BRAASCH, H. (1998). Identification of *Bursaphelenchus* species (Nematoda: Aphelenchoididae) on the basis of amplified ribosomal DNA (ITS-RFLP). *Nachrichtenblatt des Deutschen Pflanzenschutzdienstes* 50: 273–277.

MAMIYA, Y. (1988). History of pine wilt disease in Japan. *Journal of Nematology* 20: 219–226.

MAMIYA, Y. (1998). Review on the pathogenicity of *Bursaphelenchus mucronatus*. In *Sustainability of Pine Forests in Relation to Pine Wilt and Decline. Proceedings of an International Symposium, Tokyo, Japan, 26–30 October 1998* (eds. K. Futai, K. Togashi and T. Ikeda), pp. 57–64. Shokado Shote, Kyoto, Japan.

MAMIYA, Y. AND ENDA, N. (1979). *Bursaphelenchus mucronatus* n. sp. (Nematoda: Aphelenchoididae) from pinewood and its biology and pathogenicity to pine trees. *Nematologica* 25: 353–361.

MOTA, M., BRAASCH, H., BRAVO, M.A., PENAS, A.C., BURGERMEISTER, W., METGE, K. AND SOUSA, E. (1999). First report of *Bursaphelenchus xylophilus* in Portugal and Europe. *Nematology* 1: 727–734.

NAVES, P., CAMACHO, S., SOUSA, E. AND QUARTAU, J.A. (2007). Transmission of the PWN *Bursaphelenchus xylophilus* through feeding activity of *Monochamus galloprovincialis* (Coleoptera: Cerambycidae). *Journal of Applied Entomology* 131: 21–25.

NAVES, P., SOUSA, E. AND QUARTAU, J.A. (2006a). Reproductive traits of *Monochamus galloprovincialis* (Coleoptera:Cerambycidae) under laboratory conditions. *Bulletin of Entomological Research* 96: 289–294.

NAVES, P., SOUSA, E. AND QUARTAU, J.A. (2006b) Feeding and oviposition preferences of *Monochamus galloprovincialis* for certain conifers under laboratory conditions. *Entomologia Experimentalis et Applicata* 120: 99–104.

ÖZDIKMEN, H., ÖZDEMIR, Y. AND TURGUT, S. (2005). Longhorned beetles collection of the Nazife Tuatay Plant Protection Museum, Ankara, Turkey (Coleoptera, Cerambycidae). *Journal of the Entomological Research Society* 7: 1–33.

RYSS, A. (2003). Express technique to prepare permanent slides of nematodes. *Zoosystematica Rossica* 11: 257–260.

RYSS, A., VIEIRA, P., MOTA, M. AND KULINICH, O. (2005). A synopsis of the genus *Bursaphelenchus* Fuchs, 1937 (Aphelenchida: Parasitaphelenchidae) with keys to species. *Nematology* 7: 393–458.

SOUSA, E., BRAVO, M., PIRES, J., NAVES, P., PENAS, A., BONIFACIO, L. AND MOTA, M. (2001). *Bursaphelenchus xylophilus* (Nematoda: Aphelenchoididae) associated with *Monochamus galloprovincialis* (Coleoptera: Cerambycidae) in Portugal. *Nematology* 3: 89–91.

VIEIRA, P., AKBULUT, S., MOTA, M. AND VALADAS, V. (2003). First report of *Bursaphelenchus* sp. from Turkey, associated with *Pinus nigra*. *Journal of Nematology* 35: 369.

WEBSTER, J.M. (2004). The PWN: implications of factors past and present for pine wilt disease. In *The pine wood nematode*, Bursaphelenchus xylophilus. *Nematology Monographs and Perspectives, vol. 1* (eds. M. Mota and P. Vieira), pp. 55–64. E. J. Brill, Leiden, The Netherlands.

MAMIYA, Y. AND ENDA, N. (1979). *Bursaphelenchus mucronatus* n. sp. (Nematoda: Aphelenchoididae) from pinewood and its biology and pathogenicity to pine trees. *Nematologica* 25: 353–361.
MOTA, M., BRAASCH, H., BRAVO, M.A., PENAS, A.C., BURGERMEISTER, W., METGE, K. AND SOUSA, E. (1999). First report of *Bursaphelenchus xylophilus* in Portugal and Europe. *Nematology* 1: 727–734.
[illegible]
[illegible]
[illegible]
[illegible]

# Investigations on Wood–Inhabiting Nematodes of the Genus *Bursaphelenchus* in Pine Forests in the Brandenburg Province, Germany

**Ute Schönfeld, Helen Braasch, Wolfgang Burgermeister and Helmut Bröther**

## Introduction

After detection of the pine wood nematode (*Bursaphelenchus xylophilus*) from a restricted area in Portugal in 1999 (Mota et al., 1999), the European Commission decided to have investigations to confirm the absence of the pest in pine forests of the European member states. The territory of Brandenburg, a province of Germany, is covered by one third with extensive pine forests. *Pinus sylvestris* is the main forest tree in Brandenburg (861 000 ha). It is known to be a very susceptible pine species to infection by the pine wood nematode (PWN).

In the last years (since 2000), we investigated dead or dying pine trees as well as beetles, particularly the black pine sawyer, *Monochamus galloprovincialis*, the potential vector beetle of the PWN in Brandenburg, for wood–inhabiting nematodes. In order to obtain beetles for nematological examination, we set out bait logs to attract beetles for oviposition (since 2002). The following generation of hatching beetles was examined for the presence of nematode *dauer* juveniles. Samples of wood chips and sawdust from 15 sawmills were included in the nematological examination (since 1996). Sawmills predominantly use coniferous wood of *P. sylvestris* from local origin. Thus we have obtained an overview on the occurrence of *Bursaphelenchus* species in Brandenburg.

## Methods

### *Extraction and Determination of Nematodes*

Nematodes were extracted from wood and from beetles by applying the Baermann funnel technique. Morphological features of adult nematodes were used for

U. Schönfeld
Department of Customer Protection, Agriculture and Land Consolidation, Brandenburg, Steinplatz 1, D-15806 Zossen, OT Wünsdorf, Germany
e-mail: ute.schoenfeld@lvlf.brandenburg.de

M.M. Mota, P. Vieira (eds.), *Pine Wilt Disease: A Worldwide Threat to Forest Ecosystems*, 

identifying the nematode species from wood (Braasch, 2001). *Dauer* larvae extracted from beetles or nematodes from wood were multiplied on *Botryotinia fuckeliana* cultures on malt agar in order to obtain adult nematodes for diagnosis. ITS-RFLP analysis (Burgermeister et al., 2005) was applied to differentiate juveniles or to confirm morphological species diagnosis in several cases.

### Bait Logs

Healthy pine trees were cut from May to July at 45 locations, 32 near sawmills, four near wood storage sites, three at forest fire sites and six at places where felling had taken place in the summer time. The bait logs cut from these stems were collected after several months in autumn. Logs with *Monochamus* infestation show typical symptoms with large amounts of sawdust ("frass") given off by the beetle larvae. Stem segments of 50 cm length were placed in cages in the laboratory at room temperature during the winter months to accelerate the development of the beetles.

## Results

### Nematodes in Wood

Seventeen species of *Bursaphelenchus* were found, collected from 400 wood samples examined; among them were some recently described species (Braasch et al., 2004, 2006; Schönfeld et al., 2006) (Table 1). Of all samples examined, 31% were contaminated with these nematodes. So far, the PWN has not been found in Brandenburg, neither in wood from dying pines, sawmills nor bait logs or in the beetles. The most frequently detected species was *Bursaphelenchus mucronatus*, a species closely related to the PWN.

### Nematodes in Wood and Beetles from Bait Logs

No nematodes were found in the wood of the bait logs, nor did the stems show any symptoms of blue stain fungi at the time of cutting the pines. After infestation by wood-inhabiting beetles within several weeks, *B. mucronatus* could be extracted from the logs. Especially in the warm and dry summer of 2003, up to 6000 nematodes per 100 g wood were extracted in autumn.

Five beetle species hatched from the bait logs (Table 2): *M. galloprovincialis* (143 specimens), *Acanthocinus griseus* (28), *Hylobius abietis* (1), *Pissodes notatus* (14) and *Phaenops cyanea* (2). *Dauer* juveniles of *B. mucronatus* were found in beetles of *M. galloprovincialis* only.

From the tree segments kept at room temperature during the winter season, the beetles hatched several weeks or months earlier than under open air conditions

**Table 1** *Bursaphelenchus* species found in pinewood in Brandenburg/Germany

| Bursaphelenchus species | Pine forests | Number from sawmills | Bait logs | Total |
|---|---|---|---|---|
| *mucronatus* | 57 | 34 | 15 | 106 |
| *fraudulentus* | 1 | | | 1 |
| *sexdentati* | 13 | 6 | | 19 |
| *vallesianus* | 9 | 3 | | 12 |
| *poligraphi* | | 1 | | 1 |
| *pinophilus* | 3 | | | 3 |
| *borealis* | 2 | | | 2 |
| *eggersi* | 1 | 2 | | 3 |
| *tusciae* | 2 | 1 | | 3 |
| *hildegardae* | 2 | | | 2 |
| *leoni* | 6 | 3 | | 9 |
| *silvestris* | 1 | | | 1 |
| *pinasteri* | 2 | | | 2 |
| *paracorneolus* | 1 | | | 1 |
| *teratospicularis* | 1 | | | 1 |
| *fungivorus* | | 5 | | 5 |
| *willibaldi* | | 1 | | 1 |
| contaminated samples | 84 | 44 | 15 | 143 |
| examined samples | 229 | 126 | 45 | 400 |

(Fig. 1). To prevent drying out of the stem segments, it was necessary to dip them weekly into a water bath. Under these conditions we could detect up to 40 000 juveniles per hatching beetle in the laboratory as well as in the open air.

## *Monochamus galloprovincialis* in Brandenburg

The black pine sawyer developed in bait logs of 16 from 45 locations. In 15 of these locations, *B. mucronatus* was also found in the bait logs. At one location,

**Table 2** Beetles hatched from bait logs and their contamination with *dauer* juveniles of *Bursaphelenchus mucronatus*

| | Number of hatched beetles | Number of beetles from bait logs with dauer juveniles | Beetles infested with dauer juveniles |
|---|---|---|---|
| *Monochamus galloprovincialis* | 143 | 129 | 96 |
| *Acanthocinus griseus* | 28 | 17 | 0 |
| *Hylobius abietis* | 1 | 0 | 0 |
| *Pissodes notatus* | 14 | 11 | 0 |
| *Phaenops cyanea* | 2 | 0 | 0 |

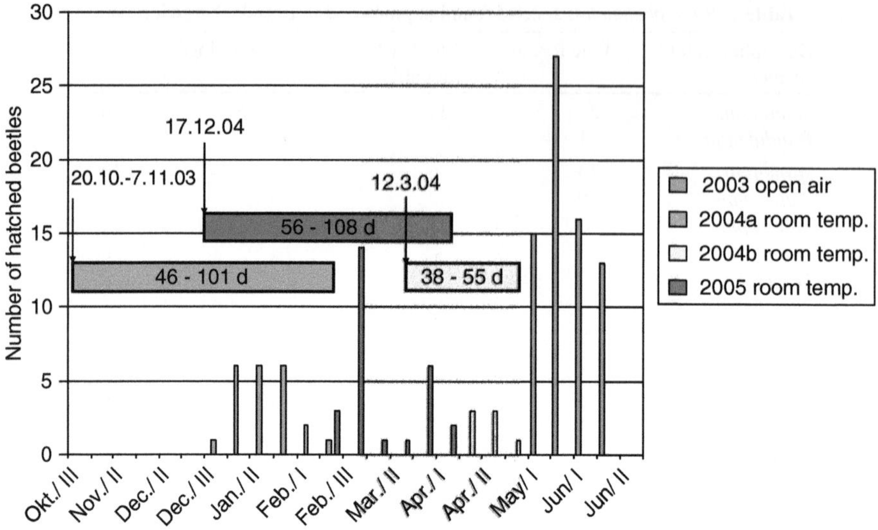


**Fig. 1** Development time and hatching of *Monochamus galloprovincialis* under open air (2003) and room temperature conditions (2004, 2005)

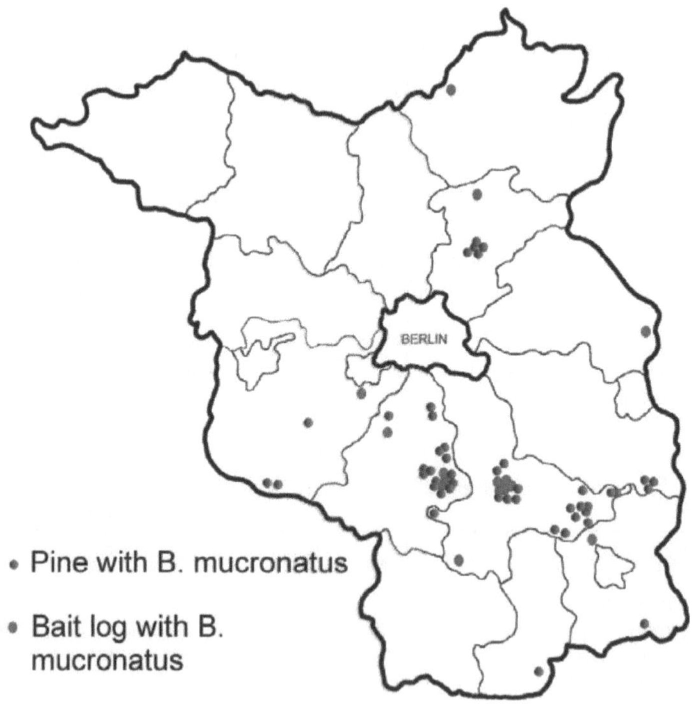


**Fig. 2** Locations with *Bursaphelenchus mucronatus* in bait logs and pine stands

*M. galloprovincialis* occurred without contamination with *B. mucronatus*. The map shows *B. mucronatus* in bait log locations and in addition in forest stands (Fig. 2).

## Conclusions

As may be concluded from our results, *B. xylophilus* is not present in Brandenburg so far. Among the 17 *Bursaphelenchus* species detected, *B. mucronatus* is the most frequent species in Brandenburg. However, we think that *B. mucronatus* is not the primary reason for dying of pines in Brandenburg. The frequent occurrence of *B. mucronatus* seems to be an indication of the wide distribution of the vector beetle *M. galloprovincialis*. Especially following forest fires or in connection with felling in the summer time, the beetles and the nematode *B. mucronatus* sometimes develop to high population densities. Dry and warm summer conditions favour this development. This indicates that the conditions for introduction and spreading of the PWN are favourable in Brandenburg.

Bait logs can successfully be used to attract beetles of the genus *Monochamus* for oviposition, during which the beetles infest the bait logs with *B. mucronatus*. Especially under conditions of latent infection with PWN, bait logs could be attractive for beetles of the genus *Monochamus* infested with *B. xylophilus*, too. Examination of beetles of the black pine sawyer and of wood from bait logs could be useful to detect an introduction of PWN in an early stage of establishment. The long-term mean summer temperatures in July/August in Brandenburg do not reach 20 °C. In warm summers, however, mean temperatures close to 20 °C or even slightly higher have been observed. An infection with PWN could remain unrecognized for some time. The use of bait logs should be an important component of the *Bursaphelenchus* survey especially in northern European countries.

## References


BRAASCH, H. (2001). *Bursaphelenchus* species in conifers in Europe: distribution and morphological relationships. *EPPO Bulletin* 31: 127–142.

BRAASCH, H., BURGERMEISTER, W., SCHÖNFELD, U., METGE, K. and BRANDSTETTER, M. (2006). *Bursaphelenchus hildegardae* sp. n. (Nematoda: Parasitaphelenchidae) – a new species of the 'eggersi' group. *Journal of Nematode Morphology and Systematics* 9: 27–38.

BRAASCH, H., SCHÖNFELD, U., POLOMSKI, J. AND BURGERMEISTER, W. (2004). *Bursaphelenchus vallesianus* sp. n. – a new species of the *Bursaphelenchus sexdentati* group (Nematoda: Parasitaphelenchoidae). *Nematologica mediterranea* 32: 71–79.

BURGERMEISTER, W., METGE, K., BRAASCH, H. AND BUCHBACH, E. (2005). ITS-RFLP patterns for differentiation of 26 *Bursaphelenchus* species (Nematoda: Parasitaphelenchidae) and observations on their distribution. *Russian Journal of Nematology* 13: 29–42.

MOTA, M.M., BRAASCH, H., BRAVO, M.A., PENAS, A.C., BURGERMEISTER, W., METGE, K. AND SOUSA, E. (1999). First report of *Bursaphelenchus xylophilus* in Portugal and in Europe. *Nematology* 1: 437–453.

SCHÖNFELD, U., BRAASCH, H. AND BURGERMEISTER, W. (2006). *Bursaphelenchus* spp. (Nematoda: Parasitaphelenchidae) in wood chips from sawmills in Brandenburg and description of *Bursaphelenchus willibaldi* sp. n. *Russian Journal of Nematology* 14: 119–126.

# Official Survey for *Bursaphelenchus xylophilus* Carried out on the Territory of the Republic of Poland

**Witold Karnkowski**

**Abstract** In Poland, the survey for the presence of *B. xylophilus* has been conducted since 2003 by inspectors of the State Plant Health and Seed Inspection Service in cooperation with the Governmental Forests. In the period 2003–2005, laboratories of the State Plant Health and Seed Inspection Service examined 5 454 samples of wood, chips, sawdust, etc. In 13 samples (0.24%) nematodes of the genus *Bursaphelenchus* were found. *B. mucronatus* was found in 11 samples (0.21%). Nematodes found in another two samples were recognized as belonging to the species *B. glochis* and *B. sexdentati* based on their morphological features.

## Introduction

The pinewood nematode (*Bursaphelenchus xylophilus*) is a serious pest of coniferous trees, especially pines. This nematode is an EU quarantine pest in accordance with the Council Directive 2000/29/EC of 8 May 2000 on protective measures against the introduction into the Community of organisms harmful to plants or plant products and against their spread within the Community. The first foci of this species in the European Union were reported in Portugal in May 1999 (Mota et al., 1999). Despite intensive control measures, its eradication in this country has not been successful. Further foci within the EU are likely to appear as this species may be introduced with coniferous wood, chips, sawdust, packaging material etc., imported from countries where it is present. Possibility of eradication of the nematode after introduction into other EU Member States will depend especially on its early detection. Therefore, each Member State is obliged to conduct on its territory surveillance for the presence of this nematode, following the EC Pinewood Nematode Survey Protocol (European Commission, 2000). In Poland, surveys for the

W. Karnkowski
Plant Health and Seed Inspection Service Central Laboratory, ul. Żwirki i Wigury 73, 87-100 Toruń, Poland
e-mail: w.karnkowski@piorin.gov.pl

presence of *B. xylophilus* have been conducted since 2003 (before accession to EU) by inspectors of the State Plant Health and Seed Inspection Service in cooperation with the Governmental Forests. It should be mentioned that the predominating tree species in Polish forests (about 70% of the whole forest area) is Scots pine (*Pinus sylvestris*), which is highly susceptible to *B. xylophilus*.

## Sampling

The survey was performed in accordance with the instructions manual elaborated in 2003 by the Central Laboratory of the Main Inspectorate of Plant Health and Seed Inspection Service on the basis of the above mentioned EC protocol. The survey was performed both in forests and in sawmills, where imported wood and wood products are stored (including pallets, packaging material, etc.). During forest inspection, special attention was paid to trees with symptoms of wilting and dieback. Samples of wood, packaging material, etc. were collected using an axe or drill (10–20 mm in diameter). Additionally, samples of wood chips, wood particles, sawdust, scrap etc. were collected in particular from sawmills, furniture producing factories, etc. From each place a sample weighing at least 300 g was collected. The samples were packed in polyethylene bags and sent to Laboratories of the Inspection (Voivodeship Laboratories or the Central Laboratory) for further examination.

## Extraction of Nematodes

Before extraction of nematodes samples were subjected to incubation for 10 days at 25 °C. The collected material (apart from chips, sawdust, etc.) was cut into small particles then placed in a 2–3 l capacity beaker and submerged with water for 3 days. Afterwards, water with debris and possibly nematodes was poured onto a nematological sieve with a proper filter in a Baermann funnel. The nematodes were extracted from the debris after 24 h using the Baermann funnel.

## Identification of Nematodes

Nematodes found in samples (if any) were identified firstly under a light (compound) microscope. The morphological identification was performed using available literature (EPPO, 2001; Ryss et al., 2005).

If the nematodes were identified as belonging to the *xylophilus* group, than each case was identified to species with a PCR-RFLP test, using the Dutch method (Dutch Plant Protection Service, 2001). The only laboratory performing this test is the Central Laboratory in Toruń.

## *Molecular Tests*

### DNA Isolation

DNA isolation from nematodes was performed in accordance with the Dutch diagnostic protocol (Dutch Plant Protection Service, 2001) using High Pure PCR template preparation kit Roche (cat. No. 1 796 828), available on the market. This is a rather simple method, which excludes the use of organic solvents, allowing for extraction and isolation of DNA simultaneously from a number of samples. The procedure was as follows:

- The sample material (at least 5 nematode specimens) was suspended in 10 μl of sterile reagent grade water.
- Homogenisation of nematodes was performed in a mixture of 200 μl tissue lysis buffer and 40 μl Proteinase K, which after immediate mixing was incubated in 55 °C for 1 h.
- After incubation, 200 μl binding buffer was added and after immediate mixing, incubation at a temperature of 72 °C was performed for 10 min.
- Next, 100 μl isopropanol was added and after mixing, the sample was pipetted into the upper reservoir of a combined high pure filter tube (so-called column).
- After centrifugation (1 min., 8 000 rpm), the flowthrough and collection tube were removed, the filter tube was combined with a new collection tube and 500 μl of removal buffer was added to the upper reservoir of the filter tube.
- After the next centrifugation (1 min., 8 000 rpm) the flowthrough and collection tube were removed, the filter tube was combined with a new collection tube and 500 μl of wash buffer was added to the upper reservoir of the filter tube, and centrifugation (1 min., 8 000 rpm) was performed.
- This step was repeated twice.
- The flowthrough was removed and the filter tube was combined with the same collection tube and centrifugation for 10 s at maximum speed (14 000 rpm) was performed to remove residual wash buffer.
- Next, the collection tube was removed and the filter tube was combined with a clean 1.5 ml reaction Eppendorf tube.
- The next step was the adding of 40 μl elution buffer, pre-warmed to 70 °C, for washing of DNA from the column.
- After centrifugation (1 min., 8 000 rpm), the reaction tube contained the eluted DNA, which was stored at −20 °C.
- DNA was diluted 1:100, if necessary.

### PCR

For every sample, the mix (50 μl) for analysis contained: 5 μl analysed DNA, 38.2 μl reagent grade water, 5 μl 10xPCR reaction buffer (Qiagen), 1 μl dNTP mix (10 mM; Promega), 0.2 μl HotStar Taq (5 U/μl; Qiagen) and 0.3 μl of each of two primers (1 mg/ml). The following primers were used: Ferris Fwd primer 5′-CGT AAC AAG GTA GCT GTA G-3′; Vrain Rev primer 5′-TTT CAC TCG CCG TTA CTA AGG-3′. Tubes with the reaction mix were placed in a MJ Research PTC-200 thermocycler. The PCR parameters were as follows: initial denaturation at 95 °C for 15 min., 35 reaction cycles at 94 °C for 15 s, 55 °C for 1 min., 72 °C for 45 s., and a final extension at 72 °C for 10 min., and 20 °C for 1 s. In each case during running of the PCR test, the negative control (not containing DNA) and positive control (DNA of known, preliminary identified species) were included.

### Electrophoresis in Agarose Gel

After completion of PCR, 5 μl aliquots of the reaction mix were resolved by gel electrophoresis in a 1% agarose gel with ethidium bromide in 1xTAE buffer at 76 V (45 min). DNA fragments were visualized under UV light using Foto/Analyst Investigator Fotodyne system.

### RFLP Analysis

Where bands of PCR products on the PCR gel indicated that the nematodes may be those belonging to *B. xylophilus*, RFLP analysis was performed. The following restriction enzymes were used: *Alu* I, *Hae* III, *Hinf* I, *Msp* I and *Rsa* I. For each restriction enzyme a mix (10 μl) was prepared, containing 8.7 μl PCR product, 0.3 μl (10 U) restriction enzyme and 1 μl 10X restriction enzyme buffer. The mixes (samples) for all restriction enzymes were then incubated for 1 h at 37 °C. The resulting products of RFLP analysis were resolved during the electrophoresis process in a 2% agarose gel in 1xTAE buffer at 80 V (0.5 h). The gel was analysed under UV light using Foto/Analyst Investigator Fotodyne system. Obtained RFLP patterns were compared with those published in the literature (EPPO, 2001; Burgermeister et al., 2005).

## Results

In the period 2003–2005, the laboratories of the State Plant Health and Seed Inspection Service examined 5 454 samples of wood, chips, sawdust, etc. (Fig. 1). In 13 samples (0.24%), nematodes of the genus *Bursaphelenchus* were found (Fig. 2). Nematodes belonging to the *xylophilus* group were found in 11 samples (0.21%) (Fig. 3). In all cases these nematodes were identified with PCR-RFLP as *Bursaphelenchus mucronatus*. Additionally, in two samples, nematodes not belonging to this group were found. The number of specimens of these nematodes in the samples was very low, therefore they were identified based on morphological features. These

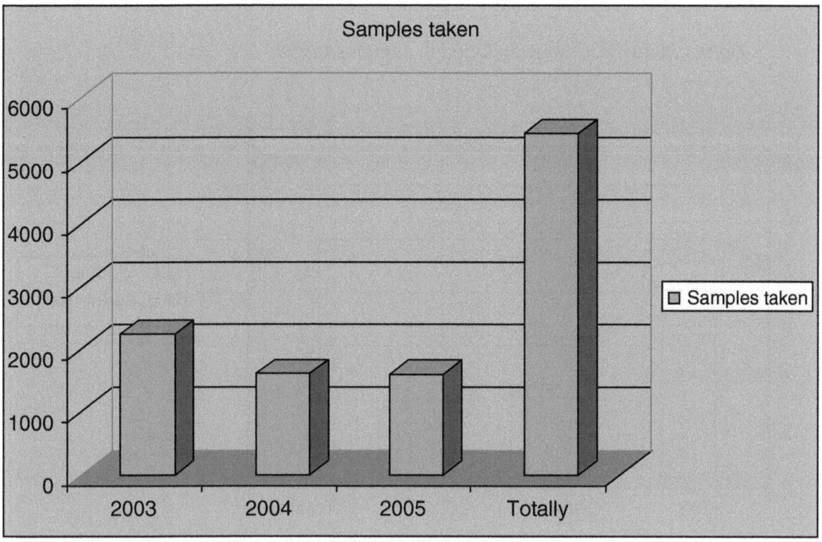


**Fig. 1** Wood samples collected in the period 2003–2005

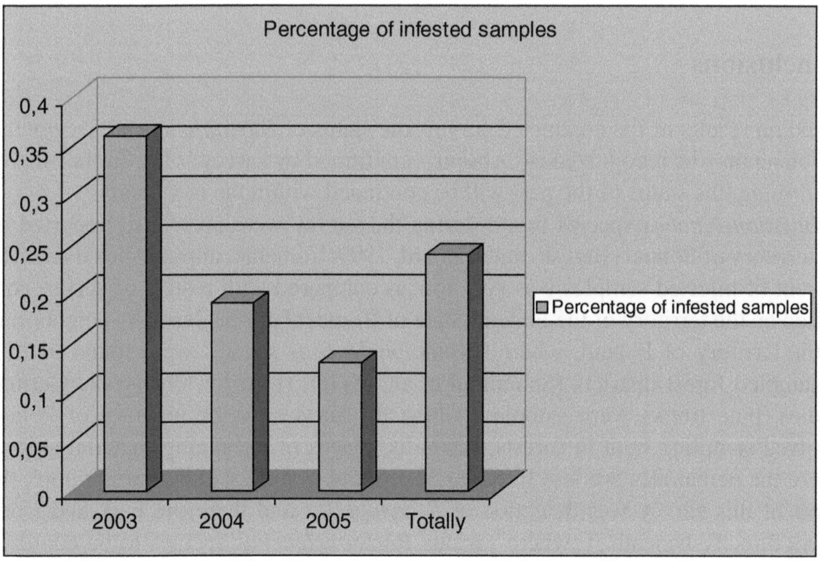


**Fig. 2** Percentage of samples infested with *Bursaphelenchus* spp

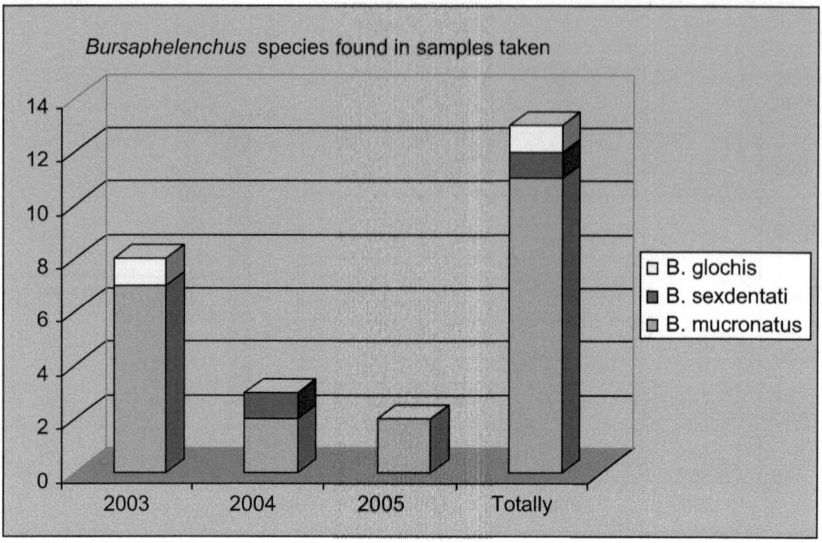


**Fig. 3** Number of samples infested with various species from the genus *Bursaphelenchus*

nematodes were recognized as belonging to the species *Bursaphelenchus glochis* and *Buraphelenchus sexdentati* (Fig. 3).

## Conclusions

Based on results of the conducted survey, the status of *Bursaphelenchus xylophilus* in Poland may be recognized as "Absent – confirmed by survey". The further survey confirming this status of the pest will be continued within the next years.

*Bursaphelenchus* species found during the survey were previously reported on the territory of Poland (Brzeski and Baujard, 1997; Tomalak, unpublished data). The percent of infested samples was very low as compared with results of survey conducted on the territory of the Federal State of Brandenburg in Germany neighboring to the territory of Poland, where 9 *Bursaphelenchus* species were found in 25% of sampled forest districts (Schönfeld et al., 2001). However, during the German studies, pine trunks were sampled, whilst the survey on the territory of Poland involved sampling both in forests, sawmills, places of packaging material storage where the nematodes are less likely to occur than in pine stands. Furthermore, the target of this survey was detection of *B. xylophilus* and therefore bark and wood layers directly beneath it, where various *Bursaphelenchus* species to be confused with the "*xylophilus*" group may occur, were generally omitted during sampling. Such sampling were proper for official PWN survey but generally did not allow to detect *Bursaphelenchus* species which are in phoretic relationship with insects occurring beneath bark, such as bark beetles.

## References

BRZESKI, M.W. AND BAUJARD, P. (1997). Morphology and morphometrics of *Bursaphelenchus* (Nematoda: Aphelenchoididae) species from pine wood of Poland. *Annales Zoologici* 47: 305–319.

BURGERMEISTER, W., METGE, K., BRAASCH, H. AND BUCHBACH, E. (2005). IST-RFLP patterns for differentiation of 26 *Bursaphelenchus* species (Nematoda: Parasitaphelenchidae) and observations on their distribution. *Russian Journal of Nematology* 13: 1, 29–42.

DUTCH PLANT PROTECTION SERVICE (2001). *Protocol Bursaphelenchus PCR-RFLP.*

EPPO (2001). Diagnostic protocols for regulated pests, *Bursaphelenchus xylophilus. Bulletin OEPP/EPPO Bulletin* 31: 61–69.

EUROPEAN COMMISSION (2000). *EC Pinewood Survey Protocol.* SANCO B/1 D(00), 18 pp.

MOTA, M. M., BRAASCH, H., BRAVO, A., PENAS, C., BURGERMEISTER, W., METGE, K. AND SOUSA, E. (1999). First report of *Bursaphelenchus xylophilus* in Portugal and in Europe. Nematology 1: 727–734.

RYSS, A., VIEIRA, P., MOTA, M. AND KULINICH, O. (2005). A synopsis of genus *Bursaphelenchus* Fuchs, 1937 (*Aphelenchida: Parasitaphelenchidae*) with keys to species. *Nematology* 7: 393–458.

SCHÖNFELD, U., APEL, K.H. AND BRAASCH, H. (2001). Nematoden der Gattung *Bursaphelenchus* (*Nematoda: Parasitaphelenchidae*) in des Landes Brandenburg – Ergebnisse eines Monitorings. *Nachrichtenbl. Deut. Pflanzenschutzd*: 180–184.

# *Bursaphelenchus* spp. in Wood Packaging Intercepted in China

**Jianfeng Gu, Jiancheng Zhang, Xianfeng Chen, Helen Braasch and Wolfgang Burgermeister**

Solid wood packaging material (including dunnage) made of unprocessed wood is known to be an important pathway for the introduction and spread of pests, but we may not know how serious this problem is. Another question is: Can a quarantine treatment certificate or the "HT" ("MB") mark on the wood really assure that the wooden packaging is free of pests?

In February of 2006, many living pinewood nematodes (PWN) were detected in wooden packaging exported from Portugal to China. The black mark "PT - 048 HT" could clearly be seen on the wood (PT representing Portugal, 048 representing the number of the heat treatment site). Was the wood cut from the PWN affected zone in Portugal and not properly treated? At Ningbo Entry-exit Inspection and Quarantine Bureau, China, almost all wooden packaging imported through this harbour have been sampled and inspected since 1997. In recent years, the quarantine pest *B. xylophilus* was detected many times in large numbers (sometimes thousands of specimens) in wood samples from different countries, and a considerable number of other *Bursaphelenchus* species, among them several undescribed species, were also found. The morphological and molecular diagnostic work was stimulated and supported by a training course for identification of *Bursaphelenchus* species given by Helen Braasch and Wolfgang Burgermeister in Shanghai, China, in October 2002.

From January, 2003 to June, 2006, wooden packaging material of 8720 batches was nematologically inspected in the Ningbo Entry-exit Inspection and Quarantine Bureau. Living nematodes were detected in 1772 batches, accounting for 20.3%. Most of them were identified as Rhabditida, Tylenchida and Aphelenchida (previously *Aphelenchoides* spp. and *Bursaphelenchus* spp., but also *Aphelenchus* spp., *Paraphelenchus* sp., *Cryptaphelenchus* spp. and *Ruehmaphelenchus* sp.). *Bursaphelenchus* spp. was detected in 343 batches from 26 different countries (Table 1). The following species were identified on the basis of their morphology and their ITS-RFLP patterns: *B. xylophilus, B. fungivorus, B. rainulfi, B. hylobianum, B. thailandae, B mucronatus, B. aberrans, B. lini, B. singaporensis, B. doui, B. conicaudatus,*

---

J. Gu
Technical Centre, Ningbo Entry-exit Inspection and Quarantine Bureau, 9 Mayuan Road, Ningbo, Zhejiang, China
e-mail: gujf@nbciq.gov.cn

**Table 1** *Bursaphelenchus* spp. detected from imported wooden packaging in Ningbo Port (2003.1–2006.6)

| *Bursaphelenchus* species | Number of detected batches | Export areas |
|---|---|---|
| *B. mucronatus* | 97 | Hong Kong (16), Taiwan (13), Germany (13), Republic of Korea (11), Japan (10), Russia (6), USA (5), France (4), Netherlands (4), Belgium (2), Spain (2), Israel (2), Saudi Arabia (1), Italy (1), Brazil (1), Canada (1), Austria (1), Norway (1), Sweden (1), Finland (1), UK (1) |
| *B. xylophilus* | 63 | USA (21), Taiwan (19), Hong Kong (7), Japan (3), Republic of Korea (3), Belgium (2), Italy (2), Netherlands (2), Thailand (1), Spain (1), Brazil (1), Portugal (1) |
| *B. rainulfi* | 37 | Republic of Korea (9), Taiwan (9), Japan (8), Hong Kong (4), USA (2), Germany (2), Belgium (2), South Africa (1) |
| *B. fungivorus* | 24 | Republic of Korea (10), Japan (5), Taiwan (4), Finland (1), Germany (2), South Africa (1), Hong Kong (1) |
| *B. thailandae* | 19 | Republic of Korea (5), Japan (3), Hong Kong (3), Taiwan (3), USA (2), Italy (1), India (1), Belgium (1) |
| *B. doui* sp.n. | 15 | Republic of Korea (6), Taiwan (4), Japan (3), South Africa (1), Malaysia (1) |
| *B. arthuri* sp.n. | 5 | Republic of Korea (2), Taiwan (1), USA (1), Japan (1) |
| *B. aberrans* | 2 | Japan (2) |
| *B. conicaudatus* | 2 | Taiwan (1), Germany (1) |
| *B. lini* | 2 | Taiwan (1), Germany (1) |
| *B. vallesianus* | 2 | Germany (1), Belgium (1) |
| *B. singaporensis* sp.n. | 1 | Singapore (1) |
| *B. pinasteri* | 1 | Italy (1) |
| *B. hofmanni* | 1 | Italy (1) |
| *B. hylobianum* | 1 | Japan (1) |
| *B. leoni* | 1 | Belgium (1) |
| *B. fraudulentus* | 1 | Belgium (1) |
| *B. paracorneolus* | 1 | White Russia (1) |
| *B. teratospicularis* | 1 | UK (1) |
| *B. africanus* sp.n. | 2 | South Africa (2) |
| Unidentified *Bursaphelechus* spp. | 65 | |

*B. vallesianus, B. pinasteri*, B. *hofmanni, B. arthuri, B. leoni, B. teratospicularis, B. fraudulentus, B. paracorneolus* and *B. africanus*. The most frequently found species were *B. mucronatus* (97 times), *B. xylophilus* (63 times), *B. rainulfi* (37 times), *B. fungivorus* (24 times), *B. thailandae* (19 times) and *B. doui* (15 times) (Table 1). The quarantine pest *B. xylophilus* was not only found in packaging wood im-

**Table 2** Annual rates of nematode detection in batches of imported packaging wood in Ningbo (2000.1–2006.8)[1]

| Year | Total | Japan | Exporting country or area USA | Republic of Korea | European Union |
|---|---|---|---|---|---|
| 2000 | 93/265 = 35.1% | 36/102 = 35.3% | 10/28 = 35.7% | 11/23 = 47.8% | 2/6 = 33.3% |
| 2001 | 76/325 = 23.4% | 22/106 = 20.8% | 15/62 = 24.2% | 12/35 = 34.3% | 10/42 = 23.8% |
| 2002 | 163/899 = 18.1% | 43/246 = 17.4% | 14/109 = 12.8% | 42/234 = 17.9% | 31/161 = 19.3% |
| 2003 | 124/719 = 17.2% | 32/204 = 15.7% | 11/93 = 11.8% | 40/144 = 27.8% | 22/100 = 22% |
| 2004 | 442/2110 = 20.9% | 50/206 = 24.3% | 42/282 = 14.9% | 74/309 = 23.9% | 86/612 = 14.1% |
| 2005[2] | 348/1567 = 22.2% | 43/157 = 27.4% | 42/259 = 16.2% | 78/526 = 14.8% | 25/146 = 17.1% |
| 2006[2] (1–8) | 150/1225 = 12.2% | 12/83 = 14.5% | 10/164 = 6% | 20/114 = 17.5% | 36/315 = 11.4% |
| Average | 1396/7110 = 19.6% | 238/1104 = 21.6% | 144/997 = 14.4% | 277/1385 = 20% | 212/1382 = 15.3% |

[1]Data are listed as follows: number of batches containing nematodes/total number of batches inspected = percent of batches containing nematodes (including *Bursaphelenchus* spp. and any other nematode species)

[2]The nematodes detected in Beilun port lab (which was established in 2005) were not included.

ported from countries/territories where it is known to occur (The United States of America, Japan, Republic of Korea, Hongkong, Taiwan, Portugal), but also from countries considered to be free of this dangerous pest (Brazil, Thailand, Belgium, The Netherlands, Italy, Spain). The records of *B. xylophilus* in packaging wood from several non-infested European countries and possibly also from Brazil can only be explained by circulation of the wood packaging material among infested and non-infested countries.

During the period from January, 2000 to August, 2006, the number of imported packaging wood batches inspected per year varied greatly by about one order of magnitude, as shown for different exporting areas in Table 2. In contrast, the percentage of batches containing any nematodes exhibited much less variation, averaging 19.6%. Typically, percentage values from about 15–25% were obtained for most of the exporting areas and years listed in Table 2. It is important to note that the percentage of batches containing *Bursaphelenchus* spp. did not show a clear tendency to decrease during these years, although the quarantine regulations are becoming more and more strict. On January 1, 2000, China implemented regulations to ensure that solid wood packaging material entering the country was not infested by the pine wood nematode. Coniferous wooden packaging material from USA and Japan had to be heat-treated. Since February 20, 2002, China required that coniferous wooden packaging material from the Republic of Korea should also be heat-treated or fumigated. Since October 1, 2002, a heat treatment or fumigation certificate is required for coniferous packing wood imported from the USA, Japan, Korea and the European Union. Since January 1, 2006, a heat treatment or fumigation (methyl bromide) mark are required for all the imported wooden packaging. Without an appropriate mark all wood must be properly treated in Chinese ports.

A global strategy is needed to address this international pest problem. As an important step, the FAO guidelines for regulating wood packaging material in international trade (ISPM 15) were issued in 2003. However, ratification of ISPM 15 in individual countries requires additional time. Strong efforts should, therefore, be made to guarantee that appropriate procedures of phytosanitary treatment are put into practice as soon as possible in all importing and exporting countries.

## References


Braasch, H., Gu, J., Burgermeister, W. and Zhang, J. (2005). *Bursaphelenchus doui* sp.n. (Nematoda: Parasitaphelenchidae) in packaging wood from Taiwan and South Korea – a new species of the "xylophilus" group. *Russian Journal of Nematology* 12: 19–27.

Burgermeister, W., Gu, J. and Braasch, H. (2005a). *Bursaphelenchus arthuri* sp. n. (Nematoda: Parasitaphelenchidae) in packaging wood from Taiwan and South Korea – a new species belonging to the "fungivorus" group. *Journal of Nematode Morphology and Systematics* 8: 7–17.

Burgermeister, W., Metge, K., Braasch, H. and Buchbach, E. (2005b) ITS-RFLP patterns for differentiation of 26 *Bursaphelenchus* species (Nematoda: Parasitaphelenchidae) and observations on their distribution. *Russian Journal of Nematology* 13: 29–42.

FAO (2003). *Guidlines for regulating wood packaging material in international trade.* International Standards for Phytosanitary Measures (ISPM). Publication No. 15, FAO, Rome, 17pp.

GU, J., ZHANG, J., BRAASCH, H. AND BURGERMEISTER, W. (2005). *Bursaphelenchus singaporensis* sp. n. (Nematoda: Parasitaphelenchidae) in packaging wood from Singapore – a new species of the *B. xylophilus* group. *Zootaxa* 988: 1–12.

HUNT, D.J. (1993). *Aphelenchida, Longidoridae and Trichodoridae – their systematics and bionomics.* CAB International, Wallingford (GB), 352pp.

KANZAKI, N., TSUDA, K. AND FUTAI, K. (2000). Description of *Bursaphelenchus conicaudatus* n. sp. (Nematoda: Aphelenchoididae), isolated from the yellow-spotted longicorn beetle, *Psacothea hilaris* (Coleoptera: Cerambycidae) and fig tree, *Ficus carica. Nematology* 2: 165–168.

MOTA, M. M., BRAASCH, H., BRAVO, A., PENAS, C., BURGERMEISTER, W., METGE, K. AND SOUSA, E. (1999). First report of *Bursaphelenchus xylophilus* in Portugal and in Europe. *Nematology* 1: 727–734.

RYSS, A., VIEIRA, P., MOTA, M. AND KULINICH, O. (2005). A synopsis of the genus *Bursaphelenchus* Fuchs, 1937 (Aphelenchida: Partasitaphelenchidae) with keys to species. *Nematology* 7: 393–458.

# Part II
# Biology and Microbial Inter-Relationships

**Kazuyoshi Futai and Manuel Mota**

## Summary

To effectively control pine wilt disease (PWD), we must understand the biology and inter-relationship of organisms involved in this epidemic disease. Among the basic biological disciplines related to development and behavior of nematodes, its regulation system and nerve network, is embryology. *Caenorhabditis elegans* is a well-known model organism in molecular embryology and has often been used for comparative purposes to understand plant parasitic nematodes. Hasegawa et al. have studied the embryonic development of *Bursaphelenchus xylophilus* in comparison with that of *C. elegans* using several methods developed for this nematode species. Among several findings are the position of the anterior-posterior axis (opposite to *C. elegans*) and the elucidation of the karyology and cytogenetics of the species, previously unknown. Reverse genetic analysis and in particular the potential of using RNAi has also been studied, albeit in a preliminary fashion. These findings should be applied to developing new tactics for controlling nematode pests such as *B. xylophilus* in the future.

It has been revealed that there is remarkable variability in the pathogenicity of *B. xylophilus* among regional strains. According to global spreading of PWD, the pathogenicity of *B. xylophilus* must have changed and evolved. Mota and his Japanese colleagues reported their recent works done in Japan, which compares various biological features of *B. xylophilus* between two Japanese and two Portugal isolates. Further comparative studies are on going also in Portugal using the same isolates of nematode and Portugal pine species. Their comparative studies stress the idea of evolution of pathogenicity in the genus *Bursaphelenchus* as well as in *B. xylophilus*. From the practical point of view, their results could be useful to discriminate between degrees of susceptibility in pine species, or to find out determinative factors of pathogenicity of *B. xylophilus*. Ongoing research is also being carried out in Portugal regarding evaluation of the susceptibility degree of maritime pine (*Pinus pinaster*), as well as elucidation of tolerance to PWN of *P. pinea* (stone pine), an important economical staple crop, which produces pineon seed, and a species being considered for a substitute of *P. pinaster* in certain areas of the affected zone.

The life cycle of *B. xylophilus* consists of two phases, the reproductive phase and the dispersal phase. In either phase, *B. xylophilus* is associated with various kinds of microorganisms and small invertebrates, and its behavior, nutrition, reproduction, and distribution in the host tree are greatly influenced by cohabiting small organisms. Sriwati et al. and Wang et al. have studied the population and distribution of *B. xylophilus* in host pine trees in relation with cohabiting fungi. They proved that the cohabiting fungi might play a role not only as food for *B. xylophilus* in dead pine trees, but also as its antagonistic cohabitant to suppress its population and distribution. Nematode density thus determined around the pupal chamber of *Monochamus* beetles, influences the number of nematodes carried by the vector beetle *Monochamus sp.*, and thereby determines the amount of damage caused to the trees.

When infected with *B. xylophilus*, and having lost its resistance ability against invading microorganisms, a pine tree becomes colonized by various species of fungi and bacteria. Zhao and his colleagues have studied the inter-relationships between *B. xylophilus* and its accompanying bacteria, and found a mutualistic relationship between them. They have also found specific toxins produced by these accompanying bacteria, and proposed a hypothesis that PWD is a complex disease, which is induced by both PWN and associated toxin-producing bacteria. Because this idea is still controversial, further studies are needed to elucidate the role of accompanying bacteria. To deepen our understanding of this serious forest disease, and thereby obtain effective control measures, further studies are needed to reveal the basic biology of *B. xylophilus* and its cohabiting microorganisms.

# Developmental Biology and Cytogenetics of *Bursaphelenchus xylophilus*

**Koichi Hasegawa, Manuel Mota, Kazuyoshi Futai and Johji Miwa**

**Abstract** The pinewood nematode, *Bursaphelenchus xylophilus*, reproduces bisexually: a haploid sperm fertilizes a haploid oocyte, and the two pronuclei rearrange, move together, fuse, and begin diploid development. Early embryonic events taking place in the *B. xylophilus* embryo are similar to those of *Caenorhabditis elegans*, although the anterior-posterior axis appears to be determined oppositely to that observed for *C. elegans*. That is, in the *B. xylophilus* embryo, the male pronucleus emerges at the future anterior end, whereas the female pronucleus appears laterally. To understand the evolution of nematode developmental systems, we cloned the full length of *Bx-tbb-1* (beta tubulin) from *B. xylophilus* cDNA and attempted to apply reverse genetics analysis to *B. xylophilus*. Several lengths of double stranded RNA (dsRNA) for the *Bx-tbb-1* gene were synthesized by *in vitro* transcription, and both *B. xylophilus* and *C. elegans* were soaked in dsRNA for RNAi. Both nematodes could suck up the dsRNA, and we could detect the abnormal phenotypes caused by *Bx-tbb-1* dsRNA in *C. elegans*, but not in *B. xylophilus*. We suspect that systemic RNAi might be suppressed in *B. xylophilus* and are attempting to establish other methods for functionally analyzing *B. xylophilus* genes.

## Introduction

The pinewood nematode (PWN), *Bursaphelenchus xylophilus*, the causal agent of pine wilt disease, which destroys millions of pine trees every year in East Asia (Mamiya, 1983), have been found recently for the first time in Europe (Mota et al., 1999). Although parasitic in nature, *B. xylophilus* is easily maintained in the laboratory and is experimentally amicable, having a large brood size with short generation time and a simple and transparent body and embryo (Hasegawa et al., 2004; Hasegawa et al., 2006). Here we show some embryological and cytological aspects of *B. xylophilus* development, as it is compared with those of the model organism

K. Hasegawa
Institute for Biological Function, Chubu University, 1200 Matsumoto, Kasugai 487-8501, Japan
e-mail: hasegawaelegans@hotmail.com

*Caenorhabditis elegans*. And we also attempted to establish methods for the functional analysis of *B. xylophilus* genes.

## Materials and Methods

### Nematodes

The *B. xylophilus* isolate S-10 (originally isolated from Shimane Prefecture, Japan) was used. Methods for culturing, handling, and observing *B. xylophilus* follow the methodology described by Hasegawa et al. (2004). *C. elegans* var. Bristol, strain N2 was used (Brenner, 1974).

### Immunostaining for Microtubules and Actin

Early *B. xylophilus* embryos were collected and fixed as described by Hasegawa et al. (2004). Mouse monoclonal anti-α -tubulin antibody (DM1α, Sigma) and rabbit monoclonal anti-actin antibody (Sigma) were used as primary antibodies, and FITC-conjugated goat anti-mouse IgG (Sigma) and Cy3-conjugated sheep anti-rabbit IgG (Sigma) were used as secondary antibodies. Antibody staining was performed as described by Hasegawa et al. (2004). Stained embryos were washed with PBS for several seconds and mounted in Vectashield (Vector Laboratories, Inc.) for viewing with a ZEISS Axiovert 200 microscope equipped with a confocal laser-scanning module (ZEISS LSM510).

### dsRNA Synthesis

The plasmid pBxTBB5R-1 was used as a PCR template. This plasmid was constructed by inserting a 5′ fragment of *Bx-tbb-1* cDNA (1079 bp) into the pPCR SCRIPT SK (+) MCS (Stratagene) (Hasegawa and Miwa, unpublished data). Three different lengths of DNA templates were prepared from pBxTBB5R-1 with the following primers: for the 1 kbp (1,079 bp) DNA template, the T7 primer (5′ -GTA ATA CGA CTC ACT ATA GGG C- 3′) and Cmo422 (5′- GCG TAA TAC GAC TCA CTA TAG GGA ACA AAA GCT GGA GCT- 3′) (Mello, personal communication); for the 500 bp (464 bp) DNA template, T7BXTBB5R_SP500 (5′ –GTA ATA CGA CTC ACT ATA GGG CCA CTC TTT CCG TC- 3′) and Cmo422; and for the 100 bp (117 bp) DNA template, T7BXTBB5R_SP100 (5′ –GTA ATA CGA CTC ACT ATA GGG CTG CTT GTG ATC CT- 3′) and Cmo422. Double-stranded RNA (dsRNA) was synthesized by in vitro transcription with T7 RNA Polymerase (Takara Bio) for each length. In addition to the three dsRNA lengths, we prepared siRNA (about 22 bp) from 1 kbp dsRNA fragment, digested with Takara siRNA Cocktail Kit (Takara Bio) for 3.5 hours at 30 °C.

## *Soaking RNAi*

Soaking RNAi of *C. elegans* L4 animals was performed following the method of Maeda et al. (2001), and *B. xylophilus* was performed following the method of Urwin et al. (2002), with some modification. For checking the soaking efficiency, the soaking buffer (Maeda et al., 2001) with or without 10 mM octopamine was mixed with FITC dye (1 μg/μL), and nematodes were dipped for 24 hours at 20 °C (*C. elegans*) or 25 °C (*B. xylophilus*) and viewed by fluorescence microscopy (Nikon E600). Double-stranded RNA for each size (1 kb, 500 bp, 100 bp) was dissolved in soaking buffer (0.5–1 μg/μL), and both *B. xylophilus* (<50 mixed stage individuals) and *C. elegans* (6–10 L4 individuals) (parental generation) were soaked in dsRNA for RNAi. After 24 hours of incubation, *C. elegans* were transferred onto a fresh NGM plate seeded with *Escherichia coli* OP50, and incubated overnight at 20 °C for recovery. After that, healthy *C. elegans* adults were individually transferred onto NGM spot plates for checking the hatching rate of F1 eggs. In case of *B. xylophilus*, nematodes were incubated in dsRNA for 24–48 hours at 25 °C. After soaking, *B. xylophilus* were transferred onto a fresh 1/10 PDA plate seeded with *Botrytis cinerea*, and incubated 1 day at 25 °C for recovery and mating. After that, mating sets of one adult female and 2 to 5 adult males were transferred onto agar plates seeded with *B. cinerea* for checking F1 hatching rate. Resulting F1 phenotypes were viewed by Nomarski optics (Nikon E600).

# Results and Discussion

## *First Cell Division and AP Axis Formation*

The *B. xylophilus* embryo is long and slender, about 25 μm wide and 60 μm long. As in *C. elegans*, the first cell division was unequal, producing the larger AB cell and the smaller $P_1$ cell (Hasegawa et al., 2004; Hasegawa et al., 2006), which were destined to a certain fate, although still reprogrammable at this stage (Gönczy and Rose, 2005). This division reveals the anterior-posterior (AP) axis of a nematode in that the AB and $P_1$ cells indicate, respectively, the future anterior and the posterior sides (Hasegawa et al., 2004). Here we analysed the relations of the male and female pronuclei, cell cytoskeleton formation, and AP axis in *B. xylophilus* embryos from fertilization to the 2-cell stage.

The cortical membrane of a newly laid egg was rumpled and active, and seemingly random cytoplasmic streaming was visible (Fig. 1A). At this time, the DAPI-stained sperm was seen as a dot positioned at the future anterior side of the embryo (Fig. 1a). Meiosis was propelled by a small seemingly acentriolar meiotic spindle positioned at the lateral mid-point position of the embryo (Fig. 1a). The 1st and 2nd polar bodies were extruded out, while vesicular male and female pronuclei were reconstituted (Fig. 1B, b). Duplicated centrosomes were on the surface of the male pronucleus, interacting with the cortical actin (Fig. 1b). The sperm entry point

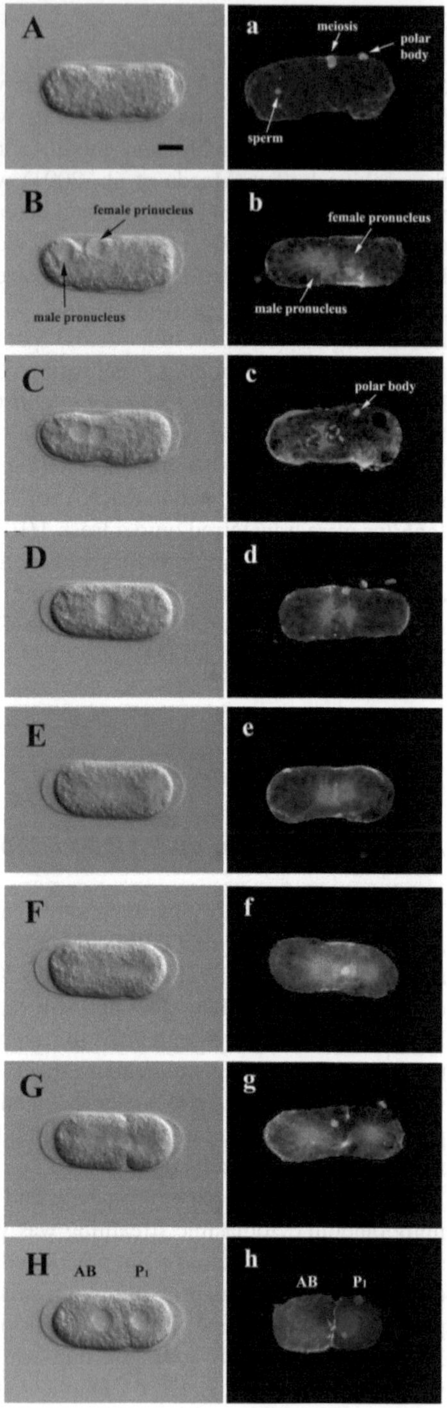


**Fig. 1** Embryonic cell divisions of *B. xylophilus*, from fertilization to the two-cell stage. **A–H**: Nomarski differential interference contrast images (DIC). **a–h**: Confocal laser-scanning microscope images, microtubules and actin cytoskeleton are visualized with antibody, and DNA with DAPI. (**A**) Just after oviposition, the entire cortical region is ruffling and one of the polar bodies is visible. (**a**) At this stage, the sperm is visible as a faint dot, and 2nd meiosis starts with microtubules elongated. (**B**) Soon after the completion of meiosis, the male pronucleus appears at the future anterior pole of the embryo, and the female pronucleus emerges at a lateral midpoint position. (**b**) Duplicate centrosomes are visible at the surface of the male pronucleus, interacting with the cortical actin. (**C**) Pronuclear meeting. (**c**) Two centrosomes are nucleating. (**D, d**) Juxtaposed pronuclei move to the center and rotate 90 degrees. (**E**) Two pronuclei are fusing. (**e**) Metaphase stage embryo, two centrosomes are located in the center of the embryo along the longitudinal axis. (**F**) Anaphase stage embryo. (**f**) Posterior centrosome moves posteriorly to pull the chromosomes posteriorly. (**G, g**) The cell membrane invaginates at the start of cell division. (**H, h**) Two-cell stage embryo. Anterior is left. Scale bar, 10 μm

appeared to become the future anterior end of the *B. xylophilus* embryo, although the *B. xylophilus* sperm also brought a centrosome into the oocyte. The rumpled cortical membrane became smooth, and male and female pronuclei moved toward each other and eventually met (Fig. 1C, c). Chromosomes became condensed in the prophase stage, and six chromosomes were visible in each pronucleus (Fig. 1c). They moved to the centre where they rotated 90° (Fig. 1D, d), and fused to become one. The pair of male and female pronuclear-derived chromatids was fused and aligned along the metaphase plate (Fig. 1E, e). Two centrosomes were located at the centre of the embryo along its longitudinal axis (Figure 1e). In anaphase, the posteriorly-positioned centrosome moved more posteriorly to divide the chromosomes antero-posteriorly (Fig. 1F, f). Subsequently, the embryo divided unequally (Fig. 1G, g) to form the larger anterior AB cell and the smaller posterior $P_1$ cell (Fig. 1H, h), thereby determining the anterior-posterior axis.

Polarity formation is an important feature for establishing many different cell types through several rounds of asymmetric cell divisions. The initial asymmetric cues such as sperm entry (Goldstein and Hird, 1996) establish the cell polarity followed by cytoskeleton reorganization and polarized localization of several cortical proteins in the *C. elegans* oocyte. A mature oocyte is fertilized as it passes through the spermatheca by sperm located there (Ward and Carrel, 1979). Compared with sperm, the oocyte is much larger and contains more cellular factors necessary for cell division, cell-fate determination, and morphogenesis. Although almost all factors required for early embryogenesis are supplied maternally (Miwa et al., 1980), paternal factors such as chromosomes, centrosome, and cytoplasm also constitute essential and important contributions (Singson, 2001). Fertilization triggers completion of the final stage of oocyte meiosis and promotes rearrangement of the cortical actin cytoskeleton, which generates cytoplasmic flows anteriorly in the cortical region and posteriorly inside of the zygote to distribute cell-fate determinants (Munro et al., 2004). At metaphase, two centrosomes are located at the centre of the embryo along its longitudinal axis, but in anaphase the posteriorly-positioned centrosome moved posteriorly to divide the chromosomes antero-posteriorly. Asymmetric distribution of cell-fate determinants should eventually lead the first cell division to produce the larger anterior AB cell and the smaller posterior $P_1$ cell, which are programmed to have different fates (Sulston et al., 1983; Miwa, 1986). Although the AP axis determination of the *B. xylophilus* embryo was entirely opposite to that of the *C. elegans* embryo, there is a question whether or not the *B. xylophilus* embryo uses the same types of molecules to generate AP polarity as in the *C. elegans* embryo? We thus need to compare the functional differences or similarities of the molecules necessary for AP axis determination between *B. xylophilus* and *C. elegans*.

## *Attempt to Induce RNA Interference in* **Bursaphelenchus xylophilus**

Next we attempted to analyse the function of *Bx-tbb-1*, a homologue of the gene encoding beta tubulin, which is an important molecule for cell division. The

*Bx-tbb-1* gene has been isolated from *B. xylophilus*, and its structure and expression patterns were analysed. Its nucleotide sequence shares 78% homology with tbb-1 of *C. elegans*. Transcriptional and functional analyses of *B. xylophilus* genes were initiated by constructing ESTs (Kikuchi et al., 2004; Kikuchi et al., 2005), and we attempted to introduce the reverse genetics technique in *B. xylophilus* for analyzing the gene function. Although conventional genetics (forward genetics) provides the most powerful tool to understand most biological phenomena, only a few organisms are endowed with this potential. The recently discovered RNAi (double strand RNA-mediated interference) is now considered by many to be a "wonder magic" for genetic analysis in non-model organisms for which classical genetics is unsuitable, and is sometimes referred to as "reverse genetics," since we first start with materials like RNA and then find a phenotype: that is the process is a reversal of classical genetics. The phenomenon of RNAi was discovered first in *C. elegans* (Fire et al., 1998). The RNAi-based "reverse genetics" in *C. elegans* offers three different ways for obtaining the interfered phenotype induced by tailor-made double stranded (ds) RNA material: by (1) microinjection of dsRNA into the nematode body (Fire et al., 1998), (2) feeding *E. coli* that expresses dsRNA (Timmons and Fire, 1998), and (3) soaking nematodes in dsRNA (Tabara et al., 1998; Maeda et al., 2001). For feeding and soaking RNAi, dsRNA is absorbed from the gut and transported to other parts of the body. Because *B. xylophilus* is very slender, it is quite difficult to perform microinjection. In this experiment we tried to perform soaking RNAi with *Bx-tbb-1*, the gene encoding β-tubulin and necessary for microtubule formation.

First, to check the soaking efficiency, the dsRNA buffer was mixed with FITC dye (1 μg/μL), and nematodes were dipped for 24 hours at 25 °C. In the presence of the neurotransmitter octopamine (10 mM), *B. xylophilus* could suck up the dsRNA more efficiently; FITC green signal was detected clearly in the pharyngeal and intestinal lumen in all nematodes (checked more than 30 individuals) (Fig. 2). Then, to apply reverse genetics to *B. xylophilus* as a functional analysis of *Bx-tbb-1*, both *B. xylophilus* and *C. elegans* animals were soaked in solutions containing several different lengths of dsRNA (1 kbp, 500 bp, 100 bp, and siRNA) for RNAi. Because it is known that introduced dsRNA is first processed into siRNA (short interfering RNA) by RNase III and siRNA with other components then forms RISC (RNA-induced silencing complex) to bind and degrade target mRNA (Hammond, 2005), we prepared *Bx-tbb-1* siRNA to facilitate the processes.

100% RNAi phenotype was detected in *C. elegans* by soaking 1 kbp and 500 bp *Bx-tbb-1* dsRNA, but none was detected in *B. xylophilus*. After *Bx-tbb-1* RNAi treatment, all *C. elegans* F1 eggs seemed to be arrested at microtubule formation and could not perform pronuclear meeting and chromosome division (Fig. 3). Although male and female pronuclei appeared, they could not move and meet each other. The embryos, nevertheless, started disordered cell division (Fig. 3). The shorter 100 bp dsRNA and siRNA were less effective for RNAi in *C. elegans*; RNAi efficiencies decreased to 67% and 22%, respectively (Table 1). We could not detect knock-down phenotypes in *B. xylophilus*, however, at any of the various dsRNA lengths (Table 1). We also examined soaking RNAi in *B. xylophilus* with several other genes *(Bx-par-1, Bx-mex-3, Bx-tbb-1, Bx-unc-22)* in several lengths (from 1.5 kb to siRNA), but could not observe knock-down phenotypes (data not shown). It might be possible

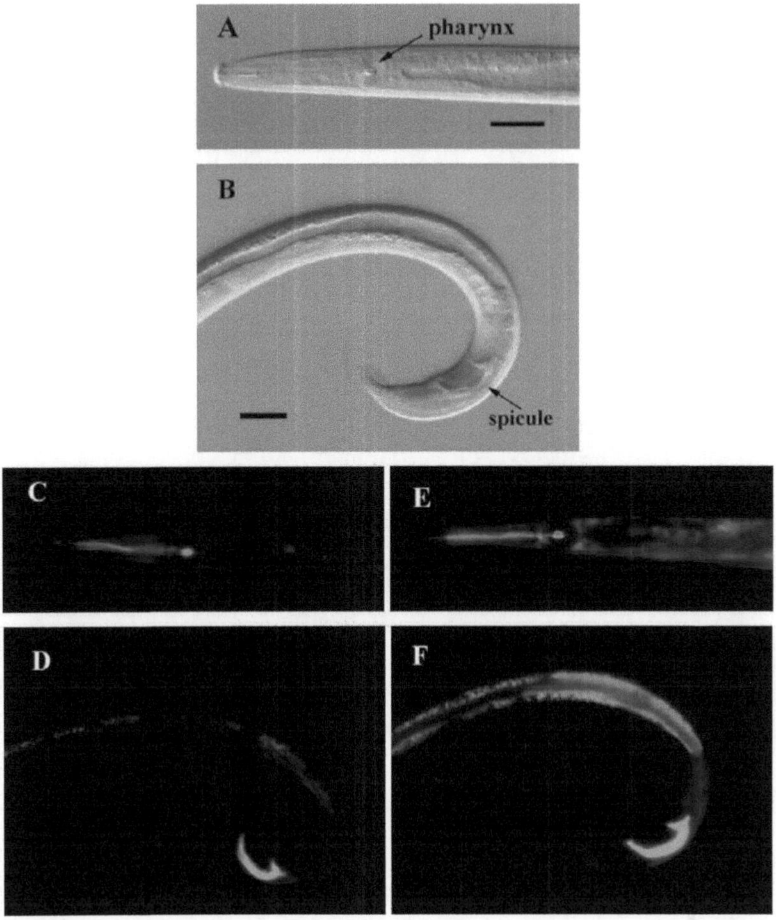


**Fig. 2** Soaking efficiency of *B. xylophilus*, checked by fluorescence dye FITC (1 μg/μL). (**A**) Head region, and (**B**) tail region of adult male, viewed by DIC. (**C**) Head region and (**D**) tail region of adult male, soaked in FITC soak buffer for 24 hours at 25 °C without octopamine, viewed by fluorescence microscopy. Only pharyngeal lumen from stylet to terminal bulb was stained with FITC green fluorescence. Blue autofluorescence signals are detected from gut and spicule. (**E**) Head region, and (**F**) tail region of adult male, soaked by FITC buffer for 24 hours at 25 °C with octopamine (10 mM), viewed by fluorescence microscopy. FITC green signal is detected from both pharyngeal and intestinal lumen. Scale bar, 20 μm

that systemic RNAi silencing, whereby mobile dsRNA is taken up by gut cells or transferred among different tissues, was suppressed in *B. xylophilus*, and thus soaking RNAi was not effective.

Successful results of RNAi in other parasitic nematodes have been reported; dsRNA was delivered by soaking (for example, Urwin et al., 2002; Fanelli et al., 2005), or by electroporation (Issa et al., 2005). *C. elegans* and perhaps also these parasitic nematodes have a systemic nature of RNAi; that is, delivered dsRNA moves from one introduced cell to the others, and that has not been reported so far in

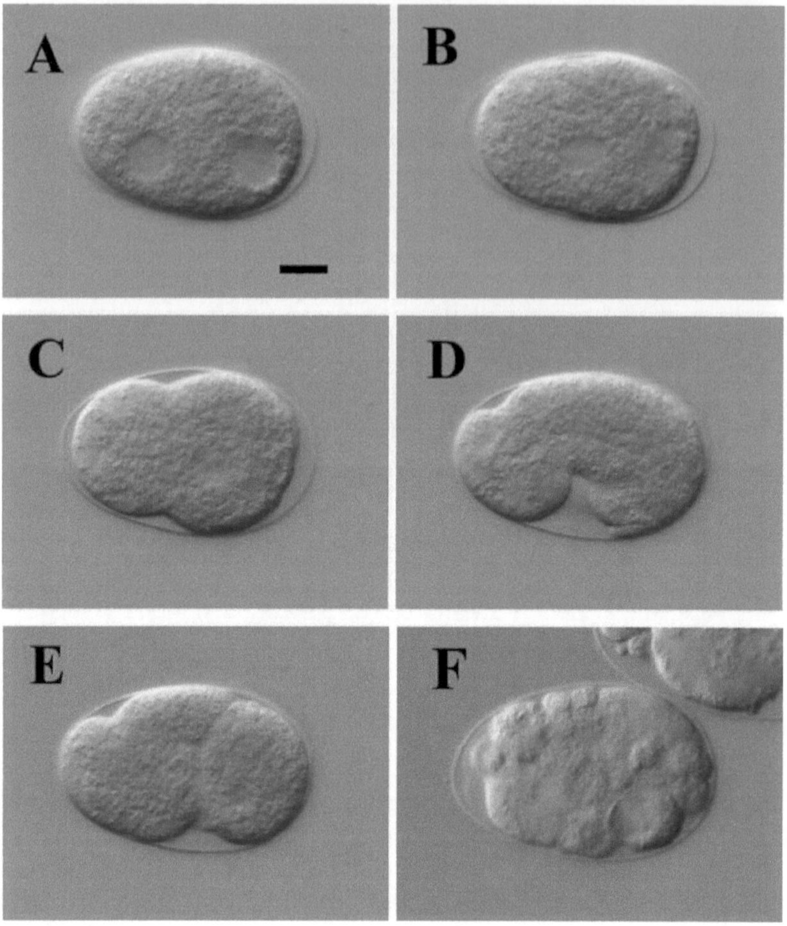

**Fig. 3** *Bx-tbb-1* RNAi knock-down phenotypes in *C. elegans* embryo. Just fertilized eggs were collected from *Bx-tbb-1* dsRNA (1 kbp) treated adults. Male and female pronuclei appeared (**A**). Two pronuclei could not move and meet (**B**), but embryo, nevertheless, started disorderly cell division (**D, E**), and became a lump of disorganized cells at 24 hours after fertilization (**F**). Scale bar, 10 μm

**Table 1** Percentage of abnormal RNAi phenotype

| dsRNA length | *B. xylophilus* | *C. elegans* |
|---|---|---|
| Control | – | 0% (14) |
| *Bx-tbb-1* (1 kbp) | 0% (41) | 100% (36) |
| (500bp) | 0% (55) | 100% (25) |
| (1000bp) | 0% (60) | 67% (24) |
| (siRNA) | 0% (29) | 22% (9) |

All abnormal phenotypes were embryogenesis arrest. Control was soaking buffer only. Numbers in parenthesis indicate the number of examined animals.

other organisms except for plants (Grishok, 2005). We need to continue seeking to establish other methods for functionally analyzing *B. xylophilus* genes to understand its developmental program and the evolution of nematode developmental systems by comparing similarities and differences between the two nematodes *B. xylophilus* and *C. elegans*.

**Acknowledgments** We thank Ms. Satsuki Miwa, Viva Informatica, for useful comments and careful reading of this manuscript. This work was supported by a Special Research Fund of Chubu University, plus a research grant from the Chubu University Institute for Biological Function to JM and a JSPS research fellowship to KH.

## References

BRENNER, S. (1974). The genetics of *Caenorhabditis elegans*. *Genetics* 77: 71–94.

FANELLI, E., DI VITO, M., JONES, J. T. AND DE GIORGI, C. (2005). Analysis of chitin synthase function in a plant parasitic nematode, *Meloidogyne artiellia*, using RNAi. *Gene* 349: 87–95.

FIRE, A., XU, S., MONTGOMERY, M.K., KOSTAS, S.A., DRIVER, S.E. AND MELLO, C.C. (1998). Potent and specific interference by double-stranded RNA in *Caenorhabditis elegans. Nature* 391: 806–811.

GOLDSTEIN, B. AND HIRD, S.N. (1996). Specification of the anteroposterior axis in *Caenorhabditis elegans. Development* 122: 1467–1474.

GÖNCZY, P. AND ROSE, L.S. (2005). Asymmetric cell division and axis formation in the embryo. In *WormBook* (ed. The *C. elegans* Research Community), doi/10.1895/wormbook.1.30.1, http://www.wormbook.org (October 15, 2005).

GRISHOK, A. (2005). RNAi mechanisms in *Caenorhabditis elegans. FEBS Letters* 579: 5932–5939.

HAMMOND, S.M. (2005). Dicing and silencing. The core machinery of the RNA interference pathway. *FEBS Letters* 579: 5822–5829.

HASEGAWA, K., MIWA, S., FUTAI, K. AND MIWA, J. (2004). Early embryogenesis of the pinewood nematode *Bursaphelenchus xylophilus. Dev. Growth Differ*. 46: 153–161.

HASEGAWA, K., MOTA, M., FUTAI, K. AND MIWA, J. (2006). Chromosome structure and behavior in *Bursaphelenchus xylophilus* (Nematoda: Parasitaphelenchidae) germ cells and early embryo. *Nematology* 8: 425–434.

ISSA, Z., GRANT, W.N., STASIUK, S. AND SHOEMAKER, C.B. (2005). Development of methods for RNA interference in the sheep gastrointestinal parasite, *Trichostrongylus colubriformis. International Journal of Parasitology* 35: 935–940.

KIKUCHI, T., JONES, J.T., AIKAWA, T., KOSAKA, H. AND OGURA, N. (2004). A family of glycosyl hydrolase family 45 cellulases from the pine wood nematode *Bursaphelenchus xylophilus. FEBS Letters* 572: 201–205.

KIKUCHI, T., SHIBUYA, H. AND JONES, J.T. (2005). Molecular and biochemical characterization of an endo-beta-1,3-glucanase from the pinewood nematode *Bursaphelenchus xylophilus* acquired by horizontal gene transfer from bacteria. *Biochemical Journal* 389: 117–125.

MAEDA, I., KOHARA, Y., YAMAMOTO, Y. AND SUGIMOTO, A. (2001). Large scale analysis of gene function in *Caenorhabditis elegans* by high-throughput RNAi. *Current Biology* 11: 171–176.

MAMIYA, Y. (1983). Pathology of the pine with disease caused by *Bursaphelenchus xylophilus. Annual Reviews of Phutopathology* 21: 201–220.

MIWA, J., SCHIERENBERG, E., MIWA, S., VON EHRENSTEIN, G. (1980). Genetics and mode of expression of temperature-sensitive mutations arresting embryonic development in *Caenorhabditis elegans. Developmental Biology* 76: 160–174.

MIWA, J. (1986). Making of the roundworm II. *Kagaku* 56: 162–170 (in Japanese)
MOTA, M., BRAASCH, H., BRAVO, M.A., PENAS, A.C., BURGERMEISTER, W., METGE, K. AND SOUSA, E. (1999). First report of *Bursaphelenchus xylophilus* in Portugal and in Europe. *Nematology* 1: 727–734.
MUNRO, E., NANCE, J. AND PRIESS, J. R. (2004). Cortical flow powered by asymmetrical contraction tansport PAR protein to establish and maintain anterior-posterior polarity in the early *C. elegans* embryo. *Development Cell* 7: 413–424.
SINGSON, A. (2001). Every sperm is sacred: Fertilization in *Caenorhabditis elegans. Development Biology* 230: 101–109.
SULSTON, J.E., SCHIERENBERG, E., WHITE, J.G. AND THOMSON, J.N. (1983). The embryonic cell lineage of the nematode *Caenorhabditis elegans. Development Biology* 100: 64–119.
TABARA, H., GRISHOK, A. AND MELLO, C.C. (1998). RNAi in *C. elegans*: soaking in the genome sequence. *Science* 282: 430–431.
TIMMONS, L. AND FIRE, A. (1998). Specific interference by ingested dsRNA. *Nature* 395: 854.
URWIN, P.E., LILLEY, C.J. AND ATKINSON, H.J. (2002). Ingestion of double-stranded RNA by preparasitic juvenile cyst nematodes leads to RNA interference. *MPMI* 15: 747–752.
WARD, S. AND CARREL, J.S. (1979). Fertilization and sperm competition in the nematode *Caenorhabditis elegans. Development Biology* 73: 304–321.

# The Relationship Between the Pinewood Nematode (PWN) and Fungi Cohabiting in Pine Trees Inoculated with the PWN

**Rina Sriwati, Shuhei Takemoto and Kazuyoshi Futai**

**Abstract** The relationship between the pinewood nematode (PWN) and fungi cohabiting with the nematodes in 15-year-old Japanese black pine (*Pinus thunbergii*) was examined bimonthly over a year after inoculation with PWN. The population of PWN in the trees was high in August, but slightly decrease in December then increased again in February. From wood samples of the pine trees examined, 18 species of fungi have been isolated. Among the 18 fungi detected, *Phialophora repens, Sphaeropsis sapinea, Pestalotiopsis* sp., *Rhizoctonia* sp. were the most frequently isolated in every season. All of these fungi had positive effects on the increase of nematode population, though the population of PWN on *Rhizoctonia* sp. was less than those on the other three dominant fungi. Under laboratory conditions, 19 species of fungi cultured on potato dextrose agar (PDA) served for PWN as food source, and the PWN's population built up on each fungus was compared at 20 °C. PWN dramatically increased on *Pestalotiopsis* sp. 1, *Pestalotiopsis* sp. 2, *Sphaeropsis sapinea, Phialophora repens*, and *Botrytis cinerea* (control), from 10 to 15 days after inoculation. From the point of view in terms of the food quality and their cohabitating ability we conclude that the species of fungi that are dominant in pine trees, except *Rhizoctonia* sp., have a compatible relationship with the PWN, while *Rhizoctonia* sp. and *Penicillium* sp. proved to be neutral, and *Trichoderma* sp. an incompatible relationship with PWN.

## Introduction

Pine wilt disease, caused by the pinewood nematode (PWN), *Bursaphelenchus xylophilus*, is a major threat to pine forests in Japan and causes serious damage to the most common native pines, *Pinus densiflora, P. thunbergii* and *P. luchuensis* (Mamiya, 1988). Since the first occurrence in Nagasaki, Kyushu, in 1905 (Yano,

R. Sriwati
Laboratory of Environmental Mycoscience, Graduate School of Agriculture, Kyoto University, Sakyo-ku, Kyoto 606-8502, Japan
Present address: Plant Protection Department, Agriculture Faculty, Syiah Kuala University, Banda Aceh Indonesia
e-mail: rin_aceh@yahoo.com

M.M. Mota, P. Vieira (eds.), *Pine Wilt Disease: A Worldwide Threat to Forest Ecosystems*, 

1913), this devastating epidemic disease has rapidly spread throughout Japan, except for the two northernmost prefectures, Aomori and Hokkaido, out of 47 prefectures. In 2000, infested areas were estimated to be 27.6% of the Japanese pine forest (2.1 million ha) (Mamiya, 2004). This nematode, presumably originating in North America (De Guiran and Bruguier, 1989), has also a rapid spread in other Asian countries (Yang, 2004). Moreover, recently it has been found also in Portugal (Mota et al., 1999).

The PWN is transmitted mainly by the Japanese pine sawyer, *Monochamus alternatus*, from wilt-killed to other healthy pine trees (Mamiya and Enda, 1972; Morimoto and Iwasaki, 1972). The adult beetles of *M. alternatus* carrying a great number of PWN's in their tracheae emerge from PWN-killed pine trees in early summer. Newly-emerging adults fly to healthy trees and feed on the bark of young twigs for maturation. At that time, PWN's on the vector beetles are transmitted to healthy trees and invade them through the feeding wounds made by the beetles. A small number of PWN's disperses widely in the infected trees and causes the cessation of oleoresin flow. Thereafter PWN's propagate dramatically and the trees show wilting symptoms, releasing volatiles such as ethanol, terpenes, etc. (see Kishi, 1995). Mature beetles are attracted to these wilting trees and lay their eggs in them. The eggs hatch within a week and the larvae feed on the inner bark and outermost sapwood, then, bore into the sapwood to form pupal chambers in autumn. The number of PWN's reaches its maximum from autumn to winter, then, decreases gradually (Mamiya et al., 1973; Fukushige and Futai, 1987). The pupal chamber of *M. alternatus* beetles is one of the most important places for PWN. Maehara and Futai (2002) reported that numerous PWN's aggregated around pupal chambers of *M. alternatus* in wilt-killed pine trees and that the beetles emerging in the subsequent year harbored many nematodes in their bodies.

PWN's, which are transmitted to healthy trees, feed on the parenchyma cells of the trees and on fungi such as *Pestalotia* spp. and *Rhizosphaera* spp., sparsely distributed in living trees. When the host tree is diseased, food sources of the PWN must be replaced with various wood-inhabiting fungi such as blue-stain fungi (Kobayashi, 1975; Kobayashi et al., 1975; Iwahori and Futai, 1990; Fukushige, 1991), though such fungi as *Trichoderma* spp. also inhabiting dead pines are unsuitable for PWN propagation (Kobayashi et al., 1975; Fukushige, 1991; Maehara and Futai, 1997).

Under field conditions, dying trees are often invade by various wood-decaying microorganisms (Shigo, 1967), and intense competition among these microorganisms brings about a succession of microbial flora and fauna. Abiotic environmental conditions, especially temperature, moisture and substrates and/or biotic factors greatly affect the succession of organisms.

Previous studies demonstrate changes in fungal flora inhabiting dead pine trees (Kobayashi et al., 1974, 1975; Maehara and Futai, 2000; Wang et al., 2005). Among the fungal species isolated from wilt-killed pine trees, some have been known to be suitable food source for the PWN, e.g., *Ceratocystis* sp., *Diplodia* sp. and *Pestalotia* sp., while others were unsuitable, e.g., *Trichoderma* spp., *Verticillium* sp. and *Cephalosporium* sp. (Kobayashi et al., 1974, 1975; Fukushige, 1991). Under laboratory conditions, Maehara and Futai (1996, 1997) demonstrated that each fungal species that proliferated around the pupal chamber of *M. alternatus*, affected not only PWN multiplication but also the number of PWN carried by the vector beetle. These findings clearly indicate that fungal flora in a dead pine tree might be one of the most determinant biotic factors for the multiplication and distribution of PWN inside the tree.

In the present study, we inoculated 15-year-old *P. thunbergii* trees grown outdoors with the PWN, investigated seasonal change of the fungal flora in the trees and analyzed the effect of fungal flora on the distribution of PWN at the microhabitat level. We also examined PWN propagation on fungi isolated from the field. On the basis of the two experiments, the effect of each fungal species on the distribution and propagation of PWN are discussed.

## Materials and Methods

### *PWN Inoculation*

A virulent isolate (S10) of the PWN, was cultured on the mycelium of *Botrytis cinerea* grown on barley grain medium (unhulled barley grain 10 ml; tap water 10 ml, autoclaved at 121 °C for 20 min) for 1 month. The Baermann funnel method was used to extract the nematodes from the fungal colonies, then the number of nematodes in the suspension was adjusted to 10,000 nematodes/ml. Inoculation of the trees was conducted on 10 June of 2004, at Kamigamo Experimental Station, Field Science Education and Research Center, Kyoto University located on a slope (average inclination: 29.7°) facing east-north-east (35°04′-N, 137°31′-E, 140 m above sea level), Kyoto, Japan. Fifteen-year-old Japanese black pines, *P. thunbergii* (average diameter at breast height was 5.3 cm and standard deviation was 3.5 cm) were inoculated as follows. An aliquot nematode suspension of 0.5 ml, containing 5,000 nematodes, was injected into a hole drilled in the bark at about 2.5 m above ground line. Small cotton balls were placed in the hole and another 0.5 ml aliquot of the nematode suspension was injected, i.e. 10,000 PWN's were injected per tree. The inoculation wound was sealed with Parafilm®. Eighteen pine trees (diameter at breast height ranging from 3.6 to 7.7 cm) were inoculated with the PWN, and another three trees with diameter from 5.3 to 5.7 cm were injected with the same volume of distilled water (control treatment). The controls were needed to determine if the trees had been infected before inoculation and to compare their results with those of the PWN-inoculated trees.

### *Introduction of* M. alternatus *Larvae into Pine Stems*

*M. alternatus* adults emerging from logs of dead, Japanese black pine located at the Arid Land Research Center, Tottori University, in the Tottori sand dune, Tottori, Japan (35°32′N, 134°13′E) where natural pine forests had been suffered from the pine wilt, were trapped. A pair of beetles was reared in a small cage and fed young pine twigs for food and oviposition. After 4 days, the laid eggs were collected from the twigs, dipped in 70% ethanol for 10 s and in 0.05% benzethonium chloride for 5 min, and then rinsed three times in sterile, distilled water (Kosaka and Ogura, 1990; Kosaka and Enda, 1991). The eggs were then placed in microplate wells, each containing 500 μl of 1/10 PDA medium, and kept under aseptic conditions until they hatched.

On 6 July 2004, when the inoculated pine trees had ceased resin exudation, the stem of each tree was drilled at 8 points (each 25 cm apart) at a height of 75–250-cm above ground and using a paintbrush one 1st-stage larva of *M. alternatus* was introduced into each hole.

## Wood Sampling

On 10 August, 13 October and 16 December 2004 and 9 February, 10 April and 10 June 2005, three trees were cut down and a 10-cm-long bole wood block was arbitrarily collected from each tree containing one of the *M. alternatus* larva-introduced points where a pupal chamber (PC) had been made. Each of the bole blocks was sliced into eight 1-cm thick discs, and 2 × 2 cm lattices were drawn on the cut surfaces. The discs were hen photocopied to record the position of PCs and tunnels of *M. alternatus*, and each of the discs was cut into small pieces (2 × 2 × 1 cm) along the line of the lattice. Each piece of wood thus prepared was split into two halves (2 × 1 × 1 cm). One-half served nematode extraction to investigate PWN density and the other half was stored at 4 °C for fungal isolation.

## Fungal Isolation

Within 2 days after collection of wood samples, the remaining half-cut wood pieces (2 × 1 × 1 cm) kept at 4 °C were surface sterilized for 1 s on the flame of a burner before placing them on potato dextrose agar (PDA) in a Petri dish for fungal isolation. After 3-day incubation at 20 °C, various fungi with different appearance grew on the plate. A small piece of agar with fungal mycelium was taken from each of the colonies with different appearance grown on the plates. They were put onto malt extract agar with chloramphenicol (2.0% malt extract, 1.5% agar, 100-ppm chloramphenicol) to establish pure cultures. This procedure was repeated every other month, after collection of samples. Isolation of fungi from healthy trees was done on June 2005 following the same procedure as mentioned above.

## Fungal Identification

Fungal mycelia were picked up by randomly scrapping the surface with a scalpel from the fungal cultures, bimonthly. A chemical procedure was used to homogenize the fungal body; the fungal cell wall was first digested by adding 50 μl of 1% Westase (TaKaRa) solution and incubating at 30 °C for 2 h. The digested solution was then transferred into a new 1.5 ml tube containing 100 μl of CTAB solution and DNA was extracted and purified according to Matsuda and Hijii (1999). The internal transcribed spacer (ITS) region of the ribosomal DNA was amplified using primers ITS1-F (5′-CTT GGT CAT TTA GAG GAA GTA A-3′) and ITS4 (5′-TCC TCC GCT TAT TGA TAT GC-3′; Gardes and Bruns, 1993; White et al., 1990). The product was digested with two restriction enzymes, *Alu*-I and *Hinf*-I. The fungal samples were divided into genotypes according to their PCR-RFLP patterns. When PCR-RLFP patterns of one sample did not match those of another, the two were considered to belong to different genotypes. Frequent genotypes, those detected from more than 50% of the bimonthly samples, were sequenced according to White et al. (1990) for ITS1, ITS2 and 5.8S regions of ribosomal DNA, using the primer ITS1F and ITS4. DNA sequences were determined using a genetic analyzer (3130, Genetic Analyzer, Applied Biosystems, USA). The DNA sequence determined for each fungus was compared with that of known species in the GenBank database. Identification at the genus level was based on identities above 90%. We also confirmed morphologically by sending the pure culture to a mycologist at the Forestry and Forest Products Research Institute, Japan.

Representative cultures are maintained in the Mycological Collection of the Laboratory of Environmental Mycoscience, Graduate School of Agriculture, Kyoto University, Kyoto, Japan.

## *Detection Frequency of Fungal Species Isolated*

For the investigation of seasonal changes in fungal flora, the presence of each fungal species was determined by observing the morphology and color of the fungal colonies grown on the PDA plate where the wood samples were placed. If a designated fungal species had been detected from a half-cut wood piece, the sample was classified as ‘positive’, otherwise ‘negative’. To calculate detection frequency of fungal isolates, the number of positive wood pieces was divided by the total number of wood pieces harvested from each trees.

## *Analysis of Compatibility Between Each Fungal Species and the PWN*

To calculate the average PWN number over fungal positive pieces harvested from each tree, sum of the number of PWN extracted from them was divided by sum of their dry weight. Likewise, a harmonic average of PWN number over fungal negative wood pieces was calculated. Here we denote the two averages as Nf and N0, respectively. We also estimated the influence of each fungus on the population growth and settlement of PWN using ‘nematode population ratio (NPR)’, which is defined as the ratio of Nf to N0. This value shows relative abundance of PWN yielded on the samples with a given fungal species, and becomes greater than one when a given fungus facilitates the growth or settlement of PWN population, on the contrary it becomes less than one when a given fungus suppress the growth or settlement of PWN population.

We also calculated the 'Index of cohabitation ability (ICA)' focusing on the relationship between the presence of a given fungi and that of the PWN as follows;

$$\text{ICA} = \text{Log}\,\frac{(\text{An}+0.01)\times(\text{B0}+0.01)}{(\text{A0}+0.01)\times(\text{Bn}+0.01)},$$

where;

A0 = Total number of wood pieces which contained only the fungus, per tree,
An = Total number of wood pieces which contained both fungus and nematode, per tree,
B0 = Total number of wood pieces which contained neither fungus nor nematode, per tree,
Bn = Total number of wood pieces which contained only nematode, per tree.

This index shows the degree of co-occurrence of PWN with a given fungus, positive ICA indicating a tendency of co-occurrence, while negative one repulsion.

## Results

### *Seasonal Changes of the Pinewood Nematode Population in Standing Pine Trees*

As shown in Fig. 1, the number of nematodes was high in August, but decreased slightly in December then recovered in February, and finally decreased again until the end of the experiment (June), though the data varied among individual trees except for those of August.

### *Identification of the Fungi and Their Detection Rate*

Eighteen species of fungi in total were identified to genus level. The results were confirmed morphologically by Dr. Kubono (Forestry and Forest Products Research Institute, Japan). Possible species, which had the highest nucleotide identity with the fungal genotype detected, are *Aspergillus* sp., *Aureobasidium* sp., *Fusarium* sp. 1, *Fusarium* sp. 2, *Gliocladium* sp., *Mucor* sp., *Mortierella* sp., *Penicillium* sp. 1, *Penicillium* sp. 2, *Penicillium* sp. 3, *Pestalotiopsis* sp. 1, *Pestalotiopsis* sp. 2, *Phialophora repens*, Rhizoctonia sp., *Sphaeropsis sapinea*, *Trichoderma* sp. 1, *Trichoderma* sp. 2 and *Trichoderma* sp. 3. Fungal identification is summarized in Table 1.

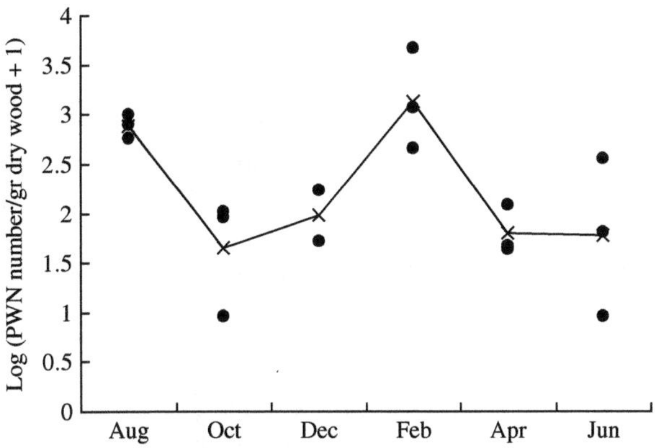


**Fig. 1** Seasonal changes of the PWN on dead pine trees (n = 3). A closed circle indicates a log-transformed average nematode density in a pine tree. To calculate the average, the total number of PWN extracted from the samples harvested from each tree was divided by their total dry weight. A cross indicates a log-transformed arithmetic average of the three average nematode densities for each sampling time

Fungal florae in the trees inoculated with PWN under natural conditions were different from tree to tree. Among the 18 fungal species identified, *P. repens, S. sapinea*, two *Pestalotiopsis* spp. and *Rhizoctonia* sp. were detected most frequently every season (Fig. 2). It was impossible to distinguish between two *Pestalotiopsis* species from morphological observation, so we treated these two species as *Pestalotiopsis* spp. We focused on these four frequent species and examined them in more detail. *Penicillium* sp. and *Trichoderma* sp. were also examined for comparison.

**Table 1** Fungi isolated from dead pine trees inoculated with PWN and healthy trees

| Isolated from dead pine trees | Isolated from healthy pine trees |
|---|---|
| *Aspergillus* sp. | *Aspergillus* sp. |
| *Aureobasidium* sp. | *Gliocladium* sp. |
| *Fusarium* sp. 1 | *Mucor* sp. |
| *Fusarium* sp. 2 | *Penicillium* sp. 1 |
| *Gliocladium* sp. | *Penicillium* sp. 2 |
| *Mucor* sp. | *Penicillium* sp. 3 |
| *Mortierella* sp. | *Pestalotiopsis* sp. 1 |
| *Penicillium* sp. 1 | *Pestalotiopsis* sp. 2 |
| *Penicillium* sp. 2 | *Phialophora repens* |
| *Penicillium* sp. 3 | *Sphaeropsis sapinea* |
| *Pestalotiopsis* sp. 1 | *Trichoderma* sp. 1 |
| *Pestalotiopsis* sp. 2 | *Trichoderma* sp. 2 |
| *Phialophora repens* | *Trichoderma* sp. 3 |
| *Rhizoctonia* sp. | |
| *Sphaeropsis sapinea* | |
| *Trichoderma* sp. 1 | |
| *Trichoderma* sp. 2 | |
| *Trichoderma* sp. 3 | |

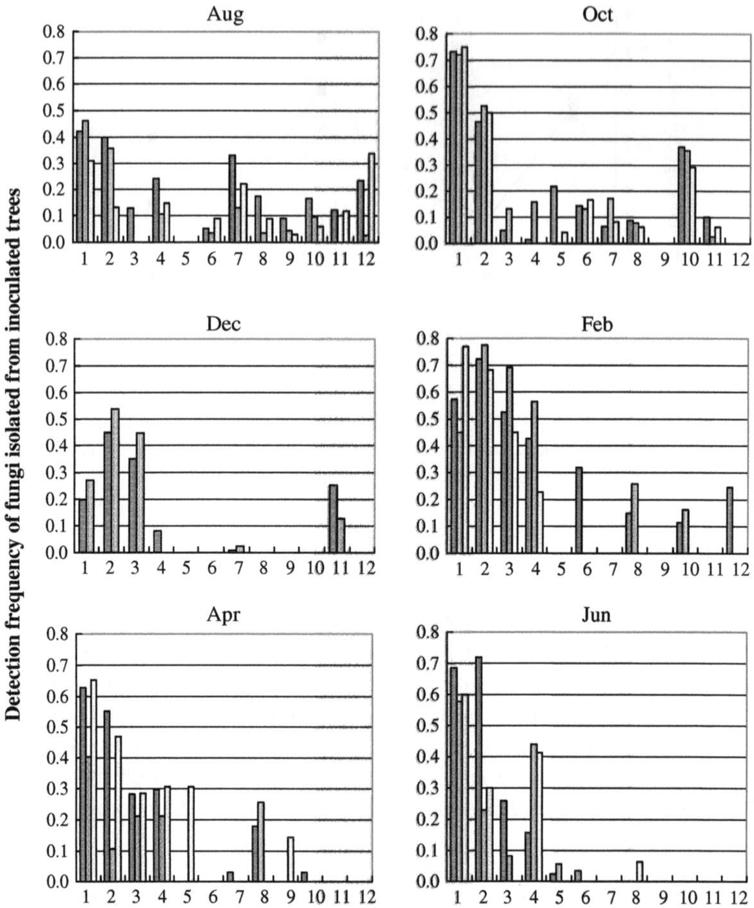


1. *Phialophora repens*
2. *Sphaeropsis* sp
3. *Pestalotiopsis* spp.
4. *Rhizoctonia* sp
5. *Fusarium* sp
6. *Penicillium* sp
7. *Aureobasidium* sp
8. *Trichoderma* sp
9. *Gliocladium* sp
10. *Mortierella* sp
11. *Mucor* sp
12. *Aspergillus* sp

**Fig. 2** Seasonal changes in frequency of fungi isolated from inoculated trees. The values is the ratio of the numbers of fungal positive wood pieces to the total number of wood pieces harvested from each trees

## *Influence of Each Fungus on the Presence and/or Propagation of PWN*

To evaluate the influence of each fungus on the presence and/or propagation of PWN, the NPR was calculated (Fig. 3). NPR of four dominant fungal species were slightly higher than one every season, suggesting that these fungi facilitate the

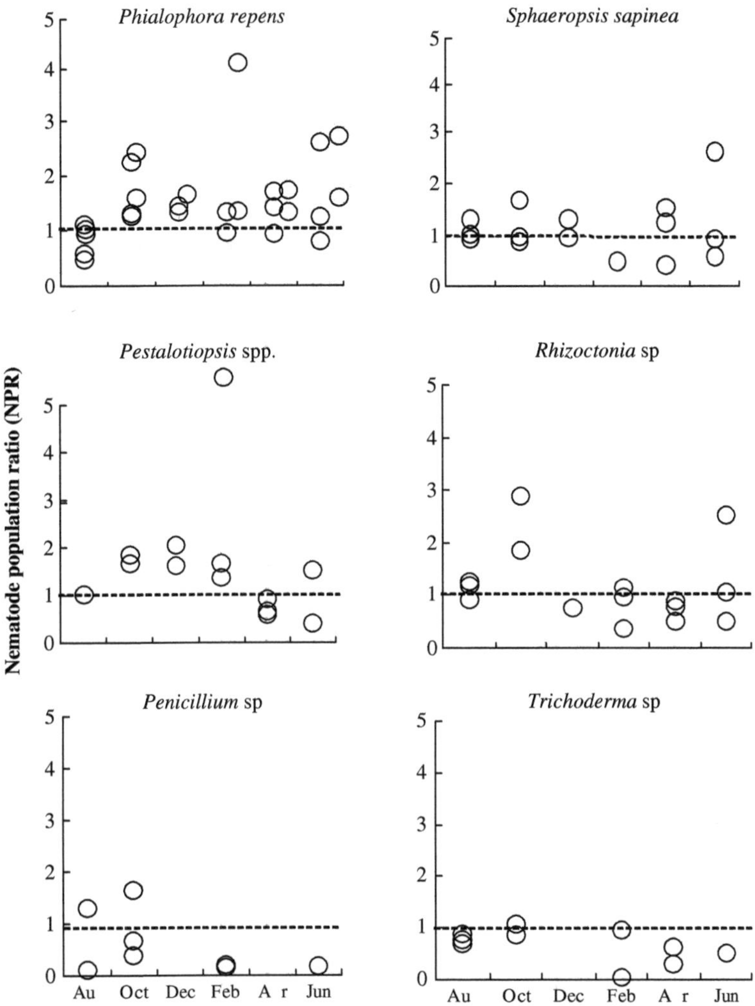


**Fig. 3** The effect of each fungal species, on the distribution of the PWN in dead pine trees. NPR is the ratio of Nf to N0, where Nf and N0 are average PWN numbers over fungal positive and negative wood pieces harvested from each pine tree, respectively (from Sriwati et al. 2007)

population growth of PWN. In contrast, the NPR of *Trichoderma* sp. and *Penicillium* sp. were lower than and equal to one, respectively. *Trichoderma* sp. suppressed the population growth of PWN, and *Penicillium* sp. influenced neither positively nor negatively on the population growth. Thus, these nematodes preferably aggregated to dominant fungi over the experimental period, but promotive effects of each fungus on the presence and/or propagation of PWN were changed from season to season and from tree to tree. The other less abundant fungal species were not analyzed in detail.

## *Cohabitation of PWN and Fungal Species in PWN-Inoculated Trees*

The ICA values which focused only on the presence or absence of fungi and the PWN were slightly higher than zero for three dominant fungal species, *P. repens, S. sapinea, Pestalotiopsis* spp., while those for *Penicillium* and *Rhizoctonia* species were around zero, and that for *Trichoderma* sp. was slightly lower than zero (Fig. 4). Thus, three dominant fungi tended to cohabit with PWN, while *Trichoderma* sp.

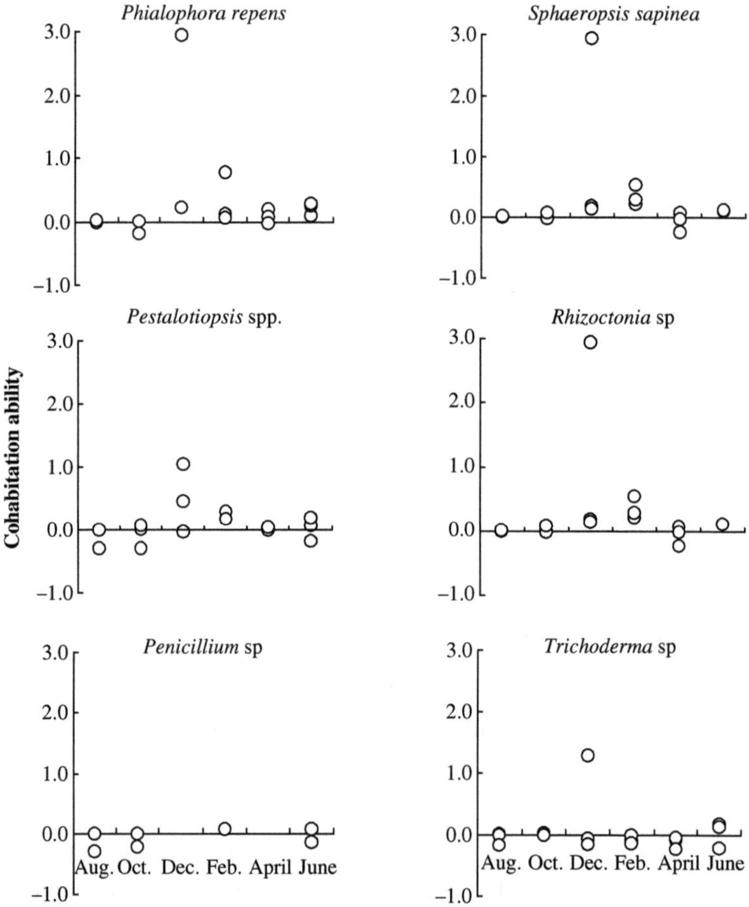


**Fig. 4** Cohabitation ability between the PWN and each fungal species. ICA = Log ((An + 0.01) × (B0 + 0.01))/((A0 + 0.01) × (Bn + 0.01)), where A0 is the total number of wood pieces which contained only the fungus per tree; An is the total number of wood pieces which contained both fungus and nematode per tree; B0 is the total number of wood pieces which contained neither fungus nor nematode per tree; Bn is the total number of wood pieces which contained only nematode per tree (from Sriwati et al. 2007)

showed repelling effects towards PWN. Both *Penicillium* sp. and *Rhizoctonia* sp. had no special relationship with PWN regarding their distribution.

## Discussion

In the present study, the number of PWN decreased until December but recovered in February. Then it decreased again until the end of the experiment (June). Fukushige and Futai (1987) reported a similar fluctuation in the number of PWN with a decrease from October to December and an increase from December to February, which did not correlate with tree water content. In the previous study, we also found no correlation between them (Sriwati et al., 2006).

From the wood samples of the pine trees examined, 18 fungal species were isolated. Among them, *P. repens, S. sapinea*, two *Pestalotiopsis* spp. and *Rhizoctonia* sp. were frequently isolated and were considered as dominant fungi. Although the dominant fungi were constantly detected over the experimental period, the composition of fungal species slightly varied among seasons as reported in previous studies (Kobayashi et al., 1974, 1975; Fukushige and Futai, 1987; Kuroda and Ito, 1992). For example, Kuroda and Ito (1992) reported that the fungal species detected from PWN-inoculated pines were the same as those in healthy trees during 4 weeks after inoculation. The species detected both from healthy and inoculated pine trees were *Pestalotiopsis* spp., *Nigrospora* spp., *Cladosporium* spp. and *Phomopsis* spp. They found a blue stain fungus, *Ceratocystis* sp., and bacteria 5 weeks after nematode inoculation. In this study, minor fungi disappeared when pine trees were completely killed in December, and fungal flora of the pine trees gradually changed until June, the end of this experiment.

Kobayashi et al. (1974) recorded one species of *Diplodia* as one of the common fungi in the wood of dead pine trees affected by *B. lignicolus* (previous name of *B. xylophilus*). They also revealed that the *B. lignicolus* multiplied well on its mycelia grown on PDA plate medium. In the present study, we frequently isolated *S. sapinea*, which is regarded as a synonym of *Diplodia* (Denman et al., 2000). De Wet et al. (2003) also suggested that *S. sapinea* should be reverted to the former name of *Diplodia pinea*. To identify the fungi detected from PWN-inoculated pine trees, we did not follow the system of Denman et al. (2000), because we depended on the GenBank data, in which the identical sequence to my data was not provided under *D. pinea* but *S. sapinea*.

A close affinity between the PWN and the blue-stain fungi has been reported in several studies (Kobayashi et al., 1974, 1975; Fukushige, 1991; Maehara and Futai, 2000; Maehara et al., 2005). Kobayashi et al. (1974, 1975) considered that *Monochamus* beetles might transmit *Ophiostoma*, when they feed on young shoots of healthy pine trees in early summer. In the present study, however, this group of fungal species was not isolated. This may be attributed to the fact that the pine trees had been killed by artificial inoculation with the PWN and thereby microbial environments might have become unsuitable for *Ophiostoma* even when it was transmitted by *Monochamus* beetle.

To evaluate the effect of fungal species and their distribution on the population density and distribution of PWN, two indices, NPR and ICA, were calculated. NPR compares the relative abundance of PWN between samples with and without a designated fungal species. ICA reflects degree of co-occurrence of a fungus and PWN. In the case of the dominant fungi except for *Rhizoctonia* sp., NPR was higher than one and ICA was higher than zero, which means that PWN had a tendency to coexist with these fungi and/or propagate well on them. In the case of the other minor fungi, e.g., *Trichoderma* spp., NPR and ICA were less than one and zero, respectively, showing an incompatible relationship to the PWN. Thus, these facts suggest that the dominant fungi in pine trees killed by PWN promoted the propagation and/or the settlement of the nematode. Since many other biotic and abiotic factors could affect the microhabitat conditions of pinewood (Shigo, 1967), further studies are needed to explore other factors that affect PWN distribution.

## References

De Guiran, G. and Bruguier, N. (1989). Hybridization and phylogeny of the pine wood nematode (*Bursaphelenchus* spp.). *Nematologica* 35: 321–330.

De Wet, J., Burgess, T., Slippers, B., Preisig, O., Wingfield, B.D. and Wingfield, M.J. (2003). Multiple gene genealogies and microsatellite marker reflect relationship between morphotypes of *Sphaeropsis sapinea* and distinguish a new species of *Diplodia*. Mycological Research 107: 557–566.

Denman, S., Crous, P.W., Tailor, J.W., Kang, J.C., Pascoe, I. and Wingfield, M.J. (2000). An overview of the taxonomic history of *Botriosphaeria*, and a re-evaluaion of its anamorphs based on morphology and ITS rDNA phylogeny. In: Molecules, morphology and classification: Towards monophyletic genera in the Ascomycetes. *Studies in Mycology* 45: 129–140.

Fukushige, H. (1991). Propagation of *Bursaphelenchus xylophilus* (Nematoda: Aphelenchoididae) on fungi growing in pine shoot segments. *Applied Entomology and Zoology* 26: 371–376.

Fukushige, H. and Futai, K. (1987). Seasonal changes in *Bursaphelenchus xylophilus* population and occurrence of fungi in *Pinus thunbergii* trees inoculated with the nematode. *Japanese Journal of Nematology* 17: 8–16.

Gardes, M. and Bruns, T.D. (1993). ITS primers with enhanced specificity for basidiomycetes – application to the identification of mycorrhizae and rusts. *Molecular Ecology* 2: 113–118.

Iwahori, H. and Futai, K. (1990). Propagation and effects of pinewood nematode on calluse of various plants. *Japanese Journal Nematology* 20: 25–36.

Kishi, Y. (1995). *The pine wood nematode and the Japanese pine sawyer*. Thomas Company limited, 302pp., Tokyo, Japan.

Kobayashi, Y. (1975). The pine wood nematode, fungi and their vector pine sawyer. *Forest Pest* 24: 199–202.

Kobayashi, T., Sasaki, K. and Mamiya, Y. (1974). Fungi associated with *Bursaphelenchus lignicolus*, the pine wood nematode (I). *Journal of the Japanese Forestry Society* 56: 136–145.

Kobayashi, T., Sasaki, K. and Mamiya, Y. (1975). Fungi associated with *Bursaphelenchus lignicolus*, the pine wood nematode (II). *Journal of the Japanese Forestry Society* 57: 184–193.

Kosaka, H. and Enda, N. (1991). Simple rearing method of larvae of the Japanese pine sawyer, *Monochamus alternatus* (Coleoptera: Cerambycidae) on artificial diets. *Forest Pests* 40: 183–187.

KOSAKA, H. AND OGURA, N. (1990). Rearing of the Japanese pine sawyer, *Monochamus alternatus* (Coleoptera: Cerambycidae), on artificial diets. *Applied Entomology and Zoology* 25: 532–535.

KURODA, K. AND ITO, S. (1992). Migration speed of pine wood nematodes and activities of other microbes during the development of pine-wilt diseases in *Pinus thunbergii. Journal of Japanese Forestry Society* 74: 383–389.

MAEHARA, N. AND FUTAI, K. (1996). Factors affecting both the numbers of pine wood nematode, *Bursaphelenchus xylophilus* (Nematoda: Aphelenchoididae), carried by the Japanese pine sawyer, *Monochamus alternatus* (Coleoptera: Cerambycidae), and the nematode's life history. *Applied Entomology and Zoology* 31: 443–452.

MAEHARA, N. AND FUTAI, K. (1997). Effect of fungal interactions on the numbers of the pine wood nematode, *Bursaphelenchus xylophilus* (Nematoda: Aphelenchoididae), carried by the Japanese pine sawyer, *Monochamus alternatus* (Coleoptera: Cerambycidae). *Fundamental and Applied Nematology* 20: 611–617.

MAEHARA, N. AND FUTAI, K. (2000). Population changes of the pinewood nematode, *Bursaphelenchus xylophilus* (Nematoda: Aphelenchoididae), on fungi growing in pine-branch segments. *Applied Entomology and Zoology* 35: 413–417.

MAEHARA, N. AND FUTAI, K. (2002). Factors affecting the numbers of *Bursaphelenchus xylophilus* (Nematoda: Aphelenchoididae), carried by several species of beetles. *Nematology* 4: 653–658.

MAEHARA, N., HATA, K. AND FUTAI, K. (2005). Effect of blue stain fungi on the number of *Bursaphelenchus xylophilus* (Nematoda: Aphelenchoididae) carried by *Monochamus alternatus* (Coleoptera: Cerambycidae). *Nematology* 7: 161–167.

MAMIYA, Y. (1988). History of pine wilt disease in Japan. *Japanese Journal of Nematology* 8: 59–67.

MAMIYA, Y. (2004) Pine wilt disease in Japan. In *The pine wood nematode*, Bursaphelenchus xylophilus. *Nematology Monographs and Perspectives, vol. 1* (eds. M. Mota and P. Vieira), pp. 9–20. E. J. Brill, Leiden, The Netherlands.

MAMIYA, Y. AND ENDA, N. (1972). Transmission of *Bursaphelenchus lignicolus* (Nematoda: Aphelenchoididae) by *Monochamus alternatus* (Coleoptera: Cerambycidae). *Nematologica* 18: 159–162.

MAMIYA, Y., KOBAYASHI, T., ZINNO, Y., ENDA, N. AND SASAKI, K. (1973). Disease development of pine trees naturally infected with *Bursaphelenchus lignicolus. Transactions of the 26th Annual Meeting of Kyushu Branch of Japanese Forestry Society*: 332–334.

MATSUDA, Y. AND HIJII, N. (1999). Characterization and identification of *Strobilomyces confuses* ectomycorrhizas on Momi fir by RFLP analysis of the PCR-amplified ITS region of the rDNA. *Journal of Forest Research* 4: 145–150.

MORIMOTO, K. AND IWASAKI, A. (1972). Role of *Monochamus alternatus* (Coleoptera: Cerambycidae) as a vector of *Bursaphelenchus lignicolus* (Nematoda: Aphelenchoididae). *Journal of the Japanese Forestry Society* 54: 177–183.

MOTA, M.M., BRAASCH, H., BRAVO, M.A., PENAS, A.C., BURGERMEISTER, W., METGE, K. AND SOUSA, E. (1999). First report of *Bursaphelenchus xylophilus* in Portugal and in Europe. *Nematology* 1: 727–734.

SHIGO, A.L. (1967). Successions of organisms in discoloration and decay of wood. In *International Review of Forestry Research, vol. 2* (eds. J.A. Romberger and P. Mikola), pp. 237–299. Academic Press, New York, London.

SRIWATI, R., TAKEMOTO, S. AND FUTAI, K. (2006). Seasonal changes of nematode fauna in pine trees killed by the pinewood nematode, *Bursaphelenchus xylophilus. Japanese Journal of Nematology* 36: 87–100.

WANG, Y., YAMADA, T., SAKAUE, D. AND SUZUKI, K. (2005). Influence of fungi on multiplication and distribution of the pine wood nematode, *Bursaphelenchus xylophilus*, in axenized *Pinus thumbergii* cuttings. *Nematology* 7: 809–817.

White, T.J., Bruns, T.D., Lee, S. and Taylor, J. (1990). Amplification and direct sequencing of fungal ribosomal RNA genes for phylogenetics. In *PCR protocols: a guide to methods and applications* (eds. M.A. Innis, D.H. Gelfand, J.J. Sninsky and T.J. White), pp. 315–322. Academic Press, New York, USA.

Yang, B. (2004). The history, dispersal and potential threat of pine wood nematode in China. In *The pine wood nematode,* Bursaphelenchus xylophilus. *Nematology Monographs and Perspectives, vol. 1* (eds. M. Mota and P. Vieira), pp. 21–24. E. J. Brill, Leiden, The Netherlands.

Yano, M. (1913). Investigation on the causes of pine mortality in Nagasaki prefecture. *Sanrin Koho* 4: 1–14.

# Influence of Fungi on Multiplication and Distribution of the Pinewood Nematode

**Yu Wang, Toshihiro Yamada, Daisuke Sakaue and Kazuo Suzuki**

**Abstract** The influence of fungi on multiplication and distribution of the pinewood nematode (PWN) was investigated in *Pinus thunbergii* cuttings. Axenized nematodes and/or one of two fungi isolated from healthy and PWN-killed *P. thunbergii* were inoculated together into autoclaved cuttings. A close relationship between existence and distribution of fungal hyphae, and the PWN's multiplication and distribution was observed. The PWN did not multiply when only axenized nematodes were inoculated in the absence of fungi. When fungi were present, PWN population size increased markedly. The number of nematodes was high at sites where fungal hyphae were distributed. It is suggested therefore, that the restriction of a large portion of the nematode population near the inoculation site during the early stage of disease development is closely related to restricted distribution of fungal hyphae.

## Introduction

The pinewood nematode (PWN), *Bursaphelenchus xylophilus*, is the causative agent of pine wilt disease (PWD), which is responsible for severe damage to pine forests in Eastern Asia (Mamiya, 1986; Enda, 1988; Enda and Taketani, 1992; Tamura, 1996). Multiplication and distribution of PWN in pine trees has been studied extensively and it is clear that multiplication and distribution of the PWN is essential for the development of pine wilt disease (Hashimoto and Kiyohara, 1973; Kiyohara and Suzuki, 1975; Kiyohara et al., 1975). The increase of population size, and distribution of PWN throughout pine trees always coincides with severe tissue damage and pine wilting, whereas a small number of PWN, restricted in distribution near the inoculation site, results in only slight damage to pines (Hashimoto and Kiyohara, 1973; Hashimoto and Dozono, 1975; Kiyohara and Suzuki, 1975; Kiyohara et al., 1975; Ishida et al., 1993; Ichihara et al., 2000). Understanding the factors that influence multiplication and distribution of PWN in pine trees will be helpful in pinpointing the mechanism(s) of disease development.

Y. Wang
Department of Forest Protection, Faculty of Forest Resources and Environment, Nanjing Forestry University, Longpan Road, Nanjing 210037, China
e-mail: njwy01@hotmail.com

M.M. Mota, P. Vieira (eds.), *Pine Wilt Disease: A Worldwide Threat to Forest Ecosystems*, 

Previous reports have shown that successful multiplication and wide distribution of the PWN resulted from infection with virulent nematodes from populations having higher multiplication rates on fungal cultures of *Botrytis cinerea* (Kiyohara, 1989; Kiyohara and Bolla, 1990; Wang et al., 2004) and in pines (Ishida et al., 1993; Ichihara et al., 2000) into susceptible pines and from some environmental factors, such as high temperature (Dozono and Yoshida, 1974; Wang et al., 2004) and water stress (Suzuki and Kiyohara, 1978).

Fungi are intimately related to the life history of the PWN because most fungi have the potential to serve as a source of nutrition for the nematode. A variety of fungi have been isolated from healthy, PWN-inoculated and PWN-infected trees (Kobayashi et al., 1974, 1975; McGawley et al. 1980; Fukushige and Futai, 1987; Fukushige, 1991; Maehara and Futai, 2000). Although the suitability of fungal species as a nutritional source for the nematode varies greatly, most of them have some ability to promote nematode population increase (Kobayashi et al., 1974, 1975; Fukushige, 1991).

Earlier results (Wang, 2004) suggests a relationship between the availability and distribution of fungi, and multiplication and distribution of PWN in pines. The restricted distribution of most of the PWN around the inoculation site in the early stage of disease development was reported by many researchers, but no studies have been done to determine why (Mamiya, 1974, 1985; Suzuki, 1984; Odani et al., 1985a). The current study was done to further demonstrate the influence of fungi on PWN's multiplication and distribution in pines. In this study, the experimental conditions were improved and simplified. Some fungi isolated from healthy or PWN-killed pine trees were inoculated together with axenized nematodes into autoclaved cuttings, and the relationship between existence and distribution of fungi, and multiplication and distribution of PWN was examined.

## Materials and Methods

### *Pine Cuttings*

Ten-cm long cuttings with the apical top were collected from 1-year-old branches of 3-year-old *Pinus thunbergii* seedlings grown in the nursery of the Experimental Station at Tanashi, the University of Tokyo. Cuttings were set upright with the basal end immersed in water in a sterile plastic plant culture jar (70 mm diam., 110 mm high) containing 30 ml ionized water. These bottles containing the cuttings were autoclaved at 121 °C for 30 min and used in the inoculation experiments. A filter capped-hole on the top was used to allow for circulation of air while keeping the aseptic condition.

### *Nematodes*

A virulent PWN isolate, S-10, was used for inoculation. Nematodes were subcultured on *Botrytis cinerea* grown on PDA in Petri dishes at 25 °C in the dark. After 10 days, the nematodes were isolated using the Baermann funnel technique at 25 °C for 24 h. Axenized nematodes were prepared by the method of Krusberg and Sardanelli (1984) as follows: nematodes were rinsed three to four times by suspension in a few milliliters of the antibiotic solution containing 20 microgrames/ml of streptomycin sulfate and 20 units/ml of penicillin G, potassium salt in a conical 15-ml centrifuge tube. The nematodes were sedimented by brief centrifugation, the supernatant discarded, and then resuspended in the antibiotic solution. After the last centrifugation, the nematodes, in about 0.5 ml of the antibiotic solution, were transferred using a sterile pipet, to the top of the antibiotic solution in an autoclaved glass chromatography column (20 mm i.d. ×300 mm long.) with 6-mm-d glass beads inside it. The nematodes were harvested from the column over a period of 4–6 h at room temperature.

## *Fungi*

Fungi were isolated on January 28, 2004 from 2 healthy and 2 PWN-killed Japanese red pine trees, *P. densiflora*. Wood chips were collected from 2 points on each tree with a drill. Then wood chips were cut into small pieces (0.2–0.3 cm long × 0.2–0.3 cm wide), surface-axenized by the normal method with 4% NaClO and cultured on PDA plates at 25 °C in the dark. Seven fungal isolates of six genera (classes) were obtained. Two out of the seven isolates were selected for the inoculation experiment: Cf. *Cryptosporiopsis* (Sutton: The Coelomycetes) sp. (F1 hereinafter), which grew very slowly; *Leptographium* sp. (F2 hereinafter), which grew relatively rapidly.

The two fungal isolates were cultured on PDA in Petri dishes at 25 °C for 10 days. Small disks (3 mm diam.) of fungal culture were made with a cork borer. The fungal discs were used in the following inoculation experiment.

## *Nematode and Fungus Inoculation*

Nematode and fungus inoculation was conducted on April 15, 2004. A 0.5 cm long cut into the xylem on the upper end (0–2 cm) of each cutting was made with sterilized scalpels. A fungal disc only, 20 μl axenized nematode suspension (containing 200 nematodes) only or combination of both was inoculated into each cut. Five different inocula were used: F1 only, F2 only, axenized nematodes only, F1 + axenized nematodes and F2 + axenized nematodes. Fifteen seedlings were inoculated per treatment. After inoculation, the cuttings in the inoculation bottles were incubated at 25 °C in the dark.

## *Distribution of Fungi and Nematodes*

At 2, 6, 11, 16 and 21 days after inoculation (DAI), three cuttings were collected for each treatment. The inoculation bottles containing the cuttings were opened in the clean bench. Two small pieces of tissue (0.2–0.3 cm long ×0.2–0.3 cm wide) containing bark and wood at every 2 cm segment from each cutting were removed with sterilized scissors and directly transferred to PDA medium in Petri dishes. The Petri dishes were then cultured at 25 °C in the dark. The presence of fungi was examined within 3 weeks. Each cutting inoculated with nematodes was cut into 5 segments of 2 cm length. Nematodes were extracted from these segments by the Baermann funnel technique. The water in the bottle was also used for extraction of nematodes. The number of nematodes as well as the percentage composition of juveniles in the population was determined for each extraction.

### *Statistical Analysis*

One-way analysis of variance (ANOVA) and Tukey-Kramer's multiple comparison test was used to determine the significance of mean differences among treatments.

## Results

### *Temporal Changes in Numbers of PWN Population*

Population size (values are means of 3 cuttings) decreased by 2 DAI for all three treatments inoculated with nematodes alone or fungus + nematodes (Fig. 1). The population number was 94, 121 and 116, respectively for the treatment of axenized nematodes only, F1 + axenized nematodes and F2 + axenized nematodes. For the treatment of axenized nematodes only, the population number increased to 145 by

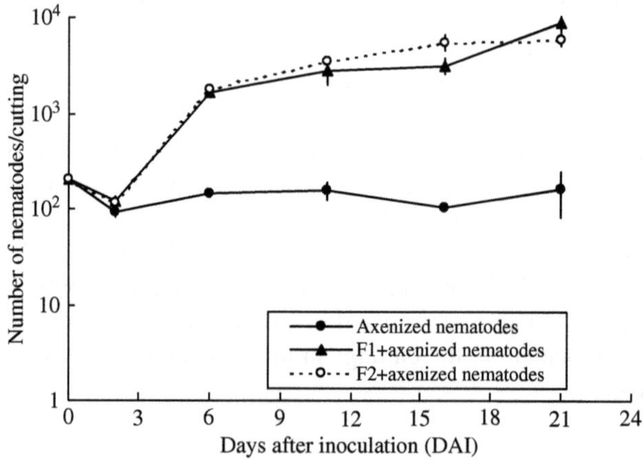


**Fig. 1** Temporal changes in numbers of PWN in autoclaved cuttings of *Pinus thunbergii* inoculated with axenized nematodes or axenized nematodes in combination with fungus. Values are means of 3 cuttings. Vertical bars indicate standard errors of means. F1: Cf. *Cryptosporiopsis* sp., F2: *Leptographium* sp

6 DAI. Thereafter the population number fluctuated around 100–165 and did not show a tendency to increase. For the other 2 treatments of fungus + nematodes, the population number steadily increased and reached 8784 (F1 + nematodes) and 5960 (F2 + nematodes) respectively by 21 DAI, significantly higher than that of the treatment of axenized nematodes only ($p \leq 0.05$). There was no significant difference of population number between the 2 treatments of fungus + nematodes at 5% level.

## *Percentage of Juveniles and Temporal Changes in the Number of Juveniles and Adults in PWN Population*

By two DAI the percentage of juveniles in the populations decreased for every treatment (Fig. 2). It decreased slightly from 62.7% to 52.0% for the treatment of axenized nematodes only but decreased greatly for the 2 treatments of axenized nematodes in combination with fungus (from 62.7% to 27.4% and 25.8%, respectively). The percentage of juveniles in the population for the treatment of axenized nematodes only was significantly higher ($p \leq 0.01$) than that of the other 2 treatments of fungus + nematodes.

The juvenile composition of the population ranged from 45.9% to 63.3% throughout the experimental period when axenized nematodes without fungi was the experimental treatment. By six DAI for the other two treatments, following the former decrease, there was an abrupt increase up to 88.1% and 94.1%. This indicated a rate of population growth that was significantly higher than that of the treatment of axenized nematodes only ($p \leq 0.05$). Thereafter, the percentage of juveniles in the population gradually decreased accompanying the decline of population growth (Fig. 1).

Temporal changes in the number of juveniles and adults in PWN population are shown in Figs. 3 and 4. For the treatment of axenized nematodes only, neither the number of juveniles nor the number of adults changed significantly. For the

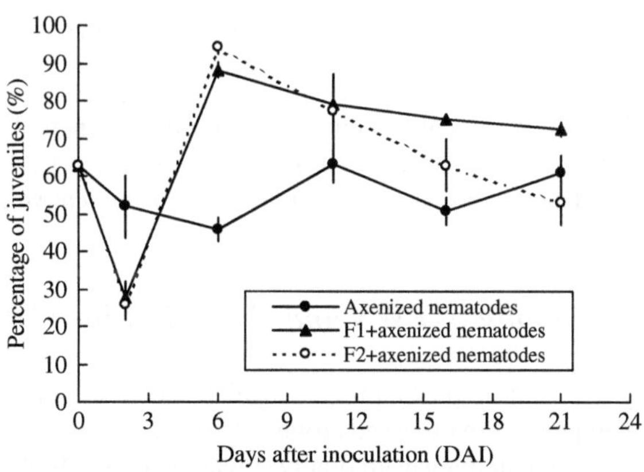


**Fig. 2** Temporal changes in percentages of juveniles in PWN population. See legends in Fig. 1

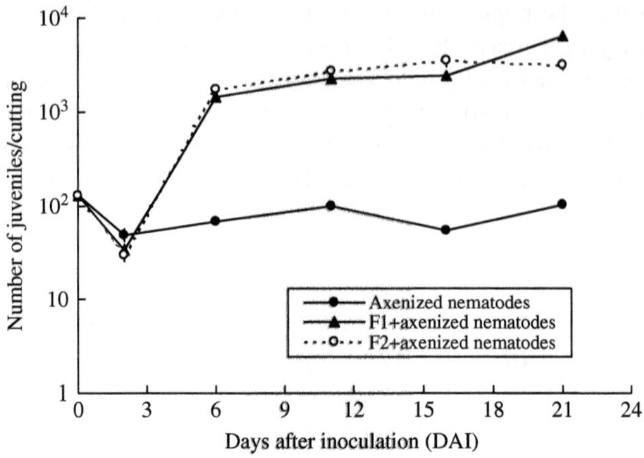


**Fig. 3** Temporal changes in numbers of juveniles in PWN population. Standard errors of means were too small to be drawn. See legends in Fig. 1

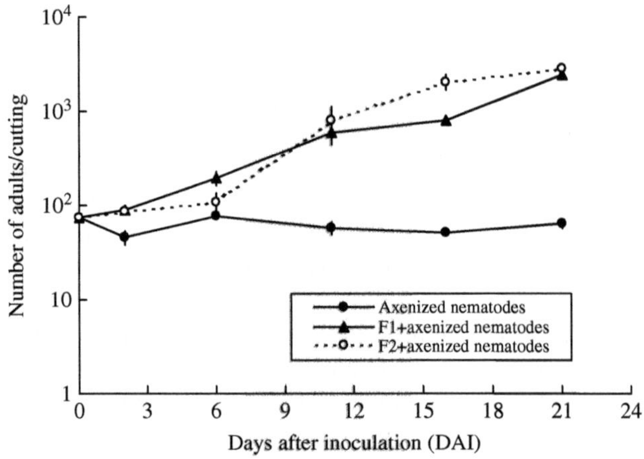


**Fig. 4** Temporal changes in numbers of adults in PWN population. See legends in Fig. 1

other two treatments inoculated with fungus + nematodes, the number of juveniles dropped by 2 DAI, then abruptly increased by 6 DAI, and increased slowly thereafter. The increase in juvenile numbers was indicative of the population growth rate.

## *Distribution of PWN and Fungi in Pine Cuttings*

By two DAI on pine cuttings inoculated with fungus only, vigorously growing fungal hyphae were observed around the inoculation site at the upper 0–2 cm. From then on, the fungal hyphae gradually spread downwards along the cutting. The spreading speed of F2 was markedly faster than that of F1 (Fig. 5). By 11 DAI, vigorously growing hyphae of F2 had spread throughout the whole cutting, whereas, until 21 DAI, hyphae of F1 only distributed at the upper 0–6 cm of the cutting.

Throughout the experimental period no fungus could be isolated from those cuttings infected only with axenized nematodes. By two DAI, on the average, most of the nematodes gathered in the upper part of the cuttings and in the water. Thereafter, most of the nematodes kept gathering in the water (Figs. 5 and 6).

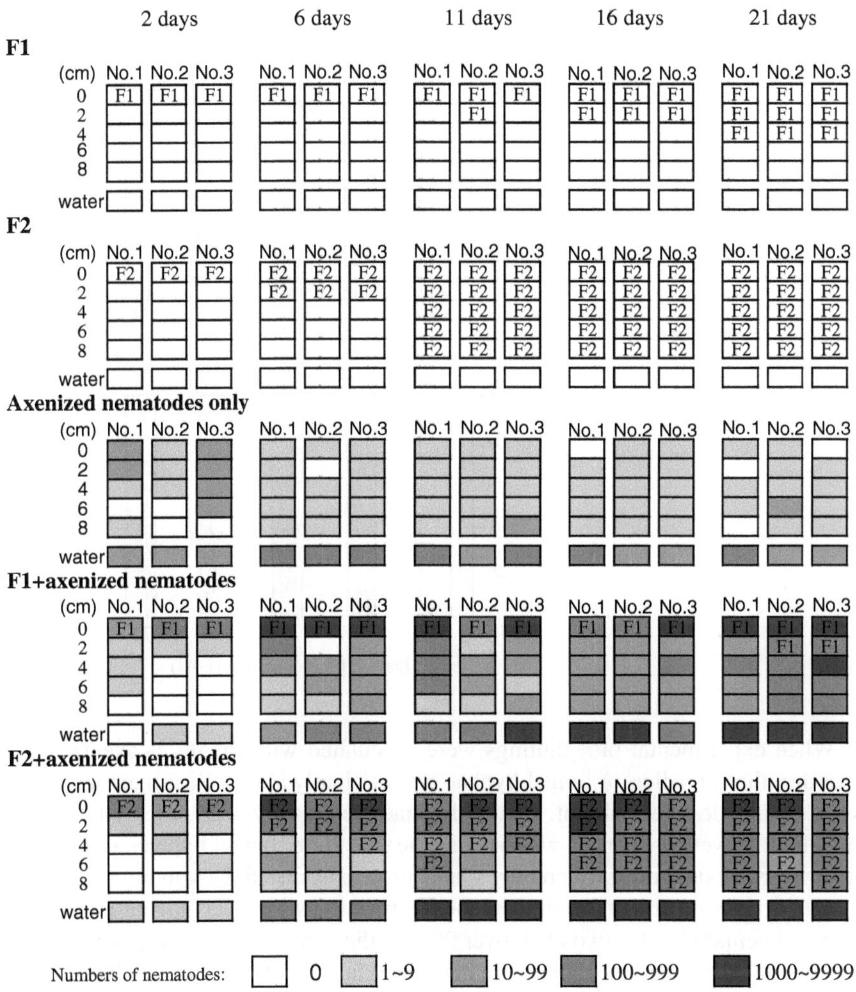


**Fig. 5** Numbers of nematodes, results of fungal and bacterial isolation, in and from each segment of the cuttings after inoculation. Each compartment means a segment of the cuttings, and the numbers on the left side indicate distance from the top end of the cutting to where the segment locates. Alphabetical letters in the figure indicate the results of fungal and bacterial isolation. F1: Cf. *Cryptosporiopsis* sp., F2: *Leptographium* sp

**Fig. 6** Temporal changes in percentages of nematodes, distributing in the different segments of the cutting and in water. (**a**) Axenized nematodes only. (**b**) F1+axenized nematodes. (**c**) F2+axenized nematodes

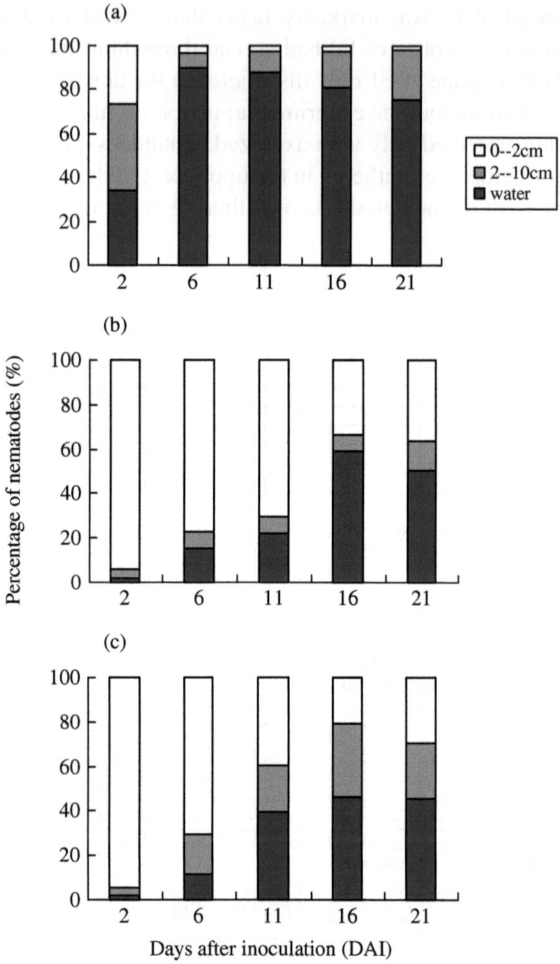


When experimental pine cuttings were inoculated with fungus + axenized nematodes, the spreading of fungal hyphae was evidently slower than their spreading when nematodes were absent. Fungal hyphae were rarely seen when F1 was the inoculum, however, when F2 was used as the inoculum, fungal hyphae grew vigorously over 2–7 days after which only sparse growth occurred. Distribution pattern of nematodes was greatly different from the treatment in which the inoculum was only axenized nematodes. By two DAI, over 90% of the nematodes gathered in the upper 0–2 cm including the inoculation site instead of in the bottom water. Subsequently, the number of nematodes increased in every part; the percentage of nematodes distributing in the 0–2 cm segment gradually decreased; the percentage of nematodes in water and in 2–10 cm segment gradually increased (Figs. 5 and 6). Throughout the experimental period, a large proportion of the nematode population distributed at the top 0–2 cm and in the water with a gradually decreasing percentage. This tendency was especially evident for the treatment of F2 + axenized nematodes, in which the percentage of nematodes distributing in 2–10 cm segment increased from the initial 3.5% up to the peak, 32.8% by 16 DAI (Fig. 6c). While this percentage only increased up to 13.4% by 21 DAI from the initial 4.2% for the treatment of F1 + axenized nematodes (Fig. 6b). Compared with the treatment of F1 + axenized nematodes, in the treatment of F2+axenized nematodes, both fungus and nematodes population showed a faster dispersal rate.

## Discussion

The present study establishes that there is a close relationship between existence and distribution of fungi, and multiplication and distribution of PWN. The nematode population failed to increase in the absence of fungi when only axenized nematodes were inoculated. When fungi were present, the PWN population increased markedly. The number of nematodes was high at sites where fungal hyphae were distributed. This has led us to infer that the vigorously growing fungal hyphae promoted the PWN's multiplication by providing a nutritional source for the nematode, and PWN preferred to gather at the place where food sources existed. Hence spreading of fungal hyphae resulted in distribution of PWN in a wider range. In living trees, wounds made by inoculation or maturation feeding of the vector beetle creates a favorable environment for fungal growth. It is suggested that in living trees, both host defenses (resin etc.) and the existence of fungi at inoculation/feeding site may contribute to the restriction of nematode distribution.

Under aseptic condition (in autoclaved cuttings inoculated with axenized nematodes), the nematodes did not multiply, and mostly remained in the water. This observation supports the authors' viewpoint that dead pine cells, in this case created by autoclaving of the cuttings, were poor food sources for PWN (Wang, 2004). By two DAI the population number of PWN decreased for all the treatments. The reason was unknown, but perhaps some of nematodes in the inoculum died during the invasion into pine tissue.

Temporal changes in the percentage of juveniles in the populaiton in cuttings inoculated with fungus + nematodes further suggested a close connection between existence of fungi and multiplication of PWN in dead pine tissue. In cuttings inoculated with axenized nematodes only, development of juveniles to adults was delayed or suppressed by poor food sources. When cuttings were inoculated with fungus + axenized nematodes, the significant drop of the percentage of juveniles in the populaiton by two DAI and the striking increase by six DAI seemed to be highly associated with existence of the fungus. By two DAI, most juveniles feed on fungal hyphae and developed into adults, which might greatly contribute to the significant decrease in percentage of juveniles in the populaiton over that in the inoculums. However, the decrease in the number of juveniles was far more than the increase in the number of adults. Considering the evident decrease in total number of nematodes and the number of juveniles in the PWN population, it is suggested that juveniles tend to die faster than adults in the presence of fungi. The reason for this is unknown, but it may be due to a negative interaction between the juveniles and the fungus. Therefore, it is suggested that both death of juveniles and growth to adults result in the observed reduction of PWN number. By six DAI adult females laid a large number of eggs which hatched into juveniles, so the percentage of juveniles in the populaiton drastically increased.

Present results were obtained in an experimental arena using autoclaved pine cuttings and simplified experimental conditions. For example, only two of seven isolates of fungi were used and these were selected at random from seven isolates. Both selected populations were favorable to the multiplication of PWN. This suggests the possibility that PWN could feed on some favorable fungi in healthy and PWN-weakened pine trees and multiply. This result supports the assumption of Kobayashi et al. (1974, 1975). In the natural environment the association of nematode and fungus in pine trees as well as, host responses and their interaction is very complicated and varies continually. Although most of the fungi related to the PWN's life history could promote nematode population increase to some degree, a few fungal species are unsuitable for the PWN's multiplication (Kobayashi et al., 1974, 1975; Fukushige, 1991). There are even some nematode-trapping fungi (Mamiya and Tamura, 1976; Tamura, 1980; Yoneda et al., 1980). Some fungi belonging to genus *Alternaria, Cephalosporium, Penicillium* and *Verticillium*, and some isolates of *Trichoderma* sp., were unsuitable for multiplication of PWN (Kobayashi et al., 1974; Fukushige, 1991; Maehara and Futai, 2000). These fungi could be used to evaluate the effect of a fungus not supportive of PWN survival on establishment of the nematode in the experimental protocol described in this paper.

In the two treatments of axenized nematodes in combination with fungus, the gradual dispersion of PWN from the inoculation site to water might be due to the rapidly increasing population density and resulting shortage of food sources at inoculation site. Thereafter PWN accompanied spreading of fungal hyphae to the lower part (2–10 cm segment) of the cutting. The relative low population number at 2–10 cm segment was possibly due to the relatively low density of fungal hyphae. It was inferred that with extension of experimental period, distribution of PWN in the whole cuttings would become uniform.

It was evident that PWN distributed accompanying spreading of fungal hyphae. This close connection further suggests that the limited local distribution of PWN in the early stage of disease development is related to the inhibition of fungal growth by host defense responses. The abrupt decrease of PWN population shortly after inoculation (Mamiya, 1974, 1985; Suzuki, 1984) was considered to be related to host defense responses (Oku et al., 1989). Based upon the results mentioned above, the process of the PWN's multiplication and distribution in pine trees was assumed to occur as follows: (i) Soon after inoculation, the host plants establish a defense against both PWN and fungi, and are not favorable food sources; a large portion of the inoculated PWN dies (Suzuki, 1984). Shortage of food sources caused by inhibition of fungal growth might further lower the nematode population. (ii) Fungal growth is restricted to the inoculation site where wound weakens and kills host cells. Correspondingly most of PWN gather near the inoculation site (Mamiya,

1974, 1985; Suzuki, 1984) where density of fungal hyphae is high. Even though the movement ability of PWN is strong (40–50 cm in one day, Suzuki, 1984) and some can distribute throughout the whole tree within 1 week (Hashimoto and Kiyohara, 1973), the number distributing at most parts is too scarce to be detected (Hashimoto and Dozono, 1975; Kuroda and Ito, 1992). (iii) After the abrupt decrease, virulent PWN gathering at the inoculation site kill pine cells by physical damage, cellulase or toxin, etc (Mamiya, 1975; Odani et al., 1985b; Oku et al., 1985; Yamamoto et al., 1986; Kojima et al., 1994). (iv) Thereafter host defense responses decline and the environment becomes favorable to fungal growth. Fungal growth promotes PWN's multiplication and therefore promotes disease development. From this assumption, it is suggested that fungus flora in pine trees may play an important role in pine wilt disease development.

**Acknowledgments** The authors thank Dr. Hayato Masuya, Forestry and Forest Products Research Institute, Japan, for his identifying the fungi.

## References

DOZONO, Y. AND YOSHIDA, N. (1974). Application of the logistic curve for the population growth of pine wood nematode, *Bursaphelenchus lignicolus*, on the cultures of *Botrytis cinerea. Journal of Japanese Forestry Society* 56: 146–148.

ENDA, N. (1988). The damage of pine wilt disease in Taiwan. *Forest Pests* 37: 161–166.

ENDA, N. AND TAKETANI, A. (1992). The damage of pine wilt disease and control in China. *Forest Pests* 41: 106–111.

FUKUSHIGE, H. (1991). Propagation of *Bursaphelenchus xylophilus* (Nematoda: Aphelenchoididae) on fungi growing in pine shoot segments. *Applied Entomology and Zoology* 26: 371–376.

FUKUSHIGE, H. AND FUTAI, K. (1987). Seasonal changes in *Bursaphelenchus xylophilus* populations and occurrence of fungi in *Pinus thunbergii* trees inoculated with the nematode. *Japanese Journal of nematology* 17: 8–16.

HASHIMOTO, H. AND DOZONO, Y. (1975). Migration and reproduction of the pine wood nematode in resistant and susceptible species of pine. *Transactions of the Annual Meeting of the Japanese Forestry Society* 86: 301–302.

HASHIMOTO, H. AND KIYOHARA, T. (1973). Migration of the pinewood nematode in the pine trees (III). *Transactions of the Annual Meeting of the Japanese Forestry Society* 84: 330–332.

ICHIHARA, Y., FUKUDA, K. AND SUZUKI, K. (2000). Early symptom development and histological changes associated with migration of *Bursaphelenchus xylophilus* in seedling tissues of *Pinus thunbergii. Plant Disease* 84: 675–680.

ISHIDA, K., HOGETSU, T., FUKUDA, K. AND SUZUKI, K. (1993). Cortical responses in Japanese black pine to attack by the pinewood nematode. *Canadian Journal of Botany* 71: 1399–1405.

KIYOHARA, T. (1989) Etiological study of pine wilt disease. *Bulletin of the Forestry and Forest Products Research Institute* 353: 127–176.

KIYOHARA, T. AND BOLLA, R.I. (1990). Pathogenic variability among populations of the pinewood nematode, *Bursaphelenchus xylophilus. Forest Science* 36: 1061–1076.

KIYOHARA, T. AND SUZUKI, K. (1975). Seasonal changes in population of *Bursaphelenchus lignicolus* in the wood of Japanese black pine. *Transactions of the Annual Meeting of the Japanese Forestry Society* 86: 296–298.

KIYOHARA, T., SUZUKI, K. AND HASHIMOTO, H. (1975). Population dynamics of *Bursaphelenchus lignicolus* at the early stage after inoculation. *Transactions of the Annual Meeting of the Japanese Forestry Society* 86: 299–300.

Kobayashi, T., Sasaki, K. and Mamiya, Y. (1974). Fungi associated with *Bursaphelenchus lignicolus*, the pine wood nematode (I). *Journal of Japanese Forestry Society* 56: 136–145.

Kobayashi, T., Sasaki, K. and Mamiya, Y. (1975). Fungi associated with *Bursaphelenchus lignicolus*, the pine wood nematode (II). *Journal of Japanese Forestry Society* 57: 184–193.

Kojima, K., Kamijyo, A., Masumori, M. and Sasaki, S. (1994). Cellulase activities of pinewood nematode isolates with different virulence. *Journal of Japanese Forestry Society* 76: 258–262.

Krusberg, L. R. and Sardanelli, S. (1984). Technique for axenizing nematodes. *Journal of Nematology* 16: 348.

Kuroda, K. and Ito, S. (1992). Migration speed of pine wood nematodes and activities of other microbes during the development of pine-wilt disease in *Pinus thunbergii. Journal of Japanese Forestry Society* 74: 383–389.

Maehara, N. and Futai, K. (2000). Population changes of the pinewood nematode, *Bursaphelenchus xylophilus* (Nematoda: Aphelenchoididae), on fungi growing in pine-branch segments. *Applied Entomology and Zoology* 35: 413–417.

Mamiya, Y. (1974). Reproduction of the pine wood nematode in seedlings of *Pinus densiflora* and *P. thunbergii. Transactions of the Annual Meeting of the Japanese Forestry Society* 85: 249–251.

Mamiya, Y. (1975). Behavior of *Bursaphelenchus lignicolus* in the wood of pine seedlings and pathological responses of pine to nematode infection. *Transactions of the Annual Meeting of the Japanese Forestry Society* 86: 285–286.

Mamiya, Y. (1985). Initial pathological changes and disease development in pine trees induced by the pinewood nematode, *Bursaphelenchus xylophilus. Annals of the Phytopathological Society of Nematology* 51: 546–555.

Mamiya, Y. (1986). Current status of pinewilt disease in China. *Forst Pests* 35: 145–150.

Mamiya, Y. and Tamura, H. (1976). A kind of nematode-trapping fungi, *Dactylella leptospora*, found in wood around pupal chambers of *Monochamus alternatus. Forest Pests* 25: 147–149.

McGawley, E., Jones, J. and Birchfield, W. (1980). Reproduction of *Bursaphelenchus lignicolus* on fungi isolated from *B. lignicolus*-infected pine trees. *Journal of Nematology* 12: 231.

Odani, K., Sasaki, S., Yamamoto, N., Nishiyam, Y. and Tamura, H. (1985a). Differences in dispersal and multiplication of two associated nematodes, *Bursaphelenchus xylophilus* and *Bursaphelenchus mucronatus* in pine seedlings in relation to the pine wilt disease development. *Journal of Japanese Forestry Society* 67: 398–403.

Odani, K., Sasaki, S., Yamamoto, N., Nishiyam, Y. and Tamura, H. (1985b). Early symptom of pine wilt disease by hydrolytic enzymes produced by pine wood nematodes – Cellulase as a possible candidate of the pathogen. *Journal of Japanese Forestry Society* 67: 366–372.

Oku, H., Shiraishi, T. and Chikawatsu, K. (1989). Active defense as a mechanism of resistance in pine against pine wilt disease. *Annals of the Phytopathological Society of Japan* 55: 603–608.

Oku, H., Yamamoto, H., Ohta, H. and Shiraishi, T. (1985). Effect of abnormal metabolites isolated from nematode-infected pine on pine seedling and pine wood nematodes. *Annals of the Phytopathological Society of Japan* 51: 303–311.

Suzuki, K. (1984). General effect of water stress on the development of pine wilting disease caused by *Bursaphelenchus xylophilus. Bulletin of the Forestry and Forest Products Research Institute* 325: 97–126.

Suzuki, K. and Kiyohara, T. (1978). Influence of water stress on development of pine wilting disease caused by *Bursaphelenchus lignicolus. European Journal of Forest Pathology* 8: 97–107.

Tamura, H. (1980). Some nematode-trapping fungi in the pupal chambers of *Monochamus alternatus. Forest Pests* 29: 39–42.

Tamura, H. (1996). After occurrence of Pine wilt disease in Korea. *Forest Pests* 45: 28–31.

WANG, Y. (2004). Life history and virulence of *Bursaphelenchus xylophilus*. PhD thesis, The University of Tokyo, Japan, 112pp.

WANG, Y., SAKAUE, D., YAMADA, T. AND SUZUKI, K. (2004). Relationship between multiplication rate on cultures of *Botrytis cinerea* and virulence of *Bursaphelenchus xylophilus* and *B. mucronatus*. *Transactions of the Annual Meeting of Kanto Branch of the Japanese Forestry Society* 55: 197–200.

YAMAMOTO, N., ODANI, K., SASAKI, S. AND NISHIYAMA, Y. (1986). Cellulase exudation by the pine wood nematode – Detection of activity in its crawling track. *Journal of Japanese Forestry Society* 68: 237–240.

YONEDA, K., KATSUMATA, J., SAIKI, H., TSUBAKI, K. AND TATSUMI, S. (1980). A nematode-trapping fungus detected in pine sap. *Journal of Japanese Forestry Society* 62: 227–229.

# Part III
# PWN Taxonomy and Detection Methods

**Alexander Ryss and Wolfgang Burgermeister**

## Summary

The session deals with both the routine and innovative approaches for pinewood nematode (PWN) survey and species identification. Firstly, two presented papers used the database and statistical tools for the species identification and taxonomical grouping within the genus *Bursaphelenchus*. In "Taxonomic databases for *Bursaphelenchus* and other aphelenchoid nematodes" (Eisenback, J., Vieira P., Ryss A. & M. Mota), the authors solve the most important problem of taxonomic resources, namely accessibility of the taxonomic publications and especially the species descriptions, and re-descriptions. The team participates in a project to create the complete electronic collection of all publications on aphelenchoid taxonomy, including papers on *Bursaphelenchus* systematics. The PDF format with OCR recognition gives an opportunity of active search within literature databases on CD ROMs.

Braasch reviewed all the data on the most practically important "*xylophilus*" species group within the genus *Bursaphelenchus*. The group has been recently enlarged with new descriptions of several species from SE Asia, the region considered now as the rich biodiversity area of the "*xylophilus* group". New detailed diagnosis of the group is also given, and changes in morphological characteristics, depending on geographic distribution, were demonstrated. New conclusions on the possible reproductive isolation between European and East Asian populations of *B. mucronatus* were drawn, with an interesting discussion whether these populations should be, in the future, considered as separate subspecies.

Metge reported on "Interspecific variation in ITS rDNA of *Bursaphelenchus* species of different groups" as opening communication on molecular biology methods. The rDNA partial sequences, including both ITS regions of 23 isolates representing 17 *Bursaphelenchus* species including five recently described species (*B. doui, B. singaporensis, B. willibaldi, B. arthuri, B. yongensis* and the rediscovered *B. eremus*), were used for studying phylogenetic relationships. The separation of isolates into different clusters of the phylogenetic trees confirmed their association to species groups which had been previously postulated on the basis of morphological features: *B. doui* and *B. singaporensis* were affiliated to the "xylophilus" group, *B. willibaldi* and *B. arthuri* to the "fungivorus" group, *B. yongensis* and *B. eremus* to the "hofmanni" group. The number of sequence differences between species was highest in ITS2, lower in ITS1 and lowest in the 5.8S rDNA. The high sequence polymorphism of the ITS2 limits phylogenetic studies among species using this sequence alone. However, the combined sequence alignment of ITS1, 5.8S rDNA and ITS2 has proven suitable for studying interspecies relationships including very closely related species. Results are in agreement with phylogenetic studies by the group of Robin Giblin-Davis on *Bursaphelenchus* species using sequences of 18S rDNA, the D2/D3 region of 28S rDNA and the gene for mitochondrial cytochrome oxidase subunit I (COI). In these studies, too, a combined evaluation of the three different types of sequences resulted in the most convincing phylogenetic arrangement, supported by generally high bootstrap values.

Application of ITS rDNA sequences to phylogeny was also the subject of the next presentation by Burgermeister on "Molecular characterisation of isolates of the *Bursaphelenchus sexdentati* group using ITS-RFLP and ribosomal DNA sequences". Alignment studies of the combined ITS1, 5.8S rDNA and ITS2 sequences of 17 isolates belonging to five species of the "sexdentati" group (*B. sexdentati, B. vallesianus, B. pinophilus, B. poligraphi* and *B. borealis*) were carried out. Branching of the phylogenetic tree in separate clusters confirmed the morphological affiliation of the isolates to their respective species. Further subclustering was observed among the *B. sexdentati* isolates suggesting the existence of a central European and a south European subtype of *B. sexdentati*, in agreement with small morphological differences among these isolates. In some *B. sexdentati* isolates, micro-heterogeneity of ITS2 sequences within the same specimen was demonstrated by sequence analysis of cloned ITS2 rDNA. This type of micro-heterogeneity had also been shown earlier for other *Bursaphelenchus* species and species of other nematode genera and seems to be a general feature of ITS rDNA sequences.

Metge further reported on the "Analysis of *Bursaphelenchus xylophilus* provenances using ISSR and RAPD fingerprints". The main objective of this study was to obtain information on the origin of the PWN population introduced in Portugal prior to 1999. By application of the two independent DNA fingerprinting techniques to 30 *B. xylophilus* populations representing all countries where the pest occurs (except Taiwan), dendrograms of genetic distances were constructed. Separate large clusters were obtained for the American (USA, Canada) and the East Asian (Japan, China, Korea) isolates. In both the RAPD and the ISSR analysis, the three Portuguese *B. xylophilus* isolates examined were grouped within the cluster of Asian isolates. The results indicated that the Portuguese *B. xylophilus* originated from East Asia and not from the native habitats of the species in North America. In a more recent investigation of RAPD analysis of 24 Portuguese *B. xylophilus* isolates, Vieira has demonstrated remarkable genetic homogeneity among the isolates, suggesting their origin from a single introduction.

In the following presentation, Leal reported on "An effective PCR-based diagnostic method for the detection of *Bursaphelenchus xylophilus* in wood samples from lodgepole pine". Utilizing sequence information from the heat shock protein 70 (hsp70) gene, PCR primer systems were developed, allowing species-specific identification of *B. xylophilus* and *B. mucronatus*. It was shown that the hsp70 PCR assay is sensitive enough to detect DNA from less than one nematode, although the hsp70 gene is present in fewer copies than ribosomal RNA genes in the *Bursaphelenchus* genome. The assay also worked well in the presence of potential inhibitors associated with wood, and thus eliminates the need to produce clean nematode samples through further culturing. The speed and sensitivity of this new diagnostic method was further increased by application of real-time PCR using a TaqMan probe.

In the following presentation by Castagnone-Sereno on "Satellite DNA as a versatile genetic marker for *Bursaphelenchus xylophilus*", applications of the Msp I satellite DNA for diagnostics and phylogenetic analyses were presented. Due to the extremely high number of satellite repeats corresponding to 30% of the total genomic DNA of *B. xylophilus*, highly sensitive identification tests have been developed, based on squash blot hybridization or PCR, including real-time PCR with a TaqMan probe. Sequencing of cloned satellite monomers has revealed the existence of highly conserved regions, but also 18.5% variable positions. Phylogenetic analyses of *B. xylophilus* isolates sampled worldwide and within the infested area in Portugal revealed considerable diversity. The apparently higher satellite sequence variability, as compared to RAPD and ISSR markers, will be exploited for more accurate analysis of the origin, evolution and pathways of spreading of the Portuguese isolates.

In the final presentation on "Port Check: the potential of real-time PCR as a valuable tool for the detection of the pine wood nematode, *Bursaphelenchus xylophilus*", Mota reported on the objectives and achievements of the EU, 6th Framework project "Port Check". In this Specific Targeted Research Project (STREP), a large consortium cooperates to develop and evaluate real-time PCR for a number of key quarantine organisms. The satDNA-based TaqMan PCR assay outlined in Castagnone-Sereno's presentation has been set up for identification of single PWN individuals, using a field-portable real-time thermocycler. To speed up the total time required for PWN detection, a fast DNA extraction method for small samples of nematode-infested wood was also developed, in order to avoid the time-consuming extraction of live nematodes from wood using the traditional Baermann funnel technique.

The contributions on molecular diagnostics presented during Part III of the symposium have demonstrated an advanced state of methodology. Molecular species identification has become a routine method in some laboratories due to the availability of specific PCR primer systems addressing different parts of the genomic DNA: satellite DNA, ITS1/2 rDNA and hsp70 gene. The speed and sensitivity of diagnosis has been further increased by introduction of several real-time PCR assays and improving extraction methods for nematode-infested wood. However, considerable effort is still necessary for introduction of molecular identification techniques to additional plant health laboratories and inspection services. Phylogenetic classification of *Bursaphelenchus* species on the basis of aligned ITS1/2 sequences of their rDNA has confirmed genetic relationships of subspecies, species and species groups which had been postulated before on the basis of their morphological features. Thus, sequencing of ITS1/2 has proved to be a very useful tool in the molecular taxonomy of the genus *Bursaphelenchus*. Other scientists not represented in this symposium have successfully used other sequences of genomic (18S rDNA and the D2/D3 region of 28S rDNA) and mitochondrial DNA (cytochrome oxidase subunit I gene, COI) for taxonomic classification of *Bursaphelenchus* species. It can be expected that cluster analysis based on the combined information from all available sequences including ITS1/2 will provide the most reliable results on the phylogeny of species. Phylogenetic analyses of PWN populations have been carried out utilizing either RAPD/ISSR fingerprints or satellite DNA sequence heterogeneity. Experience with RAPD and ISSR fingerprints has shown that patterns of the same isolates do not change substantially during ten years of culturing, but different isolates show pattern differences which could be correlated to the geographic distances of their origins. Satellite DNA heterogeneity appears to trace even short-time evolutionary changes, as was shown for PWN isolates collected in the infested area of Portugal. Therefore, RAPD/ISSR- and satellite DNA-based phylogenetic studies should be combined for pathway analysis of PWN introduction to Portugal and for tracing its movement within the affected area.

# Electronic Taxonomic Databases for *Bursaphelenchus* and Other Aphelenchid Nematodes

**Jonathan D. Eisenback, Paulo Vieira, Manuel Mota and Alexander Ryss**

**Abstract** Undoubtedly, the most important resources for nematode taxonomy are the original species descriptions. Unfortunately, nematode descriptions have been published since the middle of the eighth century in a variety of forms including many relatively obscure journals and proceedings and sometimes in lengthy special publications. As a result, the task of collecting all of the descriptions of a single genus is daunting and has been repeated by numerous nematode taxonomists around the world. Furthermore, many of the papers were printed on acid containing paper and are rapidly deteriorating by yellowing and becoming brittle. The purpose of the present project is to collect all of the original species descriptions of the Aphelenchs including the genus *Bursaphelenchus*. The original papers are digitized with a flatbed scanner and converted to the PDF documents. The image is converted into text with optical character recognition software that is built into Adobe® Acrobat; thus each document becomes a fully searchable text that includes all of the photographs and drawings that are the same quality as the original. Theses collections are put together into monographs that are published as individual CD ROMs.

## Introduction

Nematodes are the most abundant metazoans, comprising more than 80% of all animals alive today. Since 1743, when Needham described the first nematode (Needham, 1743), approximately 20,000–30,000 species have been named, with estimates of species remaining to be described ranging from 100,000 to 1 million (Blaxter, 2004; De Ley, 2000). Unfortunately, the taxonomic community is woefully inadequate for this task. The number of taxonomists currently describing new species of nematodes around the world is less than 100, and significant increases are not expected. If each of these taxonomists were able to describe 10 new species every year, it would take between 100 and 1,000 years to name these yet to be described species.

J.D. Eisenback
Dept. Plant Pathology, Physiology and Weed Science, Virginia Polytechnic Institute and State University, 103 Prince Hall, Blacksburg, VA 24061, USA
e-mail: jon@vt.edu

M.M. Mota, P. Vieira (eds.), *Pine Wilt Disease: A Worldwide Threat to Forest Ecosystems*, 

## Type Specimens

Original descriptions of nematodes provide the working database that drives nematode taxonomy. Theoretically, the taxonomy of animals is based on type specimens (Ride et al., 1999) that are selected and illustrated in the original description. In order to satisfy the rules for naming species, each is designated by a type; thus, the type becomes the definition of the species. Whenever doubts arise about the identification of a particular population of nematodes, the taxonomist can request to see the type, or at least a paratype, from the museum in which it was deposited by the original author. If the specimen is identical to the type, then it belongs to that species; however, if it is significantly different, then it belongs to another species, perhaps even one that has not been described. Obviously, this assessment is subjective, but all species descriptions can be further scrutinized when a taxonomist critically examines a particular group of species. Although complicated rules govern the change in classification and how specimens are allocated to species, subsequent authors can propose corrections to species descriptions through clarifications, amendments, and by making synonyms.

Unfortunately, type specimens of nematodes are often inadequate resources to satisfy the requirements that they are supposed to fulfill. Many type specimens are suitable at the initial time of designation, but they slowly deteriorate over time until they become nearly useless for their original intent. Because these specimens are so fragile, many museums are reluctant to send them out to taxonomists who make requests to view them. Taxonomists that go through the pains to request and review these type specimens are often frustrated because the specimens may have been poorly preserved and mounted onto glass slides. As a result of these limitations of the types, often the original publication, with its description of the type, remains the sole source that is used to evaluate specimens in order to make an identification, to justify the description of a new species, or to review the status of various taxonomic units.

## Original Descriptions

Original descriptions contain a description of the type that often replaces the use of the real type, except in very unusual circumstances that warrant its request; therefore, published species descriptions are the primary source for classification and identification. Therefore, each taxonomist who intends to describe a new species, or to evaluate a particular taxonomic unit, must begin the arduous task of collecting all of the descriptions or every species within that group. Unfortunately, this task is often difficult and time consuming since many were published in obscure journals that may be difficult to find. Some descriptions may be available as reprints; others must be photocopied from journal collections in the local library, and many may have to be requested from colleagues or other lending libraries. Even for genera with few species, the task of collecting all of the descriptions is difficult and time consuming. Genera containing more than 100 species may require several years to obtain all of the described species.

Unfortunately, original descriptions have been, and continue to be published in a wide variety of journals in numerous countries around the world. As the numbers of species descriptions increase every year, the difficulty of collecting all of the descriptions increases as well. Some are widely available and well known, but others are difficult to obtain and often obscure. Furthermore, many of these journals were published on inexpensive paper that has become brittle, yellowed, and sometimes, even faded. Like the type specimens, the publication is also subject to deterioration. Proper storage of these materials requires the use of very expensive facilities where humidity, temperature, and exposure to the ultra-violet light can be carefully controlled.

Most descriptions were published in English, but many were published in French, German, Spanish, Portuguese, Chinese, Russian, Japanese, and other less well-known languages. Translation of these descriptions into English is often difficult and costly. Furthermore, few of these translations warrant publication and are, therefore, not available except for the taxonomist who initiated the translation. Theoretically, some species descriptions have been translated many times in the past and are likely to be translated again in the future.

As physical objects, these descriptions can be damaged by the elements, lost or misplaced, require space for their storage, and are not readily shared with others. Each taxonomist has to collect all of these descriptions; therefore these collections are very personal and only known and valued by the original collector. Often they are discarded or forgotten after the collector retires.

## Electronic Databases

Fortunately the desktop computer has made it possible to store large amounts of information electronically (Eisenback, 1997). These electronic databases can be archived and stored, duplicated, and shared. If they are made available on the Internet, they can be utilized by anyone who has access to the World Wide Web. Unfortunately, if these works are archived improperly, they can rapidly become inaccessible because of changes in proprietary format. The solution to making an electronic database that will be long-lived is to choose a format that will last into the future.

## Materials and Methods

The portable document format (PDF) popularized by Adobe Systems Incorporated has revolutionized paper documents into an electronic format that makes it possible to reliably create, combine, and control documents for easy distribution and data collection. Adobe® Acrobat PDF documents are easily accessible and require minimal storage space. The Adobe Acrobat PDF format is recognized internationally and has standards that insure that documents will remain readily assessable in the future. This file format has been useful for ten years despite continuous changes in operating systems and hardware platforms.

The most difficult part of making an electronic taxonomic database is the collection of all of the original descriptions. After all of the originals have been gathered, the electronic taxonomic databases are put together by scanning the original publication on a flatbed scanner in black and white mode for text at 300 dpi resolution. Pages containing photographs are scanned in either grayscale or color at 300 dpi. The images of the scans are cleaned and sometimes enhanced with a photo editing software program like Adobe® Photoshop and saved as PDF documents. The individual pages of PDF documents are merged together into one document and the images of text are recognized with the optical character recognition (OCR) software that is included with Adobe Acrobat. This step is an important part of making the document fully searchable; otherwise, the pages with text are actually just copies of the images of the text that is on the page.

The individual descriptions are named with the genus and species name and put together into one folder with the name of the genus, the taxonomic authority of that genus, and the date of publication. If one publication contains the description of more than one species, then that PDF is copied and given another name until all of the new species in that publication are duplicated. Therefore, one publication with 100 new species will be copied 100 times. Likewise, the name that was used in the original description is the name that is used to place it into the file folder containing the name of the genus. If a species has been transferred to another genus or considered to be a synonym or placed into another taxonomic category, one copy will remain in the folder of the genus as it was originally named, and another copy will be renamed and placed in the folder for the new category. For example, the original description of *Bursaphelenchus cocophilus* can be found in the folder with the species descriptions of *Aphelenchus* as it was originally described (Cobb, 1919), *Rhadinaphelenchus*, as it was later renamed (Goodey, 1960), and finally *Bursaphelenchus* to which it was transferred (Baujard, 1989).

## Results

All of the original descriptions of the pine wood nematodes (*Bursaphelenchus* spp.) have been digitized and made available as an electronic taxonomic database (Vieira et al., 2003). In addition to the species descriptions, the major taxonomic literature on this group was included in the collection of descriptions. In addition to the pine wood nematodes, all of the descriptions of the aphelenchid nematodes are currently scheduled to be digitized. Hopefully, some day in the near future, all of the descriptions of all of the species in the entire phylum of Nematoda will be digitized and available on the Internet.

## Conclusions

Electronic taxonomic databases provide a powerful resource for nematode taxonomists. With these databases the original descriptions of nematode species are beginning to be preserved in a digital format that is searchable, shareable, storable and easily upgradeable. As collections of all of the groups become available, the drudgery of collecting and maintaining all of the species descriptions in a particular group will be greatly reduced to the task of adding new species as they are described.

Most taxonomic databases on the Internet are a simple listing of names with perhaps a few of the most important morphological features (Gewin, 2002; Page, 2005; Paterson, 2003). If all of the original descriptions can be collected and digitized, the nematodes will be the first group of organisms to be preserved and placed into a single database. This database will be a tremendous resource that will have a long-lasting and major impact on the taxonomy of the nematodes.

## References


BAUJARD, P. (1989). Remarques sur les genres des sous-familles Bursaphelenchidae Paramanov, 1964 et Radinaphelenchinae Pramanov, 1964 (Nematoda: Aphelenchoididae). *Revue de Nématologie* 12: 323–324.

BLAXTER, M.L. (2004). The promise of DNA taxonomy. *Philosophical Transactions Royal Society London B* 359: 669–679.

COBB, N.A. (1919). A newly discovered nematode *Aphelenchus cocophilus* n. sp. connected with a serious disease of the coconut palm. *West Indian Bulletin* 17: 203–210.

DE LEY, P. (2000). Lost in worm space: phylogeny and morphology as road map to nematode diversity. *Nematology* 2: 9–16.

EISENBACK, J.D. (1997). *Root-knot nematode taxonomic database*. CAB International: Wallingford, UK.

GEWIN, V. (2002). Taxonomy: All living things, online. *Nature* 418: 362–363.

GOODEY, J.B. (1960). The classification of the Aphelenchoidea Fuchs,1937. *Nematologica* 5: 111–126.

NEEDHAM, T. (1743). A Letter from Mr. Turbevil Needham, to the President; Concerning Certain Chalky Tubulous Concretions, Called Malm: With Some Microscopical Observations on the Farina of the Red Lily, and of Worms Discovered in Smutty Corn. *Philosophical Transactions* 42: 634–641

PAGE, R.D.M. (2005). A taxonomic search engine: Federating taxonomic databases using web services. *BMC Bioinformatics* 6: 48.

PATERSON, D. (2003). Progressing towards a biological names register. *Nature* 422: 661.

RIDE, W.D.L., COGGER, H.G., DUPUIS, C., KRAUS, O., MINELLI, A., THOMPSON, F.C. AND TUBBS, P.K. (1999). *International Code of Zoological Nomenclature, 4th ed.* The International Trust for Zoological Nomenclature: The Natural History Museum, London.

VIEIRA, P., MOTA, M. AND EISENBACK, J.D. (2003). *Pine wood nematode taxonomic database.* Mactode Publications: Blacksburg, VA.

# The Enlargement of the *xylophilus* Group in the Genus *Bursaphelenchus*

**Helen Braasch**

**Abstract** The *xylophilus* group of the genus *Bursaphelenchus* can be clearly distinguished from other species of the genus by the presence of four lateral lines, the presence of a vulval flap in females, a characteristic shape of the male spicules and the arrangement of the seven caudal papillae. The identification of the species within this group was previously mainly based on the female tail shape. The description of new *xylophilus* group species complicates their identification. Whereas the widely distributed species *B. xylophilus*, *B. fraudulentus*, *B. mucronatus* and *B. kolymensis* have been known for a long time, five other species of the *xylophilus* group have been described since 2000: *B. conicaudatus, B. luxuriosae, B. doui, B. singaporensis* and *B. baujardi*. They were mainly detected in East Asia and South East Asia. This seems to indicate that this region has a special enrichment of species in this group of the genus *Bursaphelenchus*. The morphological characters of the *xylophilus* group species are shown and their differentiation is depicted. An identification key of the nine species of the *xylophilus* group known so far is presented. Recording of a mucronate strain of *B. xylophilus* in packaging wood brings up a new uncertainty in morphological identification. *B. kolymensis* is considered to be the European type of *B. mucronatus*. This assumption is supported by morphological studies and former genetic results.

## Introduction

The progressive geographic spread of the pinewood nematode (PWN), *Bursaphelenchus xylophilus*, as well as national and international regulations (for example, ISPM 15: FAO, 2003) were accompanied by intensive sampling and laboratory investigations for the presence of *Bursaphelenchus* spp. in imported wood worldwide in order to significantly reduce the risk of the pest's spread. These studies revealed the presence of several previously unknown *Bursaphelenchus* species of the

H. Braasch
Kantstr. 5, D- 14471 Potsdam, Germany
e-mail: h.braasch@t-online.de

  

*xylophilus* group (*sensu* Braasch, 2001) beside *B. xylophilus*, particularly in packaging wood (Gu et al., 2006). Like the PWN, they are transported from country to country and can possibly be confused with the dangerous pest. Therefore, the species of the *xylophilus* group are listed, their essential morphological characters are described and an identification key has been elaborated.

## Material and Methods

The morphology of seven species of the *xylophilus* group and the two types of *B. mucronatus* was studied by means of a Zeiss Axioskop microscope and SEM micrographs produced by M. Brandstetter and J. Gu. Specimens of *B. kolymensis* and *B. baujardi* were not available, and their characters were studied from the literature.

## Results

The following species of the *xylophilus* group are known so far:

*Bursaphelenchus xylophilus* (Steiner & Buhrer, 1934) Nickle, 1970
*B. fraudulentus* Rühm, 1956
*B. mucronatus* Mamiya & Enda, 1979
*B. kolymensis* Korentchenko, 1980
*B. conicaudatus* Kanzaki, Tsuda & Futai, 2000
*B. baujardi* Walia, Negi, Bajaj & Kalia, 2003
*B. luxuriosae* Kanzaki & Futai, 2003
*B. doui* Braasch, Burgermeister & Zhang, 2004
*B. singaporensis* Gu, Zhang, Braasch & Burgermeister, 2005

Whereas the widely distributed species *B. xylophilus*, *B. fraudulentus*, *B. mucronatus* and *B. kolymensis* have been known for a long time, the five remaining species have been described since 2000.

Ryss et al. (2005) included three further species in the *xylophilus* group, but these are herein considered not to belong to this group due to the following reasons: *B. crenati* has a different position of caudal papillae (the double pair in front of the bursa is missing), the presence of a vulval flap is questionable, and the spicules do not show a cucullus. Additionally, it is transmitted by a bark beetle, a scenario not typical for the *xylophilus* group. *B. eroshenkii* has five incisures in the lateral field (four in the *xylophilus* group), only five caudal papillae (seven in the *xylophilus* group) and no vulval flap (Kolossova, 1998). *B. abruptus* shows spicules which are very similar to those of species of the *xylophilus* group and also a large vulval flap (Giblin-Davis et al., 1993). However, the position of caudal papillae is different from that of other members of the *xylophilus* group. Sequencing results revealed a possible basal position of *B. abruptus* at the root of the genus (Giblin-Davis et al., 2005). In contrast to the association with cerambycid beetles of the species of the *xylophilus* group, *B. abruptus* is an associate of a digger bee (Giblin-Davis et al., 1993).

The species of the *xylophilus* group are characterized by the features of the family Parasitaphelenchidae and the genus *Bursaphelenchus* according to Hunt (1993), i.e., the presence of four lateral lines and three pairs of caudal papillae (a pair adanal and a double pair in front of the bursa) and a single papilla anterior to the anus (Fig. 1), a characteristic shape of the spicules (large, strongly arcuate, bluntly rounded apex, pointed rostrum, cucullus on the distal tip, see Fig. 4), large vulval flap (Fig. 2), and long postuterine branch. Other four-lined *Bursaphelenchus* species are the *fungivorus* group, which lacks a vulval flap, and the *sexdentati* group, which shows a smaller vulval flap, but a distinctly different position of caudal papillae. Both groups display a shape of spicule which is completely different from that of the *xylophilus* group species.

Characters useful for identification of the species of the *xylophilus* group are the female tail shape (round-tailed or digitate, conoid, with or without mucro), the shape of spicules, which are, however, very similar and only slightly variable, the size of spicules (especially large in *B. singaporensis* and *B. doui*), the ratio c' of females and the position of the excretory pore. The shape of the female tail is a

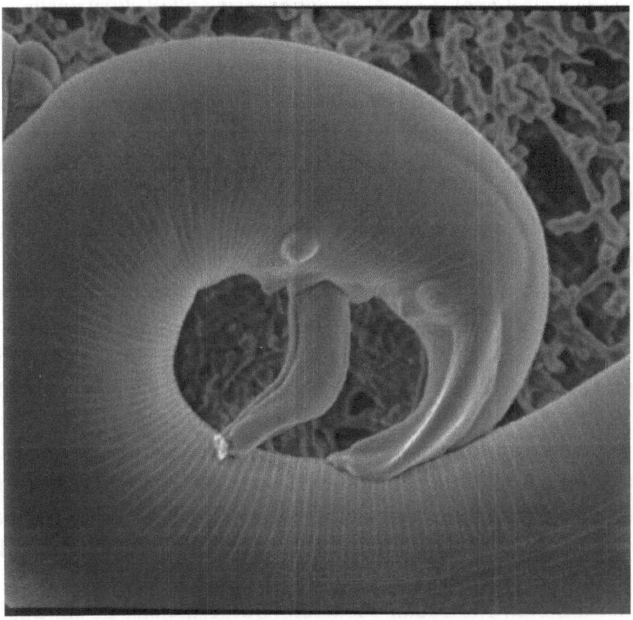

**Fig. 1** *Bursaphelenchus fraudulentus*: typical characters of the species of the *xylophilus* group: four lateral lines, bursa, spicules with cucullus, seven caudal papillae (single pre-anal papilla, first pair adanal, a double pair in front of bursa). (SEM photograph by M. Brandstetter, Vienna)

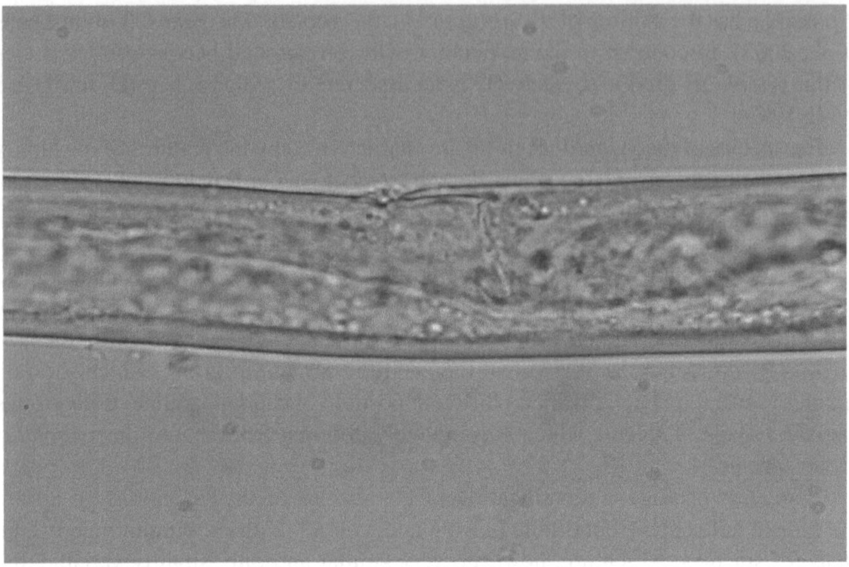

**Fig. 2** Vulval flap of females on species of the *xylophilus* group

very important character for identification of the species of the *xylophilus* group (Table 1, Fig. 3). The females of *B. singaporensis* and *B. luxuriosae* show a conical tail without mucro, and the tail of *B. luxuriosae* is strongly ventrally bent. *B. conicaudatus* and *B. baujardi* have a conical tail with a small mucro, whereas *B. xylophilus, B. mucronatus, B. fraudulentus* and *B. kolymensis* have a more or less cylindrical or digitate female tail. The females of the latter three species have a mucro of various lengths. The female tail shape of *B. doui* shows also a small mucro, but its shape is intermediate between conical and digitate. Although the most widespread form of the dangerous pest *B. xylophilus*, the form with the round-tailed females, can relatively easily be distinguished from other species of the *xylophilus* group and also from other *Bursaphelenchus* species; identification becomes more difficult in the case of the mucronate form of *B. xylophilus*. Some populations in North America and a population found in packaging wood imported from Taiwan to China (Prof. M. Lin, Nanjing, *personal communication*) exhibit this character. Whereas the females in mucronate populations generally show a mucro on the female tail end, a protuberation or a mucro can also be present in some of the females in round-tailed populations, particularly after extraction from wood. However, round-tailed females are usually also present in those populations, and their percentage decreases after multiplication on fungi in cultures. Morphological differentiation of mucronate *B. xylophilus* from *B. mucronatus* or *B. fraudulentus* is often difficult. Molecular methods (Burgermeister et al., 2005) are very helpful.

The spicule shape varies only slightly between the species with regard to the following features: the angle between the line along the capitulum and the line

**Table 1** Various female tail shapes of species of the *xylophilus* group

| | | |
|---|---|---|
| **Conoid female tail without mucro** | *B. luxuriosae*<br>*B. singaporensis* | 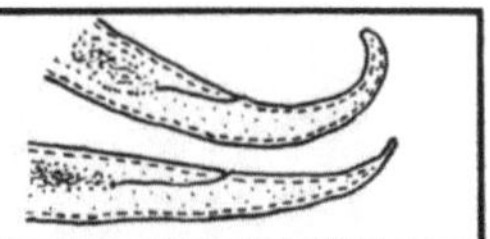 |
| **Conoid female tail with small mucro** | *B. conicaudatus*<br>*B. baujardi* | 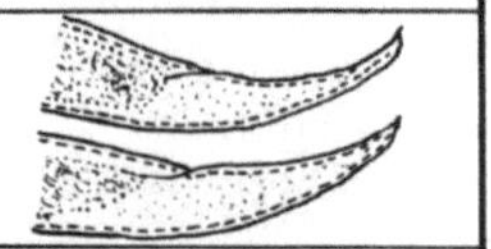 |
| **Intermediate** | *B. doui* | 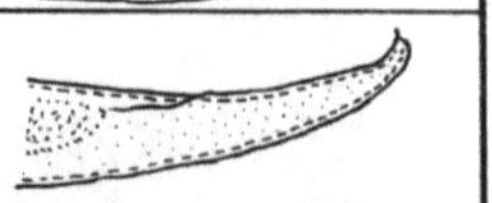 |
| **Cylindrical or digitate female tail with or without mucro** | *B. xylophilus*<br>*B. fraudulentus*<br>*B. mucronatus* (EA*)<br>*B. kolymensis*<br>*B.mucronatus* (E*) | 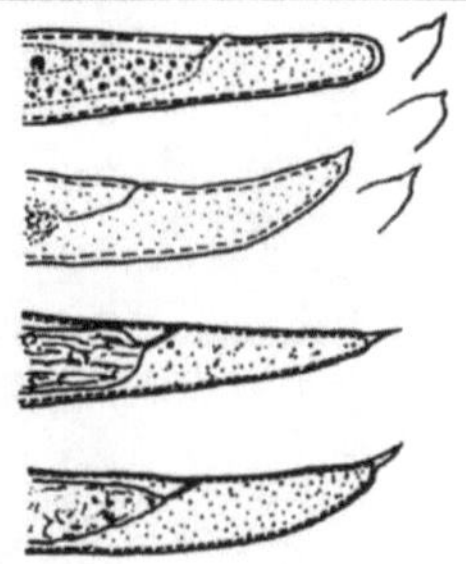 |

*EA=East Asian type
**E=European type

extending to the spicule tip; the presence (depth) of a capitulum depression; the position of the condylus; and the overall curve of the spicules. Variation within a single species makes the use of these characters difficult. Two species can be separated from the other species by the size of the spicules: *B. singaporensis* has spicules of more than 40 μm length; *B. doui* has spicules of more than 34 μm length, whereas the remaining species usually have shorter spicules. The shape of the bursa cannot be used for species differentiation in the *xylophilus* group. It varies from oval and angular to three or four-pointed within a single species.

The excretory pore is located at median bulb level or more anterior (*B. kolymensis*, European type of *B. mucronatus, B. baujardi, B. fraudulentus, B. doui*), at the level of median bulb or slightly posterior to it (*B. singaporensis, B. luxuriosae*) or posterior to median bulb (East Asian type of *B. mucronatus, B. xylophilus)*. The female c'-index (tail length related to anal body diameter) is especially high (>4) in the three species with conoid tails: *B. singaporensis, B. luxuriosae* and *B. conicaudatus*. Other species of the *xylophilus* group have c'= 2.5–4.0.

The following dichotomous key of species of the *xylophilus* group is based on the characters mentioned above. The unity of the *xylophilus* group has been confirmed by sequencing results (Kanzaki & Futai, 2003; Giblin-Davis et al., 2005).

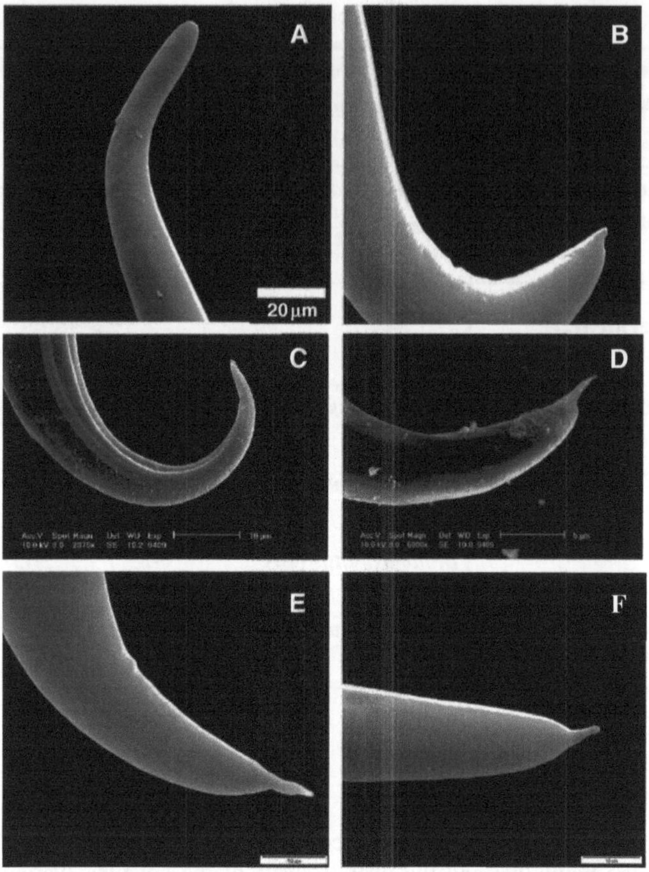

**Fig. 3** Female tails of species of the *xylophilus* group. A: *B. xylophilus*; B: *B.fraudulentus*; C: *B. singaporensis*; D: *B. doui*; E: *B. mucronatus* East Asian type; F: *B. mucronatus* European type. (SEM photographs by M. Brandstetter, Vienna, Austria [A, B, E, F] and J. Gu, Ningbo, China [C, D])

## *Key to the Species of the* xylophilus *Group*

1. Female tail conoid, c' > 4 (average), without mucro or with small hair-like mucro ..................................................................2

- Female tail cylindrical or subcylindrical, c' < 4 (average), with or without ..................................................................4

2. Female tail without mucro ..................................................3

- Female tail with small pointed mucro, spicule length < 30 μm.................... ..................................................................***B. conicaudatus***

(possibly synonymous with ***B. baujardi****, c'= 3–4)

3. Spicule length > 40 μm, spicules with prolonged medium part.................. ..................................................................***B. singaporensis***

- Spicule length 27–30 μm, female tail ventrally bent and with irregular dorsal contour near terminus. . . . . . . . . . . . . . . . . . . . . . . . . . . . . . . . . . . . . . . . . . . . . . . . ***B. luxuriosae***

4. Female tail cylindrical, terminus broadly rounded, without mucro, a part of the population may possess a short process at the tail terminus . . . . . . . . . . . . ***B. xylophilus***

- Female tail subcylindrical, terminus with mucro. . . . . . . . . . . . . . . . . . . . . . . . . . . . . . . 5

5. Spicule length < 34 μ. . . . . . . . . . . . . . . . . . . . . . . . . . . . . . . . . . . . . . . . . . . . . . . . . . . . 6

- Spicule length 34–44 μm, excretory pore at level of median bulb. . . . . . . . . ***B. doui***

6. Spicule condylus dorsally not offset, capitulum distinctly concave, mucro 4–7 μm long, often pointed, usually a continuation of the sometimes slightly conoid tail, excretory pore posterior to the median bulb. . . . . . . . . ***B. mucronatus*** (East Asian type)

- Spicule condylus dorsally offset, capitulum not distinctly concave, mucro ≤ 5 μm . . . . . . . . . . . . . . . . . . . . . . . . . . . . . . . . . . . . . . . . . . . . . . . . . . . . . . . . . . . . . . . . . . . . . . . . . .7

7. Excretory pore at or in front of the median bulb. . . . . . . . . . . . . . . . . . . . . . . . . . . . . . . 8

- Excretory pore behind median bulb, small mucro hair-like, up to 4 μm long, usually offset from tail. . . . . . . . . . . . . . . . . . . . . . . . . . . . . . .***B. xylophilus*** (mucronate form)

8. Mucro short and broad with a wide base, usually continuous with tail line. . . . . . . . . . . . . . . . . . . . . . . . . . . . . . . . . . . . . . . . . . . . . . . . . . . . . . . . . . . . . . . . ***B. fraudulentus***

- Mucro usually not continuous with tail and variably shaped, often digitate. . . . . . . . . . . . . . . . . . . . . . . . . . . . . . . . . . . . . . . . . .***B. mucronatus*** (European type)/***B. kolymensis***

* Type material of *B. baujardi* has not been studied.

## *Synonymy of* B. kolymensis *and the European Type of* B. mucronatus

Magnusson and Kulinich (1996) examined the type specimens of *B. kolymensis* and came to the conclusion that they are most similar to *B. mucronatus* populations from Russia and France, but differ from Japanese populations. *B. mucronatus* populations from France and most of the Russian populations of *B. mucronatus* belong to the European type of this species, which is wide-spread in Europe and Siberia and can be differentiated from the East Asian type of *B. mucronatus* by morphology and molecular methods (Braasch, 1991; Braasch et al., 1995; Braasch et al., 1998; Braasch et al., 1999; Hoyer et al., 1998; Burgermeister et al., 2005). The East Asian type of *B. mucronatus* is mainly distributed in East Asia, but a few populations were also found in Russian wood from Siberia (Braasch et al., 2001) and in Europe (Germany) (Braasch et al., 1999). This gives the impression that the East Asian type is spreading westwards and the European type is spreading eastwards, where a few populations have been found in Japan and China. Although many samples from Russian wood have been examined, a third type, possibly representing *B. kolymensis*, has never been found.

Morphological investigations of both types of *B. mucronatus* revealed differences in the shape of female tail terminus, length of mucro, position of excretory pore

(Table 2) and shape of spicules (Fig. 4). *B. kolymensis* resembles, in all these features, the European type of *B. mucronatus*. Usually, the female tail terminus of the East Asian type of *B. mucronatus* is conical with a mucro as a continuation of the tail, whereas the European type and *B. kolymensis* show a mucro well set off against the subcylindrical tail. The mucro is longer in the East Asian type (ca. 4–7 μm) and shorter in the European type (3–4, sometimes 5 μm) and *B. kolymensis* (type specimens 3.5–4 μm, according to Magnusson and Kulinich, 1996). The excretory pore is located behind the median bulb in the East Asian type of *B. mucronatus*, at or before the bulb in the European type, and at the level of median bulb in *B. kolymensis* (Korentchenko, 1980). The spicules of the East Asian type of *B. mucronatus* show a distinctly concave capitulum and a condylus continuous with the dorsal spicule line, whereas the capitulum of the spicules of the European type and of *B. kolymensis* is usually not concave, and the condylus is set off by a crease (Figs. 4 and 5). Korentchenko (1980) reported one pair of preanal papillae and one pair of postanal papillae, whereas the presence of seven caudal papillae is typical for the species of the *xylophilus* group. However, he used only light microscopy observations, and the single preanal papilla and the two pairs of adjacent subventral postanal papillae are distinctly recognizable only with SEM. Based on these comparisons, *B. kolymensis* is considered as a synonym of the European type of *B. mucronatus*.

**Table 2** Comparison of morphological characters of *B. mucronatus* and *B. kolymensis*

| Morphological character | *B. mucronatus* Mamiya & Enda, 1979 (East Asian type) | *B. mucronatus* European type | *B. kolymensis* Korentchenko, 1980 |
|---|---|---|---|
| Female tail shape, with mucro | Tail terminus conoid, mucro pointed and almost continually with tail | Sub-cylindrical, mucro often digitate, variable, well offset from tail | Sub-cylindrical, mucro variable |
| Length of mucro | 4–7 μm | 3–5 μm | 3.5–4.4 μm |
| Position of excretory pore | Behind median bulb | At level of median bulb, sometimes anterior to median bulb | At level of or anterior to median bulb |
| Spicules | Capitulum distinctly concave, condylus not offset from the dorsal spicule line | Capitulum not concave, condylus offset from the dorsal spicule line | Capitulum not concave, condylus offset from the dorsal spicule line |

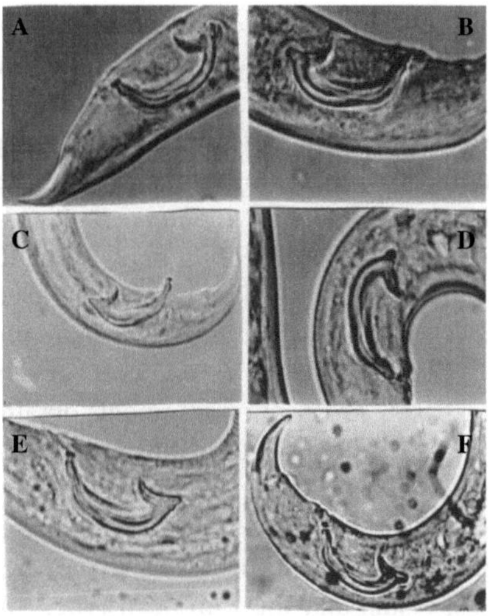


**Fig. 4** Spicules of *B. mucronatus* from various provenances. Left: East Asian type; right: European type. A, B: Germany; C: Japan; D: Canada; E, F: Russia

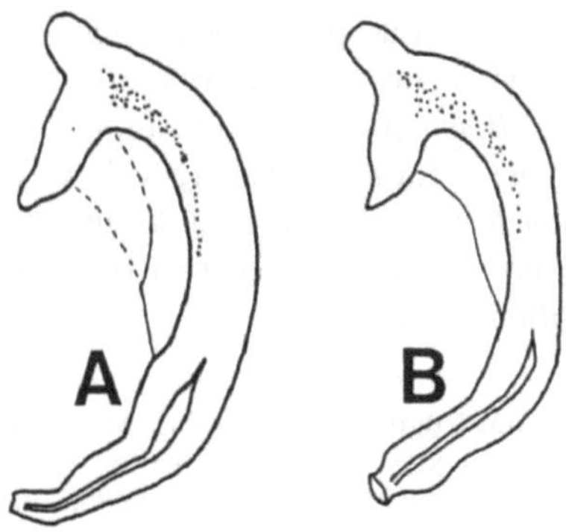


**Fig. 5** Spicules of type specimens of *B. kolymensis* (extracted from Figure 1 of Magnusson and Kulinich, 1996), showing the condylus dorsally off-set and the non-concave capitulum

## Discussion and Conclusions

Of the nine species of the *xylophilus* group known so far, three are widely distributed (*B. xylophilus, B. mucronatus, B. fraudulentus*) and occur in Asia, Europe and North America. The remaining species are distributed in East Asia, and it seems that this region has a special enrichment of species in this group of the genus *Bursaphelenchus*. The species *B. eroshenkii, B. abruptus* and *B. crenati* do not belong to the *xylophilus* group, due to morphological features as explained above.

Differentiation of species in the *xylophilus* group is sometimes difficult. Fortunately, the female tail shape of *B. xylophilus* makes recognition of this species relatively easy, provided the characteristic features of the *xylophilus* group are well known. It is important to know that other *Bursaphelenchus* species with round-tailed females and a vulval flap also exist, for instance *B. sexdentati*, but their spicules are different from those of the *xylophilus* group. However, the existence of *B. xylophilus* populations with a small mucro makes their differentiation from other mucronate species of the *xylophilus* group difficult. They were previously believed to occur only in North America, but recently, a mucronate strain of *B. xylophilus* was detected in packaging wood with imports from Taiwan to China. The key presented above may assist the differentiation of these species. A valuable and sometimes necessary aid is also the use of the ITS-RFLP method. Burgermeister et al. (2005) published the patterns of *B. xylophilus, B. fraudulentus, B. conicaudatus, B. luxuriosae* and both types of *B. mucronatus*. The patterns of *B. singaporensis* and *B. doui* can be seen in their original descriptions (Gu et al., 2005; Braasch et al., 2004). The patterns of *B. kolymensis* and *B. baujardi* are unknown.

The two types of *B. mucronatus* can be distinguished by ITS-RFLP analysis on the basis of restriction fragments obtained with *Rsa*I and *Hae*III, whereas, with both types, a PCR product of the same size of 950 bp and identical restriction fragment patterns with *Msp*I, *Hinf*I and *Alu*I were obtained (Burgermeister et al., 2005). Sequencing results of various provenances of *B. mucronatus* distinctly show two groups within the species (Zheng et al., 2003; Iwahori et al., 2004) representing the East Asian and the European type. The morphological and molecular differences between the two types of *B.mucronatus* and the morphological correspondence between the European type of *B. mucronatus* and *B. kolymensis* lead to the conclusion of synonymy of *B. kolymensis* with a part of *B. mucronatus*. Therefore, most of the '*B. mucronatus*' populations found in Europe and Russia and a few populations found in East Asia belong to *B. kolymensis*, whereas only a few populations of the real *B. mucronatus* (East Asian type) have been found in Europe and Siberia, but many populations occur in East Asia. *B. kolymensis* and *B. mucronatus* are two different species.

**Acknowledgments** I thank Ing. M. Brandstetter, Vienna, Austria and J. Gu, Ningbo, China for providing SEM micrographs, Dr. W. Burgermeister, Braunschweig, Germany for support in identification of *Bursaphelenchus* isolates by the ITS-RFLP technique, Dr. T. Schröder, Braunschweig, Germany for culturing of *Bursaphelenchus* isolates and Dr. David McNamara, U.K. for reading the manuscript.

## References

Braasch, H. (1991). Erster Nachweis von *Bursaphelenchus mucronatus* Mamiya und Enda, 1979 in der DDR und sein Vorkommen in Holzimporten aus der UdSSR nebst Ergänzungen zur Beschreibung dieser Art. *Archiv für Phytopathologie und Pflanzenschutz* 27: 209–218.

Braasch, H. (2001). *Bursaphelenchus* species in conifers in Europe: distribution and morphological relationships. *OEPP/EPPO Bulletin* 31: 127–142.

Braasch, H., Burgermeister, W. and Hoyer, U. (1998). Diversity within the species *Bursaphelenchus mucronatus* worldwide. *Nematologica* 44: 465–466.

Braasch, H., Burgermeister, W. and Pastrik, K.H. (1995). Differentiation of three *Bursaphelenchus* species by means of RAPD-PCR. *Nachrichtenblatt für den Deutschen Pflanzenschutzdienst* 47: 310–314.

Braasch, H., Gu, J., Burgermeister, W. and Zhang, J. (2004). *Bursaphelenchus doui* sp.n. (Nematoda: Parasitaphelenchidae) in packaging wood from Taiwan and South Korea – a new species of the *xylophilus* group. *Russian Journal of Nematology* 12: 19–27.

BRAASCH, H., METGE, K. AND BURGERMEISTER, W. (1999). *Bursaphelenchus*-Arten (Nematoda, Parasitaphelenchidae) in Nadelgehölzen in Deutschland und ihre ITS-RFLP-Muster. *Nachrichtenblatt für den Deutschen Pflanzenschutzdienst* 51: 312–320.

BRAASCH, H., TOMICZEK, C., METGE, K., HOYER, U., BURGERMEISTER, W., WULFERT, I. AND SCHÖNFELD, U. (2001). Records of *Bursaphelenchus* spp. (Nematoda, Parasitaphelenchidae) in coniferous timber imported from the Asian part of Russia. *Forest Pathology* 31: 129–140.

BURGERMEISTER, W., METGE, K., BRAASCH,H. AND BUCHBACH, E. (2005). ITS-RFLP patterns for differentiation of 26 *Bursaphelenchus* species (Nematoda: Parasitaphelenchidae) and observations on their distribution. *Russian Journal of Nematology* 13: 29–42.

FAO (2003). *Guidlines for regulating wood packaging material in international trade*. International Standards for Phytosanitary Measures (ISPM). Publication No. 15, FAO, Rome, 17 pp.

GIBLIN-DAVIS, R.M., HAZIR, S., CENTER, B.J., YE, W., KESKIN, N., THORP, R.W. AND THOMAS, K. (2005). *Bursaphelenchus anatolius* n. sp. (Nematoda: Parasitaphelenchidae), an associate of bees in the genus Halictus. *Journal of Nematology* 37: 336–342.

GIBLIN-DAVIS, R.M., MUNDO-OCAMPO, M., BALDWIN, J.G., NORDEN, B.B. AND BATRA, S.W.T. (1993). Description of *Bursaphelenchus abruptus* n. sp. (Nemata: Aphelenchoididae), an associate of digger bee. *Journal of Nematology* 25: 161–172.

GU, J., BRAASCH, H., BURGERMEISTER, W. AND ZHANG, J. (2006). Records of *Bursaphelenchus* spp. intercepted in imported packaging wood at Ningbo, China. *Forest Pathology* 36: 323–333.

GU, J., ZHANG, J., BRAASCH, H. AND BURGERMEISTER, W. (2005). *Bursaphelenchus singaporensis* sp. n. (Nematoda: Parasitaphelenchidae) in packaging wood from Singapore – a new species of the *B. xylophilus* group. *Zootaxa* 988: 1–12.

HOYER, U., BURGERMEISTER, W. AND BRAASCH, H. (1998). Identification of *Bursaphelenchus* species (Nematoda, Aphelenchoididae) on the basis of amplified ribosomal DNA (ITS-RFLP). *Nachrichtenblatt für den Deutschen Pflanzenschutzdienst* 50: 273–277.

HUNT, D.J. (1993). *Aphelenchida, Longidoridae and Trichodoridae – their systematics and bionomics*. CAB International, Wallingford (GB), 352 pp.

IWAHORI, H., KANZAKI, N. AND FUTAI, K. 2004. *Bursaphelenchus xylophilus* and *B. mucronatus* in Japan: where are they from? In *Nematology Monographs & Perspectives, vol.* 2 (eds. R.C. Cook and D.J. Hunt), pp. 793–803. E.J. Brill, Leiden, The Netherlands.

KANZAKI, N. AND FUTAI, K. (2003). Application and moleculkar phylogenetic analysis to the evolution and co-speciation of entomophilic nematodes. *Russian Journal of Nematology* 11: 107–117.

KOLLOSOVA, N.V. (1998). *Bursaphelenchus eroshenkii* sp. n. (Nematoda: Aphelenchoididae) from the Russian Far East, with a key to some species of *Bursaphelenchus* Fuchs, 1937. *Russian Journal of Nematology* 6: 161–164.

KORENTCHENKO, E.A. (1980). [New species from the family Aphelenchoididae, parasites of stem pests of the Dahurian larch]. *Zoologicheskii Zhurnal* 59: 1768–1780. (in Russian).

MAGNUSSON, C. AND KULINICH, O.A. (1996). A taxonomic appraisal of the original description, morphology and status of *Bursaphelenchus kolymensis* Korentschenko, 1980 (Aphelenchida: Aphelenchoididae). *Russian Journal of Nematology* 4: 155–161.

RÜHM, W. (1956). Die Nematoden der Ipiden. *Parasitologische Schriftenreihe* 6: 1–435.

RYSS., A., VIEIRA, P., MOTA, M. AND KULINICH, O. (2005). A synopsis of the genus *Bursaphelenchus* Fuchs, 1937 (Aphelenchida: Parasitaphelenchidae) with keys to species. *Nematology* 7: 393–458.

ZHENG, J., SUBBOTIN, S.A., HE, S., GU, J. AND MOENS, M. (2003). Molecular characterization of some Asian isolates of *Bursaphelenchus xylophilus* and *B. mucronatus* using PCR-FLPs and sequences of ribosomal DNA. *Russian Journal of Nematology* 11: 17–22.

# Variation in ITS and 28S rDNA of *Bursaphelenchus* Species (Nematoda: Parasitaphelenchidae)

**Kai Metge, Helen Braasch, Jianfeng Gu and Wolfgang Burgermeister**

**Abstract** Internal transcribed spacers and 28S D2/D3 domain of rDNA were used to infer the phylogenetic relationships among *Bursaphelenchus* species, with special emphasis on members of the *xylophilus* and *fungivorus* groups. Sequence alignments and phylogenetic analysis using neighbour-joining and maximum parsimony algorithms resulted in trees with similar topologies. The 17 *Bursaphelenchus* species examined can be separated into two main branches: the first includes members of the *xylophilus* group and the second includes the species of the *fungivorus* group, separated from the remaining species *B. eremus, B. hofmanni, B. rainulfi* and *B. yongensis*. As far as known, the species of the *xylophilus* group are phoretically associated with longhorn beetles. The phylogenetic analysis revealed the digger bee associated *B. abruptus* as the basal taxon of the species investigated or at least of the *xylophilus* group. The remainder of the species are, as far as known, associated with bark beetles and a soil-dwelling bee in case of *B. seani*. The significantly supported groups are largely consistent with the morphological variation within the genus *Bursaphelenchus*.

## Introduction

The main interest in species of the genus *Bursaphelenchus* is related to the phytopathogen *Bursaphelenchus xylophilus*, which is the causal agent of pine wilt disease. Since the finding of *B. xylophilus* in Portugal in 1999 (Mota et al., 1999), great efforts have been made in detecting, identifying and differentiating *Bursaphelenchus* species (Braasch et al., 1999; Kang et al., 2004; Matsunaga and Togashi, 2004; Metge and Burgermeister, 2005; Burgermeister et al., 2005). Phylogenetic studies were done to reconstruct the phylogeny of species related to the *xylophilus* group using 18S, 5.8S, internal transcribed spacer (ITS1, ITS2) of rDNA (Metge et al., 2006) and cytochrome oxidase subunit 1 (COI) (Beckenbach et al., 1999; Kanzaki and Futai, 2002). The 28S LSU and 18S SSU genetic relationships of 22 *Bursaphelenchus* species were shown by Giblin-Davis et al. (2005).

K. Metge
Institute for Biosafety of Genetically Modified Plants, Julius Kuehn-Institute (JKI), Federal Research Centre for Cultivated Plants, Quedlinburg, Germany
e-mail: kai.metge@jki.bund.de

The taxonomic grouping of *Bursaphelenchus* species outlined by Braasch (2001) based mainly on the number of lateral lines, number and position of male caudal papillae and on additional characters such as shape of spicules, presence of a vulval flap and the tail end shape of females. Giblin and Kaya (1983) and Ryss et al. (2005) used the spicule structure to separate the species into groups. Although some of these groups may reflect phylogenetic relations, such as the *xylophilus* group and the *fungivorus* group (*hunti* group of Giblin and Kaya, 1983), the species groups used by Ryss et al. (2005) are intended purely as identification units in order to facilitate species identification.

Based on the number of lines in the lateral field, three groups with four lateral lines, but characteristically different position of male caudal papillae, can be distinguished (Braasch, 2001): the *xylophilus* group, the *fungivorus* group and the *sexdentati* group. Another four-lined species, *B. abruptus*, may be a basal taxon of the genus as shown by Giblin-Davis et al. (2005). The species with reduced number of lateral lines share the biological character of phoretical association to bark beetles (Scolytidae) with the four-lined *sexdentati* group. The two-lined species are represented by a few species only (*B. abietinus, B. hylobianum, B. rainulfi*), whereas the three-lined species are numerous and can be divided into several groups (*eggersi* group, *leoni* group, *hofmanni* group and possibly others).

The aim of our study was the characterization of the ITS1, 5.8S, ITS2 and 28S rDNA sequences of rDNA of the recently described species *B. arthuri, B. doui, B. rainulfi, B. singaporensis, B. thailandae, B. willibaldi and B. yongensis*, as well as the rediscovered *B. eremus* from Germany, to determine their phylogenetic status within the genus *Bursaphelenchus* and to relate these findings to the proposed groups based on morphological data. Because of the limited number of species investigated, only three groups (*xylophilus* group, *fungivorus* group, *hofmanni* group) and a single two-lined species were considered.

## Material and Methods

The *Bursaphelenchus* species and their isolate specific data are shown in Table 1. They were maintained in petri dishes on *Botryotinia fuckeliana*/malt agar prior to DNA extraction. The nematodes were extracted from the medium using the Baermann funnel technique and washed twice with deionized water. Individuals were transferred with a needle to an Eppendorf tube containing 10 $\mu$l water and sedimented for 2 min at 9000 $\times$ g. Nematode pellets were frozen in liquid nitrogen and crushed up with an Eppendorf micro pestle. The homogenate was treated according to the protocol of the DynalBeads genomic DNA Blood Kit (Dynal Biotech, Hamburg, Germany), with omission of the blood-specific steps. DNA extracts were stored at 4 °C.

PCR was carried out employing a 50 $\mu$l reaction volume containing 2 units of Taq DNA polymerase (Fermentas, Germany), 75 mM Tris-HCl (pH 8.8), 20 mM $(NH_4)_2SO_4$, 4 mM $MgCl_2$, 0.01% Tween 20 (PCR buffer, Fermentas), 0.1 mM each dNTP, 0.6 $\mu$M K3f forward primer 5′-CCC GGG ACT GAG TTA CTT CGA GA-3′

**Table 1** *Bursaphelenchus* isolates used for sequencing of ITS-regions of rDNA. Accession numbers in brack ets refer to the complete 28 s sequences used in our study. n.s.: not submitted. [1]insect. [2]E: Europe type. EA: East Asian type

| species | first description | code at BBA | country | origin of isolate | plant or insect host | isolated by | Isolation date | EMBL. accession no. |
|---|---|---|---|---|---|---|---|---|
| *B. abruptus* | Giblin-Davis, Mundo-Ocampo, Baldwin, Norden & Batra, 1993 | Ne12/98 | USA | Maryland, Prince George County | *Anthophora abrupta*[1] | Robin M. Giblin-Davis | 1986 | (AB067756) |
| *B. arthuri* | Burgermeister, Gu & Braasch, 2005 | Ne19/04 | Taiwan | package wood from Taiwan and South Korea to China | conifers | Jianfeng Gu | 2004 | AM157742 |
| *B. conicaudatus* | Kanzaki, Tsuda & Futai, 2000 | Ne5b/05 | China | Hongkong | unknown | Maosong Lin | 2005 | AM179513 |
| *B. doui* | Braasch, Gu, Burgermeister & Zhang, 2004 | Ne26/04 | Taiwan | package wood from Taiwan to China | unknown | Jianfeng Gu | 2004 | AM157743 |
| *B. eremus* | Rühm, 1956 | Ne13/05 | Germany | Brandenburg, Eichhorst | *Scolytus intricatus*[1] | Helen Braasch | 2005 | AM180515 |
| *B. fraudulentus* | Rühm, 1956 (Goodey, 1960) | DE10w | Germany | Bavaria, Zusmarshausen | Pinus & Picea | Helen Braasch | 1997 | AM179517 |
| *B. fungivorus* | Franklin & Hooper, 1962 | Ne26/96 | Germany | Saxonia, Dresden | bark | Helen Braasch | 1996 | AM179516 |
| *B. hofmanni* | Braasch, 1998 | AT5w | Austria | Lower Austria, Maria Drei Eichen | *P. sylvestris* | Helen Braasch | 1997 | AM180516 |
| *B. luxuriosae* | Kanzaki & Futai, 2003 | Ne6/04 | Japan | Nara Prefecture, Heisaka, Gosa City | *Aclolepta luxuriosa*[1] | Natsumi Kanzaki & Kazuyoshi | 2002 | (AB097864) |
| *B. mucronatus* EA[2] | Mamiya & Enda, 1979 | DE1w | Germany | Bavaria, Zusmarshausen | *Picea abies* | Helen Braasch | 1996 | n.s. |
| *B. mucronatus* E[2] | | DE4w | Germany | Brandenburg, Templin | *Pinus sylvestris* | Helen Braasch | 1996 | AM179514 |

(continued)

**Table 1** (continued)

| species | first description | code at BBA | country | origin of isolate | plant or insect host | isolated by | Isolation date | EMBL. accession no. |
|---|---|---|---|---|---|---|---|---|
| *B. mucronatus* EA[2] | | DE5w | Germany | Thuringia, Bad Salzungen | *Picea abies* | Helen Braasch | 1996 | n.s. |
| *B. rainulfi* | Braasch & Burgermeister, 2002 | Ne27/04 | Taiwan | package wood from Taiwan to China | unknown | Maosong Lin | 2004 | AM157744 |
| *B. seani* | Giblin and Kaya, 1983 | Ne14/98 | USA | California, Fort Cronkite | *Anthophora bomboides stanfordiana*[1] | Robin M. Giblin-Davis | 1985 | AM157745 |
| *B. singaporensis* | Gu, Zhang, Braasch & Burgermeister, 2005 | Ne7/04 | Singapore | package wood from Singapore to China | unknown | Jianfeng Gu | 2004 | (AY850162) |
| *B. singaporensis* | | Ne17/05 | Malaysia | package wood from Malaysia to China | *Agathis spec.* | Jianfeng Gu | 2005 | AM180514 |
| *B. thailandae* | Braasch & Braasch-Bidasak, 2002 | Ne7b/03 | South Korea | Gyeongsangbuk-Province, Gumi | *Pinus thunbergii, densiflora* | Moon Yil Seong | 2003 | AM157746 |
| *B. willibaldi* | Schönfeld, Braasch & Burgermeister, 2006 | DE13w | Germany | Brandenburg, Beeskow | conifer wood chips | Ute Schönfeld | 2005 | AM180512 |
| *B. xylophilus* | Steiner & Buhrer, 1934 (Nickle, 1970) | Ne6/05 | Taiwan | package wood from Taiwan to China | unknown | Maosong Lin | 2005 | AM179515 |
| *B. xylophilus* | | PT1w | Portugal | Marateca, Pegoes | *Pinus pinaster* | Catarina Penas | 1999 | AM157747 |
| *B. yongensis* | Gu, Braasch, Burgermeister, Brandstetter & Zhang, (2006) | Ne14/04 | China | Zhejiang Province, Ningbo | *Pinus massoniana* | Jianfeng Gu | 2004 | AM180513 |

and D2Ap21 reverse primer 5′-GGT TTC ACG TTC TCT TGC ACT-3′ (Roth, Germany) for ITS or 28S D2D3 domain D2A ′5-ACA AGT ACC GTG AGG GAA AGT TG-3′ and D3Br 5′-TCG GAA GGA ACC AGC TAC TA-3′ and 2–10 μl of DNA template. PCR was performed with a Biometra T1 thermal cycler (Biometra, Göttingen, Germany). The PCR program consisted of an initial denaturation for 2 min 30 s at 96 °C, 35 cycles with 1 min denaturation at 94 °C, 1 min annealing at 55 °C, 2 min extension at 72 °C and a final extension for 6 min at 72 °C. After completion of the PCR, small aliquots of the samples were separated electrophoretically using a 1.8%-agarose gel and 0.5xTBE buffer.

The amplified DNA was concentrated and desalted using Microcon YM-100 centrifugal filter devices (Millipore, Schwalbach, Germany) to obtain sequencing products of good quality. Working steps were performed following the manufacturer's instructions. Additionally, the membrane was washed with 50 μl 5 mM Tris, pH 8.5. Small aliquots of each final sample were examined by electrophoresis in a 1.8%-agarose gel and 0.5xTBE buffer to estimate the concentration of desalted DNA. Gels were stained with ethidium bromide (1 μg/ml) and visualized with an UV transilluminator.

According to instructions from the sequencing company (MWG Biotech AG, Ebersberg, Germany), 20 ng/100bp of a PCR fragment were air dried and sent to the company, together with appropriate primers, to use their Value Read Service. Each fragment was sequenced using the forward and reverse primer to obtain overlapping sequences of the forward and reverse DNA strand.

The alignments were constructed with ClustalW implemented in the computer program Bioedit 7.0.5.2 (Hall, 1999). Phylogenetic trees were generated with Mega 3.1 (Kumar et al., 2004). *Aphelenchoides fragariae* (EMBL accession no. AF119049) was used as outgroup. The neighbour-joining (NJ) analysis were calculated using the Tajima-Nei distance option. The maximum parsimony (MP) algorithms were conduced using the close-neigbour-interchange (CNI) search option with level 1, starting with an initial tree generated by random addition (10 replicates). The dendrograms were tested by 1000 bootstrap replicates.

## Results

The total length of the amplified ITS rDNA regions varied from 880 bp for *B. thailandae* to 1252 bp for *B. abruptus* (Fig. 1). The 3′-end parts of 18S rDNA (42 bp) and 5′-ends of 28S rDNA (99–100 bp) were constant in length. The sequences used for analysis were 242–607 bp in ITS1, 291–399 bp in ITS2 and 157–160 bp in 5.8S rDNA excluding the outgroup *Aphelenchoides fragariae*.

The base composition of sequences differed much among the ITS regions of some species but not within the 5.8S or 28S D2/D3 domain. A very high number of insertions was shown for *B. abruptus* compared to the other species. The ITS1 sequence of *B. abruptus* (607 bp) was nearly twice as long as the median (310 bp) for all ITS1 sequences analysed.The GC content of the loci examined varied between

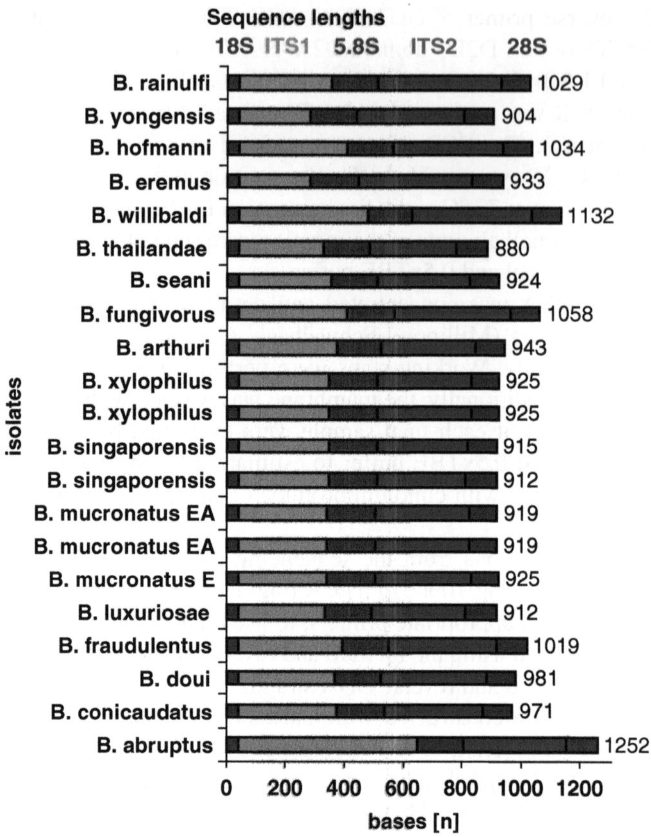


**Fig. 1** Sequence lengths of the analysed *Bursaphelenchus* rDNA. EA: East Asian type, E: European type of *B. mucronatus*

the species. ITS1 sequences exhibited the lowest GC content for *B. thailandae* (37.8%), *B. seani* (39.4%) and *B. abruptus* (41.5%). Within the *xylophilus* group the GC content was mainly higher than 51%. This percentage was also found in *B. fungivorus* and *B. rainulfi*, but within the other species it was less than 48%. The ITS2 sequences showed highest GC content for *B. fraudulentus* (60%). On the other hand, *B. thailandae* (28.9%) had the lowest GC content, followed by *B. yongensis* (36.4%) and *B. seani* (36.8%). 5.8S rDNAs were nearly constant in length, their GC content varied between 40.9% in *B. thailandae* and 44.4% in *B. abruptus*.

The global sequence alignment of the combined sequences examined including outgroup has 1549 sites, of which 325 (21%) were conserved, 981 (63%) were variable sites, 719 (46%) are parsimony-informative, 830 (54%) are parsimony-uninformative und 216 are singleton sites (Table 2). The transition/transversion ratio was 1.3 in ITS1, 1.2 in ITS2 and 2.0 in 5.8S.

The phylogenetic analysis using Neigbour Joining and Maximum Parsimonious methods carried out with the data sets yielded trees with similar topologies for the

**Table 2** Variable and constant sites of multiple alignments in *Bursaphelenchus* species. Total number of sites (n) and in % used for non-coding (ITS1, ITS2), coding (5.8S) and combined sequences (partial 18S, ITS1, 5.8S, ITS2, partial 28S).

| Primer sequence F194/5368 | ITS1 | | ITS2 | | 5.8S | | combined 18S partial, ITS1, 5.8S, ITS2, 28S partial | |
|---|---|---|---|---|---|---|---|---|
| | n | % | n | % | n | % | n | % |
| total length | 674 | | 595 | | 164 | | 1549 | |
| conserved sites | 96 | 14.2 | 42 | 7.1 | 110 | 67.1 | 325 | 21.0 |
| variable sites | 401 | 59.5 | 470 | 79.0 | 51 | 31.1 | 981 | 63.3 |
| informative sites | 258 | 38.3 | 363 | 61.0 | 30 | 18.3 | 719 | 46.4 |
| uninformative sites | 416 | 61.7 | 232 | 39.0 | 134 | 81.7 | 830 | 53.6 |
| singleton sites | 114 | 16.9 | 93 | 15.6 | 21 | 12.8 | 216 | 13.9 |

*Bursaphelenchus* groups using ITS1, 5.8S and the combined loci (Figs. 2 and 3). The 17 *Bursaphelenchus* species could be separated in two main clades: the first includes the members of the *xylophilus* group, and the second includes the species of the *fungivorus* group, separated from the remaining four species, which are related to other subgroups. The trees constructed for the D2D3 domain splitted the bark beetle branch, respectively (Fig. 4). *B. rainulfi* (*hylobianum* group) and the three species of the *hofmanni* group were found basal within the *xylophilus* branch. The 3 lined species *hofmanni* was separated from *the eremus/yongensis* branch with very low bootstrap values less than 50%.

The sequence polymorphism of the ITS2 rDNA was very high within the genus *Bursaphelenchus* (Table 2). The main branch of the *xylophilus* group was not influenced by these variations, but the tree topology was ladderized, isolates being swapped in other positions. Most subbranches were not supported by high bootstrap values (as shown in Fig. 2).

Three isolates of *B. mucronatus* and two isolates of each *B. xylophilus* and *B. singaporensis* were included in this study. Intraspecific sequence differences turned out to be very small. In the phylogenetic trees, the isolates formed subclusters of their respective species branches.

## Discussion

The phylogenetic relationship of the recently described species *B. arthuri, B. doui, B. singaporensis, B. willibaldi, B. yongensis* and the rediscovered *B. eremus* in Germany were analysed by variation of the internal transcribed spacer region of rDNA in order to compare them to species of the *xylophilus*- and *fungivorus* groups.

The sequence chromatograms obtained with the reverse primer D2Ap21r were mainly of better quality than the sequences determined with the forward primer K3f. The sizes of the amplified PCR products for the 3′- end of 18S, complete 5.8S and 28S gene were almost constant in length. They did not differ as much

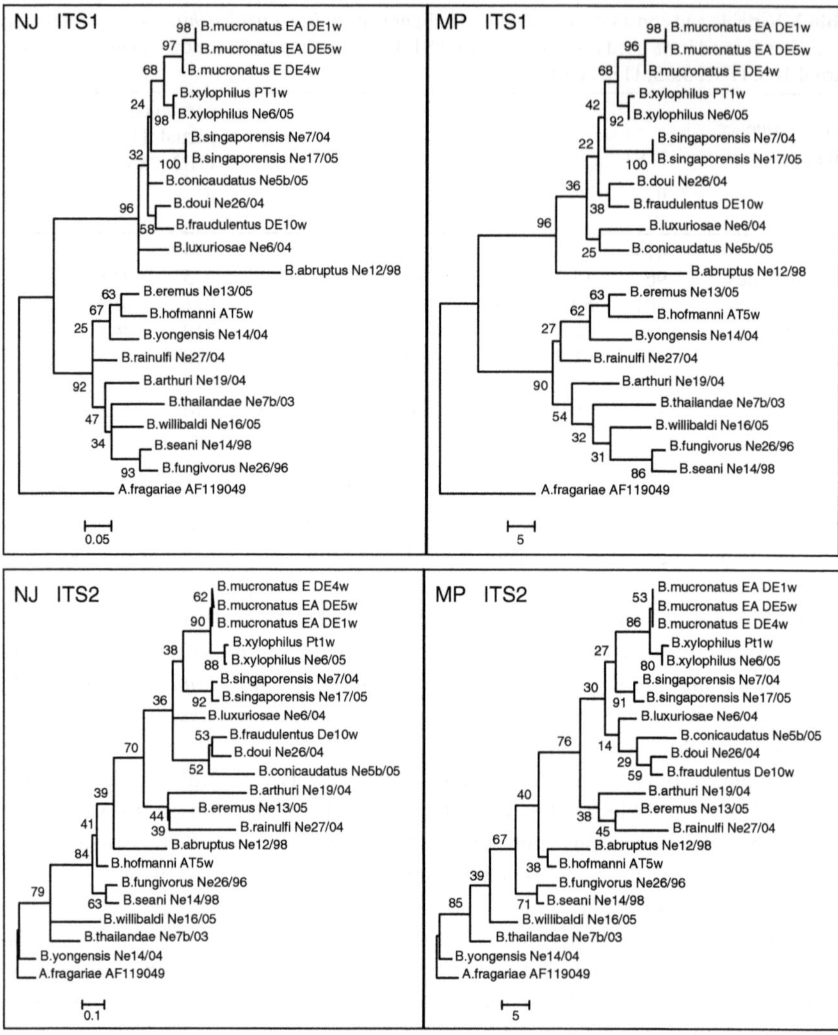


**Fig. 2** Phylogenetic relationships of *Bursaphelenchus* species. The global sequence alignments for tree constructions were calculated for ITS1 and ITS2 sequences by neighbour-joining (NJ) and maximum parsimonious (MP) algorithms. Bootstrap values (%) are given for each node. Scale bars NJ: substitutions/site; MP: tree length

as the ITS regions. The ITS2 region was more variable than the ITS1 region for the examined species. The trees calculated from the ITS region sequence alignments, 5.8S alignments and D2D3 alignments resulted in similar topologies for the *Bursaphelenchus* groups. The main branches and species correspond to the results shown by Giblin-Davis et al. (2005), however *B. arthuri, B. doui, B. eremus, B. singaporensis, B. willibaldi* and *B. yongensis* were not included there.

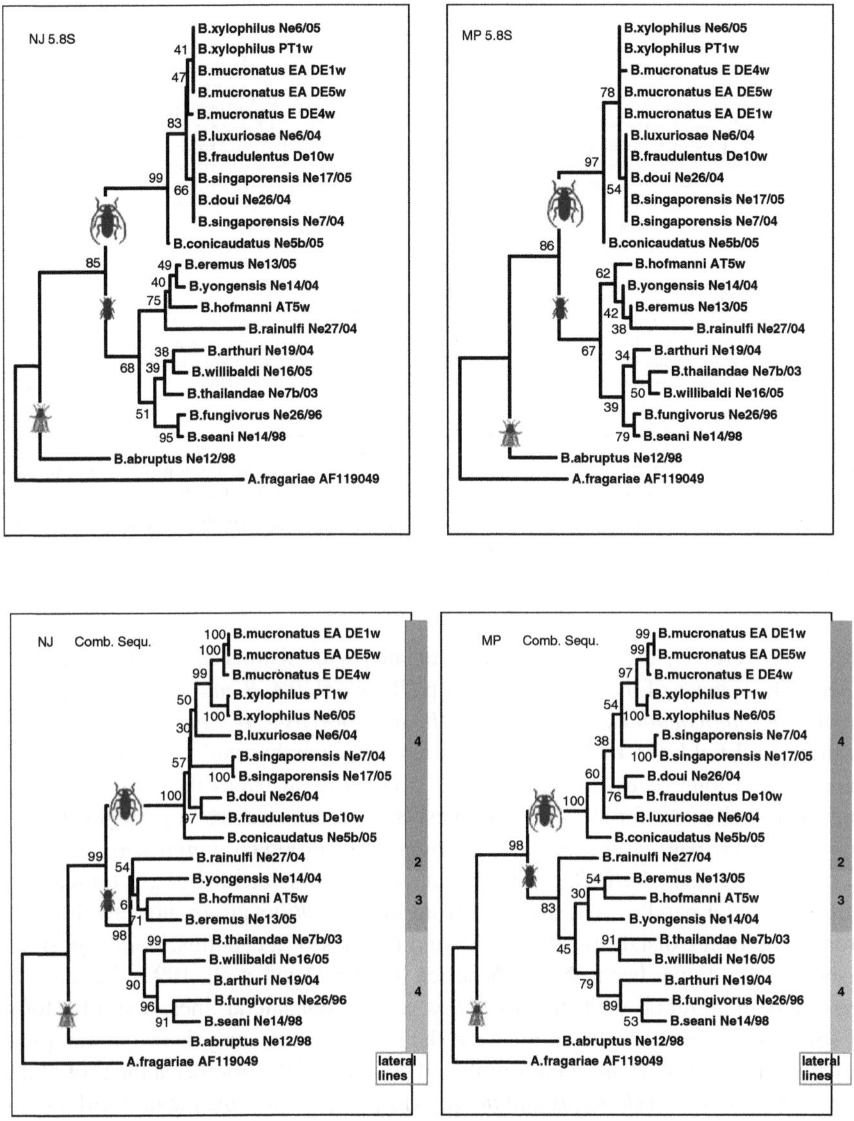


**Fig. 3** Phylogenetic relationships of *Bursaphelenchus* species and known or hypothesised nematode-insect associations with Cerambycidae, Scolytidae and Apidae. The global sequence alignments for tree constructions were calculated for 5.8S and combined sequences (18S partial, ITS1, ITS2, 5.8S and 28S partial) by NJ and MP algorithms. Bootstrap values (%) are given for each node. Scale bars NJ: substitutions/site; MP: tree length

The trees based on the DNA sequence homologies are largely consistent with the groups previously defined by morphological features (Braasch, 2001). The two main groups, the *xylophilus* group with seven species and the *fungivorus* group with five

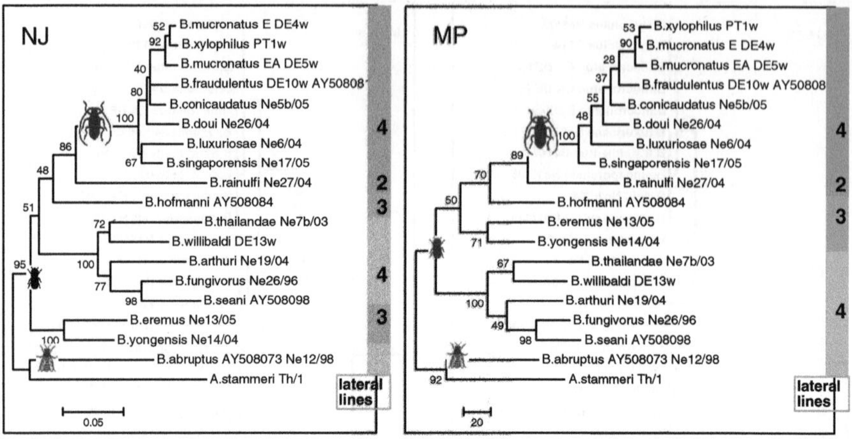


**Fig. 4** Phylogenetic relationships of *Bursaphelenchus* species and known or hypothesised nematode-insect associations with Cerambycidae, Scolytidae and Apidae. The global sequence alignments for tree constructions were calculated for D2/D3 region of the 28S by NJ and MP algorithms. Bootstrap values (%) are given for each node. Scale bars NJ: substitutions/site; MP: tree length

species, covered by this analysis, are associated with distinct morphological characters. The species of the *xylophilus* group (*B. xylophilus, B. mucronatus, B. fraudulentus, B. conicaudatus, B. luxuriosae, B. singaporensis, B. doui*) have four lines in the lateral field, the males have a characteristic number and position of caudal papillae, and long recurved spicules with a cucullus, whereas the females have a large vulval flap. As far as is known, these species are vectored by longhorn beetles (Cerambycidae). The species of the *fungivorus* group (*B. fungivorus, B. seani, B. thailandae, B. arthuri, B. willibaldi*) have also four lateral lines, a typical position of caudal papillae, but different from the *xylophilus* group, no vulval flap, a broadening of the distal tips of spicules and the lack of a cucullus. *B. abruptus*, an associate of a digger bee in North America (Giblin-Davis et al., 1993), has several characters in common with the *xylophilus* group: four lateral lines, a similar form of spicules, and a vulval flap. However, the results of the analysis of the complete sequence and of 5.8S support the significant position of *B. abruptus* at the root of the tree containing both *xylophilus* and *fungivorus* groups, whereas, the molecular trees obtained from analysis of the ITS1 region indicate a position at the base of the clade of the *xylophilus* group. The two highly supported clades (the *xylophilus* group, the clade with the *fungivorus* group and a group of several species with reduced number of lateral lines) may have developed after the divergence of *B. abruptus*. The species with reduced number of lateral lines (in this study *B. eremus, B. yongensis, B. hofmanni* and *B. rainulfi*) are most probably all associates of bark beetles, and it is suggested that they developed after the divergence of the *fungivorus* group.

Within the *xylophilus* group, *B. xylophilus* and *B. mucronatus* are closely related species, whereas *B. luxuriosae* is assumed to be close to *B. conicaudatus* and may have branched off from the ancestors of the *xylophilus* group in the early stage of speciation. Unlike the *B. xylophilus* and *B. mucronatus* branch, *B. luxuriosae*, *B. conicaudatus* and also *B. singaporensis* have long, tapering, and ventrally bent female tails. Obviously, these species branched off before the speciation to *B. xylophilus* and *B. mucronatus*. The findings correspond to the phylogenetic relationships in the *xylophilus* group outlined by Kanzaki and Futai (2002). Both *B. doui* and *B. fraudulentus* have mucronate female tails and are close to each other in all trees constructed. According to DNA base sequences, they also branched off before the speciation to *B. xylophilus* and *B. mucronatus*. These data suggest independent evolution of a mucro at the female tail terminus in *B. doui, B. fraudulentus* and *B. mucronatus*.

The *fungivorus* group defined in the molecular analysis is associated with the morphological characters given by Braasch (2001) and Burgermeister et al. (2005). The close relationship of *B. fungivorus, B. seani, B. thailandae* and *B. arthuri* was already shown by Burgermeister et al. (2005) and is confirmed by the molecular trees. The recently described species *B. willibaldi* (Schönfeld et al., 2006) is most closely related to *B. thailandae*, in accordance with its morphological description.

The clade of the species with reduced number of lateral lines (*B. eremus, B. yongensis* and *B. hofmanni* with three lateral lines, and *B. rainulfi* with two) represents only a small section of the species probably vectored by bark beetles. The tree shown in Fig. 4 and by Giblin-Davis et al. (2005) based on combined molecular data suggests two main groups after the divergence of *B. abruptus*. One clade includes the *xylophilus* and the *fungivorus* group, both with four lateral lines, beside a few other species with four lateral lines (*B. anatolius, B. kevini, B. cocophilus*). The other clade contains the species with reduced number of lateral lines and the *sexdentati* group with four lateral lines. Our results based on sequences of combined partial 18S, ITS1, 5.8S, ITS2 and partial 28S, 5.8S, ITS1 and 28S D2D3 alone suggest two main clades, i.e. the *xylophilus* group (phoretically associated with longhorn beetles) and a clade including the rest of the investigated species (mostly and as far as known phoretically associated with bark beetles). However, the *sexdentati* group and diverse subgroups of the species with reduced number of lateral lines were not considered.

The close morphological relationship of *B. eremus* and *B. yongensis* was shown by Braasch et al. (2006) and Gu et al. (2006) and is confirmed by the present investigation. These species most probably belong to the *hofmanni* group (Braasch, 2001) or are at least close to this group. The three species have three lateral lines, delicate spicules with the rostrum close to the proximal end and the same position of caudal papillae pairs, and *B. eremus* and *B. yongensis* are very similar to each other by spicule shape and a mucronate female tail. The only species with two lateral lines included in this study was *B. rainulfi*. It is well separated from the four-lined groups and close to the three-lined species in the tree.

Based on comparison of sequencing results and morphological features as well as on the number of significantly supported clades, sequencing of the ITS2 region yielded less morphologically supported results and probably a phylogenetically less informative dataset.

Studies of ITS rDNA are widely used to examine relationships between nematode species (Powers et al., 1997; Blok et al., 1998; Iwahori et al., 1998; Beckenbach et al., 1999; Subbotin et al., 2000; Kanzaki and Futai, 2002; Zheng et al., 2003; Boutsika et al., 2004; Anthoine and Mugniery, 2005). However, there are some limitations discussed by the authors that should be considered, particularly when examined isolates have great length variation like the ITS1 region of *B. abruptus* in this study. The analysis will be influenced by the high number of gaps and thus can influence the alignment and similarity results. Because of the high variation in sequence size, Kanzaki and Futai (2002) concluded, that ITS1 and ITS2 loci were unsuitable for their phylogenetic study of five *Bursaphelenchus* species proposed as members of the *xylophilus* group. We demonstrate that the very high sequence polymorphism of the ITS2 loci limits phylogenetic studies on broad relationships in the genus *Bursaphelenchus*, but it is useful to distinguish species on a closely related interspecific level. Both loci separate groups within the genus (*xylophilus/fungivorus*/other groups such as *hofmanni*) and very closely related species such as *B. xylophilus/B. mucronatus/B. singaporensis* and *B. fraudulentus/B. doui* as well as *B. fungivorus/B. seani*. The combined sequence alignment of partial 18S, ITS1, 5.8S, ITS2 and partial 28S supports the suitability of this region for the analysis of closely related species because of the high bootstrap values obtained for most nodes.

**Acknowledgments** This research was financed by the EU FP5 program, project acronym PHRAME, Contract No. QLK5-CT-2002-00672. We thank T. Schröder and U. Schönfeld for providing isolates.

## References

Anthoine, G. and Mugniery, D. (2005). Variability of the ITS rDNA and identification of *Nacobbus aberrans* (Thorne, 1935) Thorne and Allen, 1944 (Nematoda: Pratylenchidae) by rDNA amplification. *Nematology* 7: 503–516.

Beckenbach, K., Blaxter, M. and Webster, J.M. (1999). Phylogenetic of *Bursaphelenchus* species derived from analysis of ribosomal internal transcribed spacer DNA sequences. *Nematology* 1: 539–548.

Blok, V.C., Malloch, G., Harrower, B., Phillips, M.S. and Vrain T.C. (1998). Intraspecific variation in ribosomal DNA in populations of the potato cyst nematode *Globodera pallida*. *Journal of Nematology* 30: 262–273.

Boutsika, K., Brown, D.J.F., Phillips, M.S. and Blok, V.C. (2004). Molecular characterisation of the ribosomal DNA of *Paratrichodorus macrostylus, P. pachydermus, Trichodorus primitivus* and *T. similis* (Nematoda: Trichodoridae). *Nematology* 6: 641–654.

Braasch, H. (2001). *Bursaphelenchus* species in conifers in Europe: distribution and morphological relationships. *European and Mediteranean Plant Protection Organisation, OEPP/EPPO*, Bulletin 31: 127–142.

Braasch, H., Burgermeister, W., Schröder, T., Apel, K.-H. and Brandstetter, M. (2006). Occurrence of *Bursaphelenchus eremus* Rühm, 1956 (Nematoda: Parasitaphelenchidae) on oaks in State Brandenburg/Germany and supplementary species description. *Nachrichtenblatt für den Deutschen Pflanzenschutzdienst* 58: 148–153.

BRAASCH, H., METGE, K. AND BURGERMEISTER, W. (1999). *Bursaphelenchus*-Arten (Nematoda, Parasitaphelenchidae) in Nadelgehölzen in Deutschland und ihre ITS-RFLP-Muster. *Nachrichtenblatt Deutscher Pflanzenschutzdienst* 51: 312–320.

BURGERMEISTER, W., METGE, K., BRAASCH, H. AND BUCHBACH, E. (2005). ITS-RFLP patterns for differentiation of 26 *Bursaphelenchus* species (Nematoda: Parasitaphelenchidae) and observations on their distribution. *Russian Journal of Nematology* 13: 29–42.

GIBLIN-DAVIS, R.M., HAZIR, S., CENTER, B.J., YE, W., KESKIN, N., THORP, R.W. AND THOMAS, K. (2005). *Bursaphelenchus anatolius* n. sp. (Nematoda: Parasitaphelenchidae), an associate of bees in the genus *Halictus*. *Journal of Nematology* 37: 336–342.

GIBLIN-DAVIS, R.M., MUNDO-OCAMPO, M., BALDWIN, J.G., NORDEN, B.B. AND BATRA, S.W.T. (1993). Description of *Bursaphelenchus abruptus* n. sp. (Nemata: Aphelenchoididae), an associate of a digger bee. *Journal of Nematology* 25: 161–172.

GIBLIN, R.M. AND KAYA, H.K. (1983). *Bursaphelenchus seani* n. sp. (Nematoda, Aphelenchoididae), a phoretic associate of *Anthophora bomboides stanfordiana* Cockerell, 1904 (Hymenoptera, Anthophoridae). *Revue de Nématologie* 6: 39–50.

GU, J., BRAASCH, H., BURGERMEISTER, W., BRANDSTETTER, M. AND ZHANG, J. (2006). Description of *Bursaphelenchus yongensis* sp.n. (Nematoda: Parasitaphelenchidae) isolated from *Pinus massoniana* in China. *Russian Journal of Nematology* 14: 91–99.

HALL, T.A. (1999). BioEdit: a user-friendly biological sequences alignment editor and analysis program for Windows 95/98/NT. *Nucleic Acids Symposium* 41: 95–98.

IWAHORI, H., TSUDA, K., KANZAKI, N., IZUI, K. AND FUTAI, K. (1998). PCR-RFLP and sequencing analysis of ribosomal DNA of *Bursaphelenchus* nematodes related to pine wilt disease. *Fundamental and Applied Nematology* 21: 655–666.

KANG, J.S., CHOI, K.S., SHIN, S.C., MOON, I.S., LEE, S.G. AND LEE, S.H. (2004). Development of an efficient PCR-based diagnosis protocol for the identification of the pinewood nematode, *Bursaphelenchus xylophilus* (Nematoda: Aphelenchoididae). *Nematology* 6: 279–285.

KANZAKI, N. AND FUTAI, K. (2002). A PCR primer set for determination of phylogenetic relationships of *Bursaphelenchus* species within the xylophilus group. *Nematology* 4: 35–41.

KUMAR, S., TAMURA, K. AND NEI, M. (2004). MEGA3: Integrated software for molecular evolutionary genetics analysis and sequence alignment. *Briefings in Bioinformatics* 5: 150–163.

MATSUNAGA, K. AND TOGASHI, K. (2004). A simple method for discriminating *Bursaphelenchus xylophilus* and *B. mucronatus* by species-spezific polymerase chain reaction primer pairs. *Nematology* 6: 273–277.

METGE, K., BRAASCH, H., GU, J. AND BURGERMEISTER, W. (2006). Interspecific variation in ITS rDNA of *Bursaphelenchus* species (Nematoda: Parasitaphelencidae) of different groups. *Russian Journal of Nematology* 14: 147–158.

METGE, K. AND BURGERMEISTER, W. (2005). Multiple displacement amplification of DNA for ITS-RFLP analysis of *Bursaphelenchus* individual nematodes. *Nematology* 7: 253–257.

MOTA, M.M., BRAASCH, H., BRAVO, M.A., PENAS, A.C., BURGERMEISTER, W., METGE, K. AND SOUSA, E. (1999). First report of *Bursaphelenchus xylophilus* in Portugal and in Europe. *Nematology* 1: 727–734.

POWERS, T.O., TODD, T.C., BURNELL, A.M., MURRAY, P.C.B., FLEMING, C.C., SZALANSKI, A.L., ADAMS, B.A. AND HARRIS, T.S. (1997). The rDNA internal transcribed spacer region as a taxonomic marker for nematodes. *Journal of Nematology* 29: 441–450.

RYSS, A., VIEIRA, P., MOTA, M. AND KULINICH, O. (2005). A synopsis of the genus *Bursaphelenchus* Fuchs, 1937 (Aphelenchida: Parasitaphelenchidae) with keys to species. *Nematology* 7: 393–458.

SCHÖNFELD, U., BRAASCH, H. AND BURGERMEISTER, W. (2006). *Bursaphelenchus* spp. (Nematoda: Parasitaphelenchidae) in wood chips from sawmills in Brandenburg and description of *Bursaphelenchus willibaldi* sp. n. *Russian Journal of Nematology* 14: 119–126.

SUBBOTIN, S.A., HALFORD, P.D., WARRY, A. AND PERRY, R.N. (2000). Variation in ribosomal DNA sequences and phylogeny of *Globodera* parasitising solanaceous plants. *Nematology* 2: 591–604.

ZHENG, J., SUBBOTIN, S.A., HE, S., GU, J. AND MOENS, M. (2003). Molecular characterisation of some Asian isolates of *Bursaphelenchus xylophilus* and *B. mucronatus* using PCR-RFLPs and sequences of ribosomal DNA. *Russian Journal of Nematology* 11: 17–22.

# Molecular Characterization of Isolates of the *Bursaphelenchus sexdentati* Group Using Ribosomal DNA Sequences and ITS-RFLP

**Cornelia Lange, Wolfgang Burgermeister, Kai Metge and Helen Braasch**

**Abstract** Species-specific ITS-RFLP patterns have been established for more than 30 *Bursaphelenchus* species including 5 species of the 'sexdentati' group *(B. sexdentati, B. vallesianus, B. pinophilus, B. poligraphi* and *B. borealis*). Morphological species differentiation in the 'sexdentati' group is based on the shape of female tail and the spicules. However, observations on different isolates of *B. sexdentati* have revealed considerable variability of these features suggesting the existence of intraspecific genetic types which could not be differentiated by ITS-RFLP analysis using five enzymes. To improve intraspecific differentiation, we have determined ITS1/2 sequences of 17 isolates belonging to five species of the 'sexdentati' group. In some isolates, sequence heterogeneity at a few sites of ITS2 was detected. In the sequence-based phylogenetic tree, branching of clusters confirmed the affiliation of isolates to their respective species. In addition, isolates of *B. sexdentati* were separated in two groups suggesting the existence of a Central European and a South European type of this species or two separate species. This was supported by the differences in shape of female tails and spicules between the two types. The information obtained from sequencing was used to select three additional enzymes for extending the scope of ITS-RFLP analysis. In this way, improved distinction of species and differentiation of the two types of *B. sexdentati* was achieved.

## Introduction

Morphological studies by light and scanning electron microscopy have revealed common features among certain *Bursaphelenchus* species which have lead to postulation of species groups named after a typical member, i.e. the 'xylophilus', 'fungivorus', 'eggersi', 'leoni', 'sexdentati' and 'hofmanni' groups (Braasch, 2001). Species of the 'sexdentati' group include *B. sexdentati, B. vallesianus, B. naujaci, B. pinophilus, B. poligraphi, B. borealis* and probably also *B. incurvus* and

C. Lange
Max-Planck Institute for Molecular Genetics, Berlin, Germany
e-mail: lange_c@molgen.mpg.de

M.M. Mota, P. Vieira (eds.), *Pine Wilt Disease: A Worldwide Threat to Forest Ecosystems*, 

*B. piniperdae* (Braasch, 2001). They are characterized by the presence of four lateral lines, a small vulval flap, stout spicules with distinct rostrum and condylus and a typical position of caudal papillae of males. These features permit a clear differentiation from other groups. Within the 'sexdentati' group, species identification is based on the shape of female tails and of spicules. However, observations on different isolates of the same species have revealed considerable variability of these features suggesting the existence of intraspecific genetic types and complicating species affiliation of the isolates.

The intergenic transcribed spacer regions ITS1 and ITS2 of ribosomal DNA exhibit sequence polymorphisms between nematode species which can be visualized by restriction analysis of the amplified ITS1/2 region (ITS-RFLP analysis). Using five restriction enzymes, species-specific ITS-RFLP patterns have been established for 26 *Bursaphelenchus* species including five members of the 'sexdentati' group (Burgermeister et al., 2005). In most cases, however, various provenances of a species showing minor morphological differences cannot be differentiated by ITS-RFLP, since small ITS sequence differences between isolates which do not affect a restriction site remain undetected. Isolates of *B. sexdentati* exhibit two types of female tails (subcylindric with rounded tail end or conoid) and spicules (condylus rectangular or slightly hooked, absence or presence of a cucullus), whereas their ITS-RFLP patterns based on the five restriction enzymes used for differentiation of more than 30 other *Bursaphelenchus* species were identical.

Sequence analysis of the ITS1/2 region of rDNA is a more powerful method for differentiation of species and isolates of *Bursaphelenchus*. Previous ITS sequence studies have concentrated on differentiation of isolates of the pinewood nematode, *B. xylophilus* and its close relative, *B. mucronatus* (Iwahori et al., 1998; Beckenbach et al., 1999; Zheng et al., 2003). We have determined ITS1/2 sequences of 17 *Bursaphelenchus* isolates belonging to five species of the 'sexdentati' group and carried out cluster analysis of their phylogenetic relationships.

## Material and Methods

### *Bursaphelenchus Isolates*

The *Bursaphelenchus* isolates used in this study are listed in Table 1. Most isolates were obtained from the *Bursaphelenchus* culture collection maintained by Dr. Thomas Schröder at the Julius Kuehn Institute, Federal Research Centre for Cultivated Plants, Institute for National and International Plant Health, Braunschweig, Germany. They had originally been extracted from wood samples or insects by the modified Baermann funnel technique and cultured in Petri dishes with *Botryotinia fuckeliana* on malt agar. Some populations were directly isolated from wood samples which had been collected in forests in Brandenburg, Germany by chopping wood and bark material from bark beetle-infected felled pine trees. The nematodes were extracted by the modified Baermann funnel technique and identified morphologically using a Zeiss Axioscope microscope.

**Table 1** *Bursaphelenchus* isolates used in this study

| Species | Isolate | Code | Country of origin | Source | Year of isolation | Plant or insect host | Starting material | EMBL accession . No |
|---|---|---|---|---|---|---|---|---|
| *B. sexdentati* | DE-29 (w) | 42.1 | Germany | S. Schütz | 2001 | *Pinus sylvestris* | culture | AM160661 |
| | Ferch | 61 | Germany | C. Lange/ H. Braasch | 2004 | *Pinus sylvestris* | wood | AM269912 |
| | H 194/01 | 1.1 | Germany | U. Schönfeld | 2001 | *Pinus sylvestris* | wood | AM269913 |
| | ST 23 | 36.25 | Switzerland | U. Schönfeld/ J Polomski | 2002 | *Pinus sylvestris* | wood | AM269914 |
| | GR-10 (w) | 48.1 | Greece | E. Skarmoutsos | 1998 | *Pinus radiata* | culture | AM269915 |
| | IT-2 (w) | 52.3 | Italy | S. Caroppo | 1997 | *Pinus pinaster* | culture | AM269916 |
| | Portugal | 68 | Portugal | C. Penas | 2002 | *Pinus pinaster* | culture | AM269917 |
| *B. vallesianus* | GR-7 (w) | 42.2 | Greece | E. Skarmoutsos | 1997 | *Pinus nigra* | culture | AM269918 |
| | CH-1 (w) | 48.3 | Switzerland | U. Schönfeld/ J. Polomski | 2002 | *Pinus sylvestris* | culture | AM160663 |
| | CH-2 (w) | 52.1 | Switzerland | U. Schönfeld/ J. Polomski | 2002 | *Pinus sylvestris* | culture | AM269919 |
| | CH-3 (w) | 51.1 | Switzerland | U. Schönfeld/ J. Polomski | 2002 | *Pinus sylvestris* | culture | AM269920 |
| | GR-8 (w) | 52.2 | Greece | E. Skarmoutsos | 1997 | *Pinus nigra* | culture | AM269921 |
| | H 272/01A | 2.1 | Germany | U. Schönfeld | 2001 | *Pinus sylvestris* | wood | AM269922 |
| *B. pinophilus* | Ne 5/04 | 62 | Turkey | M. Mota | 2002 | *Pinus* sp. | culture | AM160664 |
| *B. borealis* | DE-8 (w) | 48.5 | Germany | H. Braasch | 1996 | *Pinus sylvestris* | culture | AM179511 |
| | DE-20 (i) | 52.6 | Germany | H. Braasch | 1998 | *Dryocoetes autographus* | culture | AM269910 |
| *B. poligraphi* | DE-17 (w) | 51.3 | Germany | H. Braasch | 1998 | *Picea abies* | culture | AM179512 |
| *B. eggersi* | DE-23 (w) | 52.4 | Germany | H. Braasch | 2000 | *Pinus sylvestris* | culture | AM160663 |
| | DE-21 (i) | 52.5 | Germany | H. Braasch | 1999 | *Hylurgops palliatus* | culture | AM269911 |

### Molecular Biological Methods

The following procedures have been described in detail by Lange et al. (2006). DNA was extracted from samples containing one to about 30 specimens. rDNA partial sequences containing either the ITS1, the ITS2 or the complete ITS1/2 region were amplified by PCR. For ITS-RFLP analysis, suitable aliquots of PCR products representing the ITS1/2 region were digested with restriction endonucleases, and the resulting restriction fragments were resolved by electrophoresis. For sequence determination, PCR products were desalted, dried and sent to MWG Biotech AG using their Value Read Service. Complete sequences of both DNA strands of the PCR products were obtained using the forward and the reverse primer, respectively. For identification of heterogeneities in the ITS2 sequence of isolate *B. sexdentati* IT-2(w), amplified ITS2 fragments of this isolate were cloned into the pCR2.1 vector and recombinant plasmids were sequenced, using the Value Read Service of MWG Biotech AG.

### Phylogenetic Analysis

Sequences were analysed and aligned using the programs BioEdit Sequencing Alignment editor version 5.0.6 (Hall, 1999) and ClustalX 1.83 (Thompson et al., 1997) and further processed for display purpose by means of DotGen (Jörg Peleikis, personal communication). Phylogenetic analyses were performed using MEGA version 2.1 (Kumar et al., 2001), applying the Tamura-Nei distance method (Tamura and Nei, 1993). Phylogenetic trees were constructed with the Neighbor Joining (NJ) method (Saitou and Nei, 1987). Bootstrap analysis was performed with 1 000 replicates.

## Results and Discussion

### Two Types of Morphological Features in the 'sexdentati' Group

The species of the 'sexdentati' group show almost uniform morphological characters, but exhibit two types of female tails (round or conical) and spicules (condylus rectangular or hooked). As shown in Fig. 1, most species of the group can be assigned unequivocally to one of these types, but *B. pinophilus* and some isolates of *B. sexdentati*, e.g. isolates IT-2(w) and GR-10(w) have intermediate features of their female tails and spicules.

### Sequencing and Phylogenetic Analysis

Complete sequences including both ITS regions were obtained for 17 *B. sexdentati* and two *B. eggersi* isolates. Phylogenetic relationships between the 19 isolates

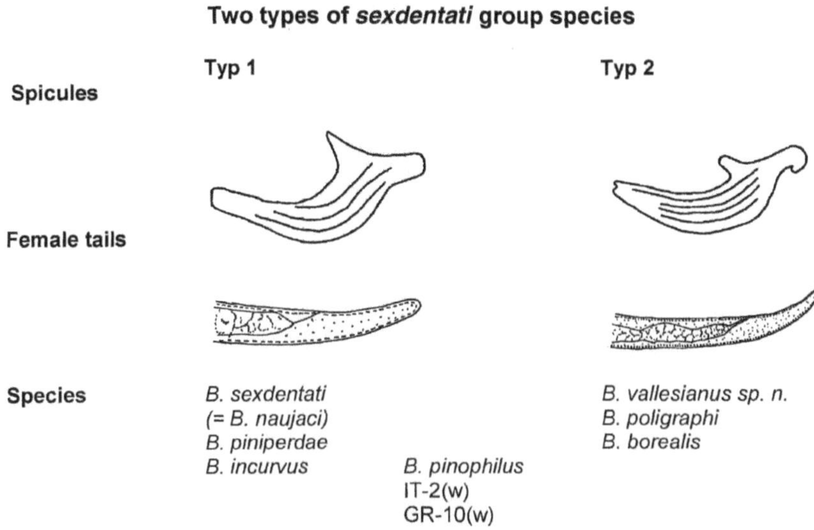


**Fig. 1** Morphological features of spicules and female tails within the *'sexdentati'* group

were examined by cluster analyses of genetic distances, using the Neighbor Joining method. Published sequences of three species of the 'xylophilus' group were included for comparison. To see whether the phylogenetic relationships based on sequence information correlate with morphological characters, drawings of the spicules and female tails were prepared for all isolates or species and shown with the sequence-based dendrogram (Fig. 2). Three main clusters branching at high genetic distance were obtained separating the species of the 'sexdentati' group *(B. sexdentati, B. vallesianus, B. pinophilus, B. poligraphi, B. borealis)*, the 'eggersi' group *(B. eggersi)* and the 'xylophilus' group *(B. xylophilus, B. mucronatus, B. fraudulentus)*. This confirms the affiliation of these species to species groups as was postulated on the basis of morphological similarities (Braasch, 2001). The cluster of the 'sexdentati' group further branched at low genetic distance, yielding a subgroup containing *B. sexdentati, B. vallesianus, B. pinophilus* and a subgroup containing *B. poligraphi* and *B. borealis*. Further subclustering was observed among the *B. sexdentati* isolates which formed two groups containing four isolates from Central Europe (Germany and Switzerland) and three isolates from South Europe (Greece, Italy and Portugal), respectively. The genetic distance between these two groups of *B. sexdentati* isolates was similar to the genetic distances between the species *B. sexdentati, B. vallesianus* and *B. pinophilus*.

The results of phylogenetic analysis are in agreement with morphological features of spicules and female tails, as shown in Fig. 2. The large genetic distances between the 'sexdentati', 'eggersi' and 'xylophilus' groups shown in the dendrogram are reflected by different types of spicule shape. Within the 'sexdentati' group species, the shapes of spicules and female tails exhibit common features, but also variability between species. The four Central European isolates of *B. sexdentati*

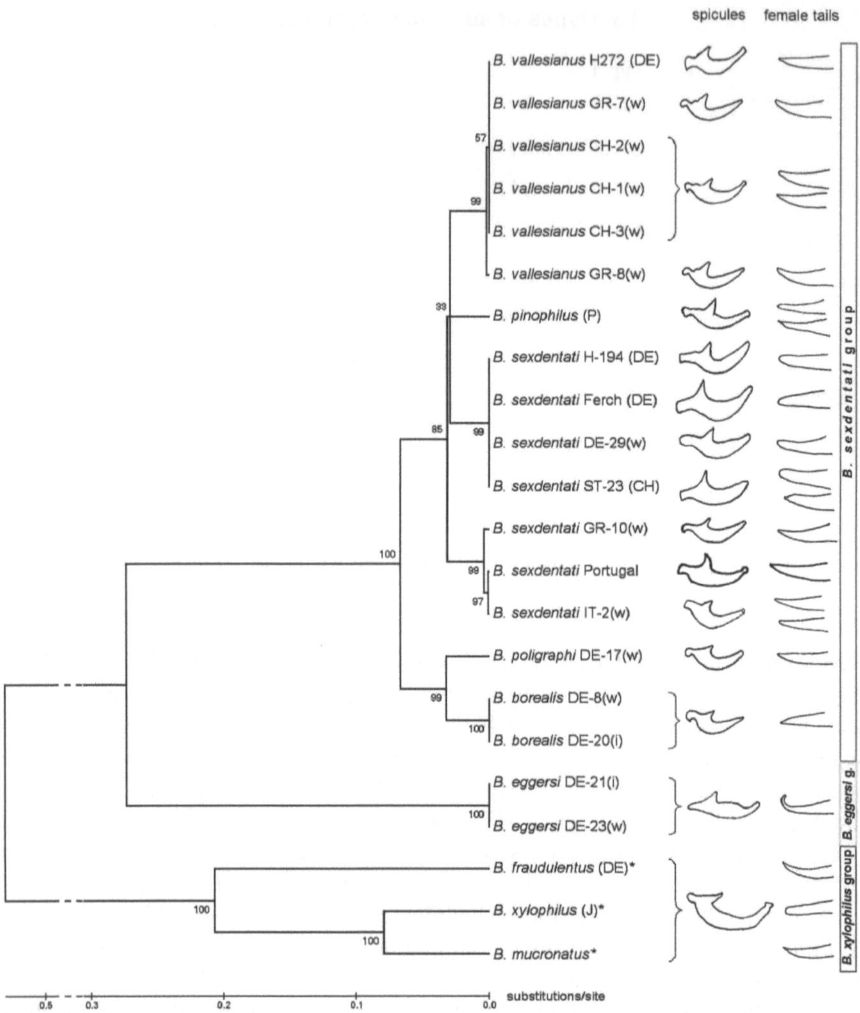


**Fig. 2** Phylogenetic tree of combined ITS1, 5.8S and ITS2 sequences of *Bursaphelenchus* isolates. The country codes used in the isolate notations are: DE: Germany; GR: Greece; CH: Switzerland; P: Portugal; IT: Italy and J: Japan. The numbers at branching points are bootstrapping values. *Sequences for these species were obtained from the EMBL database, having the IDs BXU92464 (*B. xylophilus*), BMU93554 (*B. mucronatus*) and AB067758 (*B. fraudulentus*). The following parameters were used for building the tree: Distance method: Tamura-Nei; Tree making method: Neighbor Joining; no. of bootstrap repetitions: 1000

are characterized by a rectangular condylus and a rounded female tail end (type 1 in Fig. 1), whereas the three South European isolates have a more or less hooked condylus and a conoid or even pointed female tail end (intermediates between types 1 and 2 in Fig. 1).

In conclusion, information from ITS1/2 sequencing and morphological differences support the existence of two types of *B. sexdentati* or even two species considered as *B. sexdentati* so far. Whereas the original description of *B. sexdentati* shows a subcylindrical female tail with hemispheric tail terminus, occasionally with a small digitate process (Rühm, 1960), as in the Central European isolates, the South European isolates show a conoid female tail with more or less finely rounded or blunt terminus. Additionally, a small cucullus at the spicules of the South European type has been observed (Ambrogioni and Caroppo, 1998), which is not shown in the original description of *B. sexdentati* (Rühm, 1960).

## *ITS-RFLP Analysis*

Species-specific ITS-RFLP patterns for *B. sexdentati, B. vallesianus, B. pinophilus, B. poligraphi* and *B. borealis* had been obtained earlier using the restriction enzymes *Rsa* I, *Hae* III, *Msp* I, *Hinf* I and *Alu* I (Burgermeister et al., 2005). The Central European isolates of *B. sexdentati* could not be distinguished from the South European isolates by ITS-RFLP using these restriction enzymes. Therefore, three additional restriction enzymes, *Mun* I, *Mbi* I and *Hpy* 188 I were selected based on sequence information to cut at sites where the two types of *B. sexdentati* differ in their ITS1/2 sequence. Some ITS-RFLP patterns obtained with the total set of eight enzymes are shown in Fig. 3. As expected, *B. sexdentati* DE-29(w) representing the Central European isolates and *B. sexdentati* GR-10(w), IT-2(w) and Portugal representing the South European isolates of *B. sexdentati* differ in their restriction fragments

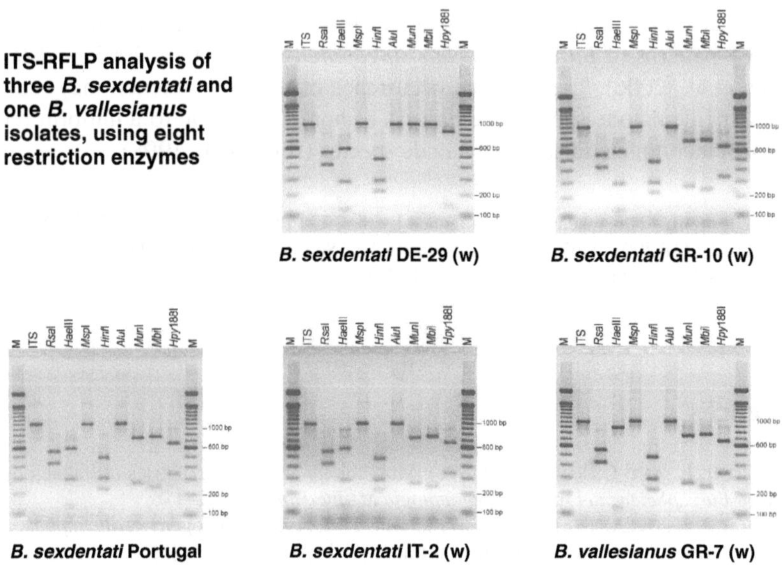


**Fig. 3** ITS-RFLP analysis of four *B. sexdentati* and one *B. vallesianus* isolates. The enzymes used for the restriction digests are indicated above the corresponding lanes. ITS: undigested PCR product containing the ITS1/2 region. M: Marker (100 bp ladder, Invitrogen)

obtained with the additional three enzymes. Isolate *B. sexdentati* IT-2(w) has the same pattern as isolates GR-10(w) and Portugal, except for an additional weaker *Hae* III fragment of approximately 860 bp which is caused by sequence heterogeneity (see below). *B. vallesianus* which can be distinguished from *B. sexdentati* by means of different *Hae* III fragments, exhibits the same pattern as the South European isolates of *B. sexdentati* on digestion with the additional three enzymes.

## *Intra-Individual Sequence Heterogeneity*

The results of ITS-RFLP analysis of *B. sexdentati* IT-2(w) suggested that two ITS2 sequences differing by the presence or absence of a *Hae* III restriction site may exist in the rDNA of this isolate. Therefore, the amplified ITS2 fragment obtained from a single specimen of *B. sexdentati* IT-2(w) was cloned (Lange et al., 2006). Sequencing of recombinant plasmids revealed the existence of two ITS2 sequences differing in four single base exchanges, one of which being responsible for the presence or absence of the *Hae* III restriction site, respectively.

Indications for ITS2 sequence heterogeneities were also obtained with other isolates, i.e. *B. sexdentati* Portugal and *B. pinophilus* Ne5/04 (Lange et al., 2006). Sequencing of cloned PCR products has shown that even the same specimen may possess slightly different ITS2 fragments within its number of rDNA tandem repeats. The presence of different ITS sequences has also been described for other nematodes, e.g. *Globodera rostochiensis* (Subbotin et al., 2000) and *Nematodirus* species (Nadler et al., 2000). Besides ITS2, ITS1 may also be affected (Zijlstra et al., 1995; Nadler et al., 2000). Indications for ITS heterogeneity were also deduced from ITS-RFLP patterns of *Bursaphelenchus singaporensis* (Gu et al., 2005) and *B. lini* (Braasch et al., 2006). The frequent occurrence of this phenomenon may indicate that in some nematode species the process of homogenization to a uniform repeat type is still incomplete in some nematode species (Blok et al., 1998).

**Acknowledgments** This research was financed by the EU FP5 program, project PHRAME, contract Nr. QLK5-CT-2002-00672. We like to thank Ute Schönfeld, Catarina Penas, Manuel Mota and Thomas Schröder for providing nematode isolates and cultures.

## References

Ambrogioni, L. and Caroppo, S. (1998). Morphology and morphometrics of Italian populations of *Bursaphelenchus* species. *Nematologia Mediterranea* 26: 97–116.

Beckenbach, K., Blaxter, M. and Webster, J.M. (1999). Phylogeny of *Bursaphelenchus* species derived from analysis of internal transcribed spacer DNA sequences. *Nematology* 1: 539–548.

Blok, V.C., Malloch, G., Harrower, B., Phillips, M.S. and Vrain, T.C. (1998). Intraspecific variation in ribosomal DNA in populations of the potato cyst nematode *Globodera pallida. Journal of Nematology* 30: 262–274.

BRAASCH, H. (2001). *Bursaphelenchus* species in conifers in Europe: distribution and morphological relationships. *Bulletin OEPP/EPPO Bulletin* 31: 127–142

BRAASCH, H., BRANDSTETTER, M. AND BURGERMEISTER, W. (2006). Supplementary characters of *Bursaphelenchus lini* Braasch, 2004 (Nematoda: Parasitaphelenchidae) and remarks on this nematode. *Zootaxa* 1141: 55–61.

BURGERMEISTER, W., METGE, K., BRAASCH, H. AND BUCHBACH, E. (2005). ITS-RFLP patterns for differentiation of 26 *Bursaphelenchus* species (Nematoda: Parasitaphelenchidae) and observations on their distribution. *Russian Journal of Nematology* 13: 29–42.

GU, J., ZHANG, J., BRAASCH, H. AND BURGERMEISTER, W. (2005). *Bursaphelenchus singaporensis* sp. n. (Nematoda: Parasitaphelenchidae) in packaging wood from Singapore – a new species of the *B. xylophilus* group. *Zootaxa* 988: 1–12.

HALL, T.A. (1999). BioEdit: a user-friendly biological sequence alignment editor and analysis program for Windows 95/98/NT. *Nucleic Acids Symposium Series* 41: 95–98.

IWAHORI, H., TSUDA, K., KANZAKI, N., IZUI, K. AND FUTAI, K. (1998). PCR-RFLP analysis of ribosomal DNA of *Bursaphelenchus* nematodes related to pine wilt disease. *Fundamental and Applied Nematology* 21: 655–666.

KUMAR, S., TAMURA, K., JAKOBSEN, I.B. AND NEI, M. (2001). MEGA2: molecular evolutionary genetics analysis software. *Bioinformatics* 17: 1244–1245.

LANGE, C., BURGERMEISTER, W., METGE, K. and BRAASCH, H. (2006). Phylogenetic analysis of isolates of the *Bursaphelenchus sexdentati* group using ribosomal intergenic transcribed spacer DNA sequences. *Journal of Nematode Morphology and Systematics* 9: 95–109.

NADLER, S.A., HOBERG, E.P., HUDSPETH, D.S.S. AND RICKARD, L.G. (2000). Relationships of *Nematodirus* species and *Nematodirus battus* isolates (Nematoda: Trichostrongyloidea) based on ribosomal DNA sequences. *Journal of Parasitology* 86: 588–601.

RÜHM, W. (1960). Ein Beitrag zur Nomenklatur und Systematik einiger mit Scolytiden vergesellschafteter Nematodenarten. *Zoologischer Anzeiger* 164: 201–213.

SAITOU, N. AND NEI, M. (1987). The neighbor-joining method: a new method for reconstructing phylogenetic trees. *Molecular Biology and Evolution* 4: 406–425.

SUBBOTIN, S.A., HALFORD, P.D., WARRY, A. AND PERRY, R.N. (2000). Variations in ribosomal DNA sequences and phylogeny of *Globodera* parasitizing solanaceous plants. *Nematology* 2: 591–604.

TAMURA, K. AND NEI, M. (1993). Estimation of the number of nucleotide substitutions in the control region of mitochondrial DNA in humans and chimpanzees. *Molecular Biology and Evolution* 10: 512–526.

THOMPSON, J.D., GIBSON, T.J., PLEWNIAK, F., JEANMOUGIN, F. AND HIGGINS, D.G. (1997). The ClustalX windows interface: flexible strategies for multiple sequence alignment aided by quality analysis tools. *Nucleic Acids Research* 25: 4876–4882.

ZHENG, J., SUBBOTIN, S.A., HE, S., GU, J. AND MOENS, M. (2003). Molecular characterization of some Asian isolates of *Bursaphelenchus xylophilus* and *B. mucronatus* using PCR-RFLPs and sequences of ribosomal DNA. *Russian Journal of Nematology* 11: 17–22.

ZIJLSTRA, C., LEVER, A.E.M., UENK, B.J. AND VAN SILFHOUT, C.H. (1995). Differences between ITS regions of isolates of root-knot nematodes *Meloidogyne hapla* and *M. chitwoodi*. *Phytopathology* 85: 1231–1237.

# Analysis of *Bursaphelenchus xylophilus* (Nematoda: Parasitaphelenchidae) Provenances Using ISSR and RAPD Fingerprints

**Kai Metge and Wolfgang Burgermeister**

**Abstract** The pinewood nematode *Bursaphelenchus xylophilus* is the causal agent of pine wilt disease. In order to trace the origin of its recently introduced Portuguese population, two PCR-based techniques, random amplified polymorphic DNA (RAPD) and inter-simple sequence repeats (ISSR), were used to determine genetic relationships among 30 *B. xylophilus* isolates from the USA, Canada, Japan, China, South Korea and Portugal. Fingerprints obtained with both methods detected a reduced genetic variation of introduced isolates as compared to native North American isolates. Cluster analyses of genetic distances between isolates were carried out and bootstrap dendrograms were constructed. The results indicated that founders of the Portuguese isolates most likely were translocated one or two times to Portugal from their colonized sites in East Asia, but not from their native habitats in North America.

## Introduction

Introducing species by global trade contributes directly to mixing of fauna from geographically separated regions. Package- and dunnage wood are of importance as transport material for the worldwide spread of economically important invasive species like *B. xylophilus* (Pfeilstetter, 2003). These low-quality wood materials were suspected to have transported the initial colonists of the Portuguese PWN population during the last decades of the twentieth century from an unknown area of distribution (Tomiczek et al., 2003). *B. xylophilus* is native in North America and was spread to China, Taiwan, South Korea and Portugal, where it was detected in 1999 in an area of 258,000 ha around Setúbal (Evans et al., 1996; Mota et al., 1999).

In spite of extension of EU quarantine regulations to phytosanitary treatments of package wood in 2001 (Commission Decision 2001/219/EC), living PWN have been recorded in 47 samples of conifer wood imported to EU countries from 2000 to

K. Metge
Institute for Biosafety of Genetically Modified Plants, Julius Kuehn-Institute (JKI), Federal Research Centre for Cultivated Plants, Quedlinburg, Germany
e-mail: kai.metge@jki.bund.de

M.M. Mota, P. Vieira (eds.), *Pine Wilt Disease: A Worldwide Threat to Forest Ecosystems*, 

2004 (Metge and Burgermeister, 2005). In order to trace the origin of the Portuguese PWN isolate, ISSR and RAPD polymorphisms were examined to determine relationships among indigenous isolates from North America and introduced isolates from various countries.

## Materials and Methods

*Bursaphelenchus* isolates were kindly provided from the *Bursaphelenchus* culture collection at our Research Centre in Braunschweig, Germany. The *B. xylophilus* isolates represent populations from the USA, Canada, Japan, China, South Korea and Portugal. *B. mucronatus* and *B. fraudulentus* from Germany were used as outgroup. The cultures were maintained in Petri dishes on *Botrytis cinerea*/malt agar prior to DNA extraction. A total number of 30 *B. xylophilus* isolates were investigated in this study (Table 1). Nineteen isolates were from infected trees of the genera *Pinus, Picea* and *Abies*, five from package wood, one from dunnage wood, three from woodchips and the remaining five isolates from hosts or products, where information was no longer available. However, no isolate was derived from an insect.

The nematodes were extracted from medium using the Baermann funnel technique and washed twice with deionized water. Genomic DNA was obtained from bulks of 2000–10000 animals of each culture without prior separation according to sex or developmental stage to obtain reproducible population-specific patterns of each isolate by ISSR- and RAPD-PCR (Schmitz et al., 1998). They were transferred in water to an Eppendorf tube and sedimented for 2 min at 9000 × g. The supernatant was discarded. Nematode pellets were frozen in liquid nitrogen and crushed with an Eppendorf micro pestle. The homogenate was then treated according to the protocol of the High Pure PCR Template Preparation Kit (Roche, Mannheim, Germany). The yield of extracted genomic DNA was determined using the fluorescent dye, Hoechst 33258 and a DyNA Quant 200 fluorometer (Amersham, Freiburg, Germany). DNA extracts were stored at 4°C.

ISSR-PCR was carried out employing a 25 μl reaction volume and a Biometra T1 thermocycler. The reaction mixture contained 2 units Taq DNA polymerase (PeqLab, Erlangen, Germany), 20 mM Tris-HCl (pH 8.5), 16 mM $(NH_4)_2SO_4$, 4 mM $MgCl_2$, 0.01% Tween 20 (Y-PCR buffer, PeqLab), 0.2 mM of each dNTP, 0.8 μM primer (Roth, Karlsruhe, Germany) and 4 ng template DNA. The PCR program consisted of an initial denaturation for 2 min, 30 s at 96°C, 35 cycles with 20 s denaturation at 94°C, 45 s annealing at 42–55°C, depending on the primer used (Table 2), 2 min extension at 72°C and a final extension for 6 min at 72°C.

RAPD-PCR was carried out employing a 25 μl reaction volume and a Perkin Elmer 9600 thermocycler. The reaction mixture contained Stoffel buffer (10 mM Tris, pH 8.3, 10 mM KCl), 5 U AmpliTaq DNA Polymerase Stoffel fragment (Applied Biosystems, Darmstadt, Germany), 4 mM $MgCl_2$, 0.2 mM of each dNTP, 0.2 μM primer (Roth; Genosys Biotechnologies, The Woodlands, TX, USA) (Table 2) and 4 ng template DNA. The PCR program consisted of an initial denaturation for 2 min, 30 s at 94°C, 40 cycles with 20 s denaturation step at 92°C,

**Table 1** List of 30 *Bursaphelenchus xylophilus*, one *B. mucronatus* and one *B. fraudulentus* populations used for ISSR and RAPD analyses

| Species | Code | BBA code country | Locality | Date of start culture at BBA |
|---|---|---|---|---|
| *B. mucronatus* | DE-4w | Germany | Templin, Brandenburg | 1996 |
| *B. fraudulentus* | DE-10w | Germany | Zusmarshausen, Bavaria | 1997 |
| *B. xylophilus* | PT-1w | Portugal1 | Marateca/Pegoes | 1999 |
| *B. xylophilus* | US-DE-1w | USA2 | unknown, packaging wood | 2002 |
| *B. xylophilus* | US-DE-2w | USA3 | unknown, packaging wood | 2002 |
| *B. xylophilus* | Ne5/00 | USA4 | Missouri | 2000 |
| *B. xylophilus* | Ne4b/00 | USA5 | Missouri | 2000 |
| *B. xylophilus* | US2 | USA6 | Burlington, Vermont | 1994 |
| *B. xylophilus* | US9 | USA7 | Tucson, Arizona | 1994 |
| *B. xylophilus* | US10 | USA8 | Cloquet Forestry Center, Minnesota | 1993 |
| *B. xylophilus* | US11 | USA9 | Vermont, New Jersey | 1994 |
| *B. xylophilus* | US15 | USA10 | Cook County, Illinois | 1993 |
| *B. xylophilus* | BxAlta | Canada11 | Smokey Lake, Alberta | 1994 |
| *B. xylophilus* | Cmz1 | Canada12 | unknown, dunnage wood | 1994 |
| *B. xylophilus* | Q52A | Canada13 | Quebec | 1993 |
| *B. xylophilus* | BxBC | Canada14 | British Columbia | 1993 |
| *B. xylophilus* | Q1426 | Canada15 | Quebec | 1993 |
| *B. xylophilus* | St.John | Canada16 | New Brunswick | 1993 |
| *B. xylophilus* | Ne3/02 | Japan17 | Tottori City, Admin. Division Tottori | 1994 |
| *B. xylophilus* | J2 | Japan18 | Izuhara, Nagasaki | 1994 |
| *B. xylophilus* | J3 | Japan19 | Ueki, Admin. Division Kumamoto | 1994 |
| *B. xylophilus* | J10 | Japan20 | Nishiaizu, Admin. Division Fukushima | 1994 |
| *B. xylophilus* | BxJP | Japan21 | Mito, Admin. Division Ibaraki | 1991 |
| *B. xylophilus* | C-14-5 | Japan22 | Chiba | 1993 |
| *B. xylophilus* | RC-DE-1w | China23 | unknown, packaging wood | 2001 |
| *B. xylophilus* | RC-DE-2w | China24 | unknown, packaging wood | 2001 |
| *B. xylophilus* | Ne12/02 | China25 | Nanjing City forest | 2002 |
| *B. xylophilus* | BxChina | China26 | unknown | 1993 |
| *B. xylophilus* | PT-3w | Portugal27 | Lezirias | 2003 |
| *B. xylophilus* | PT-4w | Portugal28 | Troia | 2003 |
| *B. xylophilus* | KR-1w | Korea29 | Jinju, Gyeongsangnam-Province | 2003 |
| *B. xylophilus* | KR-3w | Korea30 | Mokpo, Jeollanam-Province | 2003 |

**Table 2** Primer sequences, annealing temperatures and number of markers generated in ISSR-PCR's and RAPD-PCR's for *30 Bursaphelenchus xylophilus* isolates and *B. mucronatus* and *B. fraudulentus* as outgroup. Wobbles B: C, G or T; D: A, G or T; H: A, C or T; R: A or G; V: A, C or G; Y: C or T

| Primer | Primer sequence | Annealing temperature [°C] | Primer type | 30 *B. xylophilus* Populations | |
|---|---|---|---|---|---|
| | | | | with outgroup markers [n] | without outgroup markers [n] |
| 11 | $[GA]_9$-CCA | 50 | 3′-ISSR | 43 | 31 |
| 25 | $[AC]_9$-TG | 55 | 3′-ISSR | 31 | 25 |
| 26 | $[AC]_9$-GA | 55 | 3′-ISSR | 46 | 37 |
| 54 | $[TC]_9$-CG | 50 | 3′-ISSR | 26 | 16 |
| 188 | CGT-$[CA]_8$ | 55 | 5′-ISSR | 37 | 33 |
| 190 | CAG-$[GT]_9$ | 55 | 5′-ISSR | 32 | 28 |
| 841 | $[GA]_8$-YC | 50 | 3′-ISSR | 43 | 30 |
| 848 | $[CA]_8$-RG | 42 | 3′-ISSR | 47 | 40 |
| 857 | $[AC]_8$-YG | 50 | 3′-ISSR | 45 | 38 |
| 888 | BDB-$[CA]_7$ | 55 | 5′-ISSR | 39 | 31 |
| 890 | VHV-$[GT]_7$ | 55 | 5′-ISSR | 42 | 35 |
| 1423 | HVH-$[TGT]_5$ | 50 | 5′-ISSR | 42 | 34 |
| 1424 | BDB-$[CAC]_5$ | 50 | 5′-ISSR | 30 | 26 |
| 1425 | BDV-$[CAG]_5$ | 50 | 5′-ISSR | 27 | 18 |
| Total | | | | **530** | **422** |
| B07 | GGT GAC GCA G | 38 | RAPD | 49 | 40 |
| Y01 | GTG GCA TCT C | 38 | RAPD | 29 | 23 |
| Y04 | GGC TGC AAT G | 38 | RAPD | 33 | 28 |
| Y08 | AGGCAG AGC A | 38 | RAPD | 49 | 45 |
| Z01 | TCT GTG CCA C | 38 | RAPD | 34 | 32 |
| Z04 | AGG CTG TGC T | 38 | RAPD | 43 | 38 |
| Z06 | GTG CCG TTC A | 38 | RAPD | 36 | 29 |
| Z07 | CCA GGA GGA C | 38 | RAPD | 51 | 39 |
| Z08 | GGG TGG GTA A | 38 | RAPD | 53 | 42 |
| Z11 | CTC AGT CGC A | 38 | RAPD | 51 | 40 |
| Re08 | CGA TCG ATG C | 38 | RAPD | 62 | 50 |
| Re09 | GGA AGC TTC G | 38 | RAPD | 63 | 55 |
| Re10 | CCC TGC AGG C | 38 | RAPD | 58 | 48 |
| Total | | | | **611** | **509** |

15 s annealing at 38°C, 1 min extension at 72°C and a final extension for 7 min at 72°C. The rate of heating from 38°C to 72°C was regulated to 0.3°C/s. After completion of the PCR, aliquots of the samples were separated electrophoretically using a 1.8%-agarose gel and 0.5xTBE buffer. Gels were stained with ethidium bromide (1 μg/ml) and visualized with a UV transilluminator.

In repeated amplifications, DNA from the same extraction was used. We have verified the bands scored by repeated amplification of DNA from 26 isolates, using two RAPD and eight ISSR primers. Genetic relationships of isolates were assessed from bands scored by eye from the fingerprint profiles, and evaluated as presence or absence of bands. Distance data matrices were calculated (Metge and Burgermeister, 2006) using Nei & Li coeffient (Nei and Li, 1979) and dendrograms constructed with Neighbor joining (NJ) method (Saitou and Nei, 1987) and Unweighted Pair Group Method with Arithmetic Means (UPGMA) (Sneath and Sokal, 1973). Bootstrap values of trees were calculated with PAUP*4.0 (Swofford, 1998).

For testing the hypothesis that fingerprints obtained by RAPD markers and ISSR markers resulted in equal cluster analysis results, correlation between the two matrices was estimated by means of the Mantel test (Mantel, 1967), implemented in PopTools 2.6.6. (Hood, 2005).

Variation in scored bands between the group of native *B. xylophilus* isolates (n = 15) from North America and the group of isolates derived from introduced populations (n = 15) from Asia and Portugal were calculated using Chi-square test.

## Results

For comparision of genetic relationships, 14 ISSR primers and 13 RAPD decamer primers were used in this study (Table 2, Fig. 1). Amplification of genomic DNA from 32 *Bursaphelenchus* isolates yielded 530 ISSR markers and 611 RAPD markers in total. Of these, 108 ISSR and 102 RAPD markers were contributed by the outgroup of *B. mucronatus* and *B. fraudulentus*. Thus, 422 ISSR and 509 RAPD markers were derived from the 30 *B. xylophilus* isolates. Among these, only eight monomorphic RAPD and five monomorphic ISSR markers were amplified. An average of 47 RAPD markers and 40 ISSR markers were found per primer (Table 2).

When increasing numbers of bands were scored, distance values became constant above a range of 200–250 ISSR or RAPD markers. Iteration of data evaluation for 530 ISSR markers and 611 RAPD markers resulted in two distance matrices and four trees (Fig. 3) with robust branches and nodes for isolates originating from the same geographic region, with exception of the three Portuguese populations. Both distance matrices were compared using Mantel's test. The test gave a high correlation for both matrices (Fig. 2).

For the construction of trees bootstrap dendrograms were calculated (Fig. 3). *B. mucronatus and B. fraudulentus* were always positioned as outgroup. Both the NJ and the UPGMA methods were almost congruent using ISSR- and RAPD-based distances. With the UPGMA method bootstrap values of 100% support all 30

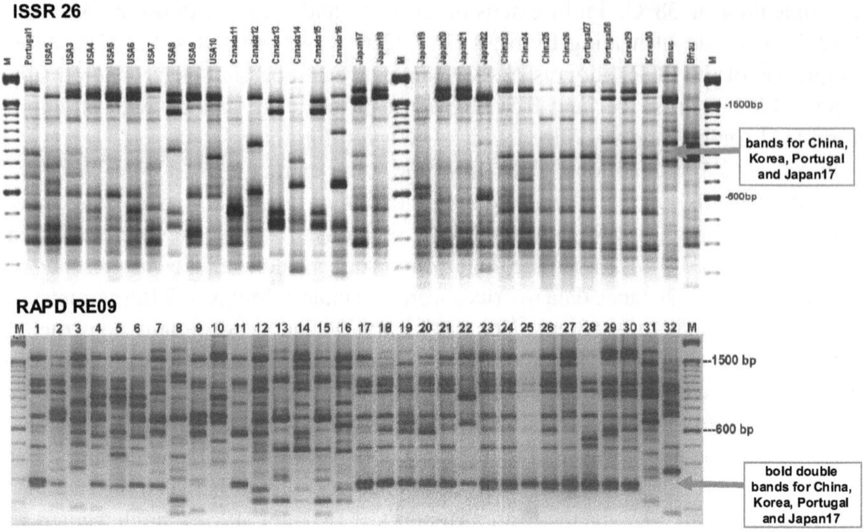


**Fig. 1** Banding profiles of amplified DNA from 32 *Bursaphelenchus* isolates using ISSR primer 26 and RAPD primer Re09. M: 100 bp ladder (Invitrogen, Karlsruhe, Germany)

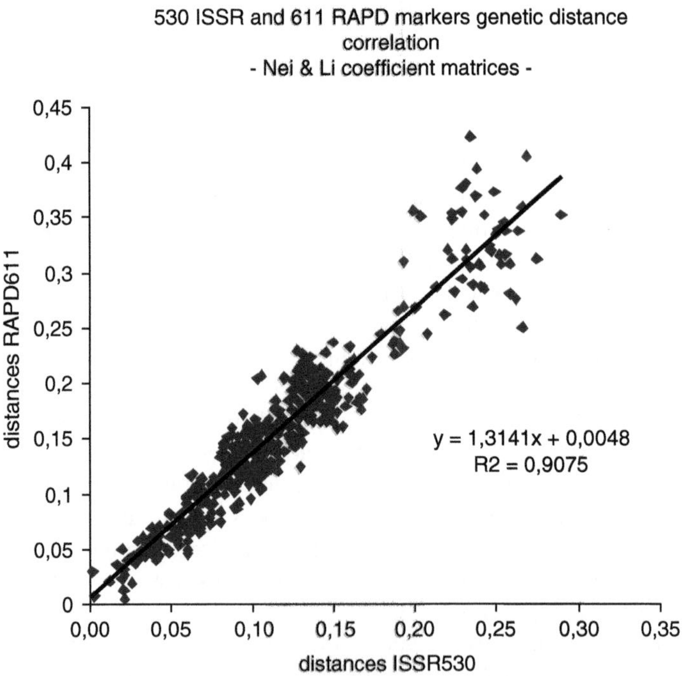


**Fig. 2** Mantel's matrix correlation test (1000 replicates) for ISSR and RAPD genetic distance matrices (Nei & Li coefficient). $R^2 = 0.9075,\ p = 0.99$

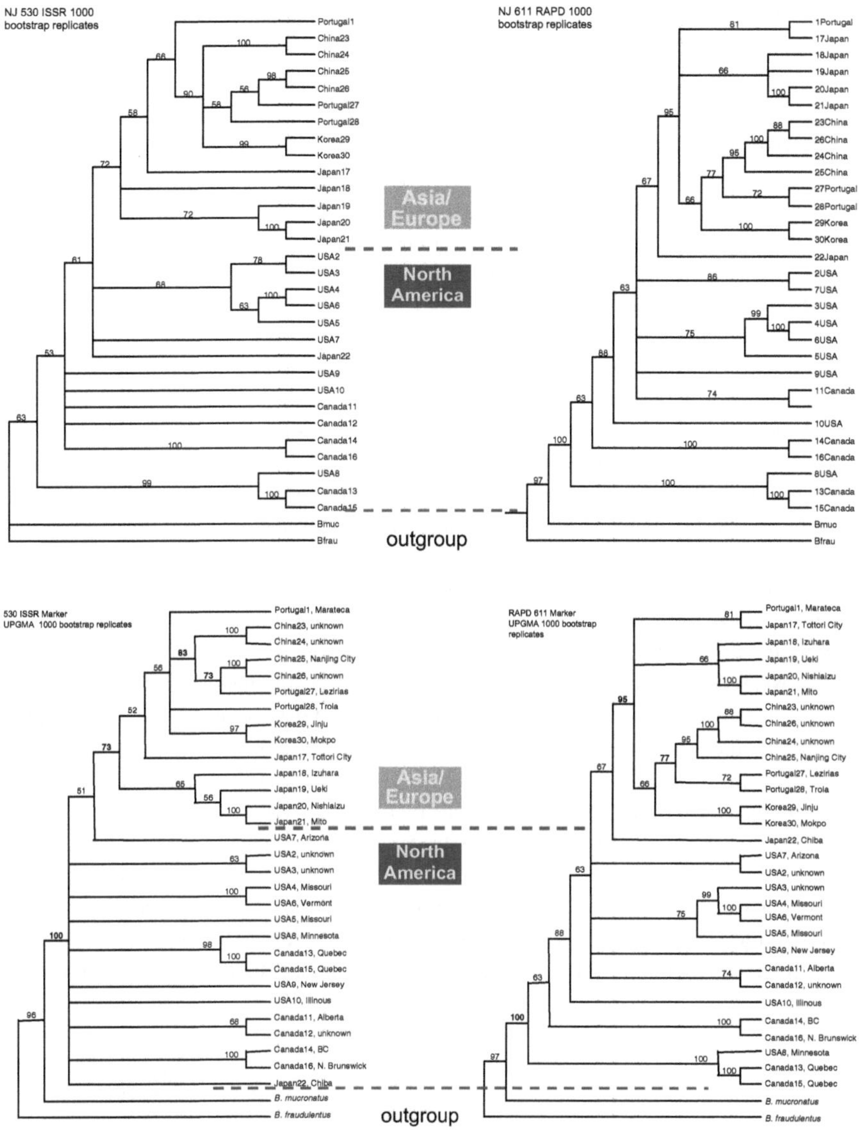


**Fig. 3** Neighbour joining and UPGMA trees inferred from 530 ISSR (**A, C**) and 611 RAPD markers (**B, D**) for 30 *Bursaphelenchus xylophilus* isolates, with *B. mucronatus* and *B. fraudulentus* isolates as outgroup. Numbers at the branches represent > 50% values of 1000 bootstrap replicates supporting the indicated subbranches and groups

*B. xylophilus* isolates as one group, the species group. An Asian/Portuguese branch including the two South Korean, four Chinese, five Japanese and three Portuguese isolates forms a separate group within the species, which is sufficiently supported by

RAPD (95%) and ISSR (73%). The nine US American and the six Canadian isolates are more or less supported by bootstrap values between those of the outgroup and the Asian/Portuguese cluster. Using the NJ method, the Asian/Portuguese branch is less strongly supported by bootstrap iterations of 72% (ISSR) and 57% (RAPD).

The five isolates USA2, USA3, Canada12, China23 and China24, which had been intercepted from package wood, could clearly be assigned to the countries as indicated in their respective accompanying documents.

Among the North American isolates, only the position of USA7 was less supported by bootstrap iterations of ISSR-based distances (51%), whereas iteration of RAPD-based distances did not support the position of USA9 from New Jersey and Japan22 from Chiba. In all dendrograms Japan22 was separated from other Japanese populations. This isolate was grouped more or less closely to the North American branch, depending on the tree constructing method used.

In general, the isolates derived from native populations of North American were genetically more diverse than isolates derived from introduced populations found in the Asian/Portuguese branch. This was demonstrated by the significantly higher number of markers of the group of 15 North American isolates as compared to the group of 15 Asian/Portuguese isolates (Fig. 4). 149 ISSR and 200 RAPD markers were found in native isolates only and not in isolates derived from introduced isolates. On the other hand, 69 ISSR and 78 RAPD markers were detected as new markers in isolates derived from introduced populations.

The Asian/Portuguese branch was separated in a subcluster built by Chinese, South Korean and Japanese isolates. The three Portuguese isolates were not grouped together. Instead they were distributed at different locations within the Asian/Portuguese branch. Portugal1 from Marateca/Pegoes and Japan17 from Tottori City built a pair when RAPD-based distances were evaluated. However, this result was not supported by ISSR dendrograms. Bootstrap iterations of 77% (RAPD, UPGMA), 56% (ISSR, UPGMA), 67% (RAPD, NJ) as well as 58% (ISSR, NJ) supported weakly closer genetic relationships of Portugal27 from Lezirias and Portugal28 from Troia to Chinese isolates. Japan18 from Izuhara, Japan19 from Ueki, Japan20 from Nishiaizu and Japan21 from Mito built a Japanese-specific subcluster which

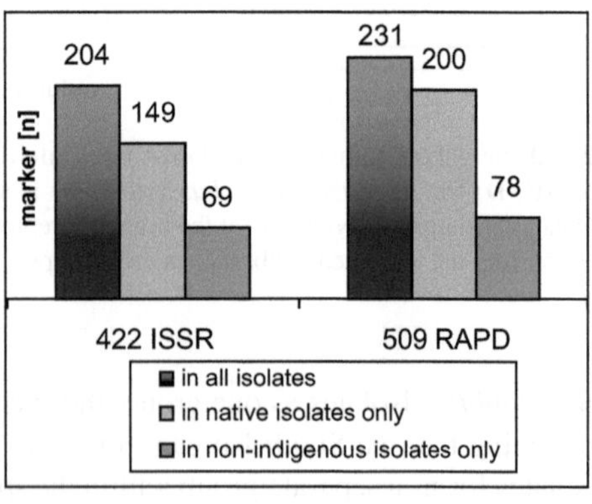


**Fig. 4** Number of genetic markers found for ISSR and RAPD fingerprints with 15 native and 15 introduced isolates of *Bursaphelenchus xylophilus*

was found in four trees with maximum bootstrap iteration of 66% (UPGMA, RAPD). Both isolates from South Korea were paired in a separate branch which was clustered in the Chinese subbranch and not in the Japanese one.

## Discussion

The results obtained in this study illustrate the applicability of ISSR and RAPD analysis for differentiation of *B. xylophilus* isolates from various countries worldwide. We have used these two independent DNA marker techniques for estimating intraspecific diversity of *B. xylophilus* isolates in order to trace the origin of the recently introduced Portuguese population. Comparing RAPD and ISSR distance matrices, Mantel's test revealed a high correlation. This supports our hypothesis that both methods are equally suited in phylogenetic studies of *Bursaphelenchus* isolates. ISSR- and RAPD-derived fingerprints have not been compared before in genetic analysis of nematodes. The cluster analyses resolved thirty *B. xylophilus* isolates into an Asian/Portuguese branch and a North American group. Concerning the Canadian isolates, the close vicinity of Canada13 and Canada15 was as expected, since both originate from Quebec. On the other hand, the subcluster calculated for Canada14 from British Columbia and Canada16 from New Brunswick was surprising concerning the large geographic distance. These two isolates were isolated before 1992 and 1993, respectively and maintained in different laboratories successively. Therefore an error in designation might have occurred.

Genetic shift during laboratory culturing may also influence the pattern of detectable markers of *B. xylophilus* reference isolates (Fig. 5). DNA samples of the same *B. xylophilus* isolates, obtained from cultures in 1994 and 2003, were examined by RAPD-PCR and cluster analysis of genetic distances (Metge et al., 2004). Slight differences between the 1994 and 2003 DNA extracts of the same culture were noted on direct comparison of electrophoretic fingerprints. Nevertheless, clustering of the 1994 DNA extracts corresponded to the 2003 DNA extracts of the same cultured isolates The isolates examined (USA6, USA7, USA8, USA9, USA10 and Canada11) were also included in this study and clustered together with additional isolates originating from North America (Figs. 2, 3).

Low genetic distance values were found within the subclusters of the isolates from China, South Korea and Japan. In contrast, the Portuguese isolates were assigned to two different branches separated at higher genetic distances. This suggests that the founders of the Portuguese reference populations could have been introduced two times to Portugal. If the three Portuguese isolates were descendants of a single introduction, they would be expected to cluster in one distinct branch as seen with the Korean isolates. In a recent RAPD analysis, 24 Portuguese isolates of *B. xylophilus* were found to be genetically homogeneous, suggesting that they were dispersed recently from a single introduction (Vieira et al., 2007).

*B. xylophilus* was introduced in Portugal before 1999. It is discussed, that the founders of this population were translocated by package wood. While the number

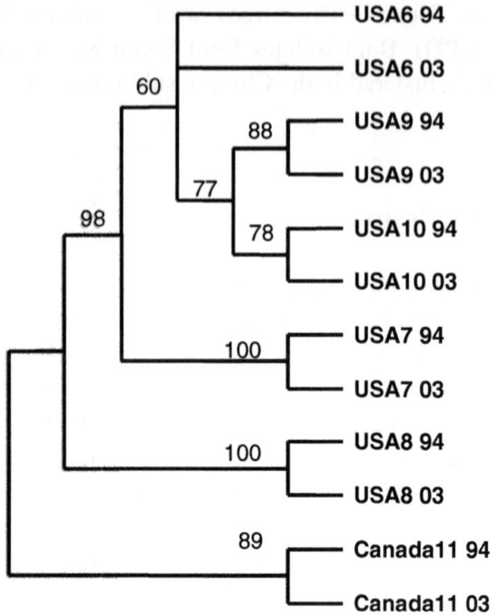


**Fig. 5** Neighbour joining tree inferred from 157 ISSR markers obtained from DNA samples isolated in 1994 and 2003 from the same nematode laboratory culture

of founder individuals translocated by package wood may have been small, they may not represent the whole genetic diversity of the species. If they were carried by their vectors, ten-thousands of nematodes originating from one tree could be translocated by one beetle (Sousa et al., 2001). Nevertheless, the variation of SSR- and RAPD-elements in the newly established population may be reduced. This effect will be strong when all founders were descendants from the same isolate, or weak when several colonization events occurred in the same new area (Sakai et al., 2001). A lower number of ISSR and RAPD fragments could be expected for recently introduced *B. xylophilus* populations. We confirmed a reduced genetic variation of RAPD and ISSR markers in 15 introduced populations as compared to 15 native populations (Metge and Burgermeister, 2006). A number of markers were found within isolates derived from native isolates only and absent in isolates derived from introduced isolates. On the other hand, a lower number of new markers were also detected in the introduced populations (Fig. 4). In general, the DNA fingerprints within the East Asian and Portuguese isolates appeared more homogeneous compared to the patterns obtained from North American isolates. This is most likely due to a higher genetic variability within native as compared to introduced populations.

ISSR and RAPD polymorphisms permitted conclusions to be drawn on the origin of the Portuguese population. Clustering of biogeographically widely separated isolates from Japan, China and Portugal is an indication for PWN introduction from East Asia to Portugal (Fig. 6). The splitting of the Portuguese PWN isolates in two different subgroups within the Asian cluster indicates that founders of the Portuguese population could have been translocated two times to Portugal from their recently colonized sites in East Asia and not from their native habitats in North America.

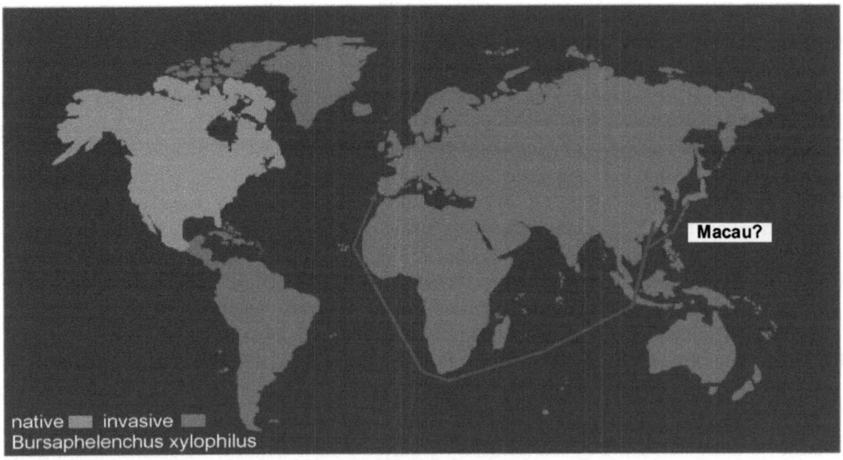


**Fig. 6** Possible pathway of introduction of *B. xylophilus* to Portugal

Translocation of *B. xylophilus* over large distances may easily be explained when the scale of international trade involving timber and package wood is taken into consideration. Four isolates (USA2, USA3, China23, China24) had been isolated from package wood. Their affiliation to the respective countries is only based on trading documents and has been confirmed by the results of our cluster analysis. This is not trivial since package wood is often used repeatedly and transported to third party countries. Until 2001 package wood was not examined officially for *B. xylophilus*. This has most probably facilitated its introduction to Portugal before 1999. Since 2001 the European Commission has ordered phytosanitary treatment of all susceptible packaging wood originating from PWN-infested countries or regions (Commission Decision 2001/219/EC). In spite of EU quarantine regulations, living PWN have been recorded in 47 samples of conifer wood (mostly package wood) imported to EU countries from 2000 to 2004 (Metge and Burgermeister, 2005). Molecular phylogenetic approaches are useful tools to trace the origin of PWN found in package wood if reference PWN material from different provenances is available.

**Acknowledgments** This research was financed by the EU FP5 program, project acronym PHRAME, Contract No. QLK5-CT-2002-00672. We would like to thank Edmundo Sousa from the Instituto Nacional de Investigação Agrária (INIA) in Oeiras and Manuel M. Mota from the University of Évora, Portugal, for providing nematode samples and Thomas Schröder at BBA for maintaining and supplying *Bursaphelenchus* cultures.

## References

EVANS, H.F., MCNAMARA, D.G., BRAASCH, H., CHADOEUF, J. AND MAGNUSSON, C. (1996). Pest Risk Analysis (PRA) for the territories of the European Union (as PRA area) on *Bursaphelenchus xylophilus* and its vectors in the genus *Monochamus*. *EPPO Bulletin* 26: 99–249.

HOOD, G.M. (2005). *PopTools version 2.6.6*. www.cse.csiro.au/poptools.

MANTEL, N. (1967). The detection of disease clustering and generalized regression approach. *Cancer Research* 27: 209–220.

METGE, K. AND BURGERMEISTER, W. (2005). Molecular identification and pathway analysis of the introduced pinewood nematode *Bursaphelenchus xylophilus*. In *Plant Protection and Plant Health in Europe: Introduction and Spread of Invasive Species. BCPC Symposium Proceedings* (eds. D.V. Alford and G.F. Backhaus), pp. 259–260.

METGE, K. AND BURGERMEISTER, W. (2006) Intraspecific variation of isolates of *Bursaphelenchus xylophilus* (Nematoda: Aphelenchoididae) revealed by ISSR and RAPD fingerprints. *Journal of Plant Disease and Protection* 6: 275–282.

METGE, K., SCHRÖDER, T. AND BURGERMEISTER, W. (2004). Biogeographische Untersuchungen zur Herkunft des nach Europa eingeschleppten Kiefernholznematoden *Bursaphelenchus xylophilus* (Nematoda: Aphelenchoididae). *Mitt. Biol. Bundesanst.* 396: 296–297.

MOTA, M., BRAASCH, H., BRAVO, M.A., PENAS, A.C., BURGERMEISTER, W., METGE, K. AND SOUSA, E. (1999). First report of *Bursaphelenchus xylophilus* in Portugal and in Europe. *Nematology* 1: 727–734.

NEI, M. AND LI, W.H. (1979). Mathematical model for studying genetic variation in terms of restriction endonucleases. *Proc. Nat. Acad. Sci. USA* 76: 5269–5273.

PFEILSTETTER, E. (2003). Monitoring invasiver Arten vor dem Hintergrund pflanzengesundheitlicher Regelungen [Monitoring of invasive species in relation to phytosanitary regulations]. *Schriftenr. BMVEL A Angew. Wiss.* 498: 233–237.

SAITOU, N. AND NEI, M. (1987). The Neighbor-joining method: a new method for reconstructing phylogenetic trees. *Mol. Biol. Evol.* 4: 406–425.

SAKAI, A.K., ALLENDORF, F.W., HOLT, J.S., LODGE, D.M., MOLOFSKY, J., WITH, K.A., BAUGHMAN, S., CABIN, R.J., COHEN, J.E., ELLSTRAND, N.C., MCCAULEY, D.E., O'NEIL, P., PARKER, J.M., THOMPSON, J.N. AND WELLER, S.G. (2001). The population biology of invasive species. *Annu. Rev. Ecol. Syst.* 32: 305–332.

SCHMITZ, B., BURGERMEISTER, W. AND BRAASCH, H. (1998). Molecular genetic classification of Central European *Meloidogyne chitwoodi* and *M. fallax* populations. *Nachrichtenbl. Deut. Pflanzenschutzd.* 50: 310–317.

SNEATH, P.H.A. AND SOKAL, R.R. (1973). *Numerical Taxonomy: The principles and Practice of Numerical Classification*. W. H. Freeman and Company, San Francisco, CA, USA.

SOUSA, E., BRAVO, M.A., PIRES, J., NAVES, P., PENAS, A.C., BONIFACIO, L. AND MOTA, M. (2001). *Bursaphelenchus xylophilus* (Nematoda; Aphelenchoididae) associated with *Monochamus galloprovincialis* (Coleoptera; Cerambycidae) in Portugal. *Nematology* 3: 89–91.

SWOFFORD, D.L. (1998). *Phylogenetic Analysis Using Parsimony (* and other methods). Version 4*. Sinauer Associates, Sunderland, MA, USA

TOMICZEK, C., BRAASCH, H., BURGERMEISTER, W., METGE, K. HOYER, U. AND BANDSTETTER, M. (2003). Identification of *Bursaphelenchus* spp. isolated from Chinese packaging wood imported to Austria. *Nematology* 5: 573–581.

VIEIRA, P., W. BURGERMEISTER, M. MOTA, K. METGE & G. SILVA (2007). Lack of genetic variation of *Bursaphelenchus xylophilus* in Portugal revealed by RAPD-PCR analyses. *Journal of Nematology* 39(2): 118–126.

# Satellite DNA as a Versatile Genetic Marker for *Bursaphelenchus xylophilus*

**Philippe Castagnone-Sereno, Chantal Castagnone, Cécile François and Pierre Abad**

**Abstract** *Bursaphelenchus xylophilus* is causal agent of the pine wilt disease, one of the major conifer diseases worldwide. This nematode is a quarantine organism in the European Union, where it has recently been discovered in a restricted area in Portugal. In our laboratory, research has been developed to characterize the genetic diversity of the nematode, and to provide a more comprehensive view of the relationships between *B. xylophilus* and the non-pathogenic *Bursaphelenchus* species. For that purpose, repetitive sequences known as satellite DNA (satDNA) have been cloned and characterized from the genome of *B. xylophilus*. Its species-specific distribution and high copy number in the genome make this sequence a very promising tool for molecular identification of the nematode, as will be illustrated with results obtained in the laboratory. In particular, the recent development of a very sensitive satDNA based real-time PCR assay will be presented. Moreover, sequencing of monomers of the satDNA was performed and phylogenetic analyses were conducted to analyse the diversity and relationships (1) between *B. xylophilus* isolates sampled worldwide; (2) among *B. xylophilus* populations all collected in the infested area from Portugal. Results will be discussed in the context of the origin and evolution of the pine wood nematode complex in Europe.

## Introduction

Pine wilt is one of the major diseases of pine (*Pinus* spp.) forests worldwide, caused by the pinewood nematode (PWN), *Bursaphelenchus xylophilus*. The PWN is indigenous to North America where it is not considered as a pathogen on native pine species. However, the nematode constitutes a severe pest in countries where it has been introduced. For example, about 30% of the total area of pine forest is estimated to be infested by this species in Japan (Mamiya, 2004), following its introduction

P. Castagnone-Sereno
UMR1301 INRA-UNSA-CNRS, Sophia Antipolis, France
e-mail: philippe.castagnone@sophia.inra.fr

M.M. Mota, P. Vieira (eds.), *Pine Wilt Disease: A Worldwide Threat to Forest Ecosystems*, 

via the timber trade. Owing to the potential threat to forest ecosystems, regulations have been introduced in many countries to prevent introduction and/or further spread of *B. xylophilus*. In Europe, the presence of PWN has recently been detected in a restricted area in Portugal (Mota et al., 1999). Because it impacts on forest health, natural ecosystem stability and international trade, the nematode has been listed as a quarantine organism in Europe (EU directive 2000/29/EC). However, to be effective, regulation steps (e.g. national surveys, special phytosanitary measures, etc.) strongly depend on accurate detection/identification of the pathogen.

The morphological identification of the more than 70 species of *Bursaphelenchus* currently recognized requires a high level of expertise, in particular when considering species belonging to the PWN species complex, such as *B. xylophilus* and *B. mucronatus* (Ryss et al., 2005). In recent years, several molecular methods have been developed for diagnostics, using a variety of targets in the genome of the nematode. Since repetitive sequences known as satellite DNA (satDNA) have been shown as excellent taxonomic markers for nematodes of agronomic interest (Grenier et al., 1997), our laboratory has been, for some time, focusing its interest on this kind of sequences in *B. xylophilus*. This paper summarizes the main results we obtained in this line of research and offers new perspectives in terms of genetic diversity and evolution in the PWN.

## The *Msp*I Satellite DNA Family

The plant and animal eukaryotic genome is constituted of both unique and repetitive sequences. Among these, satDNA represents a major component of the highly repetitive fraction of the genome. SatDNA consists of relatively short, repeated units arranged in tandemly reiterated arrays. These non-coding, repetitive DNA sequences can vary in copy number from 10 to over $10^5$ per haploid genome, and in proportion from less than 1% to more than 66% of the genome (Beridze, 1986). SatDNA is localized mainly in the constitutive heterochromatin, at both the centromeric and telomeric regions of the chromosomes. Evolution of satDNA is governed by principles of concerted evolution, in which mutations are generated and spread in a stochastic process of molecular drive (Dover, 2002). Although no defined function has yet been established, satDNA has been suggested to be involved in evolutionary processes and the stability of genome structure (Csink and Henikoff, 1998). Because it diverged rapidly during evolution, and is constantly homogenized, it often gives rise to sequences that are species- or genome-specific (Bachmann et al., 1993).

In order to characterize the PWN genome, we initiated the cloning and detailed analysis of the satDNA component in this organism. Indeed, a tandemly repeated *Msp*I satDNA has been identified in the genome of *B. xylophilus* and studied in depth (Tarès et al., 1993). It is represented as 160 bp-repetitive units organized as tandem arrays (see the typical laddern pattern observed in Southern blot experiment; Fig. 1). The relative abundance of the *Msp*I satDNA was assessed from dot blot experiments and estimated to make up to 30% of the total genomic DNA, which is

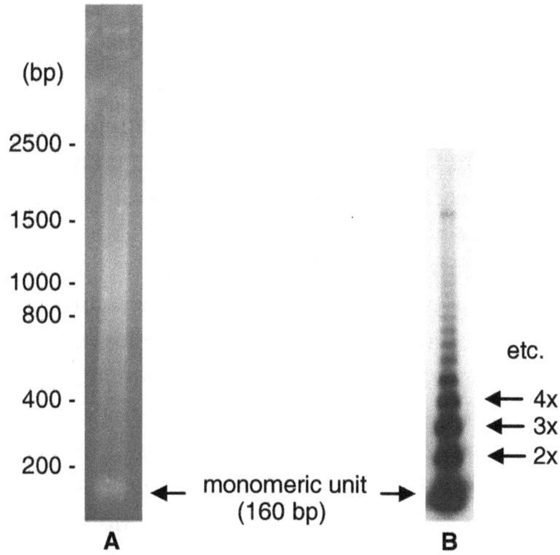


**Fig. 1** Tandem organization of *Bursaphelenchus xylophilus* satellite DNA. (**A**) Detection of the *Msp*I satellite DNA family by restriction analysis of genomic DNA. (**B**) Demonstration of the tandem organization of the repeats by Southern blot analysis. The monomer identified in (**A**) was used as a probe to reveal the typical ladder pattern

a rather high level compared to other nematodes. Sequence analysis indicated that the monomeric unit is (A + T)-rich (~ 62%) with clusters of As and Ts, which may suggest an ancient origin for this sequence. The cloning and sequence analysis of independent repeats showed a high level of similarity, with 3.9% average divergence from the consensus sequence. Such low variability suggests that some mechanism is acting upon the genome of *B. xylophilus* to maintain the sequence homogeneity of the whole family, as proposed for other satDNA sequences (Dover, 2002).

## Application to Diagnostics

As mentioned above, the *Msp*I satDNA family is highly repeated in the *B. xylophilus* genome. Recently, flux cytometry measurements indicated that the size of this genome is about 29.8 $10^6$ bp (Leroy et al., 2003), which allows to estimate the number of satDNA monomers at about 56 $10^3$ per haploid genome. This value is extremely high as compared to other repetitive components of the genome, e.g. ribosomal DNA, which is estimated to be present at a few hundred copies per haploid genome. In parallel, using the monomeric unit as an hybridization probe, the *Msp*I satDNA family proved to be specifically distributed in the *B. xylophilus* genome and absent from other closely related species of the PWN species complex, such as *B. mucronatus* and *B. fraudulentus* (Tarès et al., 1994). Therefore, these two features of the *Msp*I satDNA family (i.e. its high abundance and species-specific distribution in the genome) make it *a priori* a very good candidate as a target for diagnostics. In the following sections the two PCR methodologies we developed are presented in order to provide specific, sensitive and friendly-to-use satDNA-based diagnostic assays for the PWN.

## Conventional PCR

The introduction of the PCR technology into nematode diagnostics has revolutionized this area of research, since the small size of these organisms has no longer been considered as a serious obstacle to their identification by molecular methods. In recent years, PCR has been extensively experienced to develop differentiation tools for *Bursaphelenchus* species using ribosomal DNA (rDNA) as a specific target. Although providing valuable results for pure research purposes, such methodologies did not avoid any risk of misinterpretation of the obtained data in routine testing, since they rely on very subtle differences in rDNA primary sequences (e.g. amplified products of different size, Matsunaga and Togashi, 2004; more or less complex ITS-RFLP patterns, Burgermeister et al., 2005).

To overcome such drawback, we investigated the possibility of using satDNA as a target for PCR, based on the species-specific distribution of this repetitive element in the *B. xylophilus* genome. Specific primers were designed close to both ends of the sequence of the 160-bp monomer of the *Msp*I satDNA family previously characterized in *B. xylophilus* (Tarès et al., 1993; GenBank accession L09652), and amplification performed on individual nematodes prepared according to a single worm PCR procedure. Use of proteinase K, in combination with the alternation of high and low temperatures, proved to be efficient to make the genomic DNA of a single individual suitable as a template for PCR. Figure 2 illustrates the results obtained, with typical ladder patterns of monomers and multimers of the expected size amplified from *B. xylophilus* nematodes only, while no amplification was detected with the other *Bursaphelenchus* species tested (*B. mucronatus* in this case). The same positive or negative results were obtained with single individuals from other *B. xylophilus* isolates and from other *Bursaphelenchus* species, respectively, which confirmed the specificity and the sensitivity of the assay (Castagnone et al., 2005). The primer set developed here will be useful in positively identifying *B. xylophilus*

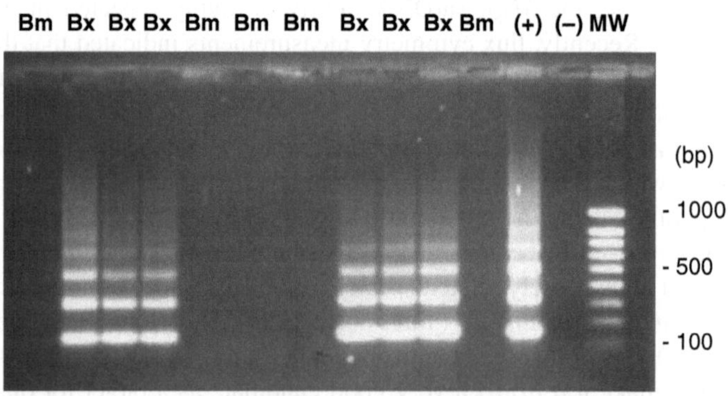


**Fig. 2** Positive identification of single *Bursaphelenchus xylophilus* individuals by conventional PCR. Bx, *B. xylophilus*; Bm, *B. mucronatus*; (+), positive control; (–), negative control; MW, molecular weight

in samples collected in the wild, and could contribute to the development of a simple 'yes/no' diagnostic procedure for this quarantine pest.

## *Real-Time PCR*

As mentioned above, conventional PCR based either on direct visualization or RFLP analysis of amplicons provided useful methods for identification of the PWN. However, the multi-step process of amplicon analysis, involving gel electrophoresis, and the risk of post-PCR contamination make conventional PCR protocols unadapted for high-throughput routine testing. To overcome such limitations, real-time PCR coupled to automate product analysis appears to be a promising alternative, since post-PCR handling for target detection is no longer required. Instead, fluorescent monitoring of amplicon accumulation allows target detection and quantification, based on the cycle threshold (Ct) value, i.e. the number of PCR cycles at which the amplicon product curve exceeds fluorescence background (Heid et al., 1996). Applicability of this new technology to the detection of the PWN has very recently been tested using rDNA as a target sequence (Cao et al., 2005).

Again, we developed and evaluated a real-time PCR assay for routine sample screening to detect and quantify the PWN, based on the specific *Msp*I satellite DNA family previously cloned in the genome of the parasite. To improve the accuracy of real-time PCR, TaqMan chemistry has been incorporated into our assay. This chemistry combines the use of a target-specific oligonucleotide probe dual-labelled with a donor fluorophore and an acceptor dye, and the 5′-exonuclease activity of Taq polymerase (Holland et al., 1991), thus associating the sensitivity of PCR and the specificity of DNA hybridization. In TaqMan real-time PCR optimized assays, genomic DNA from the *B. xylophilus* isolates was specifically amplified, while fluorescence remained below the threshold values for the negative control and the isolates of the non-target species *B. mucronatus* (Fig. 3). The sensitivity of this TaqMan PCR assay was assessed in the laboratory, and experiments showed that *B. xylophilus* target DNA was detectable in samples containing as little as 1 pg purified DNA or one

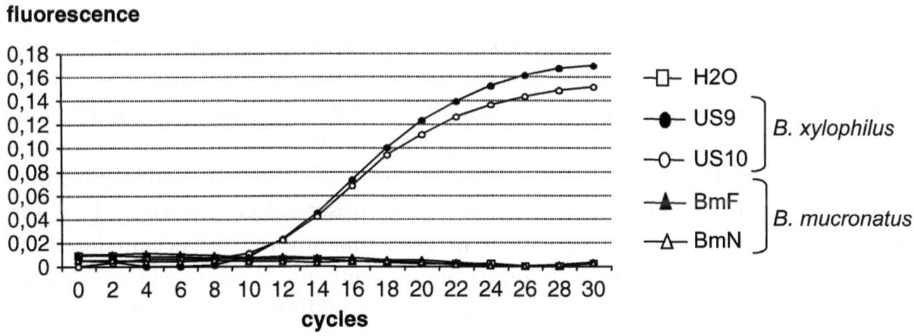


**Fig. 3** Positive identification of *Bursaphelenchus xylophilus* using TaqMan real-time PCR

single nematode. Moreover, in preliminary experiments, PWN DNA was also positively detected in a variety of PWN-infested wood samples (François et al., 2007). To this step, modifications of the method are still needed to increase throughput, increase the size of samples that can be processed, and improve sampling, which could in turn increase the advantages of routine direct testing of wood samples. However, it should be considered as a candidate methodology for diagnosing PWN infection in epidemiological surveys as well as in regulatory testing, considering the need for such a test in light of the increasing impact of this invasive pest on international trade regulation.

## Diversity and Evolution

### *Variability of SatDNA Sequences Among* B. xylophilus *Isolates*

In order to assess the genetic distances and relationships among *B. xylophilus* isolates from various geographic origins, satDNA was further used as a molecular target for comparative sequence analysis. For that purpose, we initiated the cloning and sequencing of monomeric units from *B. xylophilus* isolates from North America (Canada and the USA), East Asia (China and Japan) and Portugal, in order to study the distribution of mutations within and among isolates. Different levels of variability were detected within and among isolates, which was quite exclusively due to point mutations (within each isolate, average variability to the consensus ranged from 2.9 to 12.1%). Along with a differential structuration of this variability, this strongly suggested that the *Msp*I satDNA family exhibits differential steps of molecular evolution. However, the observation that nucleotide changes were shared among monomers supports the hypothesis that some highly effective homogenization mechanism acts upon them (such as gene conversion or unequal crossing-over; Dover, 2002), in contrast to the accumulation of independent mutational events.

In addition, alignment of consensus sequences for the overall set of isolates showed alternance of highly conserved and variable domains within the monomeric sequences (Fig. 4A). The same feature reported recently in satDNA from the root-knot nematode *Meloidogyne* spp. was interpreted as an indication of constraints imposed on particular segments of the monomer sequence (Mestrovic et al., 2006). It was suggested that the effect might be caused by functional interactions between the satellite DNA and various protein components in heterochromatin and/or in the functional centromere, although direct experimental evidence for this hypothesis is still lacking. However, the same observation performed here on *B. xylophilus* sequences reinforces the hypothesis that non-random accumulation of mutations due to selective pressure on particular segments of the monomers directs the evolution of satDNA. All together, these data highlighted the interest of satellite DNA monomer sequence analysis to reveal infraspecific genetic variability in *B. xylophilus*.

**(A)**

```
           3
      74
667        GGTGTCTAGTATAATATCAGAGTGTTTAGCCTGGTGGGGCGTAATTTGACTCCAAAGAAGCTGAGACTTGCC
601        GGTGTCTAGTATAATATCAGAGTGTTTRGCCTGGTGSGGCGTAATTTGACTCCAAAARAGCTGAGACTTGCC
Chine      GGTGTCTAGTATAATATCAGAGTTTTTCGCCCAGTGTGGGCGTAATTTGATTCCAAAAAAGCTGAAACTTGCC
US15       GGTGTCTAGTATAATATCAGAGTTTTTCGCCCAGTGTGGGCGTAATTTGATTCCAAAARAGCTGAAACTTGCC
J2         GGTGTCTAGTATAATATCAGAGTGTTTCGCCCAGTGTGGGCGTAGTTTGATTCCAAAAATGCTGAAACTTGCC
PT-3       GGTGTCTAGTATAATATCAGAGTGTTTCGCCCAGTGTGGGCGTAGTTTGATTCCAAAAAAGCTGAAACTTGCC
US2        GGTGTCTAGTATAATATCAGAGTGTTTCGCCCAGTGTGGGCGTAATTTGACTCCAAAAAAGCTGAAACTTGCC
Alta       GGTGTCTAGTATAATATCAGAGTGTTTCGCCCAGTGTGGGCGTAATTTGATTGCAAAAAAGCTGAAACTTGCC
BC         GGTGTCTAGTATAATATCAGAGTGTTTCGCCTAGTGTGGGCGTAATTTGATTGCAAAAAAGCTGAAACTTGCC
Japon      GGTGTCTAGTATAATATCAGAGTGTTTCGCCCAGTGTGGGCGTAATTTGATTGCAAAAAAGCTGAAACTTGCC
J10        GGTGTCTAGTATAATATCAGAGTGTTCGGCCTGGTRCGGGCGCARTTTDACTCCAAAAAAGGCGAGACTTGCG
US9        GGTGTCTAGTATAATATCAGAGTGTTCGGHCCGGTACGGGCGCAGTTTAVCTCCAAACAAGGCGAAACTTGCG
US10       GGTGTCTAGTATAATATCAGAGTGTTCGGCCTGATACGGGCGCARTTTTGCTCCAAAAAAGGCGAGACTTGCG
           **************************  * *   *  ***** * ***   * **** * *  ** ******
           75
     148
667        ACTCTGAAATCTCATTCAATTACGGTTCGAATGGTGTATGTCTTGTCTATTCACTCCGTCGTCACTAATTCAC
601        ACKCTGAAATCTCATTCARTTACGKTTCGAATGGTGTATGTCTTGTCTATTCACTCCGTCGTCACTAATTCAC
Chine      ATGCTAAAATCTCAGGCGATTACCTTTCGAATGGTGTATGTCTTGTCAATTCACTCCGTCGTCACTAATTCAC
US15       ATGCTAAAATCTCATGCGATTACCTTTCGAATGGTGTATGTCTTGTCAATTCACTCCGTCGTCACTAATTCAC
J2         ATGCTAAAATCTCAGGCGATTAGCTTTCGAATGGTGTATGTCTTGTCAATTCACTCCGTCGTCACTAATTCAC
PT-3       ATGCTAAAATCTCAGGCGATTACCTTTCGAATGGTGTATGTCTTGTCAATTCACTCCGTCGTCACTAATTCAC
US2        ATGCTAAAATCTCAGGCGRTTACCTTTCGAATGGTGTATGTCTTGTCAATTCACTCCGTCGTCACTAATTCAC
Alta       GTGCTGAAATCTTACGAGGTTACCTTTCGAATGGTATTAGTCCTGTCAATTCACTCCGTCGTCACTAATTCAC
BC         GTGCTRAAATCTTACGAGGTTAYSTTTCGAATGGTRYAAGTCCTGTCAATTCACTCCGTCGTCACTAATTCAC
Japon      GTGCTGAAATCTTACGAGGTTACCTTTCGAATGGTATATGTCTTGTCAATTCACTCCGTCGTCACTAATTCAC
J10        GTGTTRAAATYTCATTCAGTDACGTTTCGAATGGTATAGGTCTYGTCTATTCACTCCGTCGTCACTAATTCAC
US9        GTGTTAAAATTTCATTCGGTAACGTTTCGAATGGTATAGGTCACGTCTATTCACTCCGTCGTCACTAATTCAC
US10       GTGTTAAAATTTCATTCTGTGTTGTTTCCAATGGTATAGGTCGCGTCTATTCACTCCGTCGTCACTAATTCAC
               * **** * *     *     *** ******    ***  *** ************************
```

**(B)**

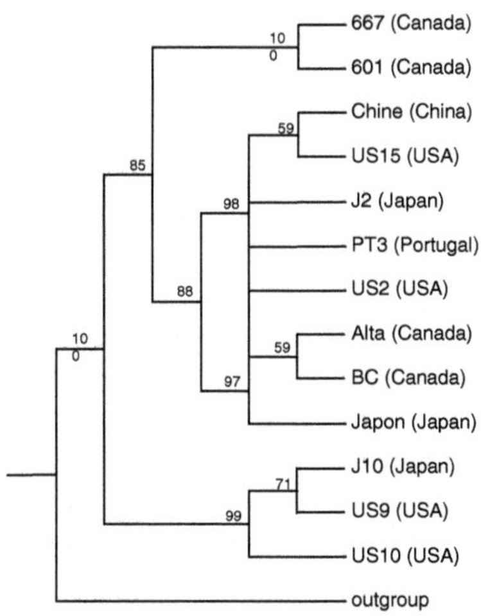


**Fig. 4** Diversity of satellite DNA monomeric sequences in *Bursaphelenchus xylophilus*. (**A**) Alignment of the consensus sequences deduced from 13 *B. xylophilus* isolates from Canada, China, Japan, Portugal and the USA. Stars indicate conserved positions. (**B**) UPGMA tree deduced from the alignment shown in (**A**). Bootstrap values > 50% based on 1000 replicates are given on each node

## *Relationships Between* B. xylophilus *Isolates*

The consensus sequences obtained previously were further used in phylogenetic analyses. As shown in Fig. 4B, no correlation could be found between the position in the tree and the geographic origin of the isolates. Since satDNAs are non-coding sequences, and thus thought to diverge rapidly during evolution, this observation is probably linked to the ancient origin of the North American and Asian isolates, and results from the accumulation of mutations in the genome of these isolates.

In contrast, a preliminary analysis of satDNA monomeric sequences from the Portuguese isolates collected in the infested region of Setúbal, tended to indicate that a significant correlation could exist between the genetic distance (based on satDNA sequences) and the location of the isolate in the area. This observation could result from the recent colonization of Portugal by the PWN, and thus from the short-time divergence of the *Msp*I satDNA family in the nematode isolate(s) involved in this colonization. If confirmed, this result would mean that the progression of the infestation could be followed in space based on sequence analysis. Further experiments, conducted on a more representative number of Portuguese *B. xylophilus* isolates, are planned to validate this hypothesis.

## Conclusion

Due to the potential threat to forest ecosystems, regulations have been introduced in many countries to prevent introduction and/or further spread of *B. xylophilus*. In Europe, the recent discovery of the PWN in a restricted area in Portugal has lead to its listing as a quarantine pest, and to the implementation of national large-scale surveys to monitor the pathogen and special phytosanitary measures for conifer wood packaging materials imported from areas where the nematode has been recorded. To be effective, such regulation steps strongly depend on accurate detection of the pathogen. The PCR-based diagnostic assays presented here, using the *Msp*I satDNA family as a target sequence, proved to be both specific and sensitive, and appeared complementary to other molecular methodologies directed against rDNA sequences. All together, they provide nowadays powerful tools to efficiently detect and identify the PWN.

Currently, very few data are available in literature concerning the genetic diversity of *B. xylophilus* isolates at various geographic scales, from a very local situation (a restricted forest) to the global distribution on earth. However, this information is clearly of outstanding importance to follow (and further anticipate?) the progression of this invasive species, and will undoubtedly constitute a new and exciting challenge for researchers in the coming years.

**Acknowledgments** These studies were financially support by the European Community (grants QLRT-2001-00672 and SSPE-CT-2004-502348). We thank all colleagues involved in the 'PHRAME' and 'PORTCHECK' projects that contributed to the research reported in this paper.

## References


BACHMANN, L., SCHIBEL, J.M., RAAB, M. AND SPERLICH, D. (1993). Satellite DNA as a taxonomic marker. *Biochemical Systematics and Ecology* 21: 3–11.

BERIDZE, T. (1986). *Satellite DNA*. Springer-Verlag, Berlin, 149 pp.

BURGERMEISTER, W., METGE, K., BRAASCH, H. AND BUCHBACH, E. (2005). ITS-RFLP patterns for differentiation of 26 *Bursaphelenchus* species (Nematoda: Parasitaphelenchidae) and observations on their distribution. *Russian Journal of Nematology* 13: 29–42.

CAO, A.X., LIU, X.Z., ZHU, S.F. AND LU, B.S. (2005). Detection of the pinewood nematode, *Bursaphelenchus xylophilus*, using a real-time polymerase chain reaction assay. *Phytopathology* 95: 566–571.

CASTAGNONE, C., ABAD, P. AND CASTAGNONE-SERENO, P. (2005). Satellite DNA-based species-specific identification of single individuals of the pinewood nematode *Bursaphelenchus xylophilus* (Nematoda: Aphelenchoididae). *European Journal of Plant Pathology* 112: 191–193.

CSINK, A.K. AND HENIKOFF, S. (1998). Something from nothing: The evolution and utility of satellite repeats. *Trends in Genetics* 14: 200–204.

DOVER, G.A. (2002). Molecular drive. *Trends in Genetics* 18: 587–589.

FRANÇOIS, C., CASTAGNONE, C., BOONHAM, N., TOMLINSON, J., LAWSON, R., HOCKLAND, S., QUILL, J., VIEIRA, P., MOTA, M. AND CASTAGNONE-SERENO, P. (2007). Satellite DNA as a target for TaqMan real.time PCR detection of the pinewood nematode, *Bursaphelenchus xylophilus*. *Molecular Plant Pathology* 8: 803–809.

GRENIER, E., CASTAGNONE-SERENO, P. AND ABAD, P. (1997). Satellite DNA sequences as taxonomic markers in nematodes of agronomic interest. *Parasitology Today* 13: 398–401.

HEID, C.A., STEVENS, J., LIVACK, K.J. AND WILLIAMS, P.M. (1996). Real time quantitative PCR. *Genome Research* 6: 986–994.

HOLLAND, P., ABRAMSON, R.D., WATSON, R. AND GELFAND, D.H. (1991). Detection of specific polymerase chain reaction product by utilizing the $5' - 3'$ exonuclease activity of *Thermus aquaticus* DNA polymerase. *Proceedings of the National Academy of Sciences USA* 88: 7276–7280.

LEROY, S., DUPERRAY, C. AND MORAND, S. (2003). Flow cytometry for parasite nematode genome size measurement. *Molecular and Biochemical Parasitology* 128: 91–93.

MAMIYA, Y. (2004) Pine wilt disease in Japan. In *The pinewood nematode, Bursaphelenchus xylophilus. Nematology monographs and perspectives, vol. 1* (eds. M. Mota and P. Vieira). Leiden, The Netherlands, E.J. Brill, pp. 9–20.

MATSUNAGA, K. AND TOGASHI, K. (2004). A simple method for discriminating *Bursaphelenchus xylophilus* and *B. mucronatus* by species-specific polymerase chain reaction primers. *Nematology* 6: 273–277.

MESTROVIC, N., CASTAGNONE-SERENO, P. AND PLOHL, M. (2006). Interplay of selective pressure and stochastic events directs evolution of the MEL172 satellite DNA library in root-knot nematodes. *Molecular Biology and Evolution* 23: 2316–2325.

MOTA, M.M., BRAASCH, H., BRAVO, M.A., PENAS, A.C., BURGERMEISTER, W., METGE, K. AND SOUSA, E. (1999). First report of *Bursaphelenchus xylophilus* in Portugal and in Europe. *Nematology* 1: 727–734.

RYSS, A., VIEIRA, P., MOTA, M. AND KULINICH, O. (2005). A synopsis of the genus *Bursaphelenchus* Fuchs, 1937 (Aphelenchida: Parasitaphelenchidae) with keys to species. *Nematology* 7: 393–458.

TARÈS, S., LEMONTEY, J.M., DE GUIRAN, G. AND ABAD, P. (1993). Cloning and characterization of a highly conserved satellite DNA sequence specific for the phytoparasitic nematode *Bursaphelenchus xylophilus*. *Gene* 129: 269–273.

TARÈS, S., LEMONTEY, J.M., DE GUIRAN, G. AND ABAD, P. (1994). Use of a species-specific satellite DNA from *Bursaphelenchus xylophilus* as a diagnostic probe. *Phytopathology* 84: 294–298.

# Application of Conventional PCR and Real-Time PCR Diagnostic Methods for Detection of the PineWood Nematode, *Bursaphelenchus xylophilus*, in Wood Samples from Lodgepole Pine

**Isabel Leal, Eric Allen, Leland Humble, Margaret Green and Michael Rott**

**Abstract** Molecular diagnostic methods have been designed for the detection and identification of the pinewood nematode, *Bursaphelenchus xylophilus*. Heat shock protein 70 (Hsp 70) gene sequences from *B. xylophilus* and the closely related *B. mucronatus*, were compared and used to design primers Bx701F and Bx701R which amplify a 171 base pair fragment from *B. xylophilus* by polymerase chain reaction (PCR). As a control, primers Bm701F and Bm701R were designed which specifically amplify a 168 base pair fragment from *B. mucronatus*. After optimization, *B. xylophilus* primers were shown to be highly sensitive and could easily detect 23 target copies, or less than 1 nematode. In addition, a real-time PCR method was developed to detect and differentiate *B. xylophilus* from other wood-inhabiting nematode species. A primer set and a specific Taqman® fluorescent probe were designed to amplify target *B. xylophilus* Hsp 70 sequences. After optimization, this real-time PCR assay was shown to be highly specific and sensitive, detecting at least 5 pg of *B. xylophilus* genomic DNA, as well as DNA extracted from individual nematodes. The species-specific detection of *B. xylophilus* was carried out directly from concentrated Baermann funnel extracts using wood samples from lodgepole pine (*Pinus contorta*, Dougl. var. *latifolia*) trees.

## Introduction

The international spread of the pinewood nematode (PWN) and its *Monochamus* vector occurs mainly through the movement of beetle-infested logs, untreated wood products and wood packaging material. To prevent new introductions and further spread, PWN is considered a regulated quarantine pest in 52 countries (mostly

I. Leal
Natural Resources Canada, Canadian Forest Service, Victoria, V8Z 1M5, British Columbia, Canada
e-mail: ileal@pfc.cfs.nrcan.gc.ca

M.M. Mota, P. Vieira (eds.), *Pine Wilt Disease: A Worldwide Threat to Forest Ecosystems*, 

from the European Plant Protection Organization, EPPO) (Liebhold et al., 1995; Evans et al., 1996; Dwinell, 1997; Mireku and Simpson, 2002). Accurate techniques for the detection and identification of PWN are required in order to comply with these quarantine regulations and to prevent the movement of PWN between countries.

Diagnosis of *B. xylophilus* has traditionally been based on morphological characteristics. However, the morphological similarity of adult PWN with *B. mucronatus* (Mamiya and Enda, 1979) and the inability to distinguish the juvenile stages of *B. xylophilus* make accurate morphological identifications difficult. Molecular detection tools, which are simple, rapid, and reliable, can be used to determine with certainty the presence of this nematode in wood.

To distinguish between *B. xylophilus* and *B. mucronatus*, multiple molecular techniques have been developed: PCR-RFLP (Hoyer et al., 1998; Braasch et al., 1999; Iwahori et al., 2000; Burgermeister et al., 2005); RAPD techniques (Braasch et al., 1995; Irdani et al., 1995), and PCR-based diagnostics with species-specific primers based on: intergenic sequences (IGS) (Kang et al., 2004), internal transcribed spacer (ITS) (Matsunaga and Togashi, 2004; Cao et al., 2005), heat shock protein 70 (Leal et al., 2005, 2007), and satellite DNA (Castagnone et al., 2005). Real-time PCR offers advantages over end-point PCR in that it is more sensitive, since it allows target quantification, and is less time-consuming (no post-PCR processing).

We report here the development of a species-specific conventional PCR assay and a real-time PCR Taqman® assay, all based on Hsp 70 gene sequences specific to *B xylophilus*, to identify and detect *B. xylophilus* and differentiate it from *B. mucronatus* in wood samples. The Hsp 70 gene family is a highly conserved gene family and consists of heat inducible or constitutively expressed proteins that act as molecular chaperones (Konstantopoulou et al., 1998). Nikolaidis and Scouras (2002) used Hsp 70 gene sequences in order to differentiate different nematode species and proposed that these sequences can be used as molecular markers.

The conventional end-point PCR and the real-time-PCR assay were evaluated using *B. xylophilus* isolates from USA, Canada, China and Europe, and other *Bursaphelenchus* species from the "xylophilus" group, such as: *B. mucronatus, B. conicaudatus, B. doui, B. fraudulentus, B. singaporensis, B. hofmanni*, and *B. thailandae*. Both assays were shown to be highly specific, while the real-time-PCR assay was also shown to be more sensitive.

Our goal was to develop a practical screening method to process hundreds of field-collected samples to detect PWN by a PCR-based method using the Baermann extraction technique. Since wood is known to contain PCR inhibitors (Lee and Cooper, 1995; Langrell and Barbara, 2001), an assay was required that was sensitive and sufficiently robust to withstand PCR inhibitors present in wood samples, and thus circumvent the need to further separate and/or culture nematodes prior to PCR analysis. Both of these PCR methods worked well in the presence of potential inhibitors associated with wood after Baermann extraction. Thus the need to isolate pure nematode samples via culturing or microscopy was eliminated. The practical applications of these diagnostic methods for the detection of *B. xylophilus* from actual wood samples of lodgepole pine containing unknown populations of nematodes, rather than pure cultures or individual samples of PWN were demonstrated.

## Conventional End-Point PCR Assay

Samples of lodgepole pine (10 cm sections from tree boles) were ground into wood chips (30 g) using a drill press, and then placed in a Kimwipe-lined Baermann funnel for nematode extraction. Tap water was added to the funnels to completely immerse the wood chips, and the samples were left for 48 h to allow the nematodes to emerge from the wood and into the water. The nematodes in water were then collected into vials and stored at 4 °C. Prior to DNA extraction, the nematodes in 20 ml of water were concentrated by centrifugation at $17,000 \times g$ for 5 min. The bulk of the supernatant was removed and the samples were spun again at $17,000 \times g$ for 5 min to further concentrate the nematodes by removing all but the last 1.0–1.5 ml of fluid. The pellet containing the nematodes was re-suspended in the remaining supernatant, transferred to a 1.5 ml microcentrifuge tube and pelleted again at $12,000 \times g$ for 5 min. The supernatant was carefully removed using a pipet, leaving no more than 50 μl of solution associated with the pellet containing the nematodes. To obtain *B. xylophilus* and *B. mucronatus* DNA, the 50 μl suspension from the nematode concentration step was briefly but thoroughly homogenised with a micro pestle (VWR, Pennsylvania). DNA isolation was carried out using the Dynal DNA Direct Universal kit (Dynal Biotech ASA, Oslo, Norway) according to the manufacturer's recommended protocol and that used by Mota et al. (1999) with the following two modifications: (1) A freeze-fracture step at −80 °C for 30 min was performed after homogenization of the nematodes, followed by centrifugation at $3,000 \times g$ for 2 min. The recovered supernatant was then added to the magnetic beads, and (2) following elution of the DNA from the beads, samples were ethanol precipitated in 2 volumes 95% Ethanol, 0.1 volume 3 M NaOAc, pH 5.4, in the presence of 1 μl GlycoBlue (Ambion), pelleted, washed in 70% ethanol, and resuspended in 20–50 μl of 10 mM Tris, pH 8.

### *End-Point PCR Parameters*

Optimized and reproducible PCR amplifications were performed using 2–5 μl of total DNA extract from wood samples. The PCR reaction contained 1X PCR buffer (20 mM Tris, pH 8.3, 50 mM KCl), 0.2 mM dNTP, 2.5 mM $MgCl_2$, 0.4 μM forward and reverse primers, and 1.25 units Fast Start *Taq* DNA polymerase (Roche) in a total reaction volume of 25 μl. PCR parameters were as follows: initial denaturation at 95 °C (10 min) to activate the *Taq* polymerase, followed by 95 °C (20 s), 60 °C (20 s), 72 °C (20 s) for 40 cycles, and a final extension at 72 °C (7 min).

## Design of Hsp 70 PCR Primers

Multiple nucleotide sequence alignments were generated using 23 isolates from North America, Japan and Europe (data not shown). A comparison of the consensus sequence from *B. xylophilus* and *B. mucronatus* is shown in Fig. 1. From the alignment, it was possible to identify conserved and non-conserved regions that could potentially be used to design universal and specific PCR primers for *B. xylophilus* and *B. mucronatus*. Three different primer pairs were designed based on this alignment. Primer pair Bm702F/Bx704R was used to universally amplify a 210 bp fragment from both *B. xylophilus* and *B. mucronatus*. Primer pairs Bx701F/R and Bm701F/R were designed to specifically amplify from *B. xylophilus* and *B. mucronatus*, respectively (sequences in Fig. 1). Purified genomic DNA from control isolates of *B. xylophilus* and *B. mucronatus* was amplified using the Bx702F/Bm704R primers, and the fragments cloned into separate plasmid vectors. Fragments were sequenced to confirm the identity of the nematode DNA. Nucleotide sequences of plasmid inserts containing the Hsp 70 fragment, amplified from the *B. xylophilus* US isolate, were 100% identical to the consensus sequence of *B. xylophilus* isolates shown in Fig. 1 (data not shown). Likewise, sequences obtained from *B. mucronatus* isolate DE-18 were identical to the consensus sequence of *B. mucronatus* also shown in Fig. 1. Two clones, pBx2-1 and pBm2-2, corresponding to *B. xylophilus* and *B. mucronatus* Hsp 70 genomic sequences, respectively were used in further experiments.

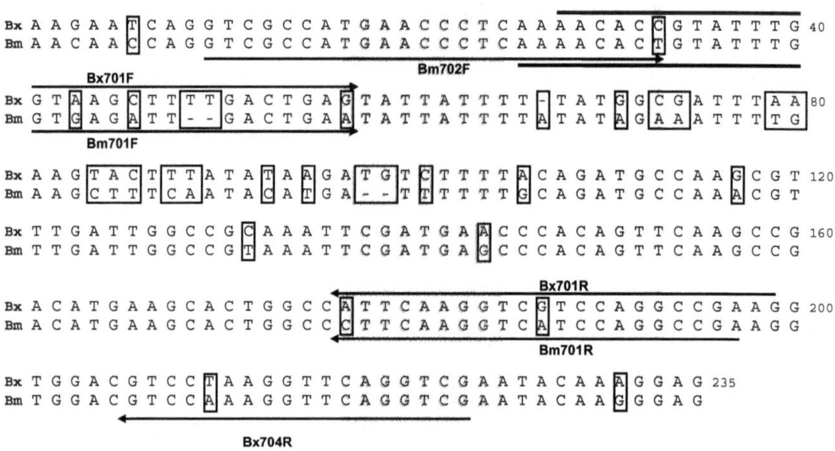


**Fig. 1** Alignment of Hsp70A consensus sequences from *B. xylophilus* (Bx) and *B. mucronatus* (Bm) isolates. Primer annealing sites are represented by arrows above (Bx primers) and below (Bm primers) the alignment. Nucleotide sequence differences are boxed. Dashes indicate gaps in the corresponding sequence and were added in for the alignment

## B. xylophilus *and B. mucronatus Primer Pair Specificity*

Due to the significant amount of nucleotide sequence difference between the Hsp 70 DNA sequences of *B xylophilus* and *B. mucronatus* (Fig. 1), it was possible to design specific primers for use in PCR amplification assays for each of these nematodes. Primer pair Bx701F/Bx701R designed to the *B. xylophilus* consensus sequence, amplified a DNA fragment of 171 bp from all six control *B. xylophilus* isolates available (Fig. 2), but not from the *B. mucronatus* isolate. Primer pair Bm701F/Bm701R designed for the *B. mucronatus* consensus sequence amplified a fragment of 168 bp from the *B. mucronatus* isolate, but not from the other six *B. xylophilus* isolates in PCR assays (Fig. 2). Bx701F and Bm701F are located in similar regions of the Hsp 70 genomic DNA sequence of *B. xylophilus* and *B. mucronatus*, but span a region that contains 6 nucleotide differences. Likewise, primers Bx701R and Bm701R span similar regions that contain 2 nucleotide mismatches. It is likely that these differences account for the specificity of the primer pairs for either *B. xylophilus* or *B. mucronatus*.

## *PCR Sensitivity and Robustness of PCR Assay*

Using the cloned Hsp 70 DNA sequences from *B. xylophilus* and *B. mucronatus* it was possible to optimize and evaluate the performance of the PCR assays. Plasmid pBx2-1 was serially diluted from 1 pg to 0.1 fg and assayed in triplicate by PCR using the primer pair Bx701F/R. All dilutions could be amplified including the lowest dilution which corresponds to approximately 23 copies of target DNA (Fig. 3A). Attempts were also made to determine the robustness of the PCR assay

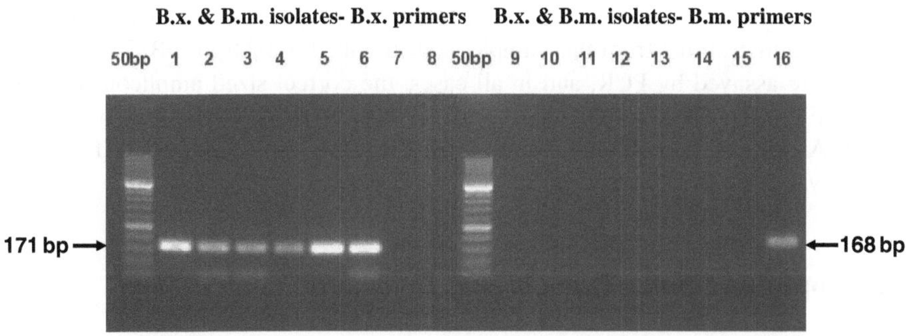


**Fig. 2** *B. xylophilus* and *B. mucronatus* primer pair specificity. Lanes 1–8 correspond to PCR amplification using the *B. xylophilus* specific primer pair (Bx701F/R) and lanes 9–16 correspond to PCR amplification using the *B. mucronatus* specific primer pair (Bm701F/R). Templates used in the PCR amplifications are as follows: *B. xylophilus* isolates: US-DE-2 (lanes 1 and 9), Ne15/03 (lanes 2 and 10) Q1426 (lanes 3 and 11), Q52-A (lanes 4 and 12), Ne12/02 (lanes 5 and 13) PT-3 (w) (lanes 6 and 14), no template control (lanes 7 and 15), and *B. mucronatus* isolate: DE-18 (lanes 8 and 16)

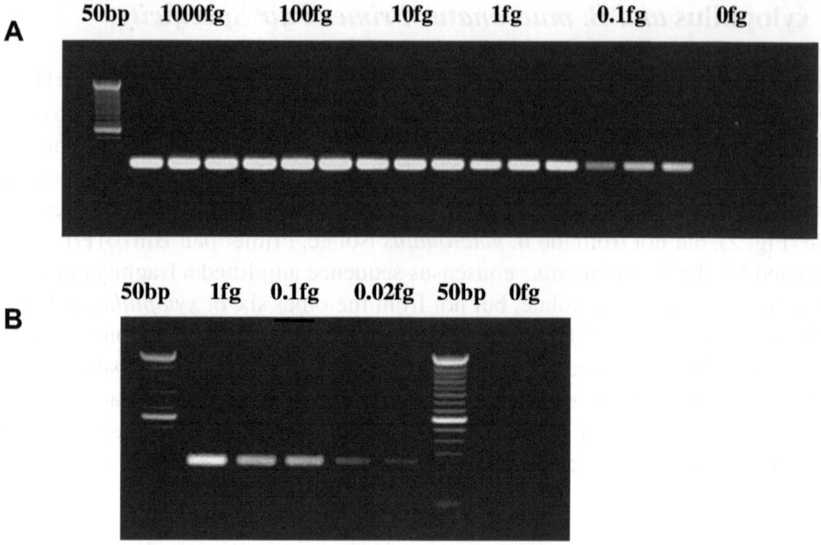


**Fig. 3** PCR sensitivity and robustness of the Bx701U and Bx701L primers. (**A**) A ten fold dilution series of pBx2-1 plasmid DNA (1 pg–0.1 fg) containing *B. xylophilus* Hsp70 sequences assayed in triplicate by PCR. All reactions containing template DNA amplified a fragment of the correct size (171 bp). No amplification was observed in control reactions that did not contain pBx2-1. (**B**) Two microliters of DNA purified using the Dynal DNA Direct Universal procedure, from Baermann funnel extracts of wood known to be free of *B. xylophilus*, were added to PCR reactions containing 1.0–0.02 fg of pBx2-1 DNA (230 to 4 target copies). The expected sized fragment was amplified in all reactions

in the presence of inhibitors in the DNA extracts obtained from the wood samples. Twenty millilitres of water collected using the Baermann funnel from wood known to be nematode free were extracted, concentrated, and eluted into a final volume of 50 μl. Two microlitres from this sample, spiked with 1–0.02 fg of pBx2-1 plasmid DNA were assayed by PCR, and in all cases, the correct sized amplicon was obtained (Fig. 3B). These results indicate that if there were PCR inhibitors present in the DNA extracts, they did not significantly affect the level of detection (LOD) of the assay.

## B. xylophilus *Primer Pair Level of Detection*

While the genomic copy number for the Hsp 70 gene in *Bursaphelenchus* species is unknown, as is the number of cells/nematode, a single *Bursaphelenchus* nematode would contain considerably more than 23 total copies of the Hsp 70 gene. This would indicate that the assay should be sensitive enough to easily detect the presence of a single nematode. This was confirmed from two-fold serial dilutions of DNA extracted from 50 cultured *B. xylophilus* nematodes (Fig. 4). Serial dilutions of total extracted *B. xylophilus* genomic DNA corresponding to 2, 1 or 0.5

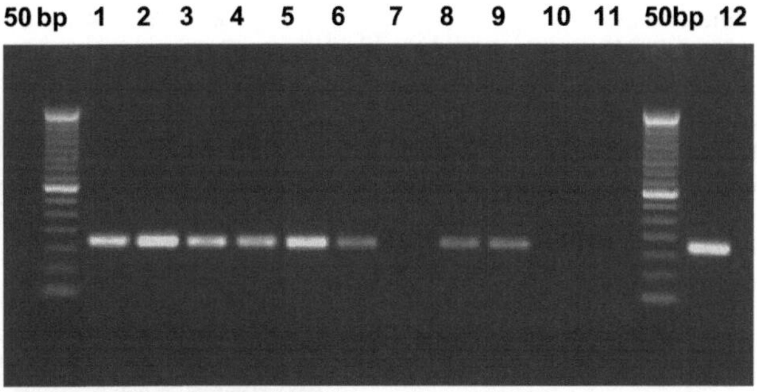


**Fig. 4** *B. xylophilus* primer pair level of detection. Lanes 1–11 represent PCR amplification using the *B. xylophilus* specific primer pair (Bx701F/R). Amount of DNA template used for PCR amplification after serial dilution corresponding to 2 nematodes (lanes 1–2), 1 nematode (lanes 3–4), 0.5 of a nematode (lanes 5–6), 0.25 of a nematode (lanes 7–8), 0.125 of a nematode (lanes 9–10), no template control (lane 11), and 0.25fg of pBx2.1 plasmid (lane 12)

*B. xylophilus* nematodes, were positive in duplicate PCRs. Reactions containing less DNA (equivalent to 0.25 or 0.125 nematode) were positive in only one of two PCRs.

## *PCR Amplification of PWN Nematode DNA from Field Samples*

*B. xylophilus* could be detected from DNA extracted from wood sampled from dead and dying lodgepole pine trees attacked by the mountain pine beetle, *Dendroctonus ponderosae* (Fig. 5). Four of these wood samples were tested, and the expected sized amplification products were obtained in 3 of the 4 samples using the Hsp 70 specific end-point PCR assay. In 2 of the samples, lower molecular weight primer/dimer bands were observed. Identity of the amplified Hsp 70 products from these 3 samples was confirmed by sequencing (data not shown). Modification of the DNA extraction method was not necessary when DNA was extracted from cultured plates of either *B. xylophilus* or *B. mucronatus*. However, for nematodes extracted from wood samples, two extra steps were included for optimum DNA purification results. It is likely that these extra steps helped to remove inhibitors co-extracted with the wood fibres that were unavoidably present when using the Baermann extraction procedure.

## Real-Time PCR Assay

Nematode samples analysed in the real-time PCR assays were also recovered from wood samples of lodgepole pine as described above. DNA from these nematodes was extracted according to the method described in Burgermeister et al. (2005) with the following changes: incubation of sample homogenate was performed at 56 °C

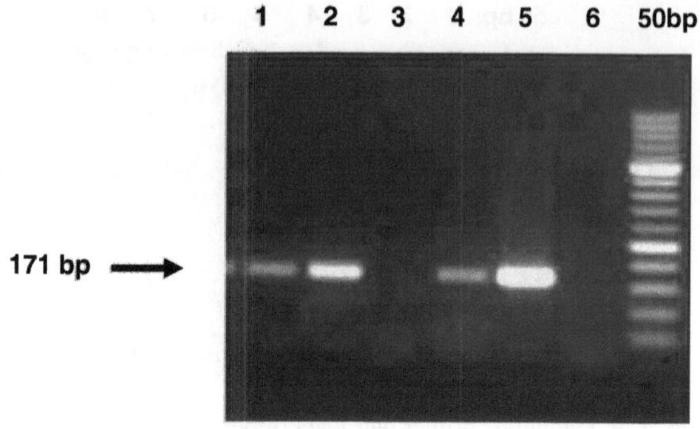


**Fig. 5** PCR amplification of PWN nematodes from wood samples using Hsp70 primers. Lane 1, wood sample 901G; lane 2, wood sample 878C; lane 3, wood samples 858G; lane 4, wood sample 883C; lane 5, *B. xylophilus* genomic DNA, and lane 6, no template control

overnight, instead of 3 h, and carrier RNA was only used when DNA was extracted from single nematodes. Samples were eluted in 20 μl (for individual nematodes), and 50 μl (for the rest of samples) using 10 mM Tris, pH 8. After elution from the mini columns, the samples were heated at 56 °C for 5 min, which resulted in improved amplification possibly due to further removal of residual ethanol from the sample DNA extracts which can inhibit PCR amplification.

## *Primers and Taqman® Probe Design*

The *B. xylophilus* Hsp 70 gene was selected as the target sequence for the design of a specific set of real-time PCR primers and a Taqman® fluorescent probe. Based on the initial sequence alignment, the regions that were most divergent between *B. xylophilus* and *B. mucronatus* were also AT rich. Consequently, the design of real-time primers and probes within these regions, with acceptable melting temperatures (Tm), would result in oligonucleotides of excessive length and poor specificity (data not shown). An alternative was to incorporate locked nucleic acid (LNA) nucleotides into the design to decrease the size of the primers and probe while increasing the Tm and the specificity of the assay.

The LNA base is a bicyclic RNA analogue which, when incorporated into a primer or probe, increases thermal duplex stability of the oligonucleotide resulting in a higher Tm and an greater specificity of hybridization to its target sequence (Demidov, 2003). The Taqman® fluorescent probe has the following sequence in which lower case letters indicate LNA nucleotides: 5′-ATtGGcCGCAAATTcGAtG AAcC-3′ (nt124 to nt146) and the primer set, forward primer: 5′-TAAGATGTcTTTt AcAGATGcCAAG-3′ (nt93 to nt117), and reverse primer: 3′-CCGgACCTGCTgGA ACTTA-5′ (nt195 to nt177). This specific set of real-time PCR primers and the Taqman® probe amplify and detect a 102 bp fragment from the same gene sequence shown in Fig. 1.

## *Specificity, Sensitivity of Real-Time PCR Assay, and Amplification of DNA from Individual Pinewood Nematodes*

Potential problems encountered with PCR inhibitors present in wood were reduced by extracting DNA using a micro version of the silica based membrane adsorption method (Burgermeister et al., 2005). DNA extraction using this method was found to be an improvement in terms of eliminating PCR inhibitors present in wood to the one used for end-point PCR with magnetic beads (data not shown). The specificity of the assay was confirmed using DNA extracted from 5 isolates of *B. xylophilus* (Bx), two isolates of *B. mucronatus* (Bm), and one isolate of each of the following species: *B. conicaudatus, B. doui* (Bd), *B. fraudulentus* (Bf), and *B. singaporensis* (Bs), *B. hofmanni* (Bh), and *B. thailandae* (Bt). DNA from all *B. xylophilus* isolates could be amplified while DNA extracted from all other seven species tested failed to give an amplification signal as shown in Fig. 6. This Figure shows single replicates, but the assay was performed in duplicate for all samples. Sensitivity of the real-time PCR assay was evaluated from a ten-fold dilution of *B. xylophilus* genomic DNA, starting with a final template amount of 50 ng down to 0.005 ng. A standard curve was plotted from the linear regression of the logarithmic amount of *B. xylophilus* DNA vs. the Ct values. The reaction was linear over a dynamic range of 50 ng – 5 pg with a correlation coefficient, $r = -1.00$, which represents six replicates at each dilution. The PCR efficiency of the assay, calculated from the slope of the linear regression curve using the equation $E = 10^{-1/\text{slope}}$ was 1.92 (Fig. 7). The lowest amount of template DNA analyzed, 5 pg, resulted in amplification in all six replicates and was therefore defined as the LOD for this assay. Agarose gel electrophoresis of the real-time PCR products indicated that all reactions containing

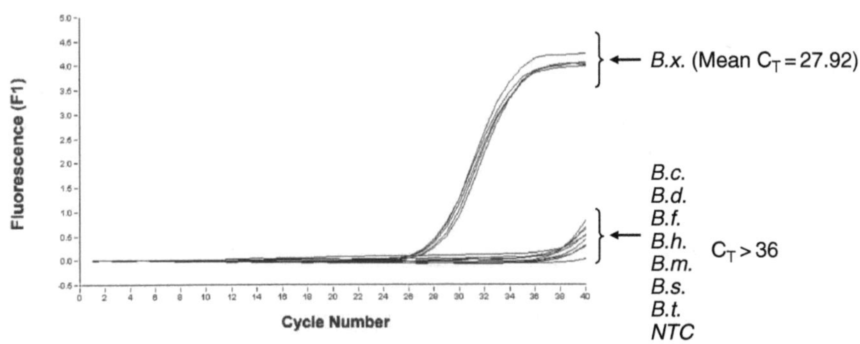


**Fig. 6** Amplification plot of real-time PCR assay showing specificity for detection of *B. xylophilus*. The amplifications were carried out with 5 ng of DNA template from 5 isolates of *B. xylophilus* (Bx): Bx(2), PT3, Q52, Ne 15, and Ne 12, and two isolates of *B. mucronatus* (Bm): DE 30, and DE 18, and one of *B. conicaudatus* (Bc): Ne5b; *B. doui* (Bd): Ne26; *B. fraudulentus* (Bf): DE10: *B. singaporensis* (Bs): Ne7; *B. hofmanni* (Bh); DE6w; *B. thailandae* (Bt): Kr2w, and NTC (no template control)

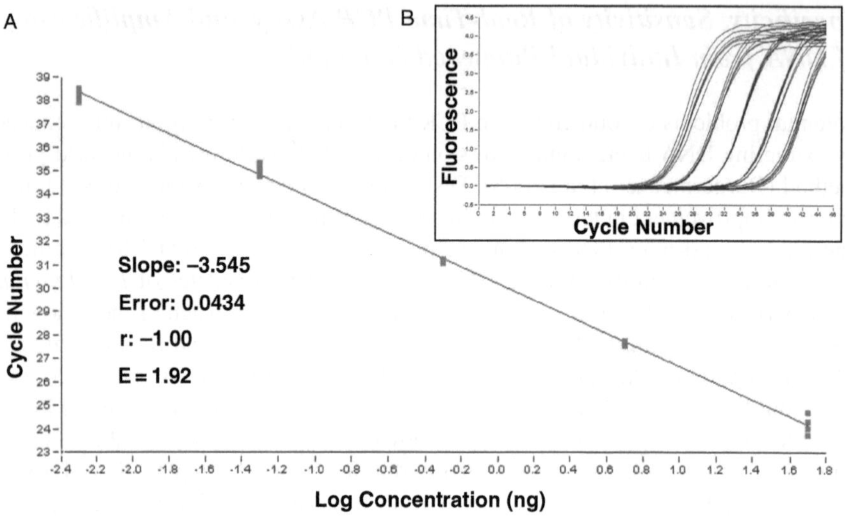


**Fig. 7** Sensitivity and linearity of the real-time PCR assay. (**A**) Standard curve for 10-fold serial dilution of *B. xylophilus* genomic DNA, based on the calculated serial dilution points and corresponding Ct values from six replicates. Standard curve parameters: slope, −3.545; intercept, 32.65; error, 0.0434; efficiency, 1.92, and r, −1.00. (**B**) Amplification plot of same experiment showing the ten fold dilution series of DNA concentrations (50 ng–5 pg) replicated 6 times. No amplification was observed in control reactions that did not contain *B. xylophilus* genomic DNA

*B. xylophilus* template DNA amplified a fragment of the correct size, 102 bp (data not shown). We have also shown that this real-time PCR assay can amplify DNA from single nematodes. We carried out ten-fold serial dilutions of DNA extracted from 100 and 10 cultured *B. xylophilus* nematodes. Serial dilutions of the total extracted genomic DNA corresponding to ten and one *B. xylophilus* nematodes were positive in all 4 replicates at each dilution, indicating that detection of a single *B. xylophilus* nematode is possible. This was confirmed by performing the Hsp 70 real-time PCR assay with DNA extracted directly from 8 individual *B. xylophilus* nematodes. The mean Ct value of individual nematodes was 33.7 with a standard deviation of 0.77 (Fig. 8).

## *Evaluation of Field Sample Extracts for Potential PCR Inhibition*

Field sample extracts were evaluated for the presence of potential PCR inhibitors. Twenty millilitres of water collected using the Baermann funnel from wood known to be nematode free were extracted, concentrated, and eluted into a final volume of 50 μl. To determine whether this extract contained PCR inhibitors, 5 μl of this wood extract was spiked into a sample containing 5 ng purified *B. xylophilus* DNA and compared to a non-spiked reaction. Both spiked and non-spiked real-time PCR assays, produced detectable fluorescence and amplified with equal efficiency resulting in comparable Ct values (Fig. 9).

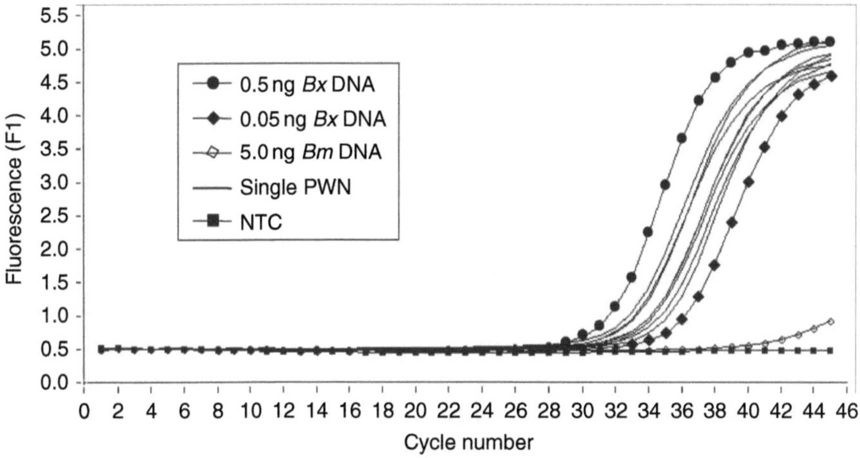


**Fig. 8** Real-time PCR amplification plot of DNA from 8 different single *B. xylophilus* nematodes (*solid lines*) flanked by 0.5 ng and 0.05 ng of purified *B. xylophilus* DNA (*circle and diamond line points*), respectively. A single reaction using 5 ng of purified *B. mucronatus* DNA (*open diamond line points*) was also performed. One microgram of carrier RNA and a NTC, no template control (*square line points*) were included as negative controls

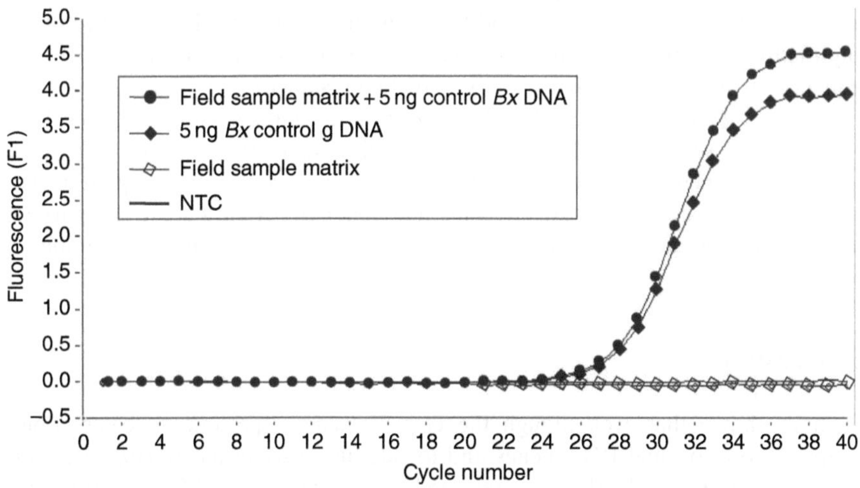


**Fig. 9** Evaluation of field sample extracts for potential PCR inhibition. Two microlitres of purified DNA from Baermann funnel extracts of wood known to be free of *B. xylophilus* were added to 5 ng of *B. xylophilus* DNA. The expected amplification curve was observed in the template sample

## *Real-Time PCR Assay Using Lodgepole Pine Wood Samples*

The sensitivity of the real-time PCR assay was evaluated using wood sampled from dead and dying lodgepole pine trees attacked by mountain pine beetle,

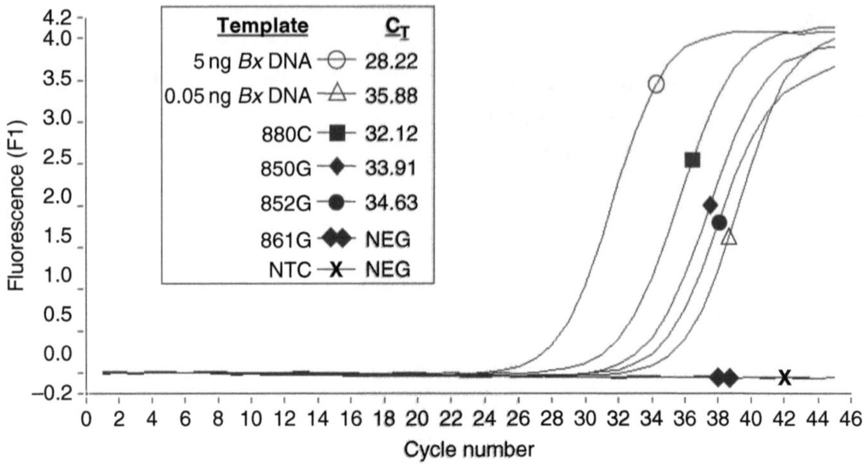


**Fig. 10** Real-time PCR amplification of PWN nematode DNA extracted from wood samples flanked by genomic DNA standards (5 ng and 0.05 ng) using Hsp70 primers and Taqman® fluorescent probe. Five microlitres of undiluted field sample extract was used as template

*D. ponderosae*. Four trees were selected for testing, and 3 out of the 4 samples produced detectable fluorescence and DNA amplification (Fig. 10). Identity of the amplified fragments was confirmed to be the expected *B. xylophilus* Hsp 70 gene sequence by nucleotide sequence analysis (data not shown).

Cao et al. (2005) also detected PWN from field samples using real-time PCR based on ITS sequences. However, after the Baermann funnel extraction, these authors performed an additional and traditional time-consuming step of isolating nematodes using a microscope. This additional step would serve the purpose of separating the nematodes from wood fibres that contain compounds inhibitory to PCR.

## Conclusion

We have shown that even though the Hsp 70 gene is present in fewer copies compared to ribosomal RNA genes and to satellite DNA in the *Bursaphelenchus* genome, both Hsp 70 PCR assays reported here (conventional PCR and real-time PCR) are very robust and sensitive assays that work well in the presence of potential PCR inhibitors associated with wood. These protocols eliminate the need to produce clean nematode samples through further culturing and/or to isolate nematodes for examination with a microscope.

In addition, real-time PCR can accurately quantify *B. xylophilus* DNA from wood samples, making it a very useful and practical method to detect the presence of *B. xylophilus*. The Baermann funnel extraction method is particularly useful, because it allows live nematodes to be extracted from wood samples. When molecular techniques are used for quarantine purposes to detect PWN in wood products, it is critical that a distinction between live and dead nematodes be made.

**Acknowledgments** We are grateful to Dr. Thomas Schröder, for providing us with isolates of *Bursaphelenchus xylophilus* and *Bursaphelenchus mucronatus*, and to Dr. Kai Metge for providing us with DNA samples of *B. conicaudatus, B. doui, B. fraudulentus, B. singaporensis, B. hofmanni, and B. thailandae*. Both above-mentioned scientists are from the Federal Biological Research Centre for Agriculture and Forestry-Department for National and International Plant Health, Germany. We are also thankful to Brett Foord and Grace Ross for their technical assistance. This study was supported by a grant from The Mountain Pine Beetle Initiative, Natural Resources Canada.

## References

Braasch, H., Burgermeister, W. and Pastrik, K.H. (1995). Differentiation of three *Bursaphelenchus* species by means of RAPD PCR. *Nachrichtenbatt des Deutschen Pflanzenschutzdienstes* 47: 310–314.

Braasch, H., Metge, K. and Burgermeister, W. (1999). *Bursaphelenchus* species (Nematoda, Parasitaphelenchidae) found in coniferous trees in Germany and their ITS-RFLP patterns. *Nachrichtenbatt des Deutschen Pflanzenschutzdienstes* 51: 312–320.

Burgermeister, W., Metge, K., Braasch, H. and Buchbach, E. (2005). ITS-RFLP patterns for differentiation of 26 Bursaphelencus species (Nematoda: Parasitaphelenchidae) and observations on their distribution. *Russian Journal of Nematology* 13: 29–42.

Cao, A.X., Liu, X.Z., Zhu, S.F. and Lu, B.S. (2005). Detection of the pinewood nematode, *Bursaphelenchus xylophilus*, using a real-time polymerase chain reaction assay. *Phytopathology* 95: 566–571.

Castagnone, C., Abad, P. and Castagnone-Sereno, P. (2005). Satellite DNA-based species-specific identification of single individuals of the pine wood nematode, *Bursaphelenchus xylophilus* (Nematoda: Aphelenchoididae). *European Journal of Plant Pathology* 112: 191–193.

Demidov, V.V. (2003). PNA and LNA throw light on DNA. *Trends in Biotechnology* 21: 4–7.

Dwinell, L.D. (1997). The pinewood nematode: Regulation and mitigation. *Annual Review of Phytopathology* 35: 153–166.

Evans, H.F., McNamara, D.G., Braasch, H., Chadoeuf, J. and Magnusson, C. (1996). Pest risk analysis (PRA) for the territories of the European Union (as PRA area) on *Bursaphelenchus xylophilus* and its vectors in the genus *Monochamus*. *EPPO Bulletin* 26: 199–240.

Hoyer, U., Burgermeister, W. and Braasch, H. (1998). Identification of Bursaphelenchus species (Nematoda: Aphelenchoidae) on the basis of amplified ribosomal DNA (ITS RFLP). *Nachrichtenbatt des Deutschen Pflanzenschutzdienstes* 50: 273–277.

Irdani, T., Marinara, A., Bogani, P., Ambrogioni, L., Caroppo, S. and Buiatti, M. (1995). Molecular diversity among pine wood *Bursaphelenchus* populations detected by RAPD analysis. *Redia* 78: 149–161.

Iwahori, H., Kanzaki, N. and Futai, K. (2000). A simple, polymerase reaction–restriction fragment length polymorphism-aided diagnosis method for pine wilt disease. *Forest Pathology* 30: 157–164.

Kang, J.S., Choi, K.S, Shin, S.H., Moon, I.S., Lee, S.G. and Lee, S.H. (2004). Development of an efficient PCR-based diagnostic protocol for the identification of the pinewood nematode, *Bursaphelenchus xylophilus* (Nematoda, Parasitaphelenchidae). *Nematology* 6: 279–285.

Konstantopoulou, I., Nikolaidis, N. and Scouras Z.G. (1998). The hsp70 locus of *Drosophila auraria* (montium subgroup) is single and contains copies in a conserved arrangement. *Chromosoma* 107: 577–586.

Langrell, S.R.H. and Barbara, D.J. (2001). Magnetic capture hybridisation for improved PCR detection of *Nectria galligena* from Lignified Apple Extracts. *Plant Molecular Biology Reporter* 19: 5–11.

LEAL, I, GREEN, M., ALLEN, E., HUMBLE L. AND ROTT, M. (2005). An effective PCR-based diagnostic method for the detection of *Bursaphelenchus xylophilus* (Nematoda: Aphelenchoididae) in wood samples from lodgepole pine. *Nematology* 7: 833–842.

LEAL, I., GREEN, M., ALLEN, E., HUMBLE L. AND ROTT, M. (2007). Detection of pinewood nematode, *Bursaphelenchus xylophilus* (Nematoda: Aphelenchoididae) using a Real-Time PCR assay in wood samples from lodgepole pine. *Nematology* 9: 351–362.

LEE, A.B. AND COOPER, T.A. (1995). Improved direct PCR screen for bacterial colonies-wooden toothpicks inhibit PCR amplification. *Biotechniques* 18: 225–226.

LIEBHOLD, A.M., MACDONALD, W.M., BERGDAHL, D. AND MASTRO, V.C. (1995). Invasion by exotic forest pests: A threat to forest ecosystems. *Forest Science Monographs* 30: 49 pp.

MAMIYA, Y. AND ENDA, N. (1979). *Bursaphelenchus mucronatus* n.sp. (Nematoda: Aphelenchoidae) from pine wood and its biology and pathogenicity to pine trees. *Nematologica* 25: 353–361.

MATSUNAGA, K. AND TOGASHI, K. (2004). A simple method for discriminating *Bursaphelenchus xylophilus* and *B. mucronatus* by species-specific polymerase chain reaction primer pairs. *Nematology* 6: 273–277.

MIREKU, E. AND SIMPSON, J.A. (2002). Fungal and nematode threats to Australian forests and amenity trees from importation of wood and wood products. *Canadian Journal of Plant Pathology* 24: 117–123.

MOTA, M.M., BRAASCH, H., BRAVO, M.A., PENAS, A.C., BURGERMEISTER, W., METGE, K. AND SOUSA, E. (1999). First report of *Bursaphelenchus xylophilus* in Portugal and in Europe. *Nematology* 1: 727–734.

NIKOLAIDIS, N. AND SCOURAS Z.G. (2002). A polymerase chain reaction (PCR) application for free-living nematodes (Rhabditida) discrimination. *Molecular Ecology Notes* 2: 248–249.

# Part IV
# The Insect Vectors: Biology and Ecology

**Marc Linit and Süleyman Akbulut**

## Summary

The association of the pinewood nematode with insects in the genus *Monochamus* (Cerambycidae) appears to have a long evolutionary history but it was not described until the widespread occurrence in Japan of nematode associated tree death now known as pine wilt. Japanese scientists discovered the association around 1900 in response to the pine wilt epidemic in their pine forests. Since then much research on the association between the nematode and its insect vectors has followed the interest in pine wilt in North America and most recently in Europe. The association is believed to be long standing in North America where the nematode survived as a secondary pathogen on stressed pines. It's emergence as a primary pathogen coincided with the widespread distribution of exotic and hence susceptible pine species such as *Pinus sylvestris*.

Japanese scientists documented the community of bark and phloem inhabiting insects associated with the nematode in dying trees and determined which of these associates were vectors of the nematode. Further studies documented the life cycle of the nematode and its vector, the number of nematodes carried by vectors, the methods of nematode transmission to new host trees, and other insect related aspects of the pine wilt disease cycle. When the pinewood nematode was identified in Columbia, Missouri, USA in 1979, similar studies were undertaken to determine if the nematode-beetle relationships were similar to those reported from Japan.

The interspecific association between the pinewood nematode and its insect vectors was determined to be similar in North America and in Japan and is an obligatory component of the disease cycle. Transmission of the nematode from nematode-infested pines to healthy, non-infested pines is believed to be limited to the feeding and oviposition activities of *Monochamus* beetles. Thus an understanding of this relationship as well as the population ecology of the beetle is essential.

*Monochamus* alternatus is the principal vector of the pinewood nematode in Japan and other parts of eastern Asia. *M. carolinensis* is the principal vector in the midwestern United States. The nematode is also vectored by congeneric species in North America and eastern Asia. The basic relationship between the nematode vector species of *Monochamus* is consistent although certain aspects of beetle biology such as voltinism and the presence or absence of diapause alter aspects of the seasonality of the disease cycle. What is consistent is that fourth-stage dispersal juveniles (JIV) of the nematode enter the tracheal system of a newly eclosed adult beetle prior to the beetle's emergence from the tree in which it developed. Beetles transmit JIVs to healthy trees through feeding wounds and to stressed or dying trees through oviposition wounds. The nematode is the causal agent of pine wilt in parts of North America and eastern Asia where susceptible pine species occur.

The discovery of the pinewood nematode in the Setubal region of Portugal in May 1999 necessitated research to establish the basic biological relationships between the nematode and its vectors in a new ecological region of the globe. Much of the information presented in Part IV addressed ongoing efforts in Portugal to understand the biology and ecology of *Monochamus galloprovincialis* the insect assosicated the pinewood nematode-killed trees. Elsewhere in Europe research was undertaken to identify insects associated with stressed and dying pines and document the distribution of *Bursaphelenchus* species.

Although the nematode has not been document in Europe outside of Portugal it is important for potential host countries to determine the community of insects likely to interact with the nematode. Robertson et al. presented a list of insects associated with declining pines in Spain. These included species of *Monochamus*, other cerambycids, buprestids, scolydids and curculionids. This community was similar to those described from declining pines in Asia and North America. *M. galloprovincialis* is the vector of pinewood nematode in Portugal and is assumed to be the principle potential vector elsewhere on the continent. Naves et. al. presented results of a series of studies on *M. galloprovincialis* that addressed the beetle's longevity, feeding behavior, fecundity, oviposition rate and oviposition preferences. Tomiczek and Tomiczek described similar studies of *M. galloprovincialis* in Austria. These studies revealed that the overall ecology and reproductive biology of *M. galloprovincialis* is similar to that describe elsewhere for *M. alternatus* and *M. carolinensis*. It is apparent from these studies that *M. galloprovincialis* serves as an ecological equivalent of *Monochamus* species in Asia and North America that vector the nematode.

Other species of *Bursaphelenchus* are carried by *Monochamus* beetles. Lieutier et al. reported the occurrence of *B. mucronatus* in France and its association with *M. galloprovincilais*. Their interest in *B. mucronatus* and other nonpathogenic nematodes in the genus *Bursaphelenchus* was to analyze the risk of propagation and the routes of dissemination in the event of an introduction of *B. xylophilus*. He found *B. mucronatus* in several locations in France; other *Bursaphelenchus* species found included *B. hellenicus, B. leoni* and *B. sexdentati*. *B. xylophilus* was not found but clearly a vector system already exists should it become established.

Genetic studies on the pinewood nematode and closely related species have been undertaken by several researchers; however, they have not been commonly applied to the insect vectors of the nematode. Two interesting studies were presented in this session. Roux-Morabito et al. used genetic molecular techniques to study the geographic distribution and host range of *M. galloprovincialis* in France. They found a low genetic variability among populations across France and revealed no clear population differentiation patterns influenced by host plant or geographic isolation suggesting there has been extensive gene flow among populations. Shoda-Kagaya reported on the genetic structure of *M. alternatus* in northern Honshu, Japan. She used mitochondrial DNA to compare the genetic structure of apparently isolated populations from Akita and Iwate prefectures. Microsatellite markers revealed a strong genetic divergence between the two populations. These locations are currently the most northern areas of pine wilt disease in Japan and the strong population structure is hypothesized to have occurred because of different population dispersal routes separated by a mountain range.

# Biology Studies Relevant to the Vector Role of *Monochamus* Species for PineWood Nematode

**Christian Tomiczek and Ute Hoyer-Tomiczek**

**Abstract** The pinewood nematode (PWN), *Bursaphelenchus xylophilus*, is transferred by an insect vector to pine trees. In countries, where the PWN occurs, *Monochamus* spp. plays the most important role as vector of the nematode. Within the context of the EU-Project "PHRAME" *M. galloprovincialis pistor, M. sutor* and *M. sartor* were collected in different areas of Austria, cultured and reared in the lab. Lifespan, breeding and emerging behaviour as well as maturation feeding were studied for all three *Monochamus* species. The consequences from these results for the control measures are discussed.

## Introduction

The pinewood nematode (PWN) needs a vector to be transferred to pine trees. In countries, where *Bursaphelenchus xylophilus* occurs, *Monochamus* spp. plays the most important role as vector of the nematode. Therefore specific knowledge of the distribution and biology of various *Monochamus* species is important to improve control measures and indicate risk areas for possible epidemic spread of PWN.

## Material and Methods

Culturing of stem pieces of *Pinus sylvestris* with symptoms of long horned beetle attack was one method of obtaining *Monochamus* species for biological studies. Therefore stem pieces of an 80–100 years old *P. sylvestris* with obvious *Monochamus* attack, originating from a southern Austrian province and cut in Spring, were stored in outdoor insectariums until June, afterwards in the lab for

C. Tomiczek
Federal Research & Training Centre for Forests, Natural Hazards & Landscape (BFW), Department for Forest Protection, Seckendorff-Gudent-Weg 8, A-1131 Vienna, Austria
e-mail: christian.tomiczek@bfw.gv.at

M.M. Mota, P. Vieira (eds.), *Pine Wilt Disease: A Worldwide Threat to Forest Ecosystems*,  

emerging observation. From a *P. sylvestris* stand in bad conditions in a north-east Austrian province stem pieces from felled trees, cut in summer, were taken to the lab, over wintered in outdoor insectariums and stored in the lab from following May for observation of the emerging phase. In both cases attack by *M. galloprovincialis pistor* was suspected. To get *M. sartor* and *M. sutor* specimens for biological studies, beetles were caught from trap trees or during flight in a mountain region of Lower Austria.

For studying the biology of the different *Monochamus* species in Austria, the collected and emerged *Monochamus* beetles were cultured in cages in the lab. The beetles were caged together with feeding plants, branches and breeding material. Seedlings and branches of older *Pinus sylvestris* were offered as feeding material to *M. galloprovincialis pistor, P. sylvestris* seedlings and branches of *Picea abies* to *M. sutor* and *M. sartor*. Breeding material was *Pinus sylvestris* in five different diameters (13–15 cm, 9–13 cm, 6–9 cm, 4–6 cm, 3–4 cm) for *M. galloprovincialis pistor, Picea abies* of 13–15 cm and 6–9 cm in diameter for *M. sutor* and *M. sartor*. Altogether 43 couples and additional 41 single beetles of three *Monochamus* species were cultured within 3 years. After the end of the egg laying period, the breeding material was stored in the lab until November, in outdoor insectariums until following May and then again in the lab. The emerging period was observed and controlled in the lab.

## Results

### *Collecting of* Monochamus *Specimens for Biological Studies*

From the 80–100 years old *P. sylvestris* originating from the southern Austrian province, long horned beetles emerged from two of the seven stem pieces during the time period 12.07–26.07.2004 (7 males: 7 females = 1:1 ratio). The beetles were identified as *M. galloprovincialis pistor* according to morphological and, after death, molecular biological features.

Culturing of the stem pieces of *P. sylvestris*, originating from a *Pinus* stand in bad conditions in a north-east Austrian province, was less successful than expected due to the very intensive larvae activity over several months. Out of seven of the thirteen stem pieces altogether 16 *M. galloprovincialis pistor* beetles (3 males: 13 females ≈ 1:4 ratio) emerged mainly in June. Compared with the activity of larvae the amount of emerged beetles was surprisingly low. One explanation could be the very strong winter 2005/2006 with very low temperatures over a long time period. The end control of these *P. sylvestris* stem pieces by chopping into small pieces and strips resulted in further four living and five dead *Monochamus* larvae.

Collecting of *Monochamus* species by trap trees or during flight was very successful in the mountain region in Lower Austria of an altitude of 600–1200 m within the time period July to September of 4 years.

The following *Monochamus* specimens were caught:

| *Monochamus sutor* | *Monochamus sartor* | *M. galloprovincialis pistor* |
|---|---|---|
| 2003 : 5♂ + 3♀ = 8 | 2003 : 1♂ + 2♀ = 3 | |
| 2004 : 1♂ + 8♀ = 9 | 2004 : 2♂ + 1♀ = 3 | |
| 2005 : 4♂ + 3♀ = 7 | 2005 : 6♂ + 5♀ = 11 | |
| 2006 : 14♂ + 5♀ = 19 | 2006 : 7♂ + 2♀ = 9 | 2006 : 1♂ + 0♀ = 1 |

## *Breeding and Emerging Behaviour of* Monochamus *sp.*

Culturing of *Monochamus* sp. couples with feeding branches and breeding material was carried out as followed over 3 years:

| *M. galloprovincialis pistor* | *M. sutor* | *M. sartor* |
|---|---|---|
| 2004: 7 couples | 1 couple, 7♀ | 1 couple, 1♂ |
| 2005: 14 couples | 2 couples | 5 couples, 2♂, 1♀ |
| 2006 : 10 couples*, 8♀ | 2 couples, 12♂, 3♀ | 1 couple, 6♂, 1♀ |
| (* : 4♂ and 2♀ "double used") | | |

In the year 2004 seven couples of *M. galloprovincialis pistor* were cultured as mentioned above. Breeding material was *P. sylvestris* in four different diameters (9–13 cm, 6–9 cm, 4–6 cm, 3–4 cm). After the end of egg laying in 2004 the breeding wood was stored in the lab until November 2004, in outdoor insectariums until following May 2005 and then again in the lab. The emerging period was end of May until beginning of September 2005 with two peaks in June and July. Altogether 73 beetles emerged with a relation of 1:2 (males:females) (28:45). The preferred dimension of breeding wood was the 6–8 cm in diameter, followed by 9–12 cm in diameter. The end control of the breeding wood of one *M. galloprovincialis pistor* couple by chopping into small pieces and strips in November 2005 resulted in five living larvae which were transferred into artificial diet, three dead larvae and one dead beetle. Therefore, it seems that not all larvae completed their development within 1 year and stayed in the wood over the second winter, indicating the possibility of a 2-years-developing-period of *M. galloprovincialis pistor*, at least in colder regions. This would require adapted control measures of the PWN-vector. For this reason the breeding wood of the remaining six couples was stored again over winter in outdoor insectariums until May of the following year 2006. In the second year of development no beetles emerged from the wood, but two of the five larvae in artificial diet finished their development and emerged. The other three larvae died due to vitality decrease of unknown reasons. The breeding wood was cut and chopped finally in September 2006 for end controlling. All development stages were found. Altogether six dead larvae, two dead pupae, and seven dead beetles were detected. This indicates that most of the individuals staying over two winters finished their development in the second year, but failed to emerge finally.

In 2005, fourteen couples of *M. galloprovincialis pistor*, four couples of *M. sartor* and one couple if *M. sutor* were cultured as mentioned above. Breeding material was *Pinus sylvestris* in five different diameters (12–15 cm, 9–12 cm, 6–9 cm, 4–6 cm, 3–4 cm) for *M. galloProvincialis pistor Picea abies* of two different diameters (6–9 cm, 12–15 cm) for *M. Sartor* and *M. Sutor*. At the end of egg laying in 2005 the breeding wood was stored in the lab until November 2005, in outdoor insectariums until following May 2006 and then again in the lab. During autumn 2005, heavy larvae activity was observed, suspecting a high number of larvae and potentially emerging beetles in the next year. After returning the breeding woods from outdoor insectariums only a low level of larvae activity was detected. From twenty breeding wood pieces of all fourteen *M. galloprovincialis pistor* couples only six beetles (3 males, 3 females) emerged out of six pieces during June and July 2006. The diameters of those pieces were 6–9 cm and 9–12 cm, corresponding to the results of the year before. The end control by chopping and cutting the wood into small pieces and strips in October 2006 revealed no further larvae bored into the wood. Only eight dead larvae were found directly under the bark not being bored into the wood. These results were surprisingly negative because of the high number of emerged beetles in the year before. One explanation could be the very strong winter 2005/2006 with very low temperatures over a long time period.

From the breeding wood of *M. sutor* and *M. sartor* of the year 2004 no beetles emerged in 2005 or in 2006. The end control by chopping the wood revealed only one dead larva per *Monochamus* species. From the five pieces of breeding wood of the four couples of *M. sartor* and one piece of one *M. sutor* couple of the year 2005 no beetles emerged in 2006. The chopping of these breeding woods in October 2006 resulted in four living *M. sartor* larvae in one piece which were transferred to artificial diet, and three dead *M. sartor* larvae out of three pieces and no larvae of the *M. sutor* breeding wood. Because of the living larvae in one *M. sartor* breeding wood another wood piece was cultured further for over-wintering a second time.

## *Lifetime of* Monochamus *Species*

The whole life-span could be investigated in the lab only for *M. galloprovincialis pistor*.

*M. galloprovincialis pistor* (emerged in the lab)

| | 2004 | 2005 | 2006 | |
|---|---|---|---|---|
| ♂: | 57–76 days | 34–94 days | 42–120 days | ► 5–17 weeks |
| ♀: | 53–76 days | 39–96 days | 35*–94 days | |
| | | | * influenced by insecticide | |

For *M. sutor* and *M. sartor* only the living weeks after catch could be noticed:

*M. sutor* (caught in nature in August/September)

| | 2004 | 2005 | 2006 | |
|---|---|---|---|---|
| ♂: | 13 days | 12–15 days | 6–37 days | ► 2–5 weeks in the lab |
| ♀: | 1–34 days | 6–33 days | 7–28 days | |

*M. sartor* (caught in nature in August/September)

| | 2004 | 2005 | 2006 | |
|---|---|---|---|---|
| ♂: | 19–41 days | 8–35 days | 5–19 days | ► 2–6 weeks in the lab |
| ♀: | 12 days | 10–25 days | 6–17 days | |

Thus the life-span of *M. galloprovincialis pistor* beetles is 5–17 weeks. *M. sutor* lived 2–5 weeks and *M. sartor* 2–6 weeks in the lab. Because breeding of *M. sutor* and of *M. sartor* failed in the lab, no exact data on the life span of these two *Monochamus* species are available. It is assumed, that the lifetime does not differ very much and is more or less the same as for *M. galloprovincialis pistor*. Regarding each pair separately, the females died on average 1–2 weeks earlier, probably due to the strength consuming egg laying.

## *Maturation Feeding of* Monochamus *Species*

Another important fact of the vector biology is the maturation feeding of *Monochamus* species, especially its duration and intensity, in view of the transferring capability of nematodes.

### Maturation Feeding of *M. galloprovincialis pistor*

Within the first 2 weeks maturation feeding was very strong. The beetles fed intensively at the basis of the needles causing their complete fall. The feeding was also very strong on the needles themselves and on the bark. If only young *Pinus* seedlings were offered, those were very strongly attacked. If young *Pinus* seedlings and branches of older *Pinus* were offered together, the branches were preferred. After the first 2 weeks the maturation feeding severely decreased. Only some feeding at the basis of the needles and on the bark could be observed. The feeding was preferred at branches of older *Pinus*. Since the third or fourth week until the end of life only slight feeding on the bark took place, but twigs were completely without bark at the end.

### Maturation Feeding of *M. sutor*

*M. sutor* specimens preferred the branches of *P. abies* and fed on the bark. The feeding at the *Pinus sylvestris* seedling was done also on the bark but less intensive and partially at the basis of the needles. There was only little feeding intensity in the cage, because the very strong maturation feeding took place outdoor before catch.

### Maturation Feeding of *M. sartor*

The preferred feeding at the branches of *P. abies* was very intensive on the bark and at the basis of the needles. The feeding at *P. sylvestris* seedling was investigated only on the bark but less intensive. There was only little feeding intensity in the cage, because the very strong maturation feeding took place outdoor before catch.

## Discussion and Conclusions

All three *Monochamus species* occurring in Austria have mainly one, more seldom a 2 years development cycle, depending on temperature sums.

The lifetime seems to be similar between *M. galloprovincialis pistor, M. sutor* and *M. sartor* and amounts to 5–17 weeks. The emergence period is from June to September for *M. galloprovincialis pistor*, and from late July to September for *M. sutor* and *M. sartor* in Austria. The *Monochamus* beetles lived until October and made maturation feeding up the end of their life. This indicates that the *Monochamus* beetles could transfer nematodes until late in October to healthy *Pinus* trees which will not show pine decline symptoms in winter season although they are already infested. This may be an important result for the monitoring and detection of *B. xylophilus* infested trees.

**Acknowledgments** The studies presented in this paper were financially supported by funding of the EU research cooperation project "PHRAME – Development of improved Pest Risk Analysis technique for quarantine pests, using pinewood nematode, *Bursaphelenchus xylophilus*, in Portugal as model system" (QLK5-CT-2002-00672).

## Reference

Hoyer-Tomiczek, U. and Tomiczek, C. (2005). *"PHRAME – eine EU-Forschungskooperation: Wie gefährlich ist der Kiefernsplintholznematode für Europa?"* Forstschutz aktuell Nr. 34, November 2005, 21–25, ISSN 1815-5103; Bundesforschungs- & Ausbildungszentrum für Wald, Naturgefahren und Landschaft (BFW), Wien, November 2005.

# Potential Insect Vectors of *Bursaphelenchus* spp. (Nematoda: Parasitaphelenchidae) in Spanish Pine Forests

**Lee Robertson, A. García-Álvarez, Susana C. Arcos, M.A. Díez-Rojo, J. Pedro Mansilla, R. Sanz, C. Martínez, Miguel Escuer, L. Castresana, A. Notario, Antonio Bello and Maria Arias**

**Abstract** Potential insect vectors of *Bursaphelenchus xylophilus* (PWN) were studied. Pathways of introduction of PWN from Portugal to Europe, through Spain, were determined and traps were located in pine stands sites along the pathways. 19 Cerambycidae, 12 Scolytidae, 12 Buprestidae and 10 Curculionidae species have been found. Trapped insects were examined for the presence of nematodes under their elytra. Nematodes were found on *Arhopalus ferus, Spondylis buprestoides, Hylastes ater, Hylurgus lingniperda, Orthotomicus erosus, Pityogenes bidentatus, Tomicus piniperda, Hylobius abietis* and *Pissodes validirrostris* specimens. *Monochamus galloprovincialis* was the most important insect species, representing a risk for the introduction of the PWN in Spanish pine forests; Cerambycidae and Curculionidae species, were taken into account because they have been reported as vectors of other *Bursaphelenchus* spp.

## Introduction

Nematodes of the genus *Bursaphelenchus* are fungal feeding organisms with ectophoretic relationships with insects, some of which are found in damaged conifers and deciduous trees associated with Coleoptera, mainly within the families Cerambycidae, Curculionidae and Scolytidae. The pinewood nematode (PWN), *Bursaphelenchus xylophilus*, is the most pathogenic species. The introduction and establishment of PWN in the Setúbal Peninsula, Portugal (Mota et al., 1999) and the evidence that *Monochamus galloprovincialis* is its insect vector (Sousa et al.,, 2001) has stimulated studies on its potential vector insects in Europe. Other *Bursaphelenchus* spp. are commonly associated with stressed, dead or decaying conifers, and may be involved in pine forest decline in some European regions (Philis, 1996;

L. Robertson
Dept. Agroecologia, Instituto de Ciencias Agrarias, CCMA, Consejo Superior de Investigaciones Científicas Serrano 115 dpdo, Madrid, 28006
e-mail: lee.r@ccma.csic.es

Braasch et al., 1999; Caroppo et al., 2000; Mamiya, 1999; Skarmoutsos et al., 2000). Therefore, the distribution of these species and their potential vectors are of increasing interest.

Numerous insect species colonize weakened and dying pine trees, but the Cerambycidae, Curculionidae, Scolytidae and Buprestidae are the most important vectors of *Bursaphelenchus* spp. A worldwide list of insects associated with *B. xylophilus* includes 21 species of Cerambycidae, one genus of Buprestidae and 2 species of Curculionidae, but the main vectors of *B. xylophilus* are beetles of the genus *Monochamus* (Linit et al., 1983; Linit, 1988). Some species of the genera *Acalolepta, Acanthocinus, Amniscus, Arhopalus, Asemum, Chrysobotris, Corymbia, Hylobius, Neoacanthocinus, Pissodes, Rhagium, Spondylis, Uraecha*, and *Xylotrechus*, have also been reported as carrying the PWN (CABI-EPPO, 2003).

*Pinus pinaster* is the most representative conifer of the autochthonous Spanish vegetation, followed by *P. halepensis, P. sylvestris, P. nigra, P. pinea* and *P. radiata* (MAPA, 2001), being *P. nigra, P. pinaster* and *P. sylvestris* the most susceptible species to PWN, and *P. halepensis, P. pinea* and *P. radiata* with an intermediate susceptibility (Evans et al., 1996).

The occurrence of *Bursaphelenchus* spp. in Spain has been previously studied (Abelleira et al., 2003; Escuer et al., 2003a; Escuer et al., 2003b; Arias et al., 2004) and the distribution of insects belonging to the family Cerambycidae, Curculionidae and Buprestidae have also been studied (Cobos, 1986; Gil Sánchez and Pajares Alonso, 1986; Vives, 2000; Verdugo Páez, 2004). However, only *Orthotomicus erosus* carrying *Bursaphelenchus fungivorus* (Arias et al., 2005) is reported as a potential insect vector. The aim of this work is to determine the occurrence and distribution of the potential insect vector species in Spanish pine forests, to analyze their ability to transport nematodes as well as their relationship with pine species and environmental conditions.

## Trap Sites and Conditions

Sampling was performed during three campaigns in 2003, 2004 and 2005 (Fig. 1 and Table 1). To capture the insects, double window traps were installed in *P. pinaster* forests using ethanol (98%) and turpentine (1:1 v/v) as attractant, along the Portuguese border in June 2003, in the provinces of Pontevedra, Ourense, Zamora, Salamanca, Cáceres and Huelva; light traps were installed in the provinces of Jaén and Teruel. They were checked every two weeks until their removal in late September.

In the second year nine new sites (7–15) were characterized, taking into account altitude, rainfall, annual temperature, conifer species, and establishing the potential pathways of PWN movement into central Europe through Spain from Portugal (Fig. 1), three to five traps were located in each site, when possible, on *P. pinaster* stands. In early May of 2005 traps were located in *P. nigra* forests of Cazorla Natural Park in Jaén and in the area of Almería Sierra Nevada. Forty-nine double window

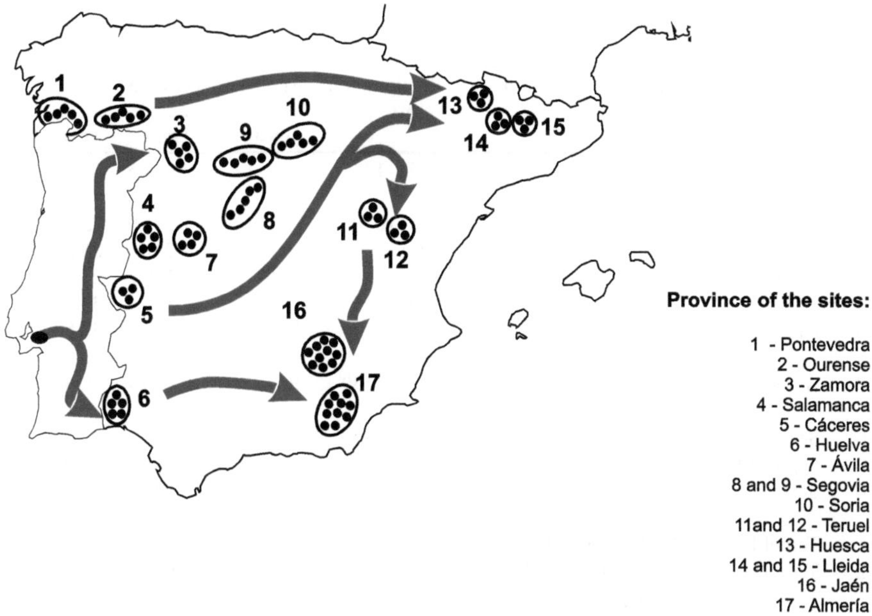


**Fig. 1** Insect trap sites 2003–2005 and potential pathways of introduction in to the Spanish Peninsula

traps were allocated to those sites using ethanol and turpentine (1/1) as an attractant and the captured insects were recovered every fifteen days from May, when the traps were allocated, until their removal at the end of October. All insects were observed under a Zeiss Stemi 2000, (Jena, Germany) stereomicroscope for the presence of nematodes under their elytrae.

## Nematode-Insect Interactions

Nematodes were found under the elytrae of about 10% of the following Cerambycidae insects: *Arhopalus ferus* from Navafría (Segovia), As Neves (Pontevedra) and Navasfrías (Salamanca); *Rhagium bifasciatum* and *R. inquisitor, Spondylis buprestoides* from As Neves and O Rosal (Pontevedra), Verín (Ourense), Santa Ana (Zamora) and several different sites in the area of Navasfrías (Salamanca) (Fig. 4); in the Scolytidae: *Hylastes ater* and *Hylurgus ligniperda* from Puerto del Pico and San Esteban del Valle (Ávila), *Ips sexdentatus* and *Orthotomicus erosus* from Cabecerán (Cáceres), Las Cumbres, Villablanca y El Granado (Huelva), El Payo and Navasfrías (Salamanca), Cuéllar (Segovia) and Cabrejas de Pinar (Soria), *Pityogenes bidentatus* from Las Cumbres (Huelva), and El Payo (Salamanca) and Cuéllar (Segovia) and *Tomicus piniperda* from Navasfrías area (Salamanca); and in the Curculionidae *Hylobius abietis* from Navasfrías (Salamanca) and Cabrejas del Pinar (Soria) and *Pissodes validirostris* from El Payo

**Table 1** Climatic data of the traps sites including pine species, altitude, average precipitation and temperature

| Site | Province | Vegetation | Altitude (m) | Precipitation (mm) | Annual temperature (°C) | | |
|---|---|---|---|---|---|---|---|
| | | | | | Tmean | Tmin | Tmax |
| 1 | Pontevedra | *P. pinaster/ P. radiata* | 154–320 | 1434–1465 | 12.7–14.1 | 2.9–4.5 | 23.4–27.7 |
| 2 | Orense | *P. pinaster* | 462–824 | 842–1050 | 12.1–12.3 | 1.1–1.9 | 25.9–28.2 |
| 3 | Zamora | *P. pinaster* | 974 | 991 | 9.8 | −2.0 | 26.6 |
| 4 | Salamanca | *P. pinaster* | 991 | 859 | 11.0 | −1.4 | 28.7 |
| 5 | Cáceres | *P. pinaster* | 424 | 619 | 15.5 | 3.2 | 33.7 |
| 6 | Huelva | *P. pinaster/ P. pinea* | 505 | 505 | 17.7 | 5.7 | 32.5 |
| 7 | Ávila | *P. sylvestris/ P. pinaster* | 661–1447 | 1072–1139 | 10.7–14.2 | 0.2–0.7 | 29.6–33.4 |
| 8 | Segovia I | *P. sylvestris* | 1471 | 643 | 9.6 | −2.2 | 27.4 |
| 9 | Segovia II | *P. pinaster* | 824 | 482 | 12.3 | −0.6 | 31.0 |
| 10 | Soria | *P. sylvestris/ P. pinaster* | 1153 | 691 | 9.3 | −3.0 | 28.2 |
| 11 | Teruel I | *P. sylvestris* | 1545 | 618 | 7.8 | −4.2 | 26.8 |
| 12 | Teruel II | *P. sylvestris/ P. nigra* | 1159 | 489 | 11.2 | −0.4 | 28.1 |
| 13 | Huesca | *P. sylvestris* | 1823 | 1073 | 6.7 | −5.9 | 21.8 |
| 14 | Lleida I | *P. nigra* | 698 | 675 | 12.5 | −1.3 | 26.9 |
| 15 | LLeida II | *P. nigra* | 860 | 779 | 11.6 | −2.1 | 25.9 |
| 16 | Cazorla | *P. nigra* | 885–1290 | 976 | 12.7 | 0 | 32 |
| 17 | Sierra Nevada | *P. sylvestris/ P. nigra* | 921–1800 | 606 | 12.7 | 1.7 | 28.4 |

(Salamanca) (Fig. 4 and Table 1). The most frequent nematodes were aphelenchids (68%) belonging to the genera *Cryptaphelenchus, Laimaphelenchus, Ektaphelenchus*, and *Seinura*, and rhabditids (30%). *B. fungivorus* appeared associated to *Orthotomicus erosus*.

## Distribution and Ecology

Among the Cerambycidae (Fig. 2) *Monochamus galloprovincialis*, the only insect captured known as vector of PWN, appears widespread in the Peninsula, especially along the Portuguese border, reaching the forests in Andalusia, the central region and even in the North-western and Eastern regions towards the Pyrenees. None of the captured specimens carried nematodes under their elytrae.

*Arhopalus ferus* as well as *Stictoleptura rubra* have been found widespread in the Peninsula, from Navafría (Segovia), O Rosal (Pontevedra) and Navasfrias (Salamanca) carrying nematodes other than *Bursaphelenchus* spp. *A. rusticus* has only been found at Las Cumbres, Villa Blanca (Huelva). Also specimens of *Spondylis buprestoides*, mainly captured along the Portuguese border but widespread in the Northern half of Spain were found in the Northwestern regions (Orense,

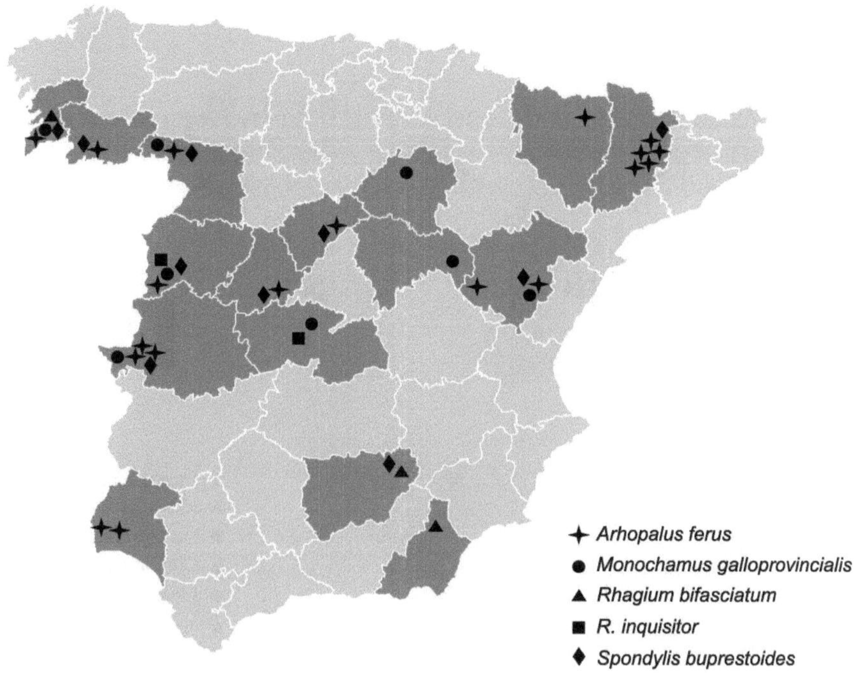


**Fig. 2** Distribution of Cerambycidae

Pontevedra, Salamanca and Zamora provinces) but carried nematodes other than *Bursaphelenchus* spp.

Other captured Cerambycidae included *Rhagium bifasciatum* in the Northwest (Pontevedra) and at Lugar Nuevo (Jaén) and Laujar (Almería), in the Southeast, and *Rhagium inquisitor* at Quintos de Mora (Toledo) and El Payo (Salamanca) on *P. pinaster* forests.

The scolytid *Hylastes ater* (Fig. 3a) appeared widespread in Andalusia and Extremadura, in the South-West, spreading towards Avila, Salamanca, and Soria and in the Northern plateau, on *P. pinaster* forests. Specimens from Puerto del Pico, San Esteban del Valle carried various nematodes. *Hylurgus ligniperda* has been trapped in the West, along the Portuguese border, carrying nematode specimens from Puerto del Pico, San Esteban del Valle (Avila), as well as *Orthotomicus erosus* which is widespread in the Spanish Peninsular, associated with damage in *P. pinaster* and frequently carrying nematodes, among them *Bursaphelenchus fungivorus* (Arias et al., 2005). Also, *Tomicus piniperda* has been found widespread in *P. pinaster* forests with the specimens from Avila, Cáceres, Huelva, Jaén, Lérida, Salamanca, Soria, Teruel and Toledo provinces carrying nematodes. Other Scolytidae found were *Ips sexdentatus* in three points at the Southeast and Southwest of Spain, *Pityogenes bidentatus* and *P. chalcographus* on *P. pinaster*.

The Curculionidae (Fig. 3b) found were *Hylobius abietis* trapped on *P. pinaster* forest from the Northwest, carrying nematode specimens from Navasfrías

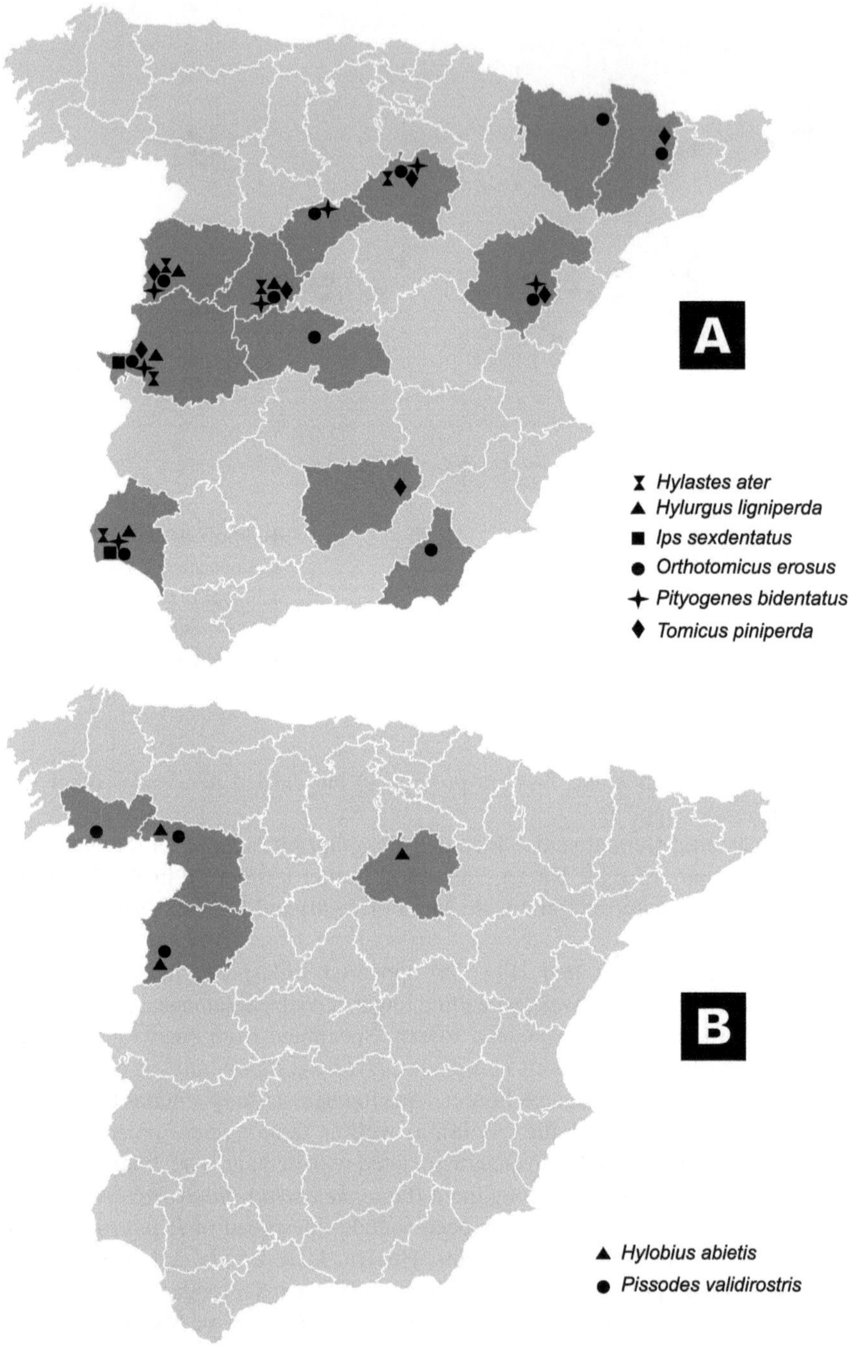


**Fig. 3** **A**: Distribution of Scolytidae. **B**: Distribution of other Curculionidae

(Salamanca) and Cabrejas de Pinar (Soria), and *Pissodes validirostris* also trapped in *P. pinaster* along the Northern part of the Portuguese border (Orense, Pontevedra and Zamora provinces).

## Discussion and Conclusions

Some of the trapped insect can produce direct damage and even death to various pine species, but more important is their potential to act as vectors of the PNW. It would appear that the role of these vector insects does not imply any specificity in their relationships with the nematodes which they transport, but are a mere vehicle as both organisms have parallel development under the same ecological conditions, therefore it could be considered that potential vectors are any insect which is able to transport nematodes, especially if the nematodes belong to the genus *Bursaphelenchus*.

In this sense the most important insect found is the cerambycid *M. galloprovincialis*, the vector of PWN in Portugal (Sousa et al., 2001), which has also been found associated to *B. mucronatus* in Finland, Germany and Italy (Magnuson and Schroeder, 1989; Tomminen et al., 1989; Ambrogioni et al., 1994; Palmisano et al., 1998; Braasch et al., 1999). It appears widespread in the Spanish Peninsula, especially along the Portuguese border, reaching the pine forests in the central region, although no nematodes were found under the elytra of any of the captured specimens. *M. galloprovincialis* was reported in Spain on the branches of dead *Pinus* spp., *Abies* spp. , *Picea* spp. and *Larix* spp. (Vives, 2000; Verdugo, 2004). It represents a risk of introducing PWN in to Central Europe through the Spanish Peninsula (Fig. 1), and the establishment of the nematode in the Southern half of Spain, as there are long periods during spring and summer with temperatures above 25 °C, low rainfall and consequently, a higher probability of trees with drought stress.

Other Cerambycidae that must be highlighted, as they have been found carrying nematodes under their elytrae, are *Arhopalus ferus* (Fig. 4), which is widespread in the Peninsula and was reported in Japan carrying *B. xylophilus*, as well as on dying shoots of *P. pinaster* in Piemonte, Italy and on *P. sylvestris* in Missouri, USA (Mamiya and Endo, 1972; Kondo et al., 1982; Caroppo et al., 1998) and *Spondylis buprestoides*, which is a very common insect species in the Spanish Peninsula on recently fallen *Pinus* spp. and *Abies* spp. trees (Vives, 2000) and has been reported in Japan as carrying *B. xylophilus* (Kobayashi et al., 1984) (Table 2).

Many Scolytidae are significant as pests to many pine species, besides being potential nematode vectors of *Bursaphelenchus* spp. as they carry nematodes under their elytrae. *Hylastes ater*, which lives under the bark of *P. nigra, P. pinaster* and *P. sylvestris*, can cause the death of young seedlings in nurseries (Gil and Pajares, 1986). *Hylurgus ligniperda* localized in the basal part and on the roots of declined or death trees, is becoming a pest in reforestation of many pine forests (Browne, 1986). This insect has been reported in Greece associated to *B. sexdentati* and in Portugal carrying *B. leoni, B. sexdentati* and *B. teratospicularis* (Skarmoutsos and Skarmoutsos, 1999; Penas et al., 2006).

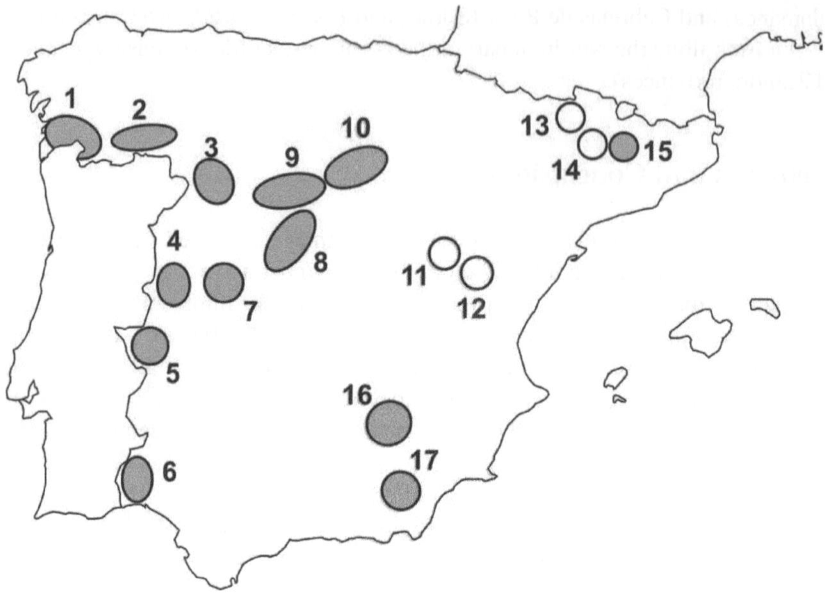


| Family | Species | Site number |
|---|---|---|
| Cerambycidae | *Arhopalus ferus* | 8 |
| | *Spondylis buprestoides* | 1, 2, 3, 5 |
| Scolytidae | *Hylastes ater* | 7 |
| | *Hylurgus ligniperda* | 7 |
| | *Ips sexdentatus (I. acuminatus)* | 16, 17 |
| | *Orthotomicus erosus* | 4, 5, 6, 7, 9, 10 |
| | *Pityogenes bidentatus* | 4, 9 |
| | *Tomicus piniperda* | 4, 15 |
| Other Curculionidae | *Hylobius abietis* | 10 |
| | *Pissodes validirrostris* | 4 |

**Fig. 4** Insects with nematodes under their elytrae and their distribution. Site designation is the same as given in Fig. 1

*Ips sexdentatus* is included in the quarantine list of the EU directive 2000/29/EC. This species is of no significance as a pest in Northern and Central Europe, where it breeds only in fresh logs or in weakened or dying trees. It has been shown to cause death of *P. sylvestris* and *P. radiata* suffering from drought stress in Central and Southern regions of France and in the Northern areas of Portugal and Spain, often in association with other pests such as *Ips accuminatus* or *Tomicus piniperda*. It is not considered to be a quarantine pest by EPPO or any other regional plant

**Table 2** Species found in traps after the 2003, 2004, and 2005 campaigns

| Family/Subfamily | Species | Frequency* | Presence of nematodes |
|---|---|---|---|
| **Cerambycidae** | *Acanthocinus hispanicus* Sama & Schurmann 1980 | 1 | − |
| | *Anaesthetis testacea* (Fabricius 1781) | 3 | − |
| | *Anoplodera sexguttata* (Fabricius 1775) | 3 | − |
| | *Arhopalus ferus* (Mulsant 1839) | 1 | + |
| | *A. rusticus* (Linnaeus 1758) | 3 | − |
| | *A. syriacus* (Reitter 1895) | 3 | − |
| | *Clytus arietis* (Linnaeus 1758) | 3 | − |
| | *Gracilia minuta* (Fabricius 1781) | 3 | − |
| | *Monochamus galloprovincialis* (Olivier 1795) | 1 | − |
| | *Nustera distigma* (Charpentier 1825) | 3 | − |
| | *Prionus (Prionus) coriarius* (Linnaeus 1758) | 3 | − |
| | *Paracorymbia fulva* (DeGeer 1775) | 2 | − |
| | *Phoracantha semipunctata* (Fabricius 1775) | 3 | − |
| | *Plagionotus arcuatus* (Linnaeus 1758) | 3 | − |
| | *Rhagium (Hagrium) bifasciatum* (Fabricius 1775) | 2 | − |
| | *R. (Rhagium) inquisitor* (Linnaeus 1758) | 3 | − |
| | *Spondylis buprestoides* (Linnaeus 1758) | 1 | + |
| | *Stictoleptura rubra* (Linnaeus 1758) | 1 | − |
| **Scolytinae** | *Hylastes ater* (Paykull 1800) | 2 | + |
| | *H. attenuatus* (Erichson 1836) | 3 | − |
| | *Hylurgus ligniperda* (Fabricius 1787) | 2 | + |
| | *Ips acuminatus* (Gyllenhal 1827) | 1 | + |
| | *I. mannsfeldi* (Wachtl 1879) | 1 | + |
| | *I. sexdentatus* (Börner 1776) | 3 | − |
| | *Orthotomicus erosus* (Wollaston 1857) | 1 | + |
| | *Pityogenes bidentatus* (Herbst 1784) | 1 | + |
| | *P. chalcographus* (Linnaeus 1761) | 1 | − |
| | *Pityokteines vorontzowi* (Jakobson 1895) | 3 | − |
| | *Tomicus piniperda* (Linnaeus 1758) | 1 | + |
| | *Xyleborinus saxesenii* (Ratzeburg 1837) | 3 | − |
| **Other Curculionidae** | *Brachyderes (Brachyderes) confusus* Viedma 1967 | 3 | − |
| | *B. (Brachylophus) lusitanicus* (Fabricius 1781) | 3 | − |
| | *B. (Brachyderes) suturalis* Graells 1851 | 3 | − |
| | *Curculio salicivorus* Paykull 1792 | 3 | − |
| | *Hylobius (Callirus) abietis* (Linnaeus 1758) | 2 | − |
| | *Magdalis (Magdalis) memnonia* (Gyllenhal 1837) | 2 | − |
| | *Pissodes (Pissodes) castaneus* (De Geer 1775) | 3 | − |
| | *P. (Pissodes) validirostris* (C.R. Sahlberg 1834) | 2 | + |

**Table 2** (continued)

| Family/Subfamily | Species | Frequency* | Presence of nematodes |
|---|---|---|---|
| | *Polydrosus (Polydrosus) tereticollis* (De Geer 1775) | 3 | – |
| | *Rhynchites bicolor* (Fabricius 1775) | 3 | – |
| **Buprestidae** | *Acmaeoderella (Carininota) flavofasciata* (Piller & Mitterpacher 1783) | 3 | – |
| | *Anthaxia (Haplanthaxia) confusa* (Gory 1841) | 3 | – |
| | *A. (Melanthaxia) quadripunctata* (Linnaeus 1758) | 3 | – |
| | *A. (Melanthaxia) rugicollis* Lucas 1849 | 3 | – |
| | *Buprestis (Buprestis) novemmaculata* Linnaeus 1758 | 2 | – |
| | *B. (Buprestis) octoguttata* Linnaeus 1758 | 3 | – |
| | *Chalcophora mariana* (Linnaeus 1758) | 3 | – |
| | *Coraebus florentinus* (Herbst 1801) | 3 | – |
| | *Phaenops cyanea* (Fabricius 1775) | 3 | – |

*1: Frequent; 2: Infrequent; 3: Occasional (Species in grey box are vectors of *Bursaphelenchus* spp.)

protection organization, because it is not generally a primary pest and is only capable of attacking trees already suffering stress, either environmental or from other pests. It is unlikely to spread naturally, so phytosanitary measures could be justified (CABBI and EPPO, 2003). *Ips sexdentatus* has been shown to reduce tree growth and weakened trees often die, but high population levels of this insect can cause healthy trees to die. It has been reported in Portugal carrying dauer larvae of *Bursaphelenchus* spp. and *B. hellenicus* (Sousa et al., 2002; Penas et al., 2006).

*Orthotomicus erosus* has been found at several sites the Central System and in Southern Spain associated to pine trees in decline. This fact must be taken into account, because the beetle although at first appears not to be a primary pathogen, infests recently fallen trees, wounded, and stressed living trees that later often die, and at high population levels massive attacks can lead to the death of healthy trees. It has been found attacking *P. pinaster* and *P. halepensis* from Tuscany (Italy), and in Greece it appeared attacking trees which contained populations of *B. leoni* and *B. sexdentati*. In Portugal it has been reported as carrying dauer larvae of *Bursaphelenchus* spp., *B. sexdentati* and *B. teratospicularis* and *B. fungivorus* in Spain (Arias et al., 2005; Caroppo et al., 1998; Penas et al., 2006; Skarmoutsos and Skarmoutsos, 1999; Sousa et al., 2002). High numbers of *O. erosus* have been found at several sites in the Central System and Southern Spain associated to pine trees in decline, although the beetle does not seem to be a primary pathogen; it infests recently fallen trees, wounded, and stressed living trees. Stressed trees are more prone to attacks. Weakened trees often die, and at high population levels, the attack can lead to the death of healthy trees.

*Tomicus piniperda*, a specific insect of pine trees associated to *P. sylvestris, P. halepensis P. brutia* and *P. nigra camaranica* has been reported in Cyprus causing malformations and loss of growth (Grüne, 1979). It is widespread in the Spanish Peninsula, usually associated with blue stain wood fungi, and represents a problem in young reforestations and in established pine forest under adverse environmental conditions (Gil and Pajares, 1986). It appeared in Greece associated with *B. eggersi, B. leoni, B. hellenicus, B. sexdentati* and *B. teratospicularis*, to *B. sexdentati* in Germany and, in Portugal it has been found transporting dauer larvae of an unidentified *Bursaphelenchus* spp. and *B. hellenicus* (Braasch et al., 1999; Penas et al., 2006; Skarmoutsos and Skarmoutsos, 1999; Sousa et al., 2002). In our study this insect appeared widespread associated mainly to *P. pinaster* and carried unidentified nematodes under their elytrae. *Hylobius abietis* must be also highlighted because some specimens have been found carrying nematodes under their elytrae and in high populations can cause damage to young pines.

Finally, the potential risk of these insects as vectors of PWN or of other *Bursaphelenchus* spp. which in themselves represent a pathological risk as it is the case of *B. sexdentati* which can cause damage to *Pinus nigra, P. pinaster* and *P. sylvestris* seedlings (Skarmoutsos et al., 2000), *B. fungivorus* to *P. sylvestris* (Caroppo et al., 2000) and *B. leoni* to *P. brutia* (Philis, 1996; Skarmoutsos et al., 2000) should be further studied.

**Acknowledgments** This work was supported by the projects: QLK5-CT-2002-00672 "Development of improved pest risk analysis techniques for quarantine pests, using pinewood nematode, *Bursaphelenchus xylophilus*, in Portugal as a model (PHRAME)".

## References

ABELLEIRA, A., ESCUER, M., ARIAS, M. AND MANSILLA, J.P. (2003). The genus *Bursaphelenchus* Fuchs (Nematoda: Aphelenchida) in North-west Spain. *Nematology* 5: 677–685.

AMBROGIONI, L., CERCHIARINI, G., IRDANI, T. AND TOSSANI, N. (1994). Preliminary data on the spread of *Bursaphelenchus* spp. in pine woods in Italy. *Redia* 77: 273–278.

ARIAS, M., ESCUER, M. AND BELLO, A. (2004). Nematodos asociados a madera y árboles de coníferas en pinares españoles. *Boletín Sanidad Vegetal Plagas* 30: 581–593.

ARIAS, M., ROBERTSON, L., GARCÍA-ÁLVAREZ, A., ARCOS, S.C., ESCUER, M., SANZ, R. AND MANSILLA, J.P. (2005). *Bursaphelenchus fungivorus* Franklin and Hooper 1962 (Nematoda: Aphelenchida) associated with *Orthotomicus erosus* (Wollaston, 1857) (Coleoptera: Scolitydae) in Spain. *Forest Pathology* 35: 375–383.

BRAASCH, H., CAROPPO, S., AMBROGIONI, L., MICHALOPOULOS, H., SKARTMOUSOS, G.C. AND TOMICZEK, H. (1999). Pathogenicity of various *Bursaphelenchus* species and implications to European forest. In *Sustainability of pine forest in relation to pine wilt and decline. Proceedings of the Symposium, Tokyo, Japan, 26–30 October 1998* (eds. K. Futai, K. Togasmi and I.T. Shokado), pp. 7–22. Shokado Shoten, Kyoto, Japan.

BRAASCH, H., METGE, K. AND BURGERMEISTER, W. (1999). *Bursaphelenchus* species in conifer woods in Germany and their ITS-RFLP patterns. *Nachrichtenblatt des Deutschen Pflanzenschuthzdienstes* 51: 312–320.

BROWNE, F.G. (1986). *Pests & diseases of forest plantation trees. An anoted list of the principal species occurring in the British Commonwealth.* Clarenton Press. London, 1330 pp.

CABBI-EPPO (2003). *Data sheets on quarantine pests: Ips sexdentatus.* CABBI and EPPO for the EU under Contract 90/399003, 3 pp.

Caroppo, S., Ambrogioni, L., Cavalli, M. and Coniglio, D. (1998). Occurrence of the pine wood nematodes, *Bursaphelenchus* spp., and their possible vectors in Italy. *Nematologia Mediterranea* 26: 87–92.

Caroppo, S., Cavalli, M., Coniglio, D. and Ambrogioni, L. (2000). Pathogenicity studies with various *Bursaphelenchus* populations on conifer seedlings under controlled and open air conditions. *Redia* 83: 61–75.

Cobos, J.M. (1986). *Fauna Ibérica de Coleópteros Buprestidae, CSIC.* Madrid, 364 pp.

de Liñán, C. (1998). *Entomología Agroforestal. Ediciones Agrotécnicas.* Madrid, 1309 pp.

Escuer, M., Arias, M. and Bello, A. (2003a). The genus *Bursaphelenchus* (Nematoda) in Spain. In *The pinewood nematode, Bursaphelenchus xylophilus. Nematology monographs and perspectives, vol. 1* (eds. M. Mota and P. Vieira), pp.93–99. E.J.Brill, Leiden, The Netherlands.

Escuer, M., Arias, M. and Bello, A. (2003b). Occurrence of the genus *Bursaphelenchus* Fuchs, 1937 (Nematoda: Aphelenchida) in the Spanish conifer forests. *Nematology* 6: 155–156.

EU (2004). *Consolidated TEXT produced by the CONSLEG system of the Office for Official Publications of the European Communities.* CONSLEG: 2000L0029, 37 pp.

Evans, H.F., McNamara, D.G., Braasch, H., Chadoeuf, J. and Magnuson, C. (1996). Pest risk analysis (PRA) for the territories of the European Union (as PRA area) on *Bursaphelenchus xylophilus* and its vectors in the genus *Monochamus. Bulletin OEPP/EPPO* 26: 199–249.

Gil Sánchez, L.A. and Pajares Alonso, J.A. (1986). *Los escolítidos de las coníferas en la Península Ibérica.* Monografias INIA, 53, Madrid, 194 pp.

Grüne, S. (1979). *Brief illustrated key to European bark beetles.* Verlag M. & H. Schaper. Hannover, 182 pp.

Kobayashi, F., Yamane, A. and Ikeda, T. (1984). The Japanese pine sawyer beetle as a vector of pine wilt disease. *Annual Review of Entomology* 29: 115–135.

Kondo, E., Foudin, A., Linit, M., Smith, M., Bolla, R., Winter, R. and Dropkin, V. (1982). *Pine wilt disease – nematological, entomological and biochemical investigations.* Columbia, Missouri, USA; Agricultural Experiment Station, Missouri University, 56 pp.

Linit, M.J. (1988). Nematode-vector relationships in the pine wilt disease system. *Journal of Nematology* 20: 227–235.

Linit, M.J., Kondo, E. and Smith, M.T. (1983). Insect associated with the pine wood nematode, *Bursaphelenchus xylophilus*, in Missouri. *Environmental Entomology* 12: 467–470.

Magnuson, C. and Schroeder, L.M. (1989). First record of a *Bursaphelenchus* species in *Monochamus* beetles in Scandinavia. *Anzeiger für Schädlingskkunde, Pflanzenschutz und Umweltschutz* 62: 53–54.

Mamiya, Y. (1999). Review on the pathogenicity of *Bursaphelenchus mucronatus*. In *Sustainability of pine forest in relation to pine wilt and decline. Proceedings of the Symposium, Tokyo, Japan, 26–30 October 1998* (eds. K. Futai, K. Togasmi and I.T. Shokado), pp. 47–51. Shokado Shoten, Kyoto, Japan.

Mamiya, Y. and Endo, N. (1972). Transmission of *Bursaphelenchis lignicolus* (Nematoda: Aphelenchoididae) by *Monochamus alternatus* (Coleoptera: Cerambycidae). *Nematologica* 18: 159–162.

MAPA (2001). *Anuario de estadística agroalimentaria 2000.* Ministerio de Agricultura Pesca y Alimentación, 691 pp.

Mota, M.M., Braasch, H., Bravo, M.A., Penas, A.C., Burgermeister, W., Metge, K. and Sousa, E. (1999). First report of *Bursaphelenchus xylophilus* in Portugal and in Europe. *Nematology* 1: 727–734.

Palmisano, A., Ambrogioni, L. and Caroppo, S. (1998) *Bursaphelenchus mucronatus* on *Pinus pinaster* in Italy. *Redia* 75: 517–527.

Penas, A.C., Bravo, M.A., Naves, P., Bonifácio, L., Sousa, E. and Mota, M.M. (2006). Species of *Bursaphelenchus* Fuchs, 1937 (Nematoda: Parasitaphelenchidae) and other nematode genera associated with insects from *Pinus Pinaster* in Portugal. *Annals of Applied Biology* 148: 121–131.

PHILIS, J. (1996). An outlook on the association of *Bursaphelenchus leoni* with wilting pines in Cyprus. *Nematologia Mediterranea* 24: 221–225.

SKARMOUTSOS, G. AND SKARMOUTSOS, H. (1999). First record of *Bursaphelenchus* nematodes from pine forest in Greece. *Plant Disease* 83: 879.

SKARMOUTSOS, H., SKARMOUTSOS, G. AND MICHALOPOULOS-SKARMOUTSOS, H. (2000). Pathogenicity of *Bursaphelenchus sexdentati, Bursaphelenchus leoni* and *Bursaphelenchus hellenicus* on European pine seddlings. *Forest Pathology* 30: 149–156.

SOUSA, E., BRAVO, M.A., PIRES, J., NAVES, P., PENAS, A.C., BONIFACIO, L. AND MOTA, M. (2001). *Bursaphelenchus xylophilus* (Nematoda: Aphelenchoididae) associated with *Monochamus galloprovincialis* (Coleoptera: Cerambycidae) in Portugal. *Nematology* 3: 89–91.

SOUSA, E., NAVES, P., BONIFÁCIO, L., BRAVO, M.A., PENAS, A.C. AND SERRÃO, M. (2002) Preliminary survey for insects associated with *Bursaphelenchus xylophilus* in Portugal. *OEPP/EPPO Bulletin* 32: 499–502.

TOMMINEN, J., NUORTEVA, M., PULKKINEN, M. AND VAKEVA, J. (1989). Ocurrence of the nematode *Bursaphelenchus mucronatus* in Finland. *Silva Fennica* 23: 271–277.

VERDUGO PÁEZ, A. (2004). *Monografías de la socieda Andaluza de Entomología (SAE). 1. Cerambicidae de Andalucía.* Diputación de Córdoba, 152 pp.

VIVES, E. (2000). *Fauna Ibérica. 12 Coleoptera Cerambycidae*. Museo Nacional de Ciencias Naturales, CSIC. Madrid, 715 pp.

# Genetic Structure of *Monochamus alternatus* in Japan

**Etsuko Shoda-Kagaya, Miho Kawai, Tadashi Maehara, Ryûtarô Iwata and Akiomi Yamane**

**Abstract** Recent progress in molecular ecological methods has enabled us to study genetic differentiation among populations. By elucidating the genetic structure, we can trace the history of populations and to infer ongoing patterns of migration. In this chapter, we review the population structure of the pine sawyer *Monochamus alternatus* detected by microsatellite markers and mitochondrial DNA. First, the process of population expansion is inferred from the genetic structure determined from populations of Japan, China and Taiwan populations by mitochondrial DNA and microsatellite markers. Secondly, microsatellite markers are used to show the pine sawyers' dispersal route in the past in the northern part of Honshu Island, Japan, which is the frontier of the area damaged by pine wilt disease. Finally we discuss further subjects in studies of the sawyers' dispersals.

## Introduction

For the last ten years or so, the biogeography and ecology of forest insects has been studied using molecular markers (e.g. Reineke et al., 1999, Salvato et al., 2002, Shoda et al., 2003). At the same time, advances in the theory of spatial genetic structure have made it possible to trace the history of populations and to infer ongoing patterns of migration over broad time scales. Molecular ecology can determine the origins of these populations in both the ecological and historical time scales by using different genome regions.

The main vector of the pinewood nematode (PWN), *Bursaphelenchus xylophilus*, in Asia is the pine sawyer, *Monochamus alternatus* (Mamiya and Enda, 1972). During the 1990s, Yamane and Iwata applied isozyme analysis involving esterase zymograms into population studies. Although the zymogram patterns showed considerable genetic variation within the species, no particular geographical pattern could be detected. Recent advances in molecular ecological techniques have made

E. Shoda-Kagaya
Department of Forest Entomology, Forestry and Forest Products Research Institute, Matsunosato 1, Tsukuba, Ibaraki 305-8687, Japan
e-mail: eteshoda@affrc.go.jp

M.M. Mota, P. Vieira (eds.), *Pine Wilt Disease: A Worldwide Threat to Forest Ecosystems*, 

it possible to use markers that contain more numerically rich traits and higher resolutions. Genetic population structure analyses using sufficient sample sizes and marker resolutions should be able to reveal the formation of the populations and their dispersal processes. Microsatellite is a very high resolution DNA marker and it is useful for detecting intra-specific genetic diversification. Because microsatellite markers for *M. alternatus* have been developed (Maehara et al., personal communication), we can evaluate fine genetic differentiations among the pine sawyers.

Whereas the nematode was introduced from North America (de Guiran and Bruguier, 1989), the pine sawyer, *M. alternatus*, is endemic to Japan (Makihara, 2004). The introduction of the nematode has likely affected the genetic population structure of its vector, but there has been little knowledge about the pine sawyer's population differentiations.

In this work, we will first explain the genetic structure of *M. alternatus* over a broad area in Asia, presenting an approximate picture of population differentiations, dispersal, and mode of colonization. Then, we will show the dispersal route of adults at a local scale. Finally, we will examine subjects for further molecular ecological study of the sawyer, e.g. expected techniques and prospects for preventive applications.

## Genetic Structure of *Monochamus alternatus* in Asia

In Japan, pine wilt disease was first recorded about 100 years ago. During the past 25 years, the disease has expanded its range into other Asian countries. Before the invasion of the PWN, the pine sawyer was a rare beetle in the pine forest community. However after the invasion, the pine sawyer's population density increased and its migration rate might have been enhanced. We researched its past colonization and dispersal process using DNA markers, mitochondrial DNA and microsatellite regions (Kawai et al., 2006). After collecting pine sawyers from 24 sites in Japan, Taiwan and China (Fig. 1), we conducted mitochondrial DNA analysis on 2–4 individuals for all populations and 18–30 individuals from Japanese populations were subject to microsatellite analysis.

Using the mitochondrial DNA CO2 region, seven haplotypes were detected in the Asian populations. The mtDNA sequence data divided the pine sawyer populations into two major groups based on the phylogenetic tree (Fig. 2): Clade A (haplotypes A-F) and Clade B (haplotype G). Clade B is monotypic, and occurs only in some populations of central Japan, while Clade A is widely distributed across Northeast Asia (Fig. 1). Calculations made using the molecular clock for the CO2 gene of Coleoptera indicate that divergence between Clades A and B began approximately 1.45 million years ago.

This haplotype distribution can be considered to been formed by two alternative scenarios. In the first scenario, the sawyer colonized Japan twice from the Eurasian continent and descendant populations have since fused and dispersed together. Clade B would have colonized Japan during the Lower Pleistocene period because it is

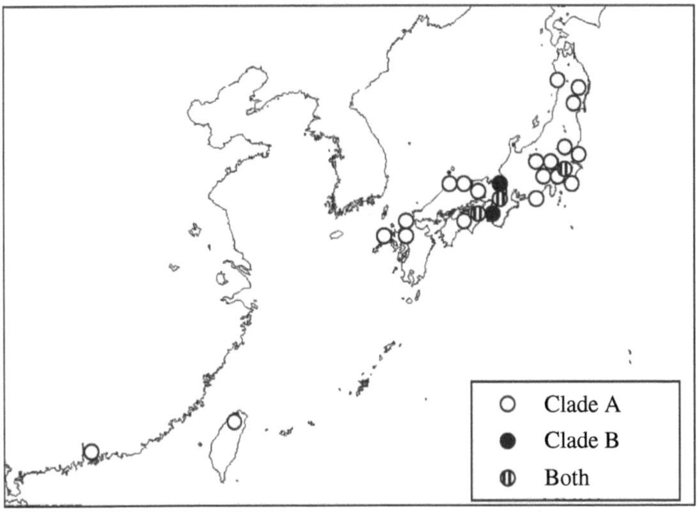


**Fig. 1** *Monochamus alternatus* sample sites, and detected haplotypes of mitochondrial DNA

monotypic and was likely formed through a genetic bottleneck in a small region, maybe in Japan. Then Clade A would have come to Japan. The second possibility is that the coalescence of genes has been extended by population subdivision. If this is the case, it is likely to have occurred by demographic chance, so-called "lineage sorting".

We also investigated population differentiations of *M. alternatus* using DNA microsatellite regions. We used five microsatellite loci (Machara et al., personal communication), developed with the dual-suppression-PCR technique (Lian and Hogetsu, 2002). There was a significant differentiation between populations, even with nearby prefectures. Thus, there seems to be little ongoing gene flow between populations between prefectures.

Using the genetic distances measured by these markers, we can determine whether or not the genetic drift and gene flow in a population is in equilibrium,

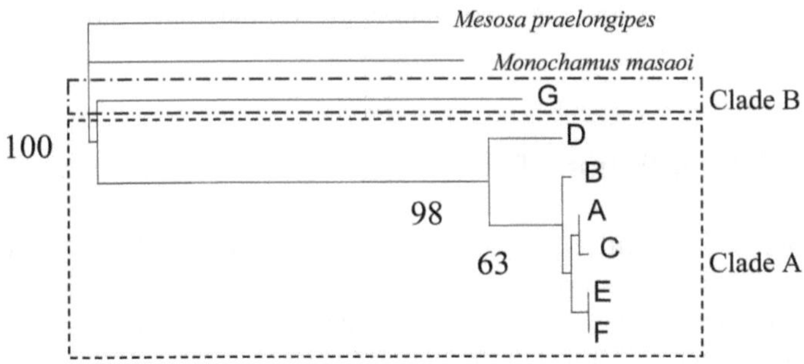


**Fig. 2** The neighbor-joining (NJ) tree of *M. alternatus* based on CO2 sequence variation. Bootstrap percentages of 1,000 replicates are noted for each branch; only bootstrap values >50% are shown

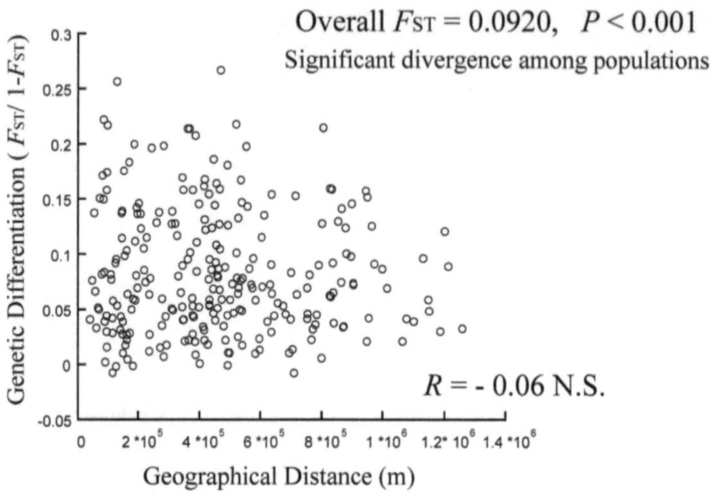


**Fig. 3** The relationship between genetic differences [$F_{ST}/(1 - F_{ST})$] and geographical distances in *M. alternatus* in Japan

by observing isolation by distance (Hutchinson and Templeton, 1999). Although significant differentiations were detected among populations, there was no correlation between geographical distances and genetic differentiation between populations (Fig. 3). In this case, these populations are not in demographic equilibrium and genetic drift may have been working upon them.

From the genetic structure, we can infer the dispersal and colonization processes of these populations. Strong genetic drift suggests intense population fluctuation. Thus, colonization into new habitats seems to be attained by a small number of adults. The sawyer population should rise rapidly in association with population increases with the PWN. Population expansion may have occurred, not only by natural dispersal on a small spatial scale, but also by long-distance dispersal because there was no isolation by distance. Relocation of infected and damaged wood has enhanced the dissemination of pine wilt disease, and might have aided in the long-distance dispersal of pine sawyer beetles. In Japan, repeated invasions of small numbers of individuals and rapid population growth may have formed the genetic structure of the sawyers.

## Genetic Differentiation of *M. alternatus* over a Mountain Range

Although we can recognize that the introduction of the PWN has influenced the genetic structure of the sawyers above the regional scale, we have not been able to explain their dispersal route or the origin of introduced populations. To apply the molecular technique into forest protection, we need to answer these questions.

Using microsatellite markers we analyzed populations in the Tohoku ("Northeast") area of Honshu Island, Japan, which are the frontier populations of pine wilt disease (Shoda-Kagaya, 2007). One of the most serious problems associated with pine wilt disease in Japan is its expansion into this area. Its occurrence and associated damage in this area has been recorded annually in administrative reports. In 1975, the first damage was reported in the southeastern part of this area and it has since spread (Fig. 4). Today, Akita Prefecture, on the western side of the Ohu Mountain Range, and Iwate Prefecture, on the eastern side, is the Northern limits of the disease. To stop the spread of pine wilt disease, it is important to elucidate the sawyer's dispersal ability and its migration dynamics.

According to the reports of the occurrence of damage, there seem to be two main routes of the invasion of the PWN (Fig. 4). The Ohu Mountain Range runs from north to south and the dispersal of the disease has occurred on both the east and west sides of the range. The disease has expanded its range in synchronism between the east and west sides, and it is hypothesized that dispersal of the sawyer over the mountain area could have enhanced its diffusion. In Akita, there seems to be another route from the southern part of the prefecture. Sudden pine wilt disease damage happened in the Oga Peninsula in 1988, and it may have been artificially introduced because there was a wood yard near the damaged pine forest. Hence, there may be

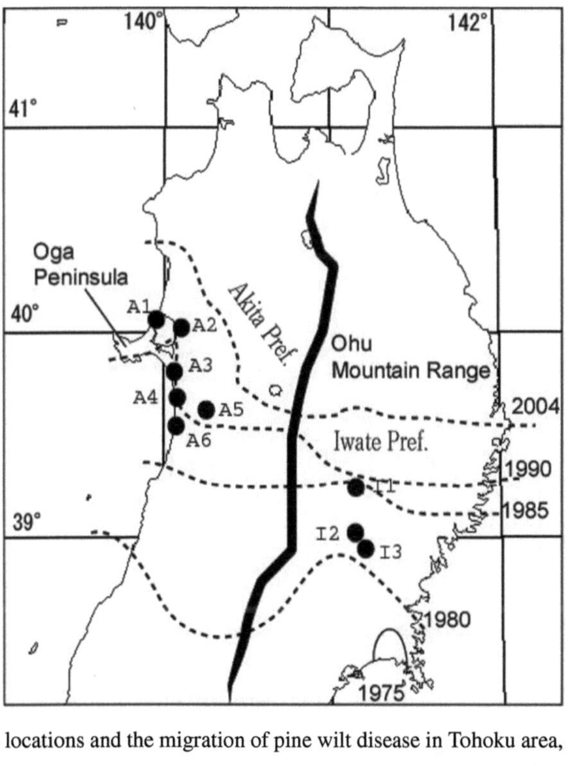


**Fig. 4** Sampling locations and the migration of pine wilt disease in Tohoku area, Japan from 1975 to 2004

two invasion routes in Akita. This study aimed to answer the following questions: (1) Do pine sawyers migrate between Iwate and Akita populations? (2) Have the two entities from the south and from the peninsula merged in Akita?

The pine sawyers used in the present survey were collected near the frontier populations. Six populations from Akita Prefecture and three from Iwate were examined (Fig. 4). Adults were trapped using lures or collected from damaged wood. Each individual was genotyped using the same five microsatellite loci used in the previous study.

The genetic relationship among populations is shown using multidimensional scaling (MDS) (Fig. 5). MDS is a class of ordination techniques that displays the complex relationships among populations in a small number of dimensions. As can be easily seen, there are two groups on the first axis. While the sawyers diverged between the prefectures, their genetic components remained similar within the prefectures. However, it is interesting that significant isolation by distance was detected within Akita Prefecture populations. This suggests that the population from the Oga Peninsula is merging with the population from the south, and/or restricted gene flow between populations is occurring in a stepping stone manner.

These results lead us make to the following conclusions: The pine sawyers scarcely disperse over the mountain range, and the synchronized expansion on both

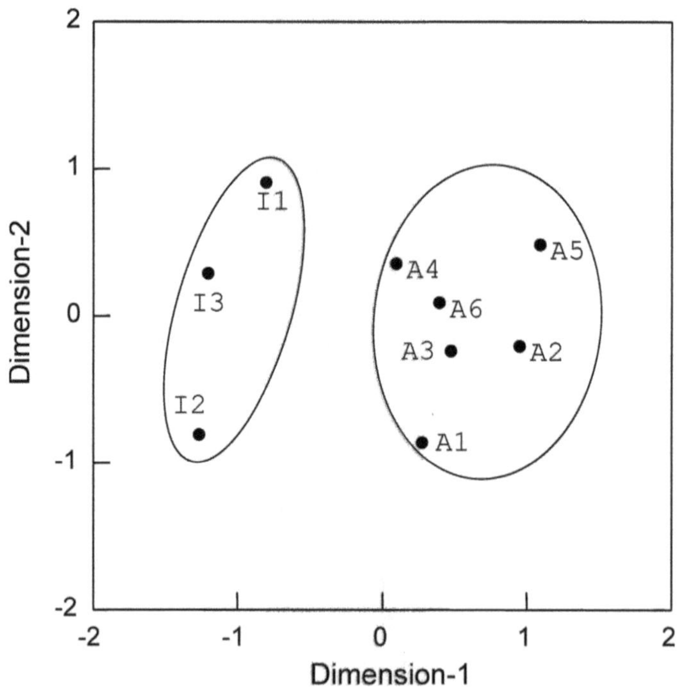


**Fig. 5** Multidimensional scaling analysis of *M. alternatus* of frontier populations in Japan based on microsatellite markers

the east and west sides of the range are most likely accidental. In Akita Prefecture, two invasion routes, from the Oga Peninsula and the southern part of the prefecture, have probably already converged.

## Subjects for Further Study

Having adopted a new methodology, we could learn about the genetic characteristics of pine sawyers in Northeast Asia and their dispersal routes in the Tohoku area of Japan. However, much remains unknown about the molecular ecology of the pine wilt disease vector and we will indicate three further subjects.

First, we should develop methods to determine whether the sawyers are endemic or derived. In undamaged areas in the north, away from the frontier populations, adult sawyers without the nematode are occasionally captured. We should elucidate whether they originally inhabited the area or not. If they are migrants from a damaged area, we should intensify protections in barrier zones to stop the spread of frontier populations.

Second, molecular ecological methods must be used to determine their dispersal ability, since this is a key factor in the expansion of the area damaged by the pine wilt disease. We should know their average movement distances and how often they engage in long distance dispersal.

Third, we need to develop new highly polymorphic markers and increase the resolution. Additional sampling near frontier populations is also necessary to obtain more accurate information about dispersal between frontier populations. Then, we should apply more sophisticated statistical methods to measure their dispersal distances.

**Acknowledgement** Thanks are due to C. Lien, Z. Zhou, T. Hogetsu, K. Kobayashi, K. Fujita, T. Goto, K. Nakamura-Matori and H. Kinuura for their cooperation with this study.

## References

De Guiran, G. and Bruguier, N. (1989). Hybridization and phylogeny of the pine wood nematode (*Bursaphelenchus* spp.). *Nematologica* 35: 321–330.

Hutchison, D.W. and Templeton, A.R. (1999). Correlation of pairwise genetic and geographical distance measures: inferring the relative influences of gene flow and drift on the distribution of genetic diversity. *Evolution* 53: 1898–1914.

Kawai, M., Shoda-Kagaya, E., Maehara, T., Zhou, Z., Lian, C., Iwata, R., Yamane, A. and Hogetsu, T. (2006). Genetic structure of pine sawyer *Monochamus alternatus* (Coleoptera: Cerambycidae) populations in northeast Asia: Consequences of the spread of pine wilt disease. *Environmental Entomology* 35: 569–579.

Lian, C. and Hogetsu, T. (2002). Development of microsatellite markers in black locust (*Robinia pseudoacacia*) using a dual-suppression-PCR technique. *Molecular Ecology Notes* 2: 211–213.

Makihara, H. (2004). Two new species and a new subspecies of Japanese Cerambycidae (Coleoptera). *Bulletin of the Forestry and Forest Products Research Institute* 390: 15–24.

Mamiya, Y. and Enda, N. (1972) Transmission of *Bursaphelenchus lingnicolus* (Nematoda: Aphelenchoididae) by *Monochamus alternatus* (Coleoptera: Cerambycidae). *Nematologica* 18: 159–162.

Reineke, A., Karlovsky, P. and Zebitz, C.P. (1999). Amplified fragment length polymorphism analysis of different geographic population of gypsy moth, *Lymantria dispar* (Lepidoptera: Lymantriidae). *Bulletin of Entomological Research* 89: 79–88.

Salvato, P., Battisti, A., Concato, S., Masutti, L., Pataranello, T. and Zane, L. (2002). Genetic differentiation in the winter pine processionary moth (*Thaumetopoea pityocampa – wilkinsoni* complex), inferred by AFLP and mitochondrial DNA markers. *Molecular Ecology* 11: 2435–2444.

Shoda-Kagaya, E. (2007). Genetic differentiation of the pine wilt disease vector, pine sawyer *Monochamus alternatus* over a mountain range – revealed from microsatellite DNA markers. *Bulletin of Entomological Research* 97: 167–174.

Shoda, E., Kubota, K. and Makihara, H. (2003). Geographical structure of mitochondrial DNA in *Semanotus japonicus* (Coleoptera: Cerambycidae). *Applied Entomology and Zoology* 38: 339–345.

# Distribution of Nematodes (*Bursaphelenchus xylophilus*) in the Beetle *Monochamus alternatus* and its Exiting Transmission Way

**Yan-Xue Lai**

**Abstract** This paper reports the distribution of the pinewood nematode (*Bursaphelenchus xylophilus*) in the beetle of *Monochamus alternatus* by dissection under a microscope and the nematode transmission way exiting the body of the beetle by its mating and feeding. The results showed a small number of nematodes in the beetle tracheae; 87.9% nematodes distributed in the coelom of the beetle; numerous nematodes distributed in the phallus. This is a previously unknown pathway of nematodes, transmitting from the phallobase to the ovipositor by beetle mating, following infection of pine trees by egg-laying. Only a few nematodes infect pines during beetle feeding on twigs for sex maturity.

## Introduction

In East Asia, *Monochamus alternatus* is the most efficient insect vector of the nematode *Bursaphelenchus xylophilus* which causes pine wilt disease – the destructive disease of pine forests, which has been shown by Japanese researchers for more than 20 years. Since then, many experiments demonstrating that the beetles of *M. alternatus* or other longhorn beetles (*Monochamus sp.*) are the most important insect vectors of *B. xylophilus* have been done in many countries, such as China, Canada and USA. Scanning electron microscopy has furthered proven that the beetle tracheae system is the organ which carries the nematodes. Since then, the idea of nematodes located in the tracheal system of the beetles transmitting nematodes to pine trees through feeding was admitted and widely accepted. Further research by dissection of beetles under the microscope and feeding, mating and egg-laying, has shown that a large number of nematodes are distributed in the coelom, the segment of thorax and abdomen and the phallus, rather than of tracheae system, and the female beetle egg-laying is the main transmission pathway of nematodes infecting pine trees instead of beetle feeding.

Y.-X. Lai
Forest and Plant Quarantine Station of Ningbo City, Ningbo, 315000 Zhejiang, China
e-mail: zhangyf@cnluye.com

M.M. Mota, P. Vieira (eds.), *Pine Wilt Disease: A Worldwide Threat to Forest Ecosystems*, 

## Materials and Methods

Equipment utilized included an XSP-103B microscope, a NTX-2B stereoscope, surgery blades and scissors, and some glassware. A Nikon Coolpix700 digital camera and analysis software were used for image capture; several healthy *M. alternatus* beetles and several fresh one- or two-year-old pine twigs were used as biological material.

Pine logs which contain borers of *M. alternatus* in the wood but no eclosion holes were placed into plastic bags for capturing the beetles when emerging from the pine logs in the following year. As soon as the beetles emerged, they were captured and divided into two groups: some for dissection research, others fed on pine twigs for sex maturation and mating. At the same time, beetles that were feeding or mating in nematode-infected pine forests were captured. The movement and distribution of nematodes on the surface of the beetles were observed under the stereoscope. With the head and shoulder fixed, the beetle was divided into two parts: the head with viscera of digestive system and reproductive system; and the shell of the thorax and abdomen with the respiratory system. The respiratory system was separated from the shell using surgery scissors, and observing under the stereoscope. The shell (body wall) without respiratory system was cut into four parts: prothorax, mesothorax, metathorax and abdomen which was cut into five segment again. Then all the above segments and systems were placed on a glass slide and observed under a stereo microscope. All nematodes found on the slide glass were counted and photographed. Two newly emerged males and three female beetles were placed inside an insect-raising box with fresh one- or two-year-old pine twigs. After feeding and mating, the female ovipositor was dissected and the nematodes counted under the microscope. The feed-damaged pine twigs in the insect-raising box were cut into small sections, and placed in a glass container with water, from which the nematodes were separated. After 24 hours, the glass container was observed under a stereoscope, the nematodes separated from the feed-damaged pine twigs and counted (nematodes per gram of pine twig). These pine twigs in the box were named type I; while those pine twigs feed-damaged by beetles from pine forests named type II.

## Results and Analysis

### *Distribution of Nematodes on the Surface of Beetles*

In four beetles observed under the stereoscope, no nematode was found on the surface, only one nematode was found on the outside of the foreleg's tibia of one female. But the segment membrane between the thorax and abdomen contained a large number of nematodes (Fig. 1).

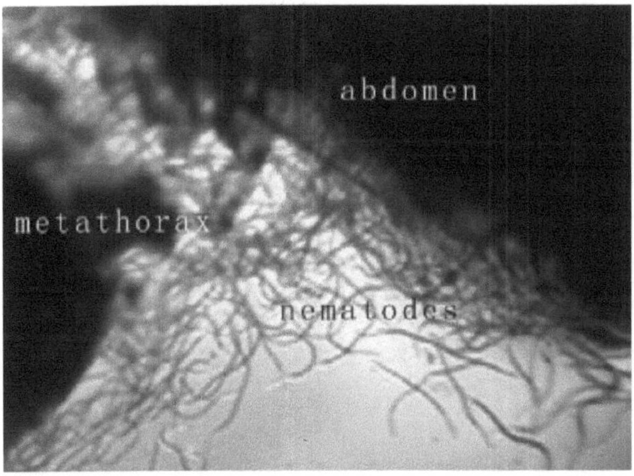


**Fig. 1** A large number of nematodes in the segment between thorax and abdomen

## *Distribution of Nematodes in the Viscera of Beetles*

The viscera consisted of digestive system, respiratory system and reproductive system. The former two systems presented a tubular structure that could be described as a "tube within a tube"; the ovipositor also presented a tubular structure. Microscope examination found nematodes distributed in the visceral organs (Table 1).

It can be seen from Table 1, that nematodes were found in the outer tube of the above three systems, however, no nematode in the inner tube of both the digestive and the male reproductive system; there were nematodes in both the respiratory and the female reproductive system, which showed that the tubular structure can carry nematodes.

It has been stated by Japanese researchers that the tracheal system is the only organ of the beetle carrying nematodes, but only few nematodes exist in the tracheal system after checking 20 beetles under the microscope, and only 25% of the beetles had nematodes distributed in the tracheal system.

**Table 1** The distribution and number of nematodes in insect adults

| adults | insect number | digestive system | | respiratory system | | reproductive system | | | |
|---|---|---|---|---|---|---|---|---|---|
| | | | | | | male | | female | |
| | | inside | outside | inside | outside | inside | outside | inside | outside |
| female | 1 | 0 | 5 | 115 | 2 | – | – | 15 | 8 |
| | 2 | 0 | 7 | 73 | 0 | – | – | 9 | 0 |
| | average | 0 | 6 | 94 | 1 | – | – | 12 | 4 |
| male | 1 | 0 | 3 | 32 | 4 | 0 | 317 | – | – |
| | 2 | 0 | 0 | 18 | 0 | 0 | 680 | – | – |
| | average | 0 | 1.5 | 25 | 2 | 0 | 498.5 | – | – |

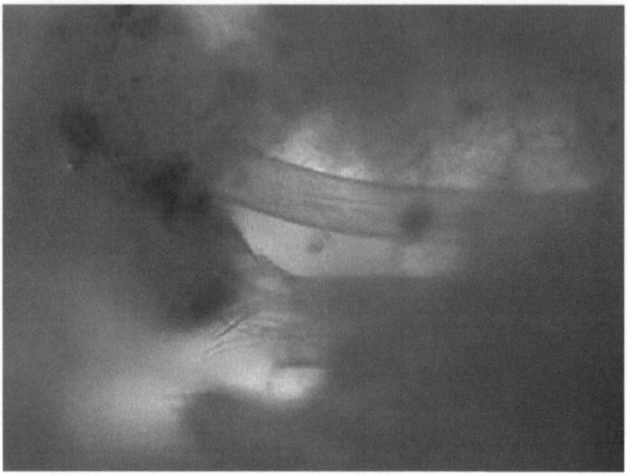

**Fig. 2** Nematodes in small trachea

The number of nematodes is related to the diameter of the tracheae. Smaller diameter tracheae contain less nematodes (Fig. 2); bigger ones contain more nematodes (Fig. 3). Nine pairs of spiracles are distributed in the thorax and abdomen of the beetle. One pair is called the prothorax spiracle, and is located on both sides of the prothorax. The spiracle orifice is located posteriorly, its long diameter being about 1.0 mm, pouch-shaped, presence of dense hair along the orifice (Fig. 4), two tracheae with a diameter of about 0.1 mm under the pouch stretched to the beetle head respectively. It is generally considered that the prothorax spiracle existed in the embryo stage and disappeared in the imago stage. However, in the beetle *M. alternatus* that spiracle still exists.

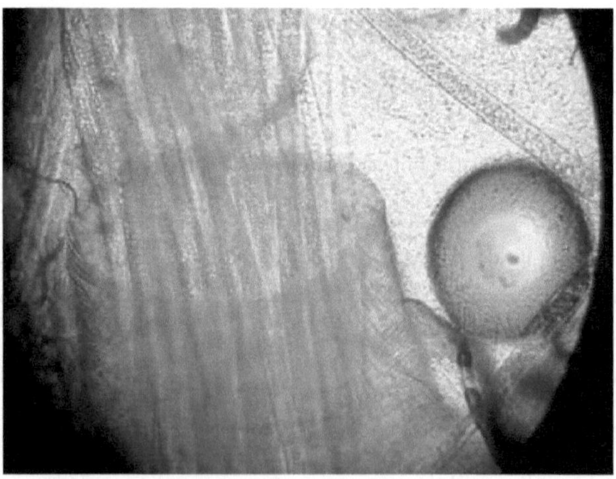

**Fig. 3** Nematodes in big trachea

**Fig. 4** Spiracle of prothorax

The mesothorax spiracle is located in the point of intersection of the mesopleuron with the metaepisternum, the spiracle orifice being outward and pouch-shaped, hair along the orifice, the long diameter is about 2.2 mm, under which over 15 tracheae which diameter was 0.12 mm connect, and open into the mesothorax and metathorax respectively. Every large trachea is divided into many branches.

The metathorax spiracle is located in the point of intersection of the metanotum and the first abdominal tergum, distributed in both sides of the metanotum, the spiracle orifice being upwards and pouch-shaped, dense hair along the orifice, the long diameter is about 2.5 mm, under which over 20 tracheae connect, and spread into the metathorax and abdomen (Fig. 5), which diameter is 0.13 mm on average. Every big trachea is divided into many branch tracheae.

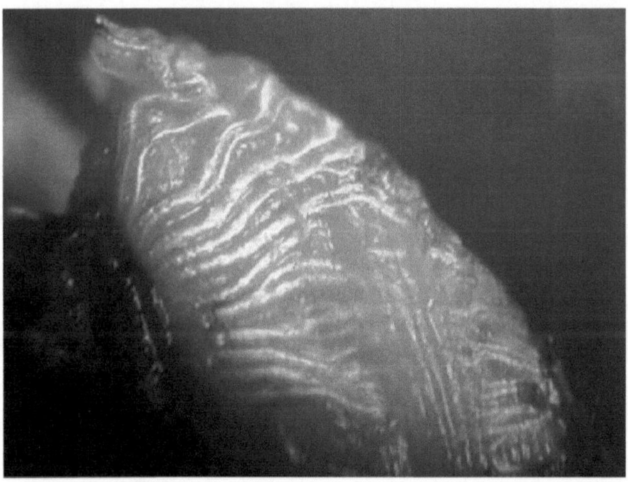

**Fig. 5** Tracheal system of metathorax

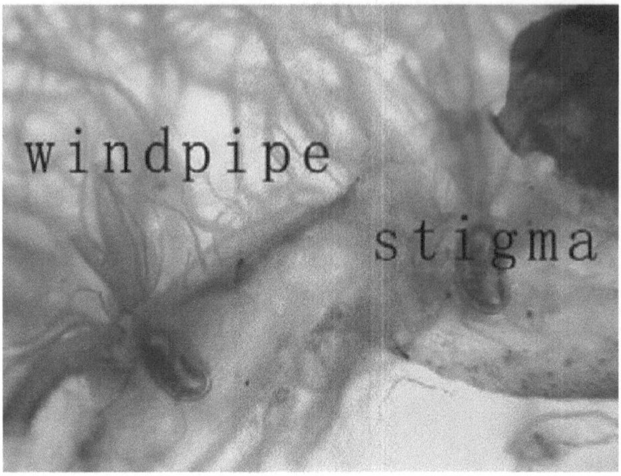


**Fig. 6** Spiracle and tracheal system of abdomen

There are 6 pair of spiracles in the abdomen, no spiracle in segment I, each pair of spiracles is distributed on both sides of the abdominal tergite from segment II to VII respectively, ellipse-shaped, dense hair along the orifice (Fig. 6), the long diameter is about 0.35 mm, under which the main tracheae are connected with divided branches like tree-roots opening into the dorsal side and ventral side of the abdomen. There is a large vessel, air-sac-shaped, connecting between spiracles lengthwise, this structure indicating that perhaps *M. alternatus* specialized in flying (Kobayshi, 1984).

There are no nematodes in the spiracles or tracheae of the prothorax and mesothorax. The metathorax spiracle is of such a trachea size that it can carry many nematodes. In the abdomen, the diameter of the trachea is so thin that it is difficult to carry

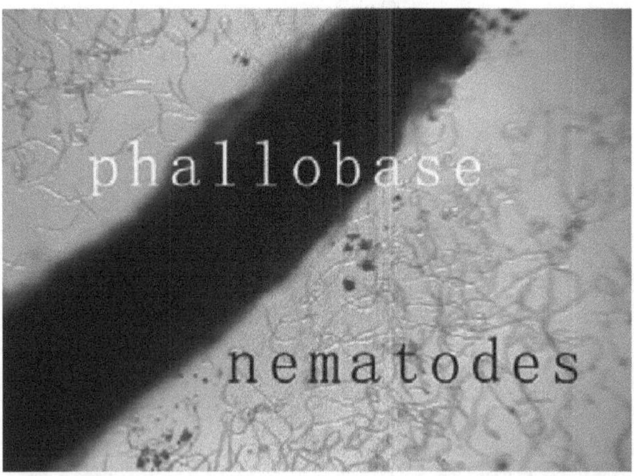


**Fig. 7** Nematodes beside phallus

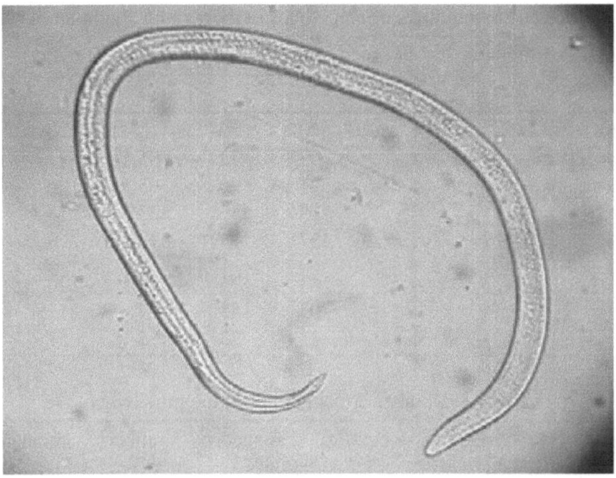

**Fig. 8** Larva shape of nematode

many nematodes; nematodes are only present in the spiracle of the last segment of the abdomen. In general, the trachea is so long, thin and winding that it is not suitable for carrying many nematodes.

The phallus was observed under the microscope, and a large number of nematodes was found in the membrane surrounding the phallus (Fig. 7). On average, 498.5 nematodes were counted in both phallus, while only four nematodes, on average, were found in the two ovipositors. The nematodes surrounding the phallus are all durable larvae with a mucronated tail (Fig. 8).

## *Distribution and Numbers of Nematodes in the Body Cavity of the Thorax and Abdomen*

The segments of the thorax and abdomen, without viscera, were dissected and then observed under the microscope, where many nematodes were found (Table 2). There were no nematodes in prothorax; in the mesothorax of the male and female, 59.5 nematodes were counted on average; in the metathorax, 87.5% nematodes were found on average. The abdomen is the main distribution area (Fig. 9), in which nematodes adhere to the adipose tissue inside the body wall and are bundle-shaped. The nematodes in the female abdomen were 87.4%, whereas in male they were 87.9%. The distribution of nematodes in the abdomen segments of males displayed little difference from the first to the last segment; while in females the nematode distribution was significantly different, 523.5 nematodes were found in the first segment, and only 18 nematodes was found in the last segment.

**Table 2** Distribution and numbers of nematodes in thorax and abdomen of insect adults

| No. | position | segment | slices | female | | | male | | |
|---|---|---|---|---|---|---|---|---|---|
| | | | | 1 | 2 | Ave. | 1 | 2 | Ave. |
| 1 | prothorax | – | 4 | 0 | 0 | 0 | 0 | 0 | 0 |
| 2 | mesothorax | – | 4 | 99 | 72 | 85.5 | 21 | 46 | 33.5 |
| 3 | metathorax | – | 4 | 91 | 96 | 93.5 | 100 | 63 | 81.5 |
| | | Seg. 1 | 4 | 415 | 632 | 523.5 | 115 | 210 | 162.5 |
| | | Seg. 2 | 4 | 327 | 218 | 272.5 | 213 | 286 | 249.5 |
| 4 | abdomen | Seg. 3 | 4 | 268 | 173 | 220.5 | 120 | 214 | 167 |
| | | Seg. 4 | 4 | 203 | 155 | 179 | 212 | 230 | 221 |
| | | telson | 4 | 13 | 25 | 18 | 57 | 16 | 36.5 |

**Table 3** Number of nematodes in the pine twigs

| type | twig age | Number/ sample | weight/g | damaged area/% | Number/ nema- todes | nematode s/g | average |
|---|---|---|---|---|---|---|---|
| I | two-year-old | 4 | 6.7 | 58 | 5 | 0.75/g | 0.62/g |
| | one-year-old | 5 | 4.6 | 50 | 2 | 0.43/g | |
| II | two-year-old | 4 | 7.3 | 43 | 2 | 0.27/g | 0.26/g |
| | one-year-old | 5 | 4.1 | 60 | 1 | 0.24/g | |

## *Pathway of Nematodes Exiting the Beetles*

The nematodes are transmitted from the male to the female ovipositor by mating, then to the phloem of pine trees by egg-laying. After dissecting three mated females, respectively 77 nematodes, 53 nematodes, 31 nematodes were found in the ovipositor, while only four nematodes were found on average in the ovipositor of

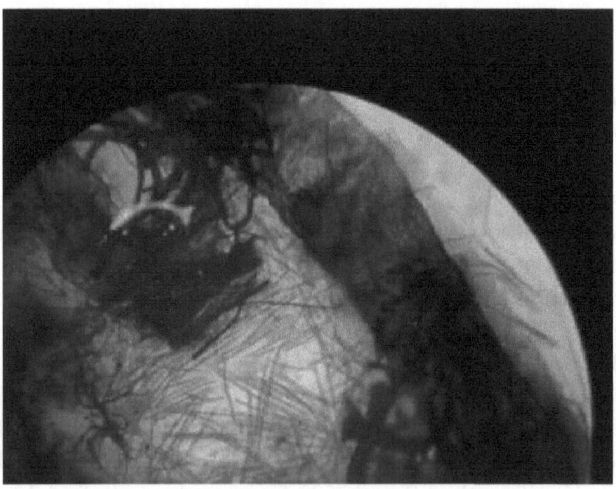

**Fig. 9** Nematodes in the body cavity of abdomen

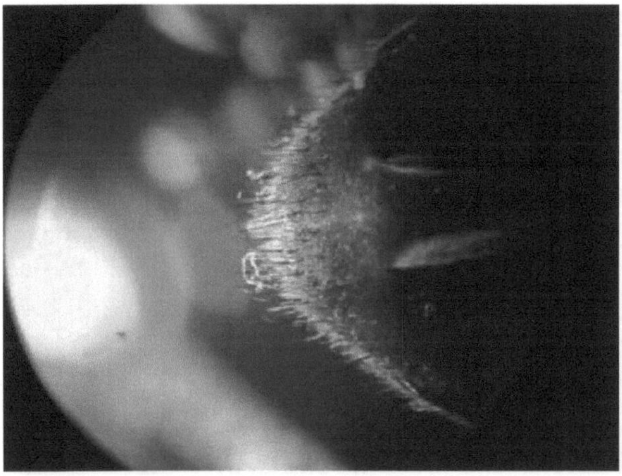

**Fig. 10** Nematodes leaving from caudal hole of the beetles

non-mated females, with nematode numbers increasing 12.4 times after mating. The results proved that the nematodes are transmitted from the phallus to the ovipositor by mating, then to the pine trees by egg-laying (Wingfield, 1983; Wingfield and Blanchette, 1983; Yanxue et al., 2000c).

Nematodes were isolated from the tail region of adults when feeding for sex maturity. It has been observed that nematodes assembled in the caudal area before infection of pine twigs in feeding for sex maturity, but no explanation has been given. Because only nematodes isolated from the spiracles were considered, when past the outside of body wall they congregated in the caudal area at that time. In fact, nematodes present in the coelom infect pine trees through the caudal hole directly, so it is difficult to find nematodes on the body wall when the beetle is feeding (Fig. 10).

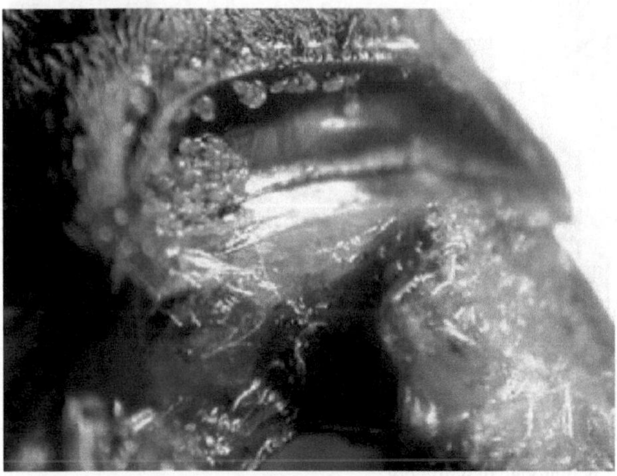

**Fig. 11** Group of acarid feeding on nematodes between the thorax and abdomen of adults

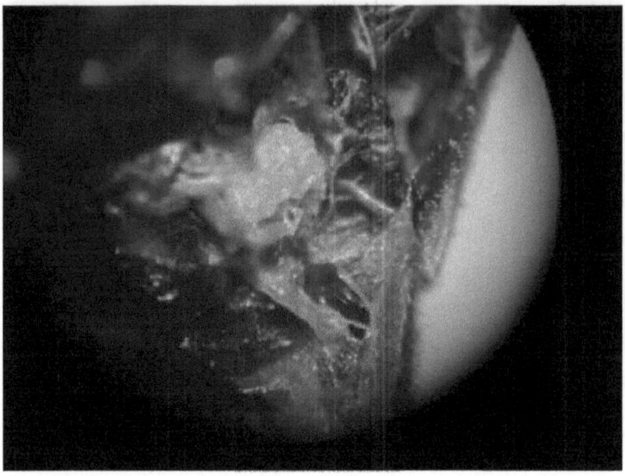

**Fig. 12** White dot in metathorax spiracle

Nematodes could hardly be transmitted effectively and isolated from the spiracle. Firstly there are many acarids feeding on nematodes on the surface of the adults; secondly because the isolated nematodes tangled into a ball and became difficult to creep. The author has once observed the adults after feeding three days, and found a "white dot" (Fig. 12) in the left and right spiracle of the metathorax respectively, which is the "nematode ball" under microscope (Fig. 13).

Nematode transmission by beetle feeding is not efficient. The type I twigs, of which 54% of the surface area was damaged by beetle feeding, were isolated and observed under the microscope, with seven nematodes being found, and only 0.62 nematodes per gram of twig on average. The type II twigs, of which 51% of the

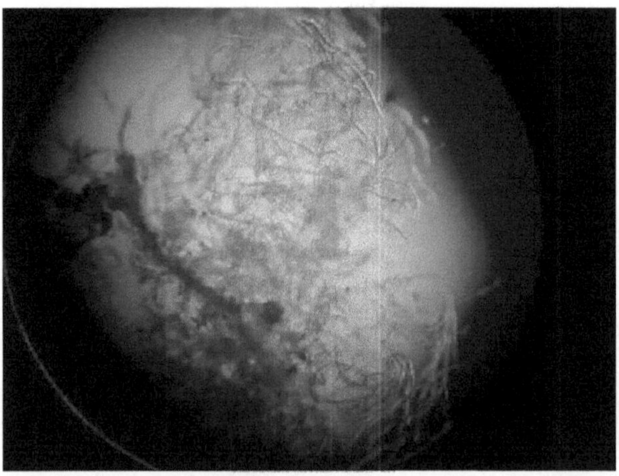

**Fig. 13** Group of nematodes in white dot

surface area was damaged, showed three nematodes, and there were only 0.26 nematodes per gram of twig on average. The nematode numbers are about equal between one-year-old twigs and two-year-old twigs.

## Discussion

The respiratory system is not the main part of the beetle carrying nematodes, since 87.5% nematodes are distributed in the abdomen and the reproductive system. This indicates the importance of two pathways in nematodes transmission by *M. alternatus*, rather than one pathway; at the same time, it has been proven that nematodes can be isolated from beetles and inspected directly and quickly by dissection under a microscope instead of the Baermann funnel method after 24 hours and then inspected under a microscope.

The theory that mating and egg-laying is the main pathway of transmission nematodes while feeding is a low effective transmission way, provided a new idea for preventing and controlling pine wilt disease. Our experience indicated that through spraying PEM on the stem except the crown of pine trees, which reject beetle egg-laying without resisting beetle feeding for sex maturity can cut off the pathway of nematodes infecting pine trees and reduce the numbers of wilted pines caused by the nematode (Yanxue et al., 2000a, b); at the same time, the number of wilted pine trees can be clearly reduced by placing bait of pine fresh log and branches to attract the beetle during egg-laying (Yanxue et al., 2001).

It is also of ecological significance that the beetle abdomen and phallus can carry a large number of nematodes which are transmitted by mating. Firstly, an unknown nematode infecting pathway is formed, which makes the nematodes capable of infecting the phloem of pine trees without exposure, so as to increase the nematode infecting efficiency. Secondly, a multiple diffusion mechanism is formed, which makes epidemic and diffusion of pine wilt disease very quickly. Because one male with nematodes fly into a new forest and cause many females to carry nematodes by mating, then the nematodes infected pine by egg-laying and could make many pine trees died; while one female with nematodes flies into a new forest could make only one pine died by egg-laying at most.

## References


BAOJUN, Y., KEGONG, Z., YUANSHENG, Z., ET AL., Control and epidemic of pine wilt disease in China. China Forestry Publishing House.

HURUI, C. (1988). The advance in finding and research on nematode disease of wilt pines. *Plant Quarantine of China* 2(1): 11–16.

IKEDA, T. (1980). Attractions for the Japanese pine sawyer Monochamus alternatus Hope. *Applied Entomology and Zoology* 15: 23–26.

JINNIAN, Z., CHANGING, Z., AND JIANCHANG, D. (1999). Studies of emigration of *Monochamus alternatus* and its ability of carrying nematode. *Forest Research* 12(6): 572–576.

KIYOHARA, T. AND TOKUSHIGE, Y. (1971). Inoculation experiments of a nematode, *Bursaphelenchus* sp., onto pine trees. *Japanese Forest Society* 53: 210–218.

KOBAYSHI, F. (1984). The Japanese pine sawyers beetle as vector of pine wilt disease. *Annual Review of Entomology* 29: 12–15.

LIAUXIANG, Z. (1986). Vector, pathology and control of pine wilt disease. *Journal of Liaoning Forestry Science and Technology* (6): 75–77.

LIU, W. AND YANG, B. (1995). Hybridization between *Bursaphelenchus xylophilus* and *B. mucronatus*. *Proceedings of international symposium on pine wilt disease caused by pine wood nematode 1995*, Beijing, 45–52.

MAMIYA, Y. AND ENDA, N. (1972). Transmission of *Bursaphelenchus lignicolus* (Nematoda: Aphelenchoididae) by *Monochamus alternatus* (Coleoptera: Cerambycidae). *Nematologica* 18: 159–162.

MANIYA, Y., TAMURA, H., AND FUTAI, K. (1992). Pine wilt disease caused by *Bursaphelenchus xylophilus*. *Japanese Nematode Research Society* (11): 201–219.

WINGFIELD, M.J. (1983). Transmission of pine wood nematode to cut timber and girled trees. *Plant Disease* 67: 35–37.

WINGFIELD, M.J. AND BLANCHETTE, R.A. (1983). The pine wood nematode, *Bursaphelenchus xylophilus* Minnesota and Wisconsin insect associate and transmission studies. *Canadian Journal of Forest Research* 13: 1068–1076.

WINGFIELD, M.J. AND BLANCHETTE, R.A. (1984). Pathogencity and insect associates of the pine wood nematode in the North Central States. *Proceedings of the United States-Japan Seminar the Resistance Mechanisms of Pine Against Pine Wilt Disease 1984, May 7–11*, 36–39.

XIMIN, C. AND ZHIHUA, H. (2000). Study on the technology of early monitor of pine wilt disease using the adult of Monochamus alternatus. *Journal of Zhejiang Forestry Science and Technology* 20(3): 1–2.

YANXUE, L., LINXIANG, Y., YONGPIN, Z., ET AL. (2000a). First aid techniques and mechanism for dying pine trees damaged by pine wilt disease. *Journal of Zhejing Forestry College* 17(4): 404–409.

YANXUE, L., YONGPIN, Z., LINXIANG, Y., ET AL. (2000b). An experiment report on controlling pine wilt disease by spraying pesticides named PEM. *Journal of Jiangsu Forestry Science and Technology* 27(5): 44–47.

YANXUE, L., YONGPING, Z., LINXIANG, Y., ET AL. (2000c). The relations the fire of pine forest and epidemic of pine wilt disease. *Forest Research* 13(2): 182–187.

YANXUE, L., LINXIANG, Y., YONGPIN, Z., ET AL. (2001). Trapping adults of *Monochamus altecnatus* with chemically-treated pine logs and branches. *Journal of Zhejiang Forestry College* 18(1): 60–65.

# Part V
# Ecology and Modeling

**Hugh Evans and Kazuyoshi Futai**

## Summary

The credentials of *Bursaphelenchus xylophilus* as a highly damaging, tree-killing pest are acknowledged universally. A fundamental requirement in relation to the potential damage that could be caused by the pinewood nematode is to improve knowledge and predictability of the key drivers leading to pine wilt disease expression in vulnerable forests. Interactions between the nematode and potential host trees are clearly central to this understanding, but the dynamics of these interactions have a strong spatial component, driven by a further level of interaction with the beetle vectors (genus *Monochamus*) of the nematode. Thus the spatial dynamics of vector, nematode and potential host trees can be regarded as the fundamental drivers in establishment and dispersal of the pinewood nematode in a given forest system. Whether transfer of nematodes to trees during maturation feeding by *Monochamus* spp vectors leads to wilt expression and death of host trees depends on a further series of factors, including host resistance and, particularly, local eco-climatic factors, of which, temperature is regarded as one of the most important.

Recent research in those regions of the world where *B. xylophilus* is not a natural component of the conifer forest system has provided insights and impetus to the need to improve prediction of both the spread and damaging potential of PWN. Particular progress has been made in understanding and modelling the interactions between PWN and host trees in Japan and, more recently, in Portugal, two countries where the nematode is an exotic component within the native and planted conifer forests. The ecological factors that determine the spatial distribution of pinewood nematode have been studied and developed into a spatial model by Pereira & Roque, who described the principal parameters that were used to develop predictions of those areas of Portugal most at risk of establishment and wilt expression. Key elements of their analysis were climate data, landscape metrics, distribution of land-use types and classifications (based on remote sensing), distributions of fire risk and current data on distribution of PWN and its vector *M. galloprovincialis*. Since precise information on eco-climatic variables was not available, the authors developed an approach that assumed that pre-disposing factors for poor tree condition (e.g. water stress, fire risk, high temperatures, low rainfall) could act as proxy variables for the likelihood of wilt expression. These data were linked to growth parameters for *Pinus pinaster* in Portugal and predictions were made of both likelihood of nematode dispersal and establishment as well as potential wilt expression. This approach is being explored and the outcomes have provided a good match to current distribution of the disease syndrome in Portugal and have also predicted a dispersal corridor from the infested zone to currently unaffected forests in the country.

The heuristic and probabilistic modelling approach adopted by Pereira & Roque was complemented by a process-model based approach by Evans, Evans & Ikegami. The latter was based on detailed tree-growth models describing tree-water-carbon relations and interactions between tree hosts and PWN. This modelling process incorporates vertical and horizontal water relations integrating soil, tree and atmosphere parameters leading to prediction of Gross Primary Productivity of host trees, with and without PWN. It, thus, uses local eco-climatic, soil and tree parameters to provide detailed site-based predictions of wilt expression. Since the model is essentially simulating photosynthesis and water-balance of the tree, presence of PWN in the cambial zone and, subsequently, in the xylem can be simulated in relation to carbon usage and, particularly, evapotranspiration. The model, consequently, predicts the gap between actual and potential water requirements of the tree, described by the degree of cavitation in the xylem and, ultimately, the likelihood of wilting and tree death. Model predictions for different sites in Portugal and, as a control, in England indicated that wilt expression would occur in the former location but would have different dynamics in different parts of the country. No wilt was predicted for sites in England. The most interesting prediction for Portuguese conditions was that in some sites, notably in the existing infested zone in the Setubal Peninsula, nematode infestation was not always likely to lead to wilt expression in the year of inoculation during maturation feeding. Thus, some trees which have nematodes present may remain asymptomatic during the winter but succumb to the disease in the following year. The likelihood of this happening is predicted by modelling which, therefore, allows for local site conditions.

The two modelling approaches have provided new insights into understanding and predicting likelihood of dispersal and subsequent wilt expression when PWN enters a new forest system. One of the key outcomes was the prediction that some trees remain asymptomatic after nematodes are introduced through maturation feeding by *Monochamus* spp vectors. This theme was elaborated by Futai & Takeuchi who described detailed work on PWN-infested trees in Japan. Central to their study was the observation that some pine trees appeared to be healthy, having a green crown and no apparent visual symptoms, but had low or nil resin flow. A novel molecular technique was used to assess for presence of PWN in these asymptomatic trees and it was possible to determine densities as low as 1 nematode in 80 g of pine wood. This nested PCR technique offers considerable potential for accurate assessment of PWN presence in host trees, even when no symptoms are evident during visual assessment. The authors also provided evidence that asymptomatic, but PWN-infested, trees were able to survive for more than one year in the field without symptom expression and suggested that these would be overlooked during eradication campaigns. Of critical importance under such circumstances was the likelihood that vector beetles would be attracted to the trees for breeding and could, therefore, result in further transmission of the nematode, even from apparently asymptomatic trees. Futai & Takeuchi assessed this in the field and found that such trees produced volatiles attractive to *M. alternatus* and so could act as future nematode carriers.

The integrated approach to improving knowledge of the transmission dynamics of PWN demonstrated by the three papers in this session provide clear evidence that campaigns to manage the nematode populations in the field must account for many factors that are not immediately apparent. However, the combined modelling approaches, supplemented by refined monitoring and detection techniques, now provide the tools for improved detection and management of this important pest in the future.

# Modeling PWN-Induced Wilt Expression: A Mechanistic Approach

**Sam Evans, Hugh Evans and M. Ikegami**

**Abstract** Pinewood nematode [PWN] (*Bursaphelenchus xylophilus*) is a saprophytic organism that usually exploits dead or dying trees. Adult feeding by its main distribution vector, longhorn beetles of the genus *Monochamus*, can introduce the organism into healthy trees, without any apparent effects on plant activity. Unusually, and in a number of geographical regions, PWN has been shown as the causal agent for the death of a mature host tree within a short time after introduction of the nematode to the vascular system of the tree. A body of evidence has demonstrated that, with significant seasonal variation in water availability combined with elevated temperatures during summer, the organism induces wilt expression in susceptible tree species. Under certain environmental conditions the severity of wilt expression can result in death of the host plant. This paper uses current understanding of plant-host interactions as the basis for assumptions incorporated into a mechanistic model describing relevant process dynamics. The model describes plant host physiological behaviour following PWN infestation that ultimately may result in death of the host plant. Several simulations are run for sites in the Iberian Peninsula, including those where fatal infestation of pines with PWN occur, indicate a high likelihood of host death, both immediately and in the year following infestation. Globally in regions where the nematode occurs and where environmental conditions do not result in significant tree stress, PWN does not result in wilting and host death. An ongoing observational experiment using PWN-infested trees, suggests a good correlation between simulated and observed results. It is proposed that, with further refinement and validation, the model may be suitable for developing a generic framework to predict the vulnerability of different hosts to the organism across a range of geographical regions.

S. Evans
Forest Research, Alice Holt Lodge, Farnham, Surrey, GU10 4LH, England, UK
e-mail: sam.evans@forestry.gsi.gov.uk

M.M. Mota, P. Vieira (eds.), *Pine Wilt Disease: A Worldwide Threat to Forest Ecosystems*, 

## Introduction

Pinewood nematode (PWN), *Bursaphelenchus xylophilus*, is a saprophytic organism, in the Parasitaphelenchidae family, usually exploiting dead or dying coniferous tree species. It is native to North America, where it is widespread, but it has been distributed internationally through trade and its present geographical distribution ranges from Japan, Korea, Taiwan to China and also Portugal.[1] A key element of its life history, which is the initial determinant of its ultimate impact on living potential host trees, is maturation feeding by adults of its principal distribution vector, longhorn beetles of the genus *Monochamus*, (Coleoptera: Cerambycidae). During the feeding phase in the crowns of living trees, the vectors can introduce *B. xylophilus* into healthy trees, which in its native range has no apparent effect on plant activity (Futai and Furuno, 1979; Rutherford and Webster, 1987). Unusually, and in a number of geographical regions, PWN has been identified as the causal agent for the death of mature, healthy host plants within a short time after introduction to the living tree following maturation feeding by adult vectors (Mamiya and Enda, 1972; Linit, 1989). The symptoms of rapid wilting of trees have led to the description of the syndrome as Pine Wilt Disease (PWD) (Kiyohara and Tokushige, 1971; Mamiya and Kiyohara, 1972; Kishi, 1995). The nematode is now known to be responsible for large-scale tree death in Japan, China, Korea and, since 1999, Portugal, and is identified as a major driver to successional processes in semi-natural stands (Fujihara et al., 2002), and a corresponding source of considerable financial loss to the timber industry as well as imposing restrictions on movement of conifer wood in international trade (OEPP/EPPO, 1986, 1989).

Although dispersal of the nematode is intimately linked to the dynamics of its only known effective vectors in the genus *Monochamus*, with considerable literature on the dispersive phase of its cycle (Mamiya and Enda, 1972; Linit, 1988), there is still much uncertainty concerning the factors that determine whether a tree that is infested through maturation feeding in the canopy is likely to succumb to PWD. The literature indicates that expression of PWD results from complex interactions within the host tree. This involves PWN movement within the tree, feeding behaviour, population dynamics, and effects on its host by disrupting physiological processes leading to destruction of cell structure and content. However, there is no mechanistic description of this process, which is essentially a 'black box'. At a generic level, the severity and time course of infestation includes visible wilt expression, accompanied by early reduction of resin flow (Mamiya, 1972), chlorosis and stem/branch cell necrosis, followed by rapid death of the tree (Mamiya, 1980). In turn, symptom severity varies according to the susceptibility of the host species (Futai and Furuno, 1979; Guiran and Bouolbria, 1985; Linit and Tamura, 1987; Yang, 1987; Baojun and Qouli, 1989). It is further apparent that certain environmental conditions leading to stress of a healthy host, primarily elevated temperatures and restricted

---

[1] http://www.eppo.org/QUARANTINE/nematodes/Bursaphelenchus_xylophilus/BURSXY_map.htm

water availability, are required for visible symptoms to occur or for death to follow. However, the internal processes of the nematode-tree interaction and their linkage to external environmental factors are poorly understood, so that prediction of wilt expression arising from PWN presence can currently only be achieved using broad environmental parameters that, inevitably, provide only general predictors of wilt expression. The ecological and host tree suitability indices used by the University of Evora in the PHRAME topic, suggest that regional variation in climatic and site factors can provide risk predictions at regional scales (M. Pires de Fonseca, personal communication). This approach fits with the predicted range of potential wilt expression in both Japan and Europe which is based on July or August isotherms exceeding 25.2°C (Yokobori, 1986). While this broad-brush approach from the literature provides useful predictors in Pest Risk Analysis for the expression of PWD (Evans et al., 1996), it does not allow more precise assessments at local and regional scales. The approach by the University of Evora, takes this a step further in integrating various datasets in correlation analysis. The final stage of interaction at the tree level is the ultimate measure in determining likelihood of tree mortality and particularly, provision of breeding resources for *Monochamus* spp. and hence acceleration of spread of the nematode and this is the subject of the current study by Forest Research, UK and Estacao Florestal Nacional, Portugal.

This paper reviews current understanding of the interactions of *B. xylophilus* and living trees and proposes a theoretical modelling framework describing the dynamics of physiological processes that can ultimately result in death of the host plant. The predictive ability of the modelling framework has been tested through a field experiment, with initial results presented here. It is further proposed that the model is suitable for developing a generic framework to predict the vulnerability of different tree hosts to PWN across a range of geographical regions.

## An Overview of Mechanisms

Although PWN is introduced to conifer hosts through maturation feeding by adult beetles, this does not always result in further development and breeding by the nematode or in death of the host tree. Normal transmission and survival of the nematode is believed to be achieved through entry to the tree during oviposition by female *Monochamus spp* (Fig. 1). This only takes place in dying or recently dead trees, which are attractive to the vectors and can support oviposition and subsequent larval development (Edwards and Linit, 1992). Both PWN and *Monochamus* larvae develop within the outer wood of the tree, the former feeding primarily on fungi. Towards the end of the *Monochamus* cycle, PWN larvae aggregate preferentially around the pupal chambers and, through responding to chemical cues (Maehara and Futai, 1997) moult to a specific stage called a 4th stage dauer larva. This stage is adapted to migrate to the callow adult stage of the vector and to enter the tracheae where it remains during subsequent flight by the newly emerged adults. The presence of other pathogens, such as blue-stain fungi, appears to increase the number

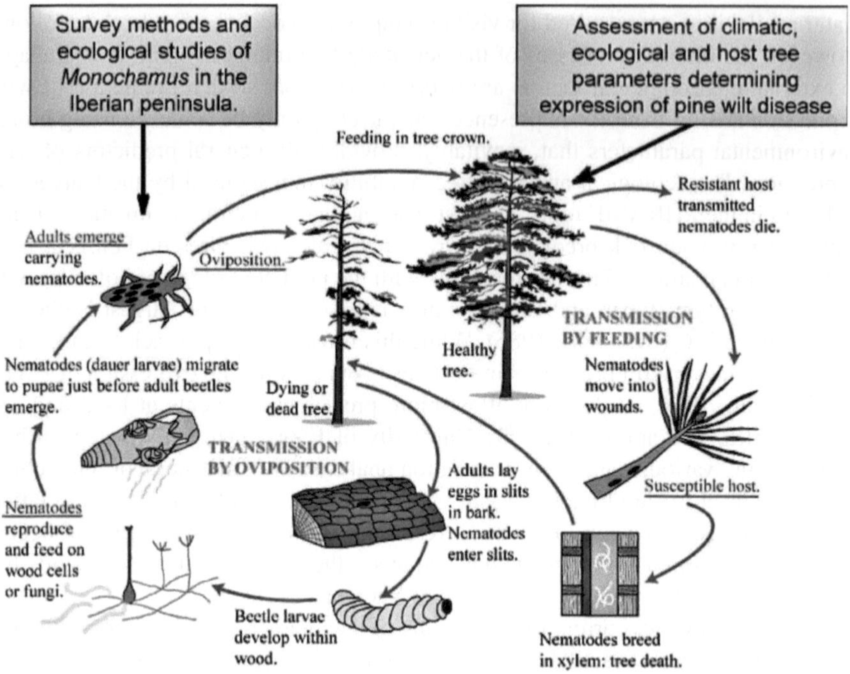


**Fig. 1** The interaction of *Bursaphelenchus xylophilus* and its *Monochamus* spp. vectors, in saprophytic and pathogenic phases of the cycle. After Wingfield (1987)

of nematodes aggregating around such chambers, by facilitating their movement through the disaggregating wood structure (Maehara and Futai, 2002). The relationship between PWN dauer larvae and *Monochamus spp.* adults appears to be co-evolved because 4th stage dauer larvae do not appear to be formed when nematodes encounter other xylophagous insects within their tree hosts. The behaviour of the 4th stage dauer larvae within the adult vectors increases their likelihood of being able to transfer to new hosts either during maturation feeding or during oviposition by the beetles. Thus, on emergence from the tree, PWN 4th stage dauer larvae are found in the tracheal system of adult beetle vectors, facing inwards to the distal end and outwards to spiracles in tracheae. After the nematodes change orientation, the proportion of tracheae with nematodes facing outward increases with beetle age (Aikawa and Togashi, 2000).

*Monochamus* spp. adult emergence tends to take place from late spring through the summer months, depending on temperature. Maturation feeding and oviposition take place throughout the adult activity period. Crucial to understanding the process of nematode entry and subsequent activity within a living tree, maturation feeding in the branches of the canopy exposes the sapwood, and PWN larvae are introduced through the feeding wound into the wood tissue (Togashi, 1985; Linit, 1989). Once introduced into the tree, the pioneer PWN population appears to decline during the first few days by up to 90% (Suzuki, 1984; Tamura, 1984; Mamiya, 1985) since most of them are captured at the feeding wound, with only a small proportion becoming established in the area immediately adjacent to the wound.

Studies on the distribution of inoculated nematodes in seedling trees, carried out within the EU PHRAME project by T. Schroeder and E. Sousa indicate that they move rapidly through the tree and gradually increase through reproduction, depending on temperature. The fate of pioneer populations of the nematodes depends on a range of host and environmental variables that are, currently, not well understood. If conditions are suitable for further development and spread within the tree, it is apparent that there are sizeable differences in PWN reproductive rates, irrespective of initial *inoculum* loading, probably influenced markedly by external environmental pressures such as temperature (Mamiya, 1975) and tree condition, making predictions of host dose responses difficult. From the literature and from studies within the PHRAME project it appears that PWN can disperse through the whole tree soon after introduction by moving along the living tissue pathway predominantly represented by the cambial zone. By feeding on this tissue the nematodes begin to break down cell content and structure, triggering local host defensive responses leading to increasing disruption of physiological processes. Evidence from the seedling studies carried out by T. Schroeder's team in the EU PHRAME project suggests a sequence of events that confirms rapid nematode migration prior to compromising the conductive system of a susceptible tree, leading to wilt expression:

(1) Formation of the initial population inside the host in a confined area around the site of entrance,
(2) Rapid migration and colonisation of all tree parts by a highly active part of the initial population of nematodes entering the tree,
(3) Establishment of the nematode inside the host through exponential growth,
(4) Retreat of the nematode into the basal parts of the tree.

These observations confirm literature data that indicate PWN migration of up to 150 cm per day in wood tissues (Kuroda and Ito, 1992) and the extent of tissue destruction appears proportional to the spread of the nematode. This suggests that PWN continues to migrate in the still living tree as increasing amounts of tissue structures collapse and, with increasing numbers, nutrient sources become scarce.

Hypotheses on the mechanisms of PWN-induced mortality in pine trees can be summarised into three broad categories of effects:

- The progressive physical destruction of living cells in wood arising from feeding by PWN. Histopathological evidence suggests that PWN tends to be found in undifferentiated, soft tissues such as cambium and living tissue (axial and radial parenchyma) as pathways of movement through the host (Myers, 1986, 1988). The presence of nematodes in resin canals is also reported frequently. Meristematic tissue and other living tissue (parenchyma and epithelium cells) in the sapwood are, therefore, exploited both as a food source and as a conduit for movement of nematodes through the tree. Compounded by its own defence responses (hypersensitive reaction), the ability of the host tree to replace damaged cambial tissues and its ability to translocate photosynthate and synthesise defence compounds is severely compromised.

- Destruction of cambial and storage tissue during feeding may also lead to progressive physical blockage of vascular tissue in the xylem, as a result of resin secretion following wounding (Fukuda et al., 1992), or embolism induced by occlusion of tracheid border pits by feeding-induced debris (Nobuchi et al., 1984a, 1984b; Kuroda et al., 1988; Kuroda, 1989).
- Toxins (i.e. cellulase) produced either by PWN (Odani et al., 1985; Yamamoto et al., 1986) or accompanying bacteria (or both) (Zhao et al., 2003), may lead to a phytotoxic reaction.

The above effects may act independently or, more likely, in concert leading to host death. Their relative contributions may also change according to local circumstance and the timing of PWN penetration in relation to the host's own phenological cycle as well as external environmental drivers. Consequently, the effects of nematodes on susceptible living host trees can be regarded as a combination of direct and indirect effects, promulgated through defence-induced biochemical reactions following release/leakage of host-synthesised volatile monoterpenes and other carbohydrates in the xylem, inducing progressive hydrophobicity and occlusion of conductive tissue, leading to cavitation (Kuroda, 1989) and, ultimately, tree wilt.

## Modelling the Likelihood of Wilt Expression at the Tree Level

The complex interactions between host, environment and pathogen would suggest recourse to a modelling solution to improve understanding of the process and to develop predictive solutions to improve risk assessment for this organism. Modelling is increasingly used to complement and integrate the hypothetico-deductive approach to experimentation, as a means of encapsulating the current, mechanistic knowledge base of process dynamics. In forests, tree growth is determined by the interactions between species and ecosystems driven by the terrestrial water, carbon and nitrogen cycles. A number of numerical models explicitly describing the interactions between trees and their environment have been developed over some years. In particular, a number of such process (or mechanistic) models are now available to predict interactions between trees and forecasted impacts of climate change (Woodward et al., 1995). Such models lend themselves well to describing the interactions between host, environment and biotic damaging organisms. The model used here, and briefly outlined below, is fully described in Evans et al. (2003).

### *The ForestETp Model*

ForestETP is an existing, fully coupled, point scale and daily time step soil-vegetation-atmosphere transfer (SVAT) model, which predicts vertical and lateral water movement through the soil-plant-atmosphere continuum as well as gross primary productivity (GPP). The model simulates relevant terrestrial hydrology processes (rainfall interception, vertical and lateral soil water movement, runoff, soil and canopy evaporation, and photosynthesis-coupled transpiration) for a forest stand of known structure, growing in locally determined soil and climate: the model structure is illustrated in Fig. 2.

At its core is the Penman-Monteith equation, extensively used in the literature to calculate free surface water evaporation and potential evapotranspiration. The equation simulates the theoretical rate (or reference evapotranspiration) at which water would be removed from the soil and plant surface by a reference crop, and does so by combining approximations of an energy balance and an aerodynamic formula, and without accounting for a structured canopy. In this model, the reference value is dynamically adjusted by the actual soil water available for plant growth, and further regulated by a number of physiological parameters such as (inter alia) species-specific stomatal conductance, air temperature, tree height and canopy size and structure. Of specific relevance in the context of potential for PWN to induce wilt, is the model's overall capacity to describe the impacts of dynamic environmental drivers on the tree's photosynthate production potential and water utilisation, determined in particular by nutrient availability (light, water and nitrogen), temperature and $CO_2$ for both current and, particularly, future climatic conditions.

The ForestETP model has been validated across a range of forest sites in Europe where, along with meteorological and soil water content, data on the water vapour and carbon dioxide content moving through the forest system are measured on a continuous basis. These data provide high-resolution values of hourly and daily changes in fluxes through the soil-plant-atmosphere system, useful for model validation.

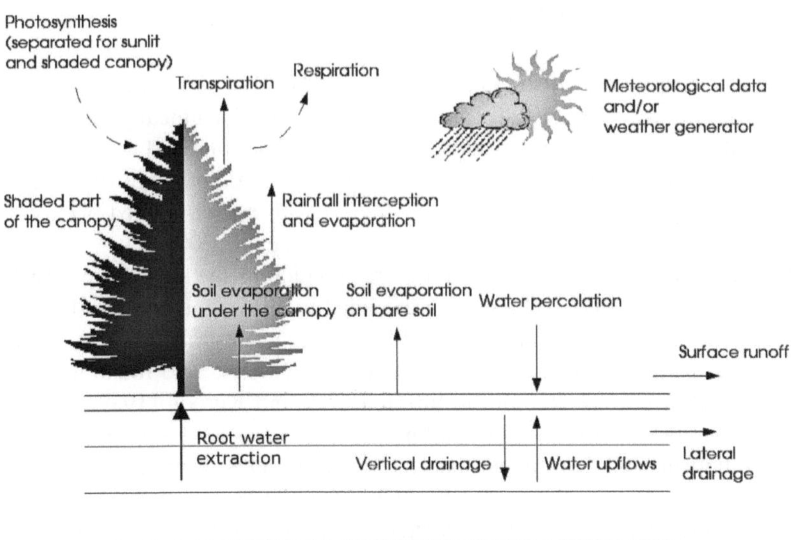


**Fig. 2** Schematic representation of water (*solid lines*) and carbon cycle (*broken lines*) processes simulated by ForestETP

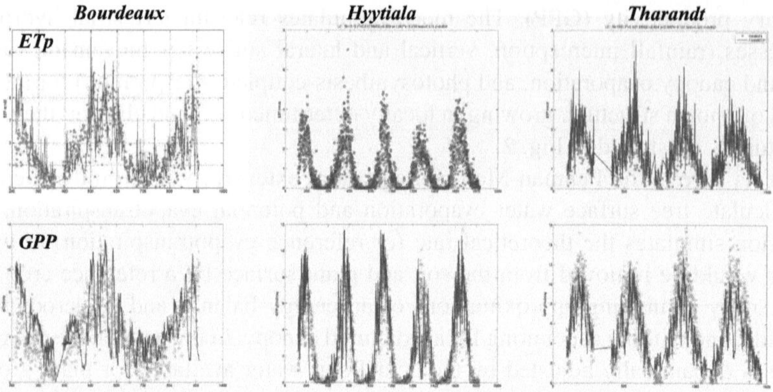


**Fig. 3** Comparison between simulated and observed values of water and carbon fluxes at stand scale, here expressed as evapotranspiration (ETp; above) and gross primary productivity (GPP; below). Green dots represent observed values from the field sites and blue lines represent estimated values from the simulation

Figure 3 illustrates the comparison between observed and simulated carbon dioxide (here expressed as gross primary productivity) and water fluxes (here expressed as evapo-transpiration or ETp) at three *Pinus sylvestris* stands in Europe. Spanning a number of growing seasons, the model appears well able to describe the fluxes of both carbon dioxide and water vapour observed in these forest systems.

## The ForestGROWTH Model

As part of a larger mechanistic model (ForestGROWTH), units of net carbon/units of water outputted by ForestETp at the tree level, are dynamically allocated to a number of tree compartments (stems, branches, foliage, fine and coarse roots). In wood, C units are further sub-divided between different tissue types (meristematic, conductive, support and storage), in accordance with a user-defined phenological scheme and a modification of pipe theory (Deckmyn et al., 2006).

In a new approach to understanding and predicting the impacts of PWN on living trees, ForestGROWTH has been extended to allow simulation of host-PWN pathogen interactions under a number of assumptions, described below. These integrate existing knowledge, principally after Ikeda (1996), who found that cavitation occurred at normal pressure in infested trees and Kuroda (1989, 1995) who indicated that the site of cavitation did not always correspond with the location of initial embolism, and determined that gaseous blockages were caused, at least in part, by volatile terpenoids; overall he found that following PWN infestation, the xylem became more vulnerable to cavitation (Kuroda, 1989, 1995; Ikeda, 1996). The model includes a number of assumptions concerning the dynamics of PWN after introduction to a susceptible tree (i.e. intrinsic susceptibility combined with environmental susceptibility).

**Assumption 1.** The PWN population will increase at known rates as a function of environmental drivers (e.g. internal tree temperature driven by air temperature).
**Assumption 2.** Once PWN is introduced into the living host tree, it immediately distributes throughout the tree.
**Assumption 3.** That following infestation the plant host will change its biomass allocation strategies; consuming reserves of non-structural carbon to activate defence responses to the PWN. The tree will divert photosynthate from growth, by increasing translocation of new or remobilised substances, to trigger local defence responses or to compensate for PWN-induced damage. In the longer term, carbon remobilisation from storage also contributes to lowering host resistance further, and to reducing the reserves available for flushing and re-growth in the year following infestation.
**Assumption 4.** The host tree becomes vulnerable to irreversible cavitation due to the direct and indirect effects of PWN, arising from PWN consumption/destruction of living cells, release and movement of cytoplasm containing molecules possibly inducing embolism (i.e. tannins) or cavitation (i.e. monoterpenoids) acting on tracheids in the xylem. When cavitation occurs, water conductance is compromised in some or all of the xylem portion where it occurs (Zimmermann, 1983); xylem water is then diverted elsewhere to avoid the affected part. Cavitation (the formation of air bubbles in conductive tissues) frequency can be measured as negative xylem water potential, approximated as the difference between the theoretical demand for soil water (potential transpiration) and the supply of available water to the canopy (calculated as actual transpiration). In dry regions, such as the Iberian Peninsula, cavitation may occur naturally in most trees during the summer and, depending on severity, can be reversible. Under extreme drought conditions, cavitation can be very extensive, as indicated by high negative xylem water potential, and may be irreversible resulting in wilt expression and tree death. When this occurs, all conductive tissue above the cavitation point, and canopy portions directly supported by affected conductive pipes, will exhibit reduced leaf water potential, lower transpiration and reduced photosynthesis. Correspondingly this lowers photosynthate availability for biomass production.
**Assumption 5.** By directly and negatively affecting tree ecophysiology in advance of visual symptoms being observed, PWN leads to host death (either in combination, or whichever occurs sooner) where:

- Net photosynthesis is less than respiration and growth maintenance costs over a threshold number of days.
- Irreversible cavitation accumulates over a threshold value above which the host tree cannot maintain the minimum necessary transpiration of water to the crown.

No assumptions are made about the number of maturation feeding events by the vector: their timing and number of nematodes are model input parameters.

## Model Definition and Parameterisation

In the model, cavitation occurs in a conductive pipe, with each pipe in turn supporting a known portion of canopy. Cavitation may be local and not always permanent but, while it occurs, all conductivity is lost in the affected pipes. Thus whole tree conductivity reduces according to the frequency of separate cavitation events. With increasing negative xylem pressure, cavitation

$$PLC = \frac{100}{(1 + e^{A(P-B)})}$$

may be reversed, allowing affected tracheids to refill with water. The impact of increasing negative xylem potential values is expressed as a Percentage Loss of Conductance (%) [PLC]. Using equation 1 from Martinez-Vilalta and Pinol (2002), PLC is expressed by a vulnerability curve proportional to xylem pressure (Fig. 4): where:

$A$ (2.0) determines the slope of the curve, and correlates with the variance of maximum size of border pits and diameter of tracheid (after Zimmermann, 1983);

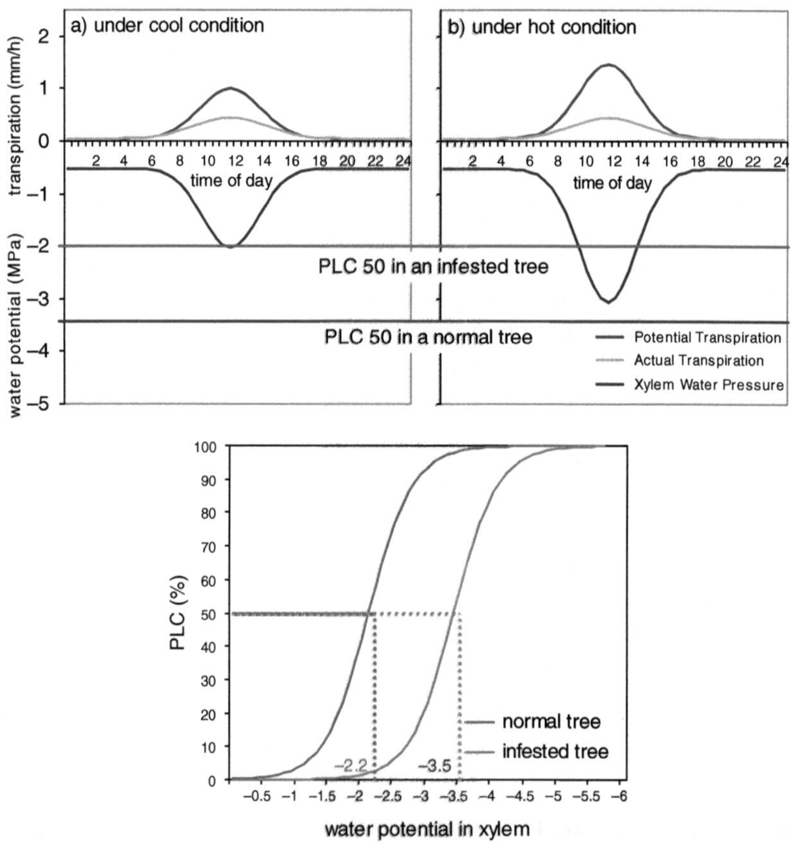


**Fig. 4** Simulated model interactions in xylem water potential between a healthy and a PWN-infested host, with interactions between climatic conditions and values of PLC50

*B* is the pressure causing a 50% loss of hydraulic conductivity (PLC50) [−3.0 to −3.9 MPa in pine species] (Martinez-Vilalta and Pinol, 2002);
*P* is xylem pressure (MPa).

There are two conditions under which xylem cavitation can occur (Fig. 4):
*Condition 1:* In *normal conditions*, expressed as the difference between the reference (potential transpiration Pt) and actual transpiration (At), as a function of climate, soils and tree characteristics. Where xylematic pressure recovers, the model assumes that tracheids affected by cavitation can be refilled overnight.
*Condition 2:* In the *presence of* PWN the range of negative xylem potential values is increased by assuming a vulnerability curve shift (set to −2.2 MPa, after Ikeda, 1996) in order to simulate the loss of conductive tissue through feeding, embolism and loss of tissue cohesion: the shift is also proportional to population size and pathogenicity of nematodes. Living cells, such as parenchymic tissue cells, are believed to have an important role in refilling xylematic tracheids (Tyree and Sperry, 1989; Canny, 1995; Tyree et al., 1999). In an infested tree, these cells are consumed by nematodes: the model therefore assumes that refilling cannot occur in the infested tree. With cavitation, and by reducing canopy water content and photosynthate production, this affects biomass productivity, the re-deployment of carbon reserves to defence and death by drought. When needle water potential drops below −5.0 MPa, irreversible damage is assumed and the needles cease to function.

## Modelling Experiments

The PWN process models have been run at the single tree level, simulating the impact on key physiological processes and on whole tree physiology following PWN inoculation. Simulations 1–2 describe conditions typical of those observed in the area of Portugal currently affected by pine wilt disease; simulation 3 is broader in scope, and extends to two sites in the Iberian Peninsula and a third UK site. In simulation 4, the model has been run for the same tree inoculated with nematodes under a range of pathogenicities; here the results were used to express the likelihood of mortality following infestation.

Species-specific model input parameters determining condition 1 are from the literature; in the absence of measurements for trees affected by PWN, input values to simulate condition 2 are approximated, and determined by the speed with which loss of leaf water potential has been observed (Sobardo et al., 1992; Salleo et al., 1996).

## Modelling Simulations

Modelling simulations have been undertaken of the trajectories of key physiological processes of a single tree following the inoculation of PWN and with increasing nematode population numbers. Simulations 1–2 are for conditions typical of those

observed in the area of Portugal currently affected by pine wilt disease; simulation 3 refers to 2 sites in the Iberian Peninsula and to a third UK site.
SIMULATION 1. The model is run for 1 year (ending 31 Dec = day 365), with PWN inoculation taking place on Julian day 180. Figure 5 provides trajectories for PWN population growth and key physiological parameters.
Three phases can be observed:

*Phase A*. The PWN population is present in low numbers. Tree physiological parameters respond primarily to local environmental conditions.

*Phase B*. Favoured by high air and, therefore, tree temperatures, the PWN population increases. In conjunction with summer drought, and as a result both of PWN movement through the host and of feeding, cavitation becomes extensive. This is expressed by significant reductions in leaf water potential, net photosynthesis and stomatal conductance. Further, the host responds by decreasing the amount of non-structural C to reserves and towards defence mechanisms.

*Phase C*. With the onset of autumn, and a significant increase in soil water content and with temperatures limiting further PWN reproduction, the host has an opportunity for some recovery in physiological processes and photosynthesis.

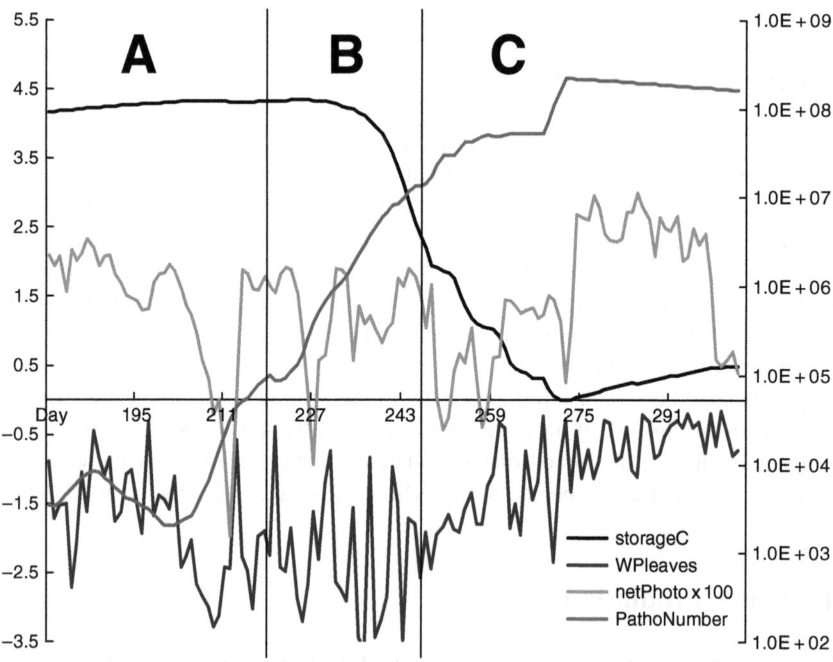


**Fig. 5** Simulation 1: simulated trajectories of physiological parameters following PWN infestation, over the period between day 180 and day 320

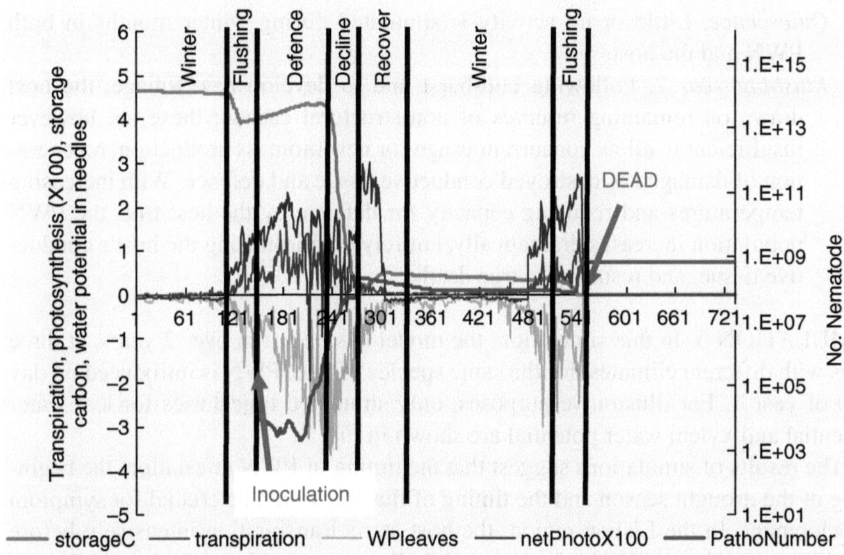


**Fig. 6** Simulation 2: simulated trajectories of physiological parameters following PWN infestation, over the period between day 1 and day 721

SIMULATION 2. Using the same conditions as Simulation 1, the model is run over a 2 year period. Figure 6 shows key model outputs over the period 1 Jan. (Day 1) to the date of tree death, occurring on day 541.

Six phases can be observed:

*Flushing year 1*. Prior to inoculation with PWN on day 165, normal tree physiological activity is simulated prior to and after the onset of flushing. Non-structural carbon reserves decrease slightly in order to support the production of new biomass and in advance of both older and new needles becoming sources of carbon.

*Defence*. Again, the PWN population is present in low numbers and does not affect tree processes adversely. Tree physiological parameters respond primarily to local environmental conditions.

*Decline*. Cavitation occurs both as a result of environmental pressure and as a result of growth in the PWN population, resulting in significant reductions in photosynthate production alongside diversions of carbon from growth and storage to defence.

*Recovery*. With increasing soil wetness and lower air temperatures in the autumn, the host tree increases photosynthesis concurrently with fixed or decreasing PWN population. This improves tree tissue conductivity and allocates modest amounts of photosynthate to parenchyma storage sites of non-structural carbon.

*Quiescence*. Little or no activity is simulated during winter months in both PWN and the host.

*Flushing year 2*. Following bud-burst and to develop new foliage, the host draws on remaining reserves of non-structural carbon: these are however insufficient to allow concurrent usage for new biomass production, regeneration of damaged or destroyed conductive tissue and defence. With increasing temperatures and reducing capacity for defence by the host tree, the PWN population increases dramatically, entirely compromising the host's conductive tissue, and resulting in tree death.

SIMULATION 3. In this simulation, the model has been run over 2 years, at three sites with different climates, for the same species of tree: PWN is introduced on day 180 of year 1. For illustrative purposes, only simulated trajectories for leaf water potential and xylem water potential are shown in Fig. 7.

The results of simulations suggest that the timing of PWN infestation, the beginning of the drought season and the timing of flushing could be crucial for symptom development. In the Lisbon region, the host starts transpiration intensively before cambium growth and PWN infestation while Bragança and London show high transpiration after cambium growth. PWN infestation occurs in the dry season in the Lisbon area where, due to soil drought, the host will tend to close stomata, thus transpiration is highly regulated. As a result, although soil water potential has high negative values, the host can, through stomatal regulation, maintain adequate xylem water potential in the needles, which acts as a protection mechanism to avoid irreversible cavitation. In contrast, PWN infestation occurs during a relatively wet season in Bragança, thus the host still transpirates intensively which results in high negative xylem water potential that can accumulate cavitation quickly leading to host mortality in the year of nematode infestation. Tree hosts in the Lisbon area can, therefore, survive until the second year but will face a high transpiration period before renewing damage tissues, which can result in wilting and death in the second year.

Where environmental pressures are less marked and conditions unfavourable to PWN population growth, as at the UK site, while a reduction is xylem conductivity after inoculation is predicted, low PWN numbers and higher soil water content result in only limited damage to the host, which appears able to localise the PWN population and to survive beyond the second year after infestation. In all simulations, the host is known to be intrinsically susceptible to PWD in areas with suitable climatic and environmental conditions.

SIMULATION 4. Given the sensitivity of the model to both host-environment and host-PWN interactions predicted by the prior simulations, this simulation was undertaken to predict the probability of tree death (as a percentage), following PWN infestation with various assumed nematode pathogenicities at selected sites in Portugal, where each has different environmental conditions. Figure 8 provides simulated values of the likelihood of host mortality occurring following inoculation.

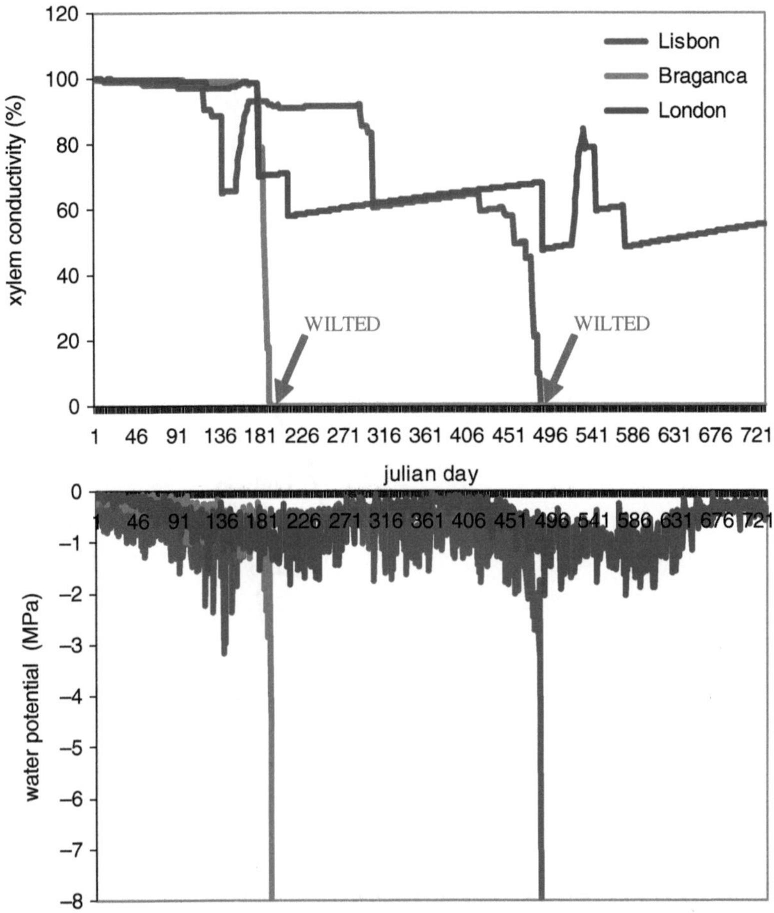


**Fig. 7** Simulation 3: simulated trajectories of leaf water potential (MPa) and xylem conductivity (as a %) at 3 sites over 2 years

Predicting a range of mortality values across the region, the model suggests that some 90% of infested tress will die in the Lisbon area, and are more likely to do so in the year following infestation. In the Bragança region the simulation suggests that around 40% of infested trees will die, and are more likely to do so in the year of infestation. Those results are dependent on the timing of high transpiration and flushing period of trees.

## *Model Validation*

To date there has been a lack of field or laboratory data to allow verification of simulated physiological processes and PWN population dynamics. Validation of

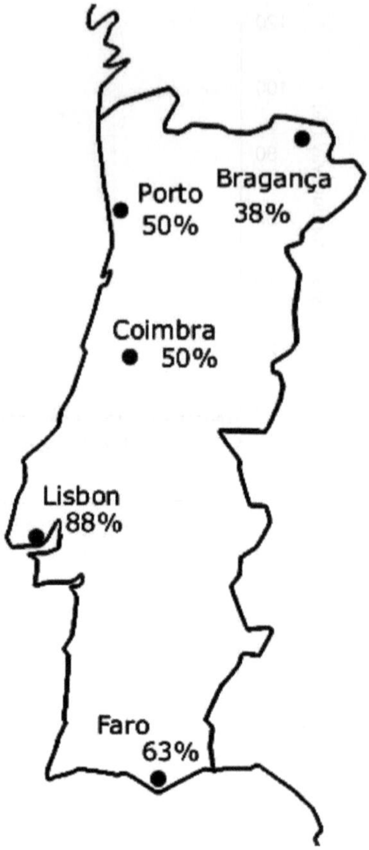


**Fig. 8** Simulated values of the likelihood of host mortality occurring following inoculation of susceptible pine trees in Portugal

carbon and water flux outputs from the ForestETp model has elsewhere shown good agreement between (inter- and intra-annual)simulated and observed values (Evans et al., 2003). Indirect comparison with mortality rates at a site in the Setubal Peninsula, Portugal where PWD is prevalent and significant environmental stresses occur indicates simulated mortality values not dissimilar to those reported across the PWN affected area.

## Conclusions

Using an existing mechanistic model of tree and stand scale carbon and water cycles that simulates tree growth dynamics under a range of conditions, here extended to describe PWN growth dynamics following host inoculation and nematode-induced cavitation, a series of model simulations have been undertaken to simulate and predict the impact of PWN infestation on host physiology in the Iberian Peninsula. The model makes a number of assumptions concerning plant host-pathogen interactions, particularly in relation to the environmental conditions of host growth.

Model results suggest that, under normal conditions, seasonal drought and/or high temperature significantly reduce host physiological activity, resulting in seasonal water stress and partial, reversible cavitation. Initially affecting the cambial zone, and with rapid extension to the conductive and reserve tissues, PWN infestation represents a significant, additional stress. This is expressed by the increasing presence of cavitation in conductive tissue that reduces canopy water supply and photosynthates to sites of meristematic activity and non-structural C storage. Ultimately, PWN activity, possibly in combination with the host's own defence mechanisms, results in degradation and destruction of cambial, conductive and storage tissues.

The model has been used to simulate tree mortality across a range of sites characterised by varying degrees of environmental pressure on tree physiology which, when coupled with the dynamics of PWN population growth, can result in tree death. Depending on local environmental conditions, the model suggests that infested trees may die either in the year of infestation or in the following year and following flushing. This result is, however, heavily dependent on the phenology and physiology of the host plant. In all simulations, the period of flushing is constant, but if the flushing period is significantly earlier than the period of high transpiration, then a tree may be able to renew damaged tissues and survive.

Initial model results are currently being compared with the results of an ongoing, observational experiment being undertaken in Portugal, using PWN extracted from a naturally infested site where significant pine wilting and mortality has been observed. Initial results suggest good agreement between observed values of host physiological parameters and those simulated by the model, as well as predicted rates of mortality. On the basis of these results it is proposed that, with further refinement and validation, the model may be suitable for developing a generic framework to predict the vulnerability of different hosts to the organism across a range of geographical regions. This represents a highly significant breakthrough in understanding the processes that result in pine wilt in some regions of the world. Conversely, it also explains why, in its natural range, despite having the same tree species and known exposure to the nematode through maturation feeding in the crowns of susceptible trees, little or no tree mortality occurs.

## References

AIKAWA, T. AND TOGASHI, K. (2000). Movement of *Bursaphelenchus xylophilus* (Nematoda: Aphelenchoididae) in tracheal system of adult *Monochamus alternatus* (Coleoptera: Cerambycidae). *Nematology* 2: 495–500.

BAOJUN, Y. AND QOULI, W. (1989). Distribution of the pinewood nematode in China and susceptibility of some Chinese and exotic pines to the nematode. *Canadian Journal of Forest Research* 19: 1527–1530.

CANNY, M.J. (1995). A new theory for the ascent of sap-cohesion supported by tissue pressure. *Annals of Botany* 75: 343–357.

DECKMYN, G., EVANS, S. AND RANDLE, T. (2006). Refined pipe theory for mechanistic modelling of wood developments. *Tree Physiology* 26: 703–718.

EDWARDS, O.R. AND LINIT, M.J. (1992). Transmission of *Bursaphelenchus xylophilus* through oviposition wounds of *Monochamus carolinensis* (Coleoptera: Cerambycidae). *Journal of Nematology 24*: 133–139.

EVANS, H.F., MCNAMARA, D.G., BRAASCH, H., CHADOEUF, J. AND MAGNUSSON, C. (1996). Pest risk analysis (PRA) for the territories of the European Union (as PRA area) on *Bursaphelenchus xylophilus* and its vectors in the genus Monochamus. *Bulletin OEPP* 26: 199–249.

EVANS, S., RANDLE, T., HENSHALL, P., ARCANGELI, C., PELLENQ, J., LAFONT, S. AND VIALS, C. (2003). Recent advances in the mechanistic modelling of forest stands and catchments. *Forest Research Annual Report and Accounts* 2003–2004: 98–111.

FUJIHARA, M., HADA, Y. AND TOYOHARA, G. (2002). Changes in the stand structure of a pine forest after rapid growth of Quercus serrata Thunb. *Forest Ecology and Management* 170: 55–65.

FUKUDA, K., HOGETSU, T. AND SUZUKI, K. (1992). Cavitation and cytological changes in xylem of pine seedlings inoculated with virulent and avirulent isolates of *Bursaphelenchus xylophilus* and *B. mucronatus*. *Journal of the Japanese Forestry Society* 74: 289–299.

FUTAI, K. AND FURUNO, T. (1979). The variety of resistances among pine-species to pine wood nematode, *Bursaphelenchus lignicolus*. *Bulletin of the Kyoto University Forests* 51: 23–36.

GUIRAN, G. AND BOUOLBRIA, A. (1985). Susceptibility of three pine species to the French and Japanese strains of pinewood nematodes (*Bursaphelenchus* spp.). Sensibilite de trois especes de pins a la souche francaise et aux souches japonaises de nematodes des pins (*Bursaphelenchus* spp.). *Meded Fac Landbouwwet Rijksuniv* 50: 809–814.

IKEDA, T. (1996). Xylem dysfunction in *Bursaphelenchus xylophilus*-infested *Pinus thunbergii* in relation to xylem cavitation and water status. *Annals of the Phytopathological Society of Japan* 62: 554–558.

KISHI, Y. (1995). *The pinewood nematode and the Japanese pine sawyer*. Thomas Company Ltd., Tokyo, Japan, 302 pp.

KIYOHARA, T. AND TOKUSHIGE, Y. (1971). Inoculation experiments of a nematode, *Bursaphelenchus* sp., onto pine trees. *Journal of the Japanese Forestry Society* 53: 210–218.

KURODA, K. (1989). Terpenoids causing tracheid-cavitation in *Pinus thunbergii* infested by the pine wood nematode (*Bursaphelenchus xylophilus*). *Annals of the Phytopathological Society of Japan* 55: 170–178.

KURODA, K. (1995). Acoustic emission technique for the detection of abnormal cavitation in pine trees infested with pine wilt disease. *Proceedings of International symposium on pine wilt disease caused by pine wood nematode*: 53–58.

KURODA, K. AND ITO, S. (1992). Migration speed of pine wood nematodes and activities of other microbes during the development of pine-wilt disease in *Pinus thunbergii*. *Journal of the Japanese Forestry Society* 74: 383–389.

KURODA, K., YAMADA, T., MINEO, K. AND TAMURA, H. (1988). Effects of cavitation on the development of pine wilt disease caused by *Bursaphelenchus xylophilus*. *Annals of the Phytopathological Society of Japan* 54: 606–615.

LINIT, M.J. (1988). Nematode-vector relationships in the pine wilt disease system. *Journal of Nematology* 20: 227–235.

LINIT, M.J. (1989). Temporal pattern of pinewood nematode exit from the insect vector *Monochamus carolinensis*. *Journal of Nematology* 21: 105–107.

LINIT, M.J. AND TAMURA, H. (1987). Relative susceptibility of four pine species to infestation by pinewood nematode. *Journal of Nematology* 19: 44–50.

MAEHARA, N. AND FUTAI, K. (1997). Effect of fungal interactions on the numbers of the pinewood nematode, *Bursaphelenchus xylophilus* (Nematoda: Aphelenchoididae), carried by the Japanese pine sawyer, *Monochamus alternatus* (Coleoptera: Cerambycidae). *Fundamental and Applied Nematology* 20: 611–617.

MAEHARA, N. AND FUTAI, K. (2002). Factors affecting the number of *Bursaphelenchus xylophilus* (Nematoda: Aphelenchoididae) carried by several species of beetles. *Nematology* 4: 653–658.

MAMIYA, Y. (1972). Abnormality of Japanese red pine and Japanese black pine inoculated with the pine wood nematode, and distribution of the nematode in wood tissue. *16th Annual Meeting of the Japanese Society of Applied Entomology and Zoology*, Shizuoka, Japan: 118.

MAMIYA, M. (1975). Pine lethal wilting disease caused by the pine wood nematode, *Bursaphelenchus lignicolus. 2nd World technical consultation on forest diseases and insects*, New Delhi, India 7 12 April, 1975: 7.

MAMIYA, Y. (1980). Inoculation of the first year pine (*Pinus densiflora*) seedlings with *Bursaphelenchus lignicolus* and the histopathology of diseased seedlings. *Journal of the Japanese Forestry Society* 62: 176–183.

MAMIYA, Y. (1985). Initial pathological changes and disease development in pine trees induced by the pine wood nematode, *Bursaphelenchus xylophilus*. *Annals of the Phytopathological Society of Japan* 51: 546–555.

MAMIYA, Y. AND ENDA, N. (1972). Transmission of *Bursaphelenchus lignicolus* (Nematoda: Aphelenchoididae) by *Monochamus alternatus* (Coleoptera: Cerambycidae). *Nematologica* 18: 159–162.

MAMIYA, Y. AND KIYOHARA, T. (1972). Description of *Bursaphelenchus lignicolus* n.sp. (Nematoda: Aphelenchoididae) from Pine wood and histopathology of nematode-infested trees. *Nematologica* 18: 120–124.

MARTINEZ-VILALTA, J. AND PINOL, J. (2002). Drought-induced mortality and hydraulic architecture in pine populations of the NE Iberian Peninsula. *Forest Ecology and Management* 161: 247–256.

MYERS, R.F. (1986). Cambium destruction in conifers caused by pinewood nematodes. *Journal of Nematology* 18: 398–402.

MYERS, R.F. (1988). Pathogenesis in pine wilt caused by pinewood nematode, *Bursaphelenchus xylophilus*. *Journal of Nematology* 20: 236–244.

NOBUCHI, T., TOMINAGA, T., FUTAI, K. AND HARADA, H. (1984a). Cytological study of pathological changes in Japanese black pine (*Pinus thunbergii*)seedlings after inoculation with pine-wood nematode (*Bursaphelenchus xylophilus*). *Bulletin of the Kyoto University Forests* 56: 224–233.

NOBUCHI, T., TOMINAGA, T., FUTAI, K., HARADA, H. AND SUDO, S. (1984b). Cytological changes of the parenchyma cells of pine seedlings after inoculation with *Bursaphelenchus xylophilus* (Nematoda: Aphelenchoidae). Proceedings, Pacific Regional Wood Anatomy October: 1–7, 1984, Tsukuba, Ibaraki, Japan.

ODANI, K., SASAKI, S., NISHIYAMA, Y. AND YAMAMOTO, N. (1985). Early symptom development of the pine wilt disease by hydrolytic enzymes produced by the pine wood nematodes – cellulase as a possible candidate of the pathogen. *Journal of the Japanese Forestry Society* 67: 366–372.

OEPP/EPPO (1986). Data sheets on quarantine organisms. List A1, No.158 – *Bursaphelenchus xylophilus* (Steiner & Buhrer) Nickle et al. Nematoda: Aphelenchoididae.*Bulletin OEPP* 16: 55–60.

OEPP/EPPO (1989). Conclusions of the EPPO ad hoc panel on pine wood nematode, Uppsala, 9–10 February, 1988. *Bulletin OEPP* 19: 4–6.

RUTHERFORD, T.A. AND WEBSTER, J.M. (1987). Distribution of pine wilt disease with respect to temperature in North America, Japan, and Europe. *Canadian Journal of Forest Research* 17: 1050–1059.

SALLEO, S.A.L.G., DE PAOLI, M. AND ZIPPO, M. (1996). Xylem recovery from cavitation-induced embolism in young plants of *Laurus nobilis*: A possible mechanism. *New Phytologist* 132: 47–56.

SOBARDO, M.A., GRACE, J. AND JARVIS, P.G. (1992). The limits to xylem embolism recovery in *Pinus sylvestris* L. *Journal of Experimental Botany* 43: 831–836.

SUZUKI, K. (1984). General effect of water stress on the development of pine wilting disease caused by *Bursaphelenchus xylophilus*. *Bulletin of Forestry and Forest Products Research Institute* 325: 97–126.

TAMURA, H. (1984). Early development of *Bursaphelenchus xylophilus* (Nematoda: Aphelenchoididae) population in the inoculated branches of pine seedings. *Applied Entomology and Zoology* 19: 125–129.

TOGASHI, K. (1985). Transmission curves of *Bursaphelenchus xylophilus* (Nematoda: Aphelenchoididae) from its vector, *Monochamus alternatus* (Coleoptera: Cerambycidae), to pine trees with reference to population performance. *Applied Entomology and Zoology* 20: 246–251.

TYREE, M.T., SALLEO, S., NARDINI, A., GULLO, M.A.L. AND MOSCA, R. (1999). Refilling of embolized vessels in young stems of laurel. Do we need a new paradigm? *Plant Physiology* 120: 11–22.

TYREE, M.T. AND SPERRY, J.S. (1989). Vulnerability of xylem to cavitation and embolism. *Annual Review of Plant Physiology* 40: 19–38.

WINGFIELD, M.J. (1987) A comparison of the mycophagous and the phytophagous phases of the pine wood nematode. In *Pathogenicity of the Pine Wood Nematode* (ed. M.J. Wingfield), pp. 81–90. APS Press, St. Paul, Minnesota.

WOODWARD, F.I., SMITH, T.M. AND EMANUEL, W.R. (1995). A global land primary productivity and phytogeography model. *Global Biogeochemical Cycles* 9: 471–490.

YAMAMOTO, N., ODANI, K., SASAKI, S. AND NISHIYAMA, Y. (1986). Cellulase exudation by the pine wood nematode – detection of activity in its crawling track. *Journal of the Japanese Forestry Society* 68: 237–240.

YANG, B.J. (1987). The resistance of pine species to pine wood nematode, *Bursaphelenchus xylophilus*. *Acta Phytopathologica Sinica* 17: 211–214.

YOKOBORI, M. (1986). Catastrophic withering of pine forests by pine wilt disease in Ibaraki Prefecture and its environmental factors. *Bulletin of the Ibaraki Prefectural Forest Experiment Station*: 26.

ZHAO, B.G., WANG, H.L., HAN, S.H. AND ZHENG, M.H. (2003). Distribution and pathogenicity of bacteria species carried by Bursaphelenchus xylophilus in China. *Nematology* 5: 899–906.

ZIMMERMANN, M.H. (1983). Xylem structure and the ascent of sap. Berlin, Springer-Verlag.

# Field Diagnosis of the Asymptomatic Carrier of Pinewood Nematode

**Kazuyoshi Futai and Yuko Takeuchi**

**Abstract** To prevent pine wilt disease (PWD) from spreading over pine forests, elimination of pine trees killed by the pinewood nematode (PWN), *Bursaphelenchus xylophilus* is desirable, though this method is very laborious and time-consuming. If such dead trees are left in the field, pathogenic nematodes and their vector, *Monochamus* beetles, could spread from tree to tree without any difficulty. In our university arboretum, where many precious foreign pine species are planted in the field, all pine trees killed by PWD have been eradicated thoroughly before the next pine wilt season. Despite intensive efforts in removing dead trees from the stands, new dead trees tend to appear in the vicinity of the stumps of trees killed in the previous year, and wilting recurs in the same pine stand every year. Why does PWD recur at the same stand even after thorough eradication of dead pine trees?

## Introduction

Following enormous control efforts for eradication of the pinewood nematode (PWN), many asymptomatic carrier trees still remain on the field, because they could not be detected as being infected. They can survive not only in the year they were infected, but also during the following years. Once resistance is overcome by pathogenic nematodes, the host tree begins to release volatiles that attract vector beetles, thereby initiating a new epidemic cycle. We confirmed the importance of asymptomatic carrier trees in spreading of pine wilt disease (PWD) to surrounding pine forests. In the first part of this paper, long-term field survey suggests the role of asymptomatic carriers in spreading the PWD. In the second part, a new nested PCR method developed to detect the PWN from asymptomatic carrier trees is described and the results of field survey using this new technique are presented. The results of these two parts have been previously published (Futai, 2003; Takeuchi et al., 2005; Takeuchi and Futai, 2007), but to grasp the scope of the important role of the

K. Futai
Laboratory of Environmental Mycoscience and Nematology, Graduate School of Agriculture, Kyoto University, Sakyo-ku, Kyoto 606-8502 Japan
futai@kais.kyoto-u.ac.jp

asymptomatic carrier trees in spreading PWD, we have described more general aspects and characteristics of asymptomatic trees.

## Materials and Methods

### *Long-Term Field Survey of PWD Spreading over a Small Stand of* **Pinus koraiensis**

Figure 1 shows a *Pinus koraiensis* stand, where we conducted a survey from 1990 to 1994. This is an arboretum of the Kamigamo experimental forest station of Kyoto University Forests. Plant material examined were 72 trees of 45-year-old, Korean pine, *P. koraiensis* planted on a slope of 25 degrees at the arboretum. To confirm the location of all test trees, a site map was made, and all 72 trees were labeled by numbering (Fig. 2).

Tree oleoresin flow degrades rapidly when pine trees are infected with PWN, so the amount of oleoresin exudation from artificial wound have been widely adopted as an assessing probe for pine wilt infection. In this study, the amount of oleoresin was surveyed by piercing a thumbtack into the bark of the trunk, then evaluating the amount of oleoresin at the following measurement. The resin exudation of all 72 trees was examined once or twice a month, 58 times in total for four years, from December 1990 to August 1994. The degree of the exuded oleoresin was classified into five ranks from 0 to 4, as shown in Fig. 3. At each measurement of resin exudation, the wilting symptoms appeared; the sawyer beetle's oviposition scars that had been marked on the trunks, and the feeding debris of the sawyer beetle larvae were inspected. To monitor the physiological condition of each test tree, all resin indices were evaluated and then were summed up at each measurement, thereby a cumulative curve was drawn for each tree. To show the standard line, the mean cumulative indices of 15 healthy pine trees were calculated and shown as a black line without a symbol (Fig. 4).

In May 1992, 21 pine trees were found with a partially wilted branch in the Korean pine stand. All of the branches were harvested, cut into small pieces, and a portion of the wood pieces (about 4 g in fresh weight) were placed overnight in a Baermann funnel containing water, to recover nematodes. The PWN was detected from 13 out of 21 branches examined. To examine disease development of the 13 nematode positive trees, the amount of oleoresin exudation was observed from February 1992 to December 1993.

### *A New Method to Detect the PWN from Asymptomatic Carrier Trees*

Because the population of the PWN in asymptomatic trees is very low, it is often difficult to detect the nematodes by the usual Baermann funnel method. We therefore

**Fig. 1** The arboretum of *Pinus koraiensis* of the Kamigamo experimental forest station of Kyoto University Forests, in May 1992 (*top*) and April 1994 (*bottom*)

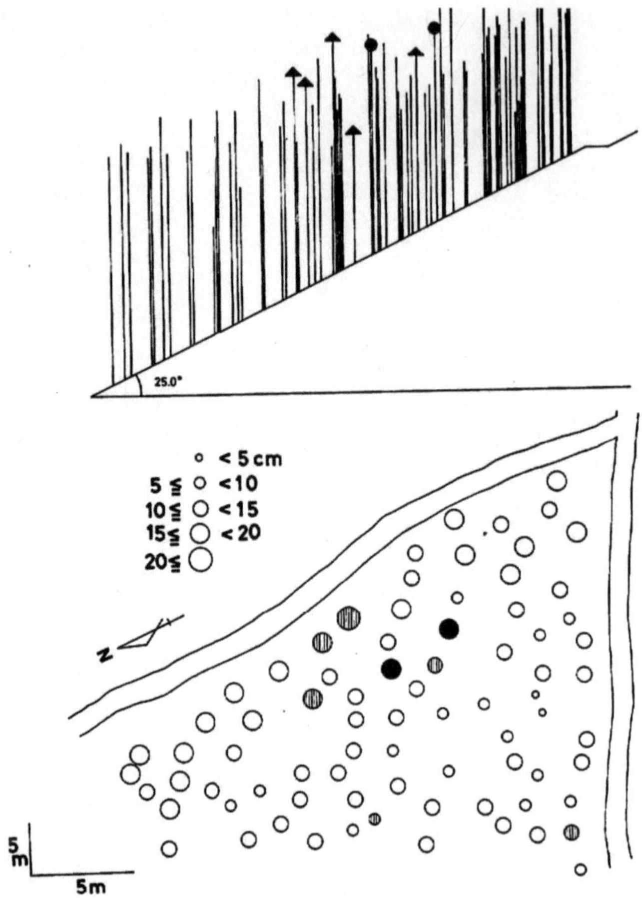


**Fig. 2** An arboretum of Korean pine, *Pinus koraiensis* locates at Kamigamo experimental forest station of Kyoto University Forests, where 72 trees of 45-year-old Korean pines were planted on a slope of 25 degrees. ***Top***: side view of the stand; ***Bottom***: overhead view

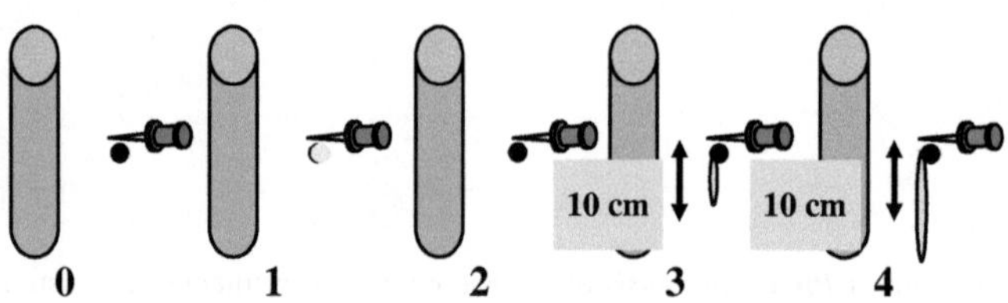


**Fig. 3** Rating of resin exudation. 0: No resin exuded. 1: The pinhole of thumbtack was filled with the resin, or the needle of the thumbtack became sticky due to the resin. 2: A small amount of resin exuded from the thumbtack pierced point to make a drop, but never flew down. 3: The resin flew down from the point shorter than 10 cm. 4: The resin vigorously flew down 10 cm or more from the point

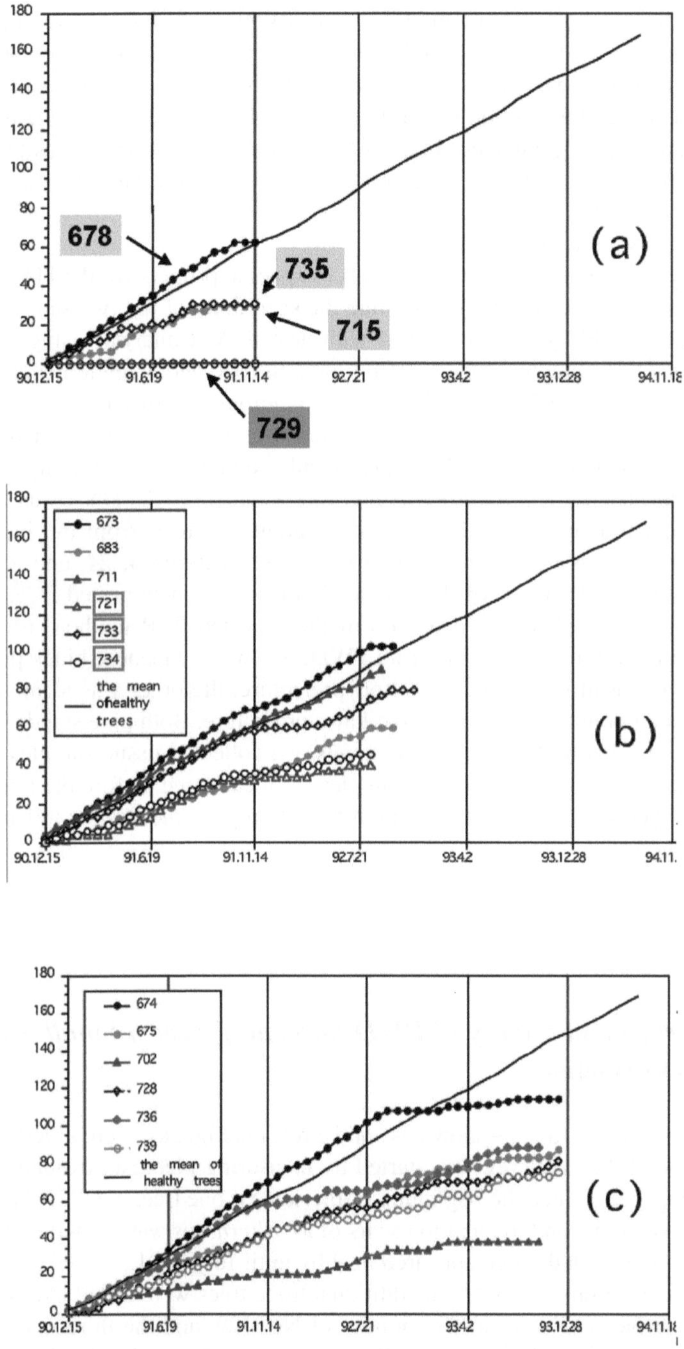


**Fig. 4** Cumulative oleoresin index curves evaluated for four trees killed before October 1991 (**a**). Curves for six trees killed in the summer of 1992 (**b**); curves for six trees on which oviposition scars were detected in July 1993 (**c**)

developed a new molecular method to detect PWN from wood tissues of asymptomatic pine trees with sparse populations. To establish the new molecular method for PWN detection, we had to overcome two problems: the first one was applicability of the general extraction method of plant DNA. We examined whether or not CTAB is applicable to Japanese black and red pine, and obtained a positive result; the CTAB method was suitable for extracting DNA from both Japanese pine species, and the resulting DNA was suitable for PCR analysis both in quality and quantity, though it was somewhat fragmented.

The next problem was to make two sets of specific primers for the ITS region of the ribosomal DNA of PWN. To examine the specificity of the two sets of primers designed for the PWN, we proceeded with nested PCR analysis using these two primer sets, DNA samples extracted from pine tissues, and a single nematode, and succeeded in distinguishing PWN's from *B. mucronatus*. To confirm the applicability, this new detection method was applied to pot pine seedlings of three pine species. A given number of PWN's were inoculated on the stem of potted seedlings of three pine species, *P. thunbergii, P. densiflora,* and *P. taeda*. DNA was extracted directly from four points of the stem at different distances from the inoculation point. Such stem tissues were obtained from each seedling 4, 24, and 72 hrs after inoculation, and served for PWN detection using the new nested PCR method. Furthermore, we applied this detection method (nested PCR method) to standing pine trees in two forests suffering from PWD; one was a Japanese black pine stand located at a coastal sand dune in Tottori prefecture, the other was a Japanese red pine stand in a mountainous area in Ishikawa prefecture. Both pine stands had been previously invaded by PWN. At both stands, we collected tissue samples from all trees in each experimental plot, five samples from each tree, before the new season of pine sawyer activity started. Apparent symptoms and oleoresin exudation were also examined.

## Results

### *Long-Term Field Survey of PWD Spreading over a Small Stand of* Pinus koraiensis

In October 1990, two trees – shown as solid circles in Fig. 5a – were killed by PWN. In December 1990, a survey was started by measuring oleoresin exudation for all 72 surviving trees. Since the beginning of the survey, one tree, No. 729, had ceased oleoresin exudation, and oviposition scars of *M. alternatus* were found on the trunk in June 1990, though the tree appeared healthy until July 1991.

During the summer of 1991, an additional three trees were killed. Among them, No. 735 and No. 715 were in the vicinity of No. 729, and the third one, No. 678 was located apart from these trees. At the beginning of September 1991, oviposition scars were found on trees 715, and 735. Before May 1992, four trees that had wilted in the previous year were cut down, and all their stems, branches, twigs and needles

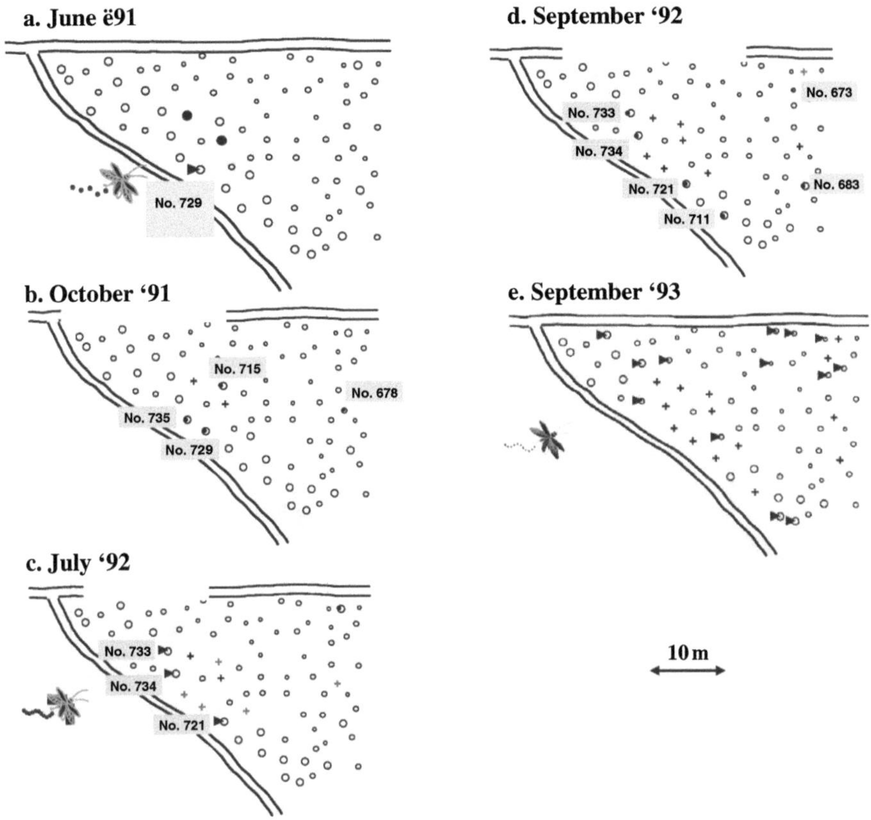


**Fig. 5** Disease development in a stand of *P. koraiensis*, where two trees shown as solid circles (**a**) were killed in October 1990

were removed from the stand. In early July 1992, however, oviposition scars were found on three trees (Nos. 721, 733, 734). All of them were located just adjacent to the stumps of the wilted trees. Before September 1992, these three trees (Nos. 721, 733, 734) and an additional three trees (Nos. 673, 683, 711) wilted, though the latter three trees had no oviposition scars.

Figure 4b shows the cumulative oleoresin index curves estimated for six *P. koraiensis* trees that were killed in the summer of 1992. Among the six trees, three (Nos. 721, 733, 734) had shown decline in oleoresin exudation since the previous year (1991), thereby selected by the *Monochamus* beetles for oviposition in 1992. In September 1992, six other trees received sawyer oviposition.

In May 1993, six pine trees that had wilted in the previous summer, and two more trees that had died thereafter, were removed from the stand before the newly emerged pine sawyer adults started their activity. In July 1993, however, sawyer oviposition scars were detected on six pine trees, each located in the vicinity of the new stumps. Among them were three trees (Nos. 702, 736, 739), which had shown

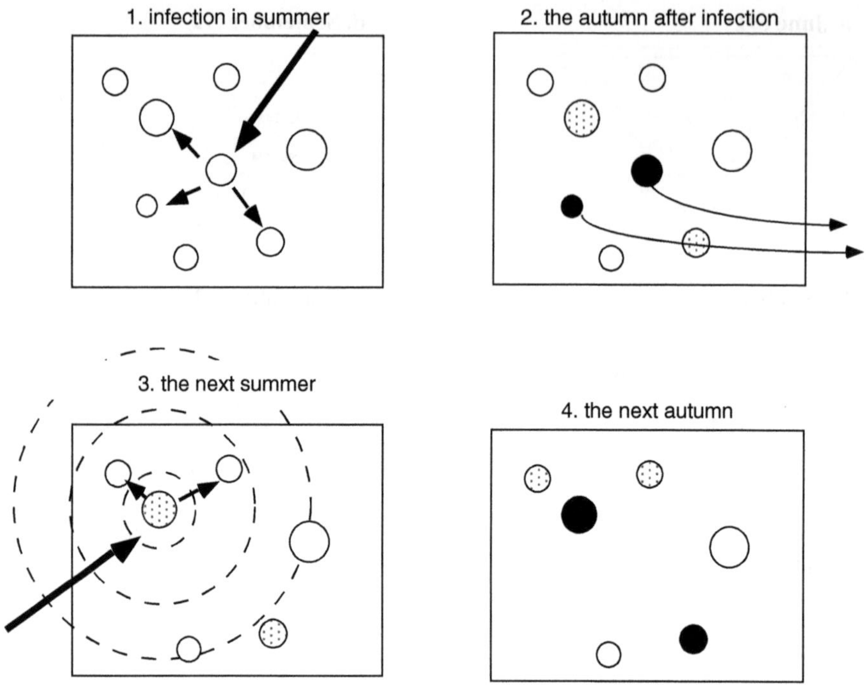


**Fig. 6** "A chain infection model", which explains the role of an asymptomatic carrier tree in spread of pine wilt in a stand

abnormal resin exudation as early as 1991. The other three (Nos. 674, 675, 728) had started to decrease their oleoresin exudation in 1992, suggesting that they already had been infected with the PWN the previous year or perhaps earlier.

The bottom picture in Fig. 1 shows the stand of *P. koraiensis* in April 1994, which had become very sparse and bright colored.

## *Fate of the Pine Trees with a Partially Wilted Branch*

Figure 7 shows the changes in oleoresin exudation of 13 asymptomatic *P. koraiensis* trees that had a partially wilted branch harboring the PWN. Seven of the 13 asymptomatic trees decline their oleoresin exudation, then died. Three others showed abnormal oleoresin exudation until the end of the observation. The remaining three trees recovered their oleoresin exudation and kept their healthy appearance.

## *Applicability of the New Detection Method of PWN*

As shown in Fig. 8, the PWN dispersed over the entire area examined in the susceptible pine, *P. thunbergii*, while dispersion of PWN was delayed both in low

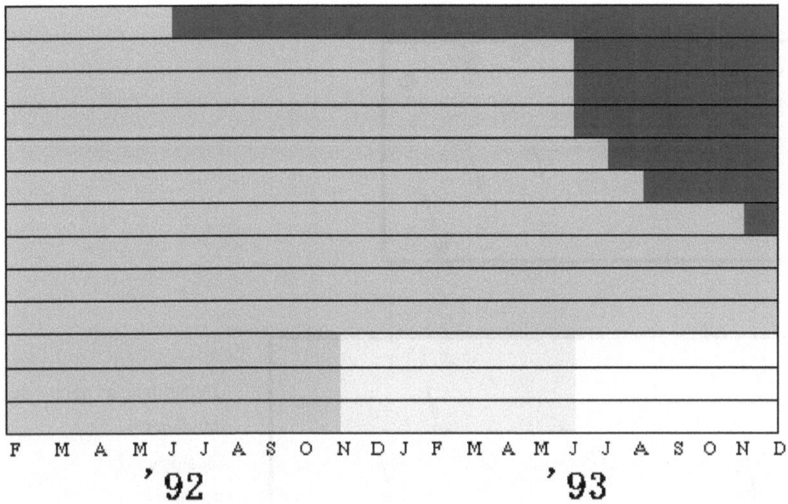


**Fig. 7** The fate of healthy-looking trees with a branch harboring PWN's

susceptible *P. densiflora* and in resistant *P. taeda*. Thus, results obtained by this new detection method reflected well the difference in host resistance, and so confirmed the applicability for potted seedlings. This result suggests the applicability of this method to potted pine trees.

In further experiments, we applied this detection method to two pine stands suffering from PWD; one is a Japanese black pine stand locating at a coastal sand dune, the other is a Japanese red pine stand in a mountainous area. Both pine stands had been previously been affected by PWD. Before the new season of pine sawyer activity started, we collected tissue samples from all trees in each experimental plot, five samples from each tree. Apparent symptoms and oleoresin exudation were also examined. As shown in Figs. 9 and 10, PWN was detected from many pine trees, both from diseased and healthy-looking ones, in either stand. Arrow mark shows asymptomatic carriers, and some trees survived another year without showing any symptoms.

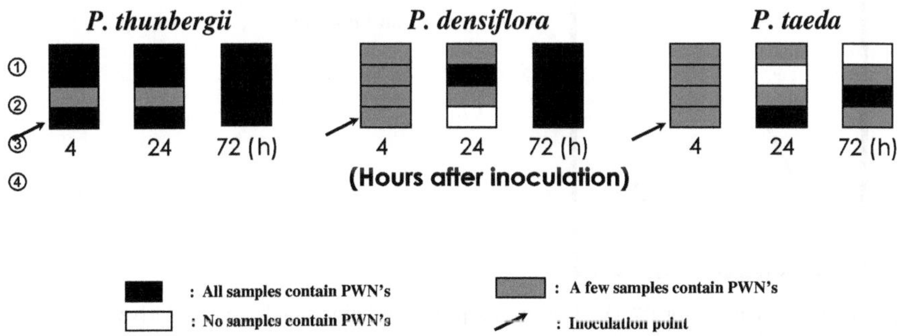


**Fig. 8** Application of nested PCR method for detecting PWN's from Potted seedling

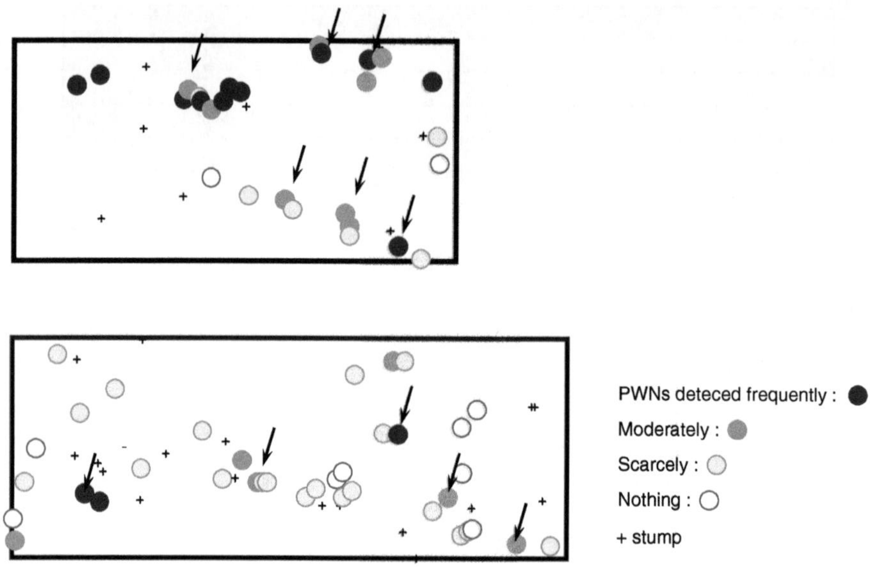


**Fig. 9** Field diagnosis for PWD infection at Japanese black pine stands, Tottori. Arrow means an asymptomatic carrier

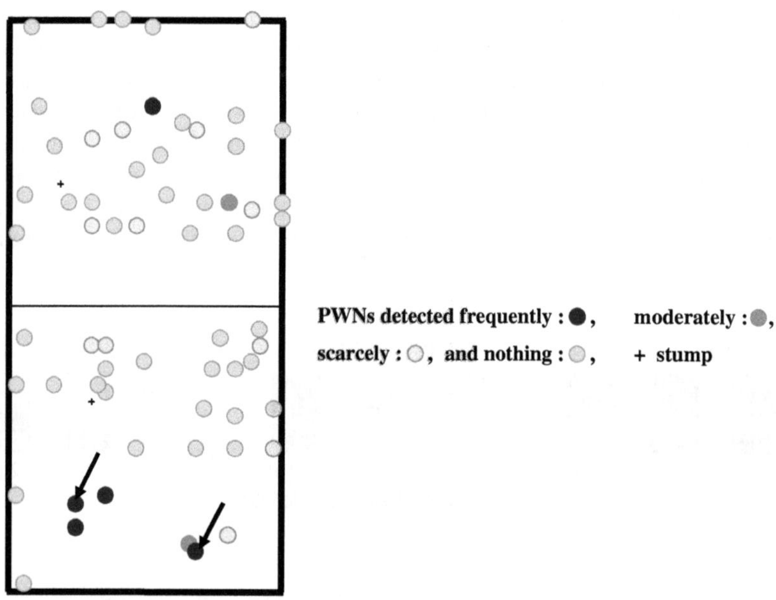


**Fig. 10** Field diagnosis for PWD infection at Japanese red pine stands, in Mt. Kariyasu, Ishikawa prefecture. Arrow means an asymptomatic carrier

## Discussion

Based on a field study at the stand of *P. koraiensis*, we made a "A chain infection model" as shown in Fig. 6 to explain the role of an asymptomatic carrier tree in spread of pine wilt in a stand. In this figure, solid and shaded circles represent dead and asymptomatic carrier trees, respectively.

1. A *Monochamus* adult firstly visits the stand to feed on the young branch of a tree and then visits surrounding trees while inoculating PWN in these trees.
2. Some of the trees infected were killed and removed from the stand, the other trees remain asymptomatic carriers.
3. In the early summer of the following year, some of the asymptomatic carriers start to emit volatiles, thereby attract newly emerged sawyer beetles.
4. Some of asymptomatic carrier trees and further newly attacked pine trees killed, and some more trees remain as another asymptomatic trees.

This model was established based on the results obtained at *P. koraiensis* stand, but may be applicable to most pine stands of any susceptible pine species. New molecular methods for PWN detection facilitate the recognition of asymptomatic pine trees. Actually, we could identify many asymptomatic trees in two pine forests by this method. If the asymptomatic carrier is an ubiquitous phenomenon in PWD, the current control method must be reviewed carefully, to sanitize infected forests by removing not only symptomatic but also asymptomatic trees.

To conclude, (1) there are more asymptomatic pine trees than previously thought, (2) such asymptomatic pine trees may emit volatiles and attract *Monochamus* beetles, thereby expanding PWD over the surrounding forests, (3) a new nested-PCR method could detect effectively asymptomatic carrier trees, and therefore may become a very promising tool for diagnosis.

## References


FUTAI, K. (2003). Role of asymptomatic carrier trees in epidemic spread of pine wilt disease. *Journal of Forest Research* 8: 253–260.

TAKEUCHI, Y. AND FUTAI, K. (2007). Asymptomatic carrier trees in pine stands naturally infected with *Bursaphelenchus xylophilus. Nematology* 9: 243–250.

TAKEUCHI, Y., KANZAKI, N. AND FUTAI, K. (2005). A nested PCR-based method for detecting the pine wood nematode, *Bursaphelenchus xylophilus,* from pine wood. *Nematology* 7: 775–782.

# Part VI
# The Tree: Physiology, Resistance and Histopathology as a Result of Pine Wilt Disease

**Keiko Kuroda and Dale Bergdahl**


## Summary

Much is still to be learned about the ecophysiology and bioecology of pine wilt disease (PWD) caused by *Bursaphelenchus xylophilus*, the pinewood nematode (PWN) in different regions of the world. Virulence of the PWN is known to vary widely from highly virulent to avirulent or less pathogenic forms. The more virulent forms appear to build their populations and migrate rapidly in susceptible host tissues, thus, quickly inciting disruption of the tree's normal physiological function, which results in rapid mortality. This disruption in function is primarily the result of the PWN's stimuli at feeding on epithelial and parenchyma cells associated with axial and horizontal resin canals in the early period of infection. This physiological dysfunction begins soon after infection, especially in trees experiencing some form of environmental stress and thus is responsible for early symptom development and rapid mortality.

The virulent PWN tends to migrate more slowly and the population growth is retarded in the trees that were selected as resistant to pine wilt from the survivors of susceptible species. Investigations on the mechanism of resistance will help to find resistant trees in the forests of susceptible pine species. It has been shown that inoculating susceptible pines with avirulent forms of the PWN may induce resistance to subsequent inoculation with known virulent forms. The nature of this induced resistance is unknown but it may hold promise as a future mechanism of biological control, especially, in areas of extensive PWD.

The PWN is present throughout the pine growing regions of North America (NA) but rarely does it cause PWD, especially, in cooler climates. Long-term inoculation studies in field grown Scots pine (*Pinus sylvestris*), indicate that low numbers of the PWN may persist in asymptomatic, living trees for many years without causing PWD. Therefore, these asymptomatic trees may complicate sanitation and other PWD management efforts and healthy appearing trees may not be any assurance they are free of the PWN.

In Portugal, the threat of the PWN is real but has yet to result in the catastrophic levels of tree mortality experienced in Asia. This is primarily the result of early detection followed by aggressive PWD management efforts with an emphasis on thorough sanitation procedures for diseased trees. However, it now appears that the older-aged, more dominant trees are being selectively removed from the forest canopy by PWD. Therefore, these pine stands are changing in age structure and composition as well as becoming less dense to the point where other tree species are beginning to occupy the open spaces. While these changes in stand structure and composition seem minor at this time, the final result may mean a drastic change in the bioecology of these PWN infested areas. This ecological change would be especially true if the natural fire regimes of the pine forests of Portugal are also suppressed and/or if the affected areas become infested by invasive exotic plant species. In addition, the threat of global warming may significantly influence the behavior, aggressiveness, and impact of the PWN and its principal vectors (*Monochamus* spp.) throughout this newly invaded region.

Primary transmission of the PWN is known to occur when the vector creates fresh feeding wounds on healthy twigs and branches of susceptible trees. This feeding activity begins soon after *Monochamus* emerges as an adult and then continues throughout its life. A secondary transmission pathway also is known to occur during the breeding and oviposition phases of the insect's life. This secondary pathway is probably the most common mode of PWN transmission in North America.

It is well known that the PWN is more aggressive in infesting and killing stressed trees, especially, those growing in the warmer and drier areas of Asia and NA. In NA, the PWN is often associated with trees that have been impacted by other biotic or abiotic stresses, such as those created by bark beetle attacks, root rots, high temperatures, drought, off-site planting, wildland fire, etc. However, these types of associations may also be the result of the increased attractiveness of stressed trees to vectors of the PWN. European species of *Monochamus* spp. are also attracted to stressed trees and so secondary transmission may become a more important pathway due to changing climatic conditions as well as other forms of environmental stress. These changing conditions may have a significant influence on our understanding of the bioecology and options for management of the PWN in the future.

The recent inoculation studies reported at this meeting indicate that other species of *Bursaphelenchus* (*B. mucronatus* and *B. vallesianus*) may also cause wilt disease of pines; therefore, it may be prudent to consider expanding the original concept of PWD to include their apparent pathogenicity. However, inoculation studies with these other *Bursaphelenchus* species were only done on seedling-size trees maintained at elevated temperatures and low moisture regimes in a greenhouse. Therefore, an inoculation study needs to be completed on older aged trees to confirm pathogenicity of these other *Bursaphelenchus* spp. in the field.

# Inoculation of Pine Trees with Avirulent Pinewood Nematode Under Experimental Conditions: Risk-Benefit Analysis

**Hajime Kosaka**

**Abstract** Virulence of the pinewood nematode, *Bursaphelenchus xylophilus*, varies widely from highly virulent to avirulent, or less pathogenic. The inoculation of pine trees with avirulent *B. xylophilus* induces resistance to subsequently inoculated virulent *B. xylophilus*. This induced resistance may constitute a biological control strategy for pine wilt disease. The characteristics of induced resistance have been previously identified in short-term experiments. In this study, the long-term and yearly repeated experiments were conducted to further understand the nature of induced resistance and the potential for control measures. Induced resistance of pine trees by avirulent *B. xylophilus* was re-confirmed, although the effects were not as strong as trunk injection with nematicides. The avirulent *B. xylophilus* remained avirulent in the long term, but occasionally caused tree mortality. The results show that the benefit, i.e., the chance of tree survival, outweighed the risk of tree mortality by avirulent *B. xylophilus* when pine trees were subsequently inoculated with virulent *B. xylophilus*. Explorative use of this resistance-inducing method will be possible in areas where pine wilt disease occurs naturally. More study is necessary to determine the effect of induced resistance against natural infection with *B. xylophilus* from vector insects.

## Introduction

The pinewood nematode (PWN), *Bursaphelenchus xylophilus*, is the causal agent of pine wilt disease (Kiyohara and Tokushige, 1971). The PWN is transmitted from infected and dead pine trees to healthy trees by longhorn beetles of the genus *Monochamus* (Mamiya and Enda, 1972; Kishi, 1995). Although control of the disease is difficult, it is possible if the strategy is carefully considered and

H. Kosaka
Hokkaido Research Center, Forestry and Forest Products Research Institute, Sapporo 062-8516, Japan
e-mail: hkosaka@ffpri.affrc.go.jp

M.M. Mota, P. Vieira (eds.), *Pine Wilt Disease: A Worldwide Threat to Forest Ecosystems*, 

pesticides are used adequately. For instance, the combination of fumigating dead and infected trees, injecting nematicides into trunks and aerial spraying of insecticides ended outbreaks of disease (Nakamura and Yoshida, 2004; Yoshida, 2006). However, pesticides cannot be used without limitation, so alternatives have to be developed.

The virulence of *B. xylophilus* varies widely, from highly virulent to avirulent, or less pathogenic (Kiyohara and Bolla, 1990). The inoculation of pine trees with avirulent *B. xylophilus* induces resistance to subsequently inoculated virulent nematodes. A series of studies on this induced resistance was conducted and appropriately reviewed (Kiyohara, 1984, 1989; Kosaka et al., 2001a). Briefly, the main characteristics of induced resistance were as follows: the degree of nematode virulence can be identified empirically through the inoculation of pine trees, but not from morphology. The effects of induced resistance were evaluated after an inoculation with virulent nematodes in comparison with a control without avirulent nematode inoculation. The mechanism of induced resistance has not been elucidated. Inoculation with living avirulent nematodes induced resistance, but not inoculation with disrupted or heat-killed nematodes. Inoculation with large numbers of avirulent nematodes (30,000 maximum) induced more robust resistance. Induced resistance is considered common in pine trees susceptible to pine wilt disease.

Induced resistance of pine trees by avirulent *B. xylophilus* might offer a way to control pine wilt disease. However, most tests of induced resistance lasted less than a year, and experimental conditions changed in each experiment. Now, long-term observations and yearly repeated experiments are necessary to develop biological control using avirulent *B. xylophilus*, which might kill trees many years after inoculation, or because the effects of induced resistance can be transient (Kosaka et al., 2001a). The aims of this study were to determine the long-term virulence of avirulent *B. xylophilus*, an adequate period in which to evaluate the effect of induced resistance, and stability of the resistance over time. The benefit and risk of inoculation of pine trees with avirulent *B. xylophilus* were then evaluated when the trees were subsequently infected with virulent *B. xylophilus*.

## Materials and Methods

### *Nematodes, Trees and Inoculation*

The experiments used *B. xylophilus* avirulent isolates, C14-5 and OKD-1, virulent isolates Ka-4 and S-10 and *B. mucronatus*, which is closely related to *B. xylophilus* but is not pathogenic (Mamiya and Enda, 1979). Test trees were the Japanese red pine, *Pinus densiflora*, and the Japanese black pine, *P. thunbergii*. The nematodes were cultured on *Botrytis cinerea* growing on potato dextrose agar or autoclaved polished barleycorns, and were then isolated in Baermann funnels. A given number of nematodes in aqueous suspension (see later) were inoculated with a pipette into cuts made by a knife, a hatchet, or an electric drill into the tree trunks.

## *Long-Term Comparison of Nematode Virulence*

Three-year-old *P. densiflora* and *P. thunbergii* growing in a nursery at the Forestry and Forest Products Research Institute (FFPRI), Ibaraki, Central Japan, were used in all experiments. The four isolates of *B. xylophilus* and *B. mucronatus* were used as inocula. A nematode suspension (10,000 nematodes per 0.05 ml) or the same amount of distilled water was inoculated into the trees on 17 July 2001. The survival of trees was observed until 23 April 2004. The number of replicates was 18 except for 17 *P. thunbergii* inoculated with Ka-4.

## *Experiment on Mature Pine Trees*

Sixteen 30-year-old trees (*P. densiflora*) near FFPRI were pre-inoculated with a suspension of avirulent *B. xylophilus* OKD-1 (50,000 nematodes per 0.5 ml) on 12 June 1997, and then inoculated with virulent Ka-4 (10,000 nematodes per 0.5 ml) on 9 July 1997. Two other experiments were conducted inoculating only with OKD-1 (16 trees) or Ka-4 (15 trees). Twenty trees remained untreated. The survival of trees was observed until 8 June 2000. Preliminary results have already been reported (Kiyohara et al., 1999; Kosaka et al., 2001a).

## *Experiments on Young Pine Trees with Yearly Repetitions*

Six-year-old *P. densiflora* and *P. thunbergii* growing at Chiyoda Experimental Station of FFPRI, Ibaraki, were used in 2002. Avirulent *B. xylophilus* C14-5 and OKD-1 were used for pre-inoculation and virulent Ka-4 was used for the challenge inoculation. The same experiments were repeated from 2002 to 2004 in the same forest stands, but in different trees. Suspensions of the avirulent nematodes (100,000 nematodes per 0.3 ml) were pre-inoculated into 3 points per trees (300,000 nematodes per tree) on 28 June 2002, 28 May 2003, and 4 June 2004. Then a suspension of virulent nematodes (6,000 in *P. densiflora* and 3,000 nematodes in *P. thunbergii* per 0.05 ml) was inoculated on 24 July 2002, 8 July 2003, and 14 July 2004. The same amount of distilled water was used as control. In the pre-inoculation in 2003 and 2004, half of the trees were inoculated with 300,000 nematodes per 0.9 ml at one point, but the data were compared with the multiple pre-inoculations because the results were similar to those of the multiple pre-inoculations. The survival of the trees inoculated with the avirulent and virulent nematodes in 2002 and 2003 was observed for 2 years until 16 April 2004 and 25 May 2005, respectively. The survival of the other trees was observed on 16 April 2003, 16 April 2004, and 25 May 2005 following inoculation. The number of replicates in 2002 was 5 except for 4 *P. thunbergii* inoculated with OKD-1 and Ka-4, and 4 *P. thunbergii* inoculated with C14-5 and Ka-4. The number of replicates in 2003 and 2004 was 10.

## Results

### *Long-Term Comparative Virulence*

The survival rates of *P. densiflora* and *P. thunbergii* inoculated with the nematodes are shown in Fig. 1. Most trees inoculated with the virulent *B. xylophilus* were dead by the following winter. The survival rates of trees inoculated with avirulent *B. xylophilus* gradually decreased, although the changes were similar to those of trees inoculated with non-pathogenic *B. mucronatus* or distilled water. These results were similar between *P. densiflora* and *P. thunbergii*. The cause of death of these trees was likely suppression, because the trees were planted densely in the nursery.

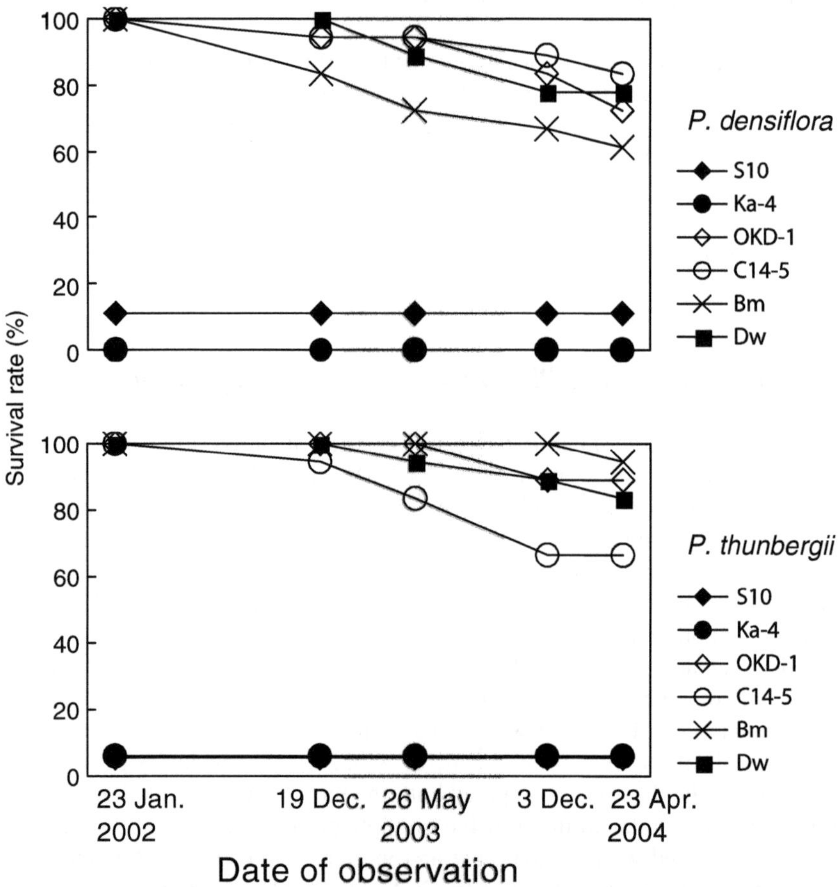


**Fig. 1** Changes in the survival of 3-year-old *Pinus densiflora* and *P. thunbergii* inoculated with *Bursaphelenchus xylophilus* or *B. mucronatus*. S-10 and Ka-4 are virulent isolates of *B. xylophilus*. OKD-1 and C14-5 are avirulent isolates of *B. xylophilus*. Bm and DW are *B. mucronatus* and distilled water, respectively. The survival rates of *P. thunbergii* inoculated with S-10 and Ka-4 were the same throughout the period

## Experiment on Mature Trees

The survival rates of 30-year-old *P. densiflora* in each treatment are shown in Fig. 2. The survival rates of trees inoculated with virulent *B. xylophilus* more or less decreased from the winter of the year of inoculation to the next spring whether the trees were pre-inoculated with avirulent *B. xylophilus* or not. The survival rate of trees inoculated with avirulent and then virulent *B. xylophilus* was about twice that of trees inoculated only with virulent nematodes. After that, the survival became nearly stable in each treatment.

## Changes in Survival of Young Trees Inoculated with Avirulent and Virulent Nematodes

The survival rates of *P. densiflora* and *P. thunbergii* inoculated with both avirulent and virulent *B. xylophilus* in Chiyoda Experimental Station decreased in some experimental plots especially in *P. thunbergii* from December in the year of inoculation to the next spring (Fig. 3). After that, survival became nearly stable.

## Yearly Replications in Young Trees

The survival rates of *P. densiflora* and *P. thunbergii* in each treatment from 2002 to 2004 in the spring following inoculation are shown in Fig. 4. Inoculation with avirulent *B. xylophilus* did not kill any tree except for one *P. thunbergii* inoculated with C14-5 in 2004. The survival rates of trees inoculated with virulent *B. xylophilus* fluctuated by year both in *P. densiflora* and *P. thunbergii* whether the trees were

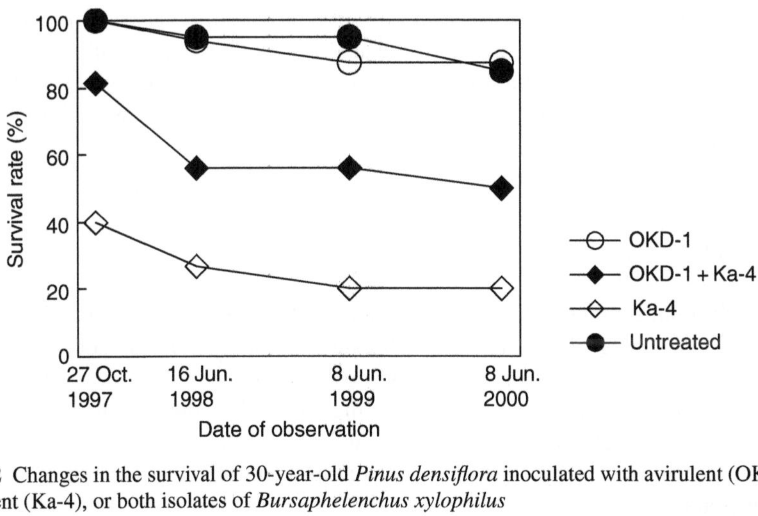


**Fig. 2** Changes in the survival of 30-year-old *Pinus densiflora* inoculated with avirulent (OKD-1), virulent (Ka-4), or both isolates of *Bursaphelenchus xylophilus*

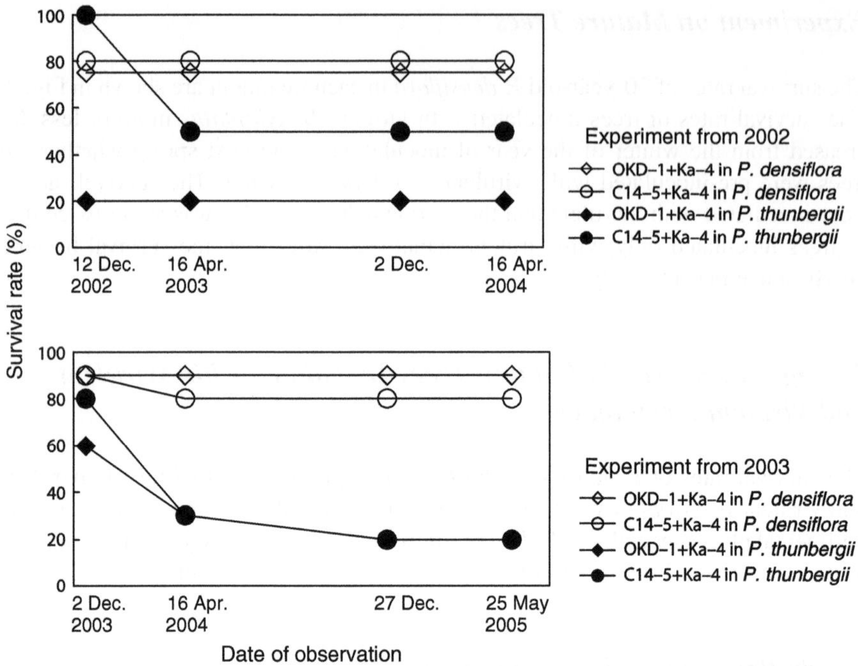


**Fig. 3** Changes in the survival of young (6 and 7-year-old) *Pinus densiflora* and *P. thunbergii* inoculated with an avirulent (OKD-1 or C14-5) and virulent isolate (Ka-4) of *Bursaphelenchus xylophilus*

pre-inoculated with the avirulent isolates or not. In most cases, the survival rates of trees inoculated with avirulent and virulent *B. xylophilus* (hatched bars in Fig. 4) were 20–40% higher than those of trees inoculated with the virulent isolate only (solid bars in Fig. 4). In 2004, however, the survival rate of *P. thunbergii* inoculated with avirulent OKD-1 and virulent Ka-4 was 0%, less than that inoculated with the virulent isolate only. Also, the survival rate of *P. thunbergii* inoculated with C14-5 and Ka-4 was only 10% higher than of that of trees inoculated with the virulent isolate only.

## Discussion

The results confirm that both avirulent isolates of *B. xylophilus*, C14-5 and OKD-1, remained avirulent in the long term. The survival rates of trees inoculated with avirulent and then virulent *B. xylophilus* sometimes decreased from the winter of the year of inoculation to the following spring. After that, they became nearly stable. Therefore, the effects of resistance induced by avirulent *B. xylophilus* should be evaluated in the spring following inoculation. The effects were not consistent among years.

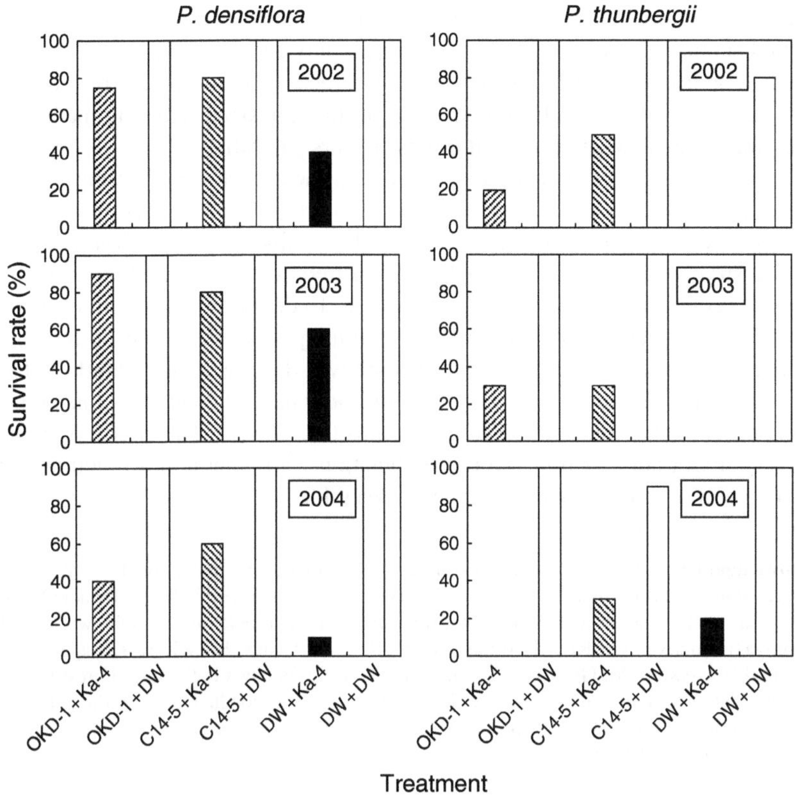


**Fig. 4** Effects of inoculation of an avirulent isolate (OKD-1 or C14-5) and/or a virulent isolate (Ka-4) of *Bursaphelenchus xylophilus* on the survival of young (6–8-year-old) *Pinus densiflora* and *P. thunbergii* the next spring (April or May) following the inoculation. The same experiments were repeated from 2002 to 2004

Several experiments on resistance induced by avirulent *B. xylophilus* have been conducted (Kiyohara, 1984, 1989; Kosaka et al., 2001a). Some showed that avirulent *B. xylophilus* did not induce resistance (Fukuda and Suzuki, 1993; Mori et al., 2007), as in *P. thunbergii* in 2004 (Fig. 4). Sometimes the resistance was temporally induced to the autumn of the year of inoculation (Kosaka et al., 2001a). Induced resistance was highly effective in Kyushu, southern Japan (Kiyohara, 1984). However, in this study conducted in Kanto region of central Japan, the survival of trees with induced resistance was about 20–40% higher than that of trees inoculated with the virulent isolate only. It was recently confirmed that avirulent *B. xylophilus* also induced resistance in the Ryukyu pine, *P. luchuensis*, growing on Okinawa Island in subtropical Japan for the first time (Sakai et al., 2007), similar to the results in this study.

The effects of induced resistance are not as great as the trunk injection with nematicides. However, the results show that although avirulent *B. xylophilus* might cause tree mortality, it still induces resistance to the virulent nematodes. This shows that the benefit, i.e., the chance of tree survival, outweighs the risk of tree mortality caused by avirulent *B. xylophilus* when pine trees were subsequently inoculated with virulent *B. xylophilus*. Explorative use of this resistance-inducing method will be possible in areas where pine wilt disease occurs naturally.

The actual effects of induced resistance as a measure for controlling pine wilt disease are unclear because, in nature, pine trees are infected with *B. xylophilus* by beetles of the genus *Monochamus* (Kishi, 1995). Unlike in inoculation tests, more than one strain of *B. xylophilus* would infest a tree in the wild, and the number of *B. xylophilus* in a tree would depend on the tree. To determine the effect of induced resistance to natural infection with *B. xylophilus*, pine trees should be inoculated with avirulent nematodes at a site where pine wilt disease occurs naturally. However, inoculation with avirulent *B. xylophilus* under natural conditions once did not prevent the spread of pine wilt disease (Kosaka et al., 2001b). The effects of induced resistance might not be strong, so its combination with other control methods such as eradication of infected and dead trees is desirable.

**Acknowledgement** I thank T. Kiyohara (formerly of FFPRI), who encouraged me to study induced resistance. Further thanks are due to T. Aikawa, T. Kikuchi (FFPRI), K. Tabata (Yashima Sangyo Co., Ltd.), and N. Ogura (Meiji University) for their help in the field work, and to Y. Mamiya (formerly of FFPRI), M. Akiba (Kyushu Research Center, FFPRI), K. Kobayashi (Akita Prefectural University), M. Gushiken, Y. Sakai, and H. Irei (Okinawa Prefectural Forest Resource Research Center) for their valuable advice. Part of this study was conducted under the guidance of a research committee organized by the Japan Forest Technology Association on commission from the Forestry Agency of Japan.

## References

Fukuda, K. and Suzuki, K. (1993). [Induced resistance in the pine wilt disease.] In *Elucidations of pathogenicity and induced resistance of pine wilt disease* (ed. K. Suzuki), pp. 71–76. Ministry of Education, Culture and Science, Japan.

Kishi, Y. (1995) *The pine wood nematode and the Japanese pine sawyer.* Tokyo, Japan, Thomas Company, 302 pp.

Kiyohara, T. (1984). Pine wilt resistance induced by prior inoculation with avirulent isolate of *Bursaphelenchus xylophilus*. In *Proceedings of the United States-Japan Seminar. The Resistance Mechanisms of Pines against Pine Wilt Disease* (ed. V. Dropkin), pp. 178–186. Honolulu, Hawaii.

Kiyohara, T. (1989). Etiological study of pine wilt disease. *Bulletin of Forestry and Forest Products Research Institute* 353: 127–176.

Kiyohara, T. and Bolla, R.I. (1990). Pathogenic variability among populations of the pinewood nematode, *Bursaphelenchus xylophilus*. *Forest Science* 36: 1061–1076.

Kiyohara, T., Kosaka, H., Aikawa, T., Ogura, N and Tabata, K. (1999). Experiments of induced resistance to pine wilt disease in pine forest. In *Sustainability of Pine Forests in Relation to Pine Wilt and Decline. Proceedings of an International Symposium, Tokyo, Japan, 26–30 October 1998* (eds. K. Futai, K. Togashi and T. Ikeda), pp. 103–104. Shokado Shoten, Kyoto, Japan.

Kiyohara, T. and Tokushige, Y. (1971). Inoculation experiments of a nematode, *Bursaphelenchus* sp., onto pine trees. *Journal of Japanese Forestry Society* 53: 210–218.

KOSAKA, H., AIKAWA, T., OGURA, N., TABATA, K. AND KIYOHARA, T. (2001a). Pine wilt disease caused by the pine wood nematode: the induced resistance of pine trees by the avirulent isolates of nematode. *European Journal of Plant Pathology* 107: 667–675.

KOSAKA, H., AIKAWA, T., OGURA, N., TABATA, K. AND KIYOHARA, T. (2001b). Induced resistance of pine trees against pine wilt disease by avirulent nematode inoculation. *IUFRO World Series Vol. 11. Protection of world forests from insect pest: Advances in research*: 181–184.

MAMIYA, Y. AND ENDA, N. (1972) Transmission of *Bursaphelenchus lignicolus* (Nematoda: Aphelenchoididae) by *Monochamus alternatus* (Coleoptera: Cerambycidae). *Nematologica* 18: 159–162.

MAMIYA, Y. AND ENDA, N. (1979) *Bursaphelenchus mucronatus* n. sp. (Nematoda: Aphelenchoididae) from pine wood and its biology and pathogenicity to pine trees. *Nematologica* 25: 353–361.

MORI, Y., MIYAHARA, F., TSUTSUMI, Y. AND KONDO, R. (2007). Experimental verification of induced resistance after pre-inoculation with avirulent pine wood nematode isolated with consideration of genetic factors in Japanese Black Pine. *The Journal of Japanese Forest Society* 89: 401–406.

SAKAI, Y., KOSAKA, H. AND AKIBA, M. (2007). Induced resistance of the Ryukyu pine, *Pinus luchuensis* to pine wilt disease by pre-inoculation with the avirulent pinewood nematode, *Bursaphelenchus xylophilus*. *The Journal of Japanese Forest Society* 89:102–106.

NAKAMURA, K. AND YOSHIDA, N. (2004). Successful control of pine wilt disease in Fukiagehama seacoast pine forest in southwestern Japan. In *The pine wood nematode*, Bursaphelenchus xylophilus. *Nematology Monographs and Perspectives, vol. 1* (eds. M. Mota and P. Vieira), pp. 247–260. E. J. Brill, Leiden, The Netherlands.

YOSHIDA, N. (2006) A strategy for controlling pine wilt disease and its application on- site. *The Journal of Japanese Forest Society* 88: 422–428.

# Rapidity of Disease Development Seems to Result in High Mortality – Insight from an Inoculation Test Using Hybridized Populations Between a Virulent and an Avirulent Isolates of *Bursaphelenchus xylophilus*

**S. Takemoto and K. Futai**

**Abstract** The mortality of pine trees caused by the pinewood nematode (PWN), *Bursaphelenchus xylophilus*, still remains insufficiently characterized. One of the reasons is that only a few isolates have been used in previous comparative studies that attempted to describe biological characteristics of the PWN in relation to its virulence. In this study we prepared 27 hybridized populations with a variety of virulence by mixing 2 isolates, S10 (virulent) and C14-5 (avirulent), at various proportions (S10 proportion in each population was 100, 99, 90, 70, 50, 30, 10, 1 or 0%) to characterize the pattern of disease progress over seedling populations. One- and two-year-old seedlings of Japanese black pine served the inoculation test. The numbers of the dead seedlings were recorded every 2 days for 45 days after inoculation. Using these data, we calculated 3 indices, rate of mortality increase, tolerance limit and mortality durability of the seedlings. The rate of mortality increase reflects rapidity of the disease progress and the tolerance limit reflects critical value of load necessary to kill a seedling. There was no correlation between the tolerance limit and the eventual mortality of the seedlings. This may indicate that the tolerance limit reflects physiological conditions of host rather than characteristics of inocula. The eventual mortality correlated closely with the rate of mortality increase. We also found that seedling death was rather durable when the rate of mortality increase was low. On the basis of these analyses, we concluded that the mortality caused by the PWN would eventually be higher when disease progress in a seedling population is faster.

---

K. Futai
Laboratory of Environmental Mycoscience, Graduate School of Agriculture, Kyoto University, Sakyo Ward, Kyoto City, Kyoto Prefecture, Japan
e-mail: futai@kais.kyoto-u.ac.jp

M.M. Mota, P. Vieira (eds.), *Pine Wilt Disease: A Worldwide Threat to Forest Ecosystems*, 

## Introduction

The pinewood nematode (PWN), ***Bursaphelenchus xylophilus*** (Steiner et Buhrer) Nickle, the causal agent of the pine wilt disease (Kiyohara and Tokushige, 1971), is supposed to originated in North America (de Guiran and Bruguir, 1989). Since 1905, when the pine wilt disease was first reported (Yano, 1913), the area damaged by this disease has been spreading throughout Japan except for the northernmost prefectures, Aomori and Hokkaido (Forestry Agency, Ministry of Agriculture, Forestry and Fisheries, 2005). The pine wilt disease has also destroyed huge areas of pine forests in other East Asian countries such as China and Korea, and has recently been found in Portugal (Mota et al., 1999), becoming a possible threat to European countries as an invasive forest pest.

Among the many PWN populations collected in Japan, several populations from different geographical origins are known to have only slight or no pathogenicity to host pine trees (e.g., Kiyohara and Bolla, 1990), and were generally called 'avirulent' isolates. Molecular phylogenetic study confirmed that some of the virulent and the avirulent isolates belong to different clades in a phylogenetic tree of *B. xylophilus* and related species (Iwahori et al., 1998), though no reproductive isolation has developed between the two groups of isolates (e.g., Kiyohara and Bolla, 1990; Aikawa et al., 2003a).

These 2 isolate groups, the virulent and the avirulent, have been used in comparative studies to investigate the nature of the virulence of this nematode species. For instance, it has been known that the avirulent isolates have lower potential to invade bark tissues of pine shoots (Asai and Futai, 2006), lower ability to disperse (Ichihara et al., 2000) and propagate (Kiyohara and Bolla, 1990) within healthy pine trees. However, so little replications within the groups were taken in most of the studies that the differences in virulence could not be exclusively attributed to the difference in such characteristics.

In this research, we established a series of 9 hybridized populations between avirulent and virulent isolates to compare the pattern of disease progress among 27 seedling groups each received any of the hybridized populations. The patterns were analyzed in relation to the eventual mortality of the seedling groups, which reflected the virulence of corresponding nematode populations.

## Materials and Methods

### *Nematodes*

A series of hybridized populations of a virulent (S10) and an avirulent isolate (C14-5) of the PWN were obtained as follows: S10 and C14-5 were mixed so that the proportion of S10 in the mixtures 0, 1, 10, 30, 50, 70, 90, 99 and 100% and the mixtures were applied to 1-month-old *Botrytis cinerea* Pers. mycelia grown on autoclaved barley grain (barley grain 10 ml; tap water 10 ml) in a 50-ml conical flask for 1 month to hybridize the parental isolates. Three replications were made for each of the different S10 proportions. Therefore, 27 populations were obtained in total. These hybridized populations were used as inoculum in the inoculation test below-mentioned. Frequency of '*S* allele' of the heat-shock protein 70a (hsp70a), a distinctive genetic marker for S10, was estimated for each of the hybridized populations following Takemoto et al. (2005). Briefly, the relative intensity of a PCR-RFLP band of the hsp70a specific to S10 against that specific to C14-5 was evaluated by image analysis and the frequency of the allele was estimated with standard lines. In this study, the frequency of *S* allele in a hybridized population was used as an index of the genetic contribution of S10 to the population.

## Inoculation Test

One- and two-year-old *P. thunbergii* seedlings planted in a plastic pot with mixed molding (Hiuga pumice:red loamy clay = 1:1) were inoculated with each of the hybridized populations from 22 to 24 June 2004. For the 1-year-old seedlings, 20-μl suspension containing 500 PWNs was pipetted onto a piece of filter paper inserted into a slit made on the hypocotyl following Asai and Futai (2002), while, for the 2-year-old ones, 500-μl suspension containing 10,000 PWNs was pipetted onto a rolled cotton put on the xylem exposed by peeling the bark (Futai and Furuno, 1979). For each hybridized population, 23 and 18 seedlings were used. After inoculation, the 1- and 2-year-old seedlings were kept in a greenhouse and in a nursery of the Kitashirakawa Experimental Station of Field Science Education and Research Center of Kyoto University. Visible symptoms were recorded every 2 days for 2 months to count the number of dead seedlings, i.e., those showing discoloration on their current-year needles.

## Statistical Analysis

Two indices, $R_{mi}$ and $T_0$, were calculated following Asai and Futai (2005) for each combination of inoculum and seedling age. They proposed the rate of mortality increase (originally named as mortality velocity, $V$, by Asai and Futai (2005)), $R_{mi}$, as an index reflecting the disease progress in a seedling population and the tolerance limit, $T_0$, as the critical value of the cumulative damage to cause seedling death. We additionally calculated the mortality durability, $D$. Percent cumulative mortality of pine seedlings, $M$, is expressed as a function of cumulative damage, $L$, as follows:

$$\begin{cases} 0 & (L < T_0) \\ M = R_{mi}(L - T_0) & (T_0 \leq L < D) \\ M_{max} & (D \leq L), \end{cases}$$

where $L$ is the product of $\log_{10}$ (inoculum dose) and (days after inoculation) (Fig. 1).

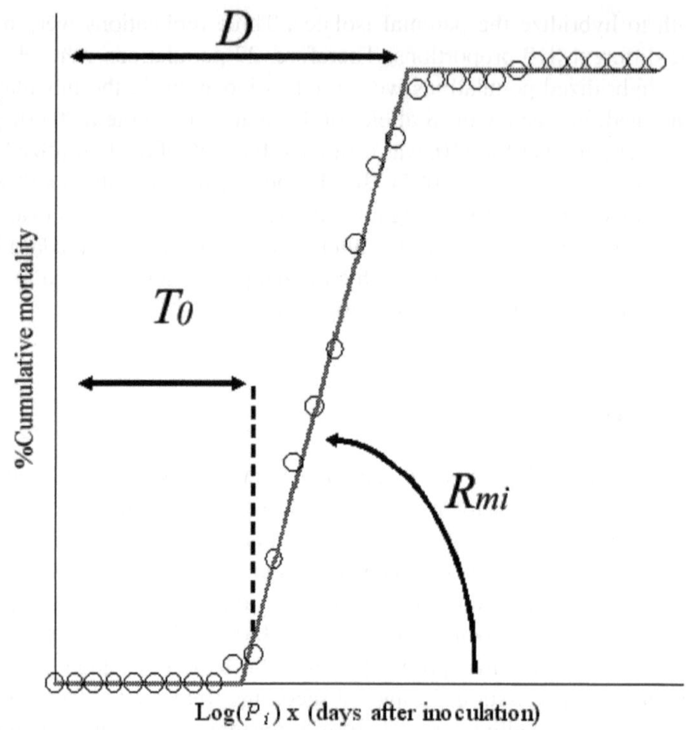


**Fig. 1** Relationship between cumulative stress [log($P_i$) × (days after inoculation)] caused by pinewood nematode and seedling mortality (modified from Asai and Futai (2005). $P_i$ is inoculum density. The slope ($R_{mi}$) and the x-intercept ($T_0$) of the regression line reflect the rate of mortality increase in seedlings and the seedling tolerance limit (critical load necessary to kill the seedlings), respectively. The value of $D$ shows mortality durability

Correlation of $R_{mi}$, $T_0$ and the cumulative mortality at 2 months after the inoculation ($M_{sat}$) with the genetic contribution of S10 to each of the inocula was analyzed. A hyperbolic curve was fitted to the data by the least-squares method and the departure from linearity was also tested following Snedecor and Cochran (1989).

## Results

Matrices of Spearman rank correlation coefficient among *S*-allele frequency in inoculum populations, tolerance limit ($T_0$), rate of mortality increase ($R_{mi}$), mortality durability ($D$) and mortality ($M_{sat}$) of the 1- and 2-year-old *P. thunbergii* seedlings inoculated with hybridized PWN populations are shown in Table 1. No significant correlation was detected between $T_0$ and other variables. However, a strong correlation was found between $R_{mi}$ and $M_{sat}$ of the seedlings. The mortality durability was

**Table 1** Matrices of Spearman rank correlation coefficient among $S$-allele frequency in inoculum populations, tolerance limit ($T_0$), rate of mortality increase ($R_{mi}$), mortality durability ($D$) and mortality ($M_{sat}$) of the 1-year-old (A) and the 2-year-old (B) *Pinus thunbergii* seedlings inoculated with hybridized PWN populations

| | %$S$-allele frequency[a] | $T_0$ | $R_{mi}$ | $D$ | $M_{sat}$[b] |
|---|---|---|---|---|---|
| **(A) 1-year-old** | | | | | |
| %$S$-allele frequency | 1.000 | | | | |
| $T_0$ | −0.004 | 1.000 | | | |
| $R_{mi}$ | *0.399 | 0.300 | 1.000 | | |
| $D$ | −0.353 | −0.226 | −0.818*** | 1.000 | |
| $M_{sat}$ | *0.453 | −0.052 | 0.755*** | −0.434* | 1.000 |
| **(B) 2-year-old** | | | | | |
| %$S$-allele frequency | 1.000 | | | | |
| $T_0$ | 0.023 | 1.000 | | | |
| $R_{mi}$ | 0.550** | 0.272 | 1.000 | | |
| $D$ | −0.078 | −0.255 | −0.671*** | 1.000 | |
| $M_{sat}$ | 0.620** | −0.0129 | 0.659*** | −0.183 | 1.000 |

*$p < 0.05$. **$p < 0.01$. ***$p < 0.001$.
[a] $S$-allele frequency in the inoculum populations.
[b] Mortality of the seedlings 2 months after inoculation.

correlated with $M_{sat}$ in the case of 1-year-old seedlings, and with $R_{mi}$ irrespective to the age of the seedlings.

Figure 2A–D shows the correlation of $S$-allele frequency in the inoculum populations with the four dependent variables ($R_{mi}$, $T_0$, $D$ and $M_{sat}$). The correlations of % $S$-allele frequency with $R_{mi}$ and $M_{sat}$ were regressed to convex hyperbolic curves and their departure from linearity was significant ($p < 0.05$), but those with $T_0$ and $D$ were not. The relationship between $M_{sat}$ and $R_{mi}$ was regressed to a convex hyperbolic curve, and that between $M_{sat}$ and $D$ a concave (Fig. 3).

## Discussion

No correlation was found between $T_0$ of *P. thunbergii* seedlings and $S$-allele frequency in the inoculum. However, there was a significant difference in the average $T_0$ of the seedlings inoculated with various hybridized populations between the 2-year-old seedlings $71.39 \pm 1.56$ (mean ± S.E.) and the 1-year-old seedlings $49.70 \pm 1.14$ (Fig. 2A). Asai and Futai (2005) treated 4-month-old *P. thunbergii* seedlings before PWN inoculation with several kinds of simulated acid rain, and reported that non-treated seedlings showed a $T_0$ of 26.47, lower than those in the present study. They also found that $T_0$ varied with the treatments. These results imply that $T_0$ might depend on the physiological status of seedlings rather than the characteristics of inoculum. In agreement with this speculation, the older seedlings showed a higher tolerance limit in the present study.

In general, the correlation coefficient between two variables dependent on one independent variable should be equal to the product of correlation coefficients

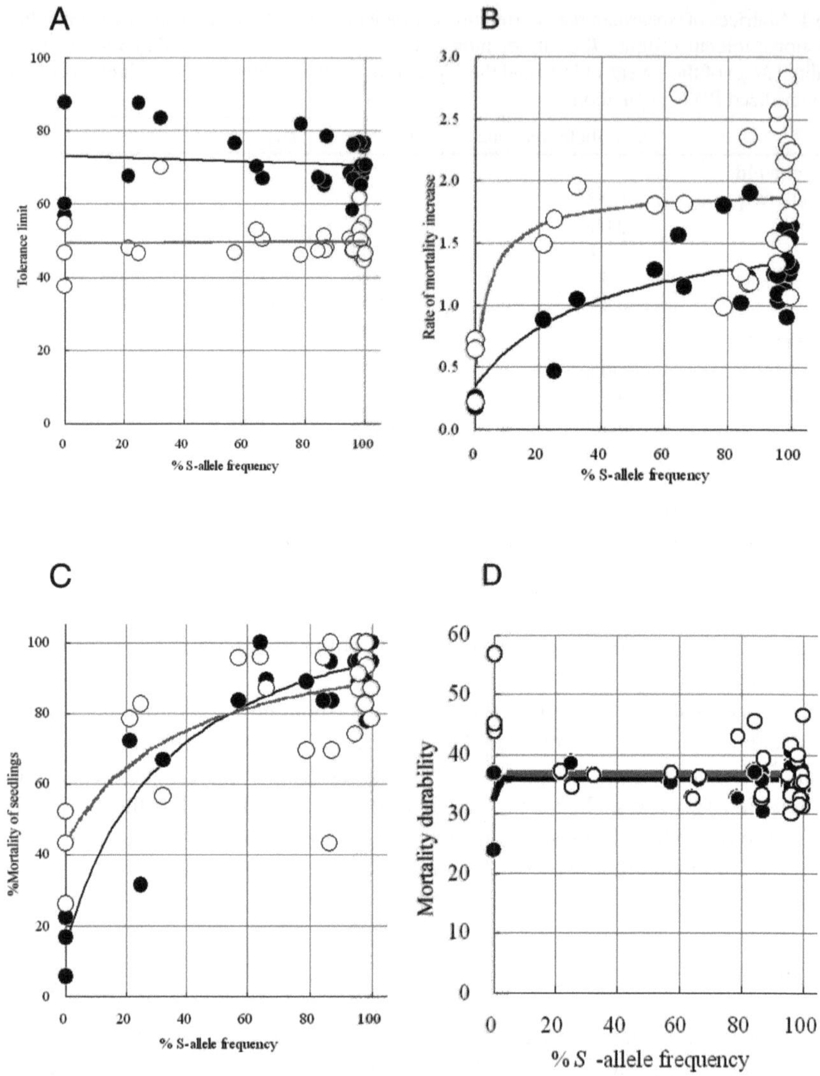


**Fig. 2** Relationship between the genetic contribution of S10 (a virulent isolate) to hybridized populations and the parameters related to their virulence. Horizontal axis is the frequency of *S*-allele of hsp70a in a population. Closed and open circles are 2- and 1-year-old seedlings, respectively. (**A**) Tolerance limit of the *P. thunbergii* seedlings inoculated. (**B**) Rate of mortality increase of the *P. thunbergii* seedlings inoculated. (**C**) Mortality of the *P. thunbergii* seedlings 2 months after inoculation. (**D**) Mortality durability of the *P. thunbergii* seedlings inoculated

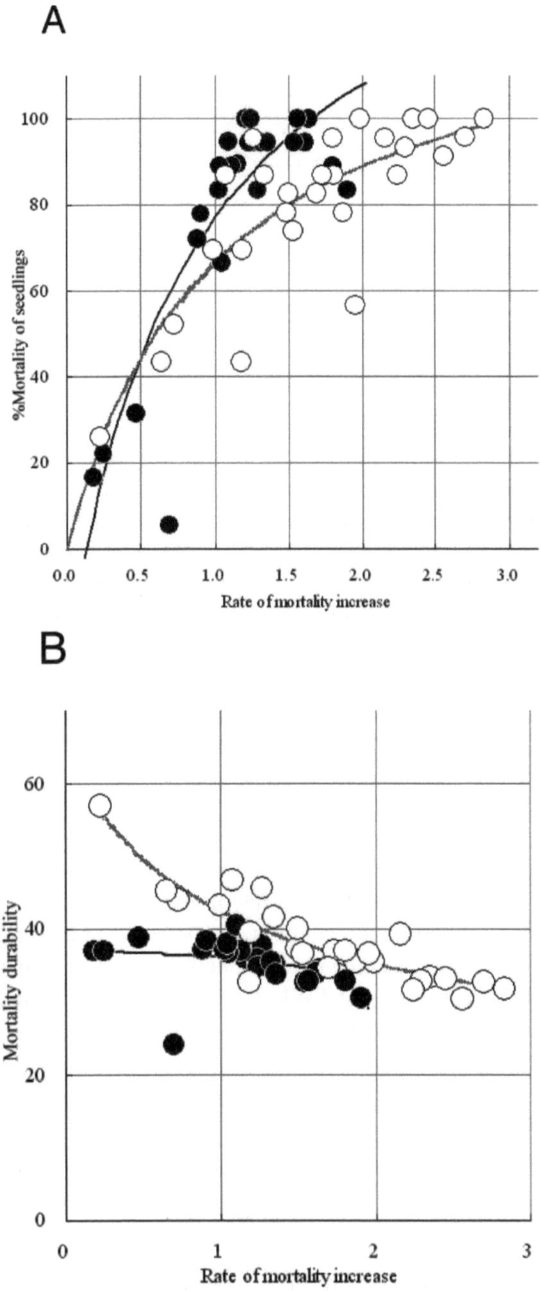


**Fig. 3** Relationship between two virulence parameters. Closed and open circles are 2- and 1-year-old *Pinus thunbergii* seedlings, respectively. The seedlings were inoculated with hybridized populations with different genetic contribution of S10 (a virulent isolate). (**A**) Mortality and the rate of mortality increase. (**B**) Mortality and the mortality durability

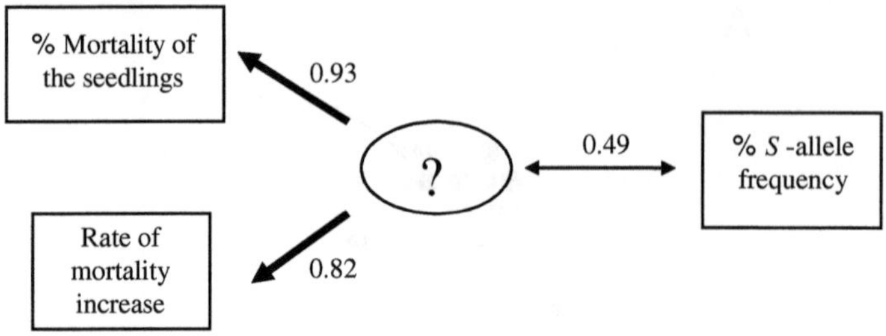


**Fig. 4** A structural model of the relationship among the mortality and the rate of mortality increase of *P. thunbergii* seedlings inoculated with hybridized populations, and the *S*-allele frequency within them. The numbers shown on the arrows are Spearman's rank correlation coefficients among the components calculated from the data of 2-year-old seedlings for instance

between the independent variable and each of the two dependent variables (Sokal and Rohlf, 1995). However, in this present study, Spearman's rank correlation coefficient between $M_{sat}$ and $R_{mi}$ was 0.659 and 0.755 for the 2- and 1-year-old seedlings (Table 1). These are much higher than the product of correlation coefficient between the *S*-allele frequency and each of the two dependent variables, $M_{sat}$ and $R_{mi}$ of the seedlings (0.341 and 0.181 for 2- and 1-year-old seedlings, respectively). This result infers a close relationship between the $M_{sat}$ and $R_{mi}$ that does not include the *S*-allele frequency as an intervening factor (Fig. 4).

The same discussion is applicable to the negative relationship between the rate of mortality increase and the mortality durability (Table 1, Fig. 3B). This means that seedling death may occur sporadically for a long time or intensively for a short time. Taking all into account, interestingly, it is expected that a PWN population with low virulence may cause slow but durative decline among pine seedlings.

In this study we clearly showed the relationship between the virulence of the hybridized PWN populations and the patterns of disease progress over the seedling populations by means of correlation analysis. The strategy of our present study may provide a reliable system to examine if some phenomenon, e.g., enzymatic production or behavioral characteristics of PWN populations or responses of host plants induced by PWN populations, is related with their virulence.

**Acknowledgement** The authors are grateful to the members of Laboratory of Environmental Mycoscience who helped us in the inoculation.

## References

AIKAWA, T., KIKUCHI, T. AND KOSAKA, H. (2003). Demonstration of interbreeding between virulent and avirulent populations of *Bursaphelenchus xylophilus* (Nematoda: Aphelenchoididae) by PCR-RFLP method. *Applied Entomology and Zoology* 38: 565–569.

ASAI, E. AND FUTAI, K. (2002). Promotion of the population growth of pinewood nematode in 4-month-old Japanese black pine seedlings by pretreatment with simulated acid rain. *Journal of Forest Research* 7: 113–116.

ASAI, E. AND FUTAI, K. (2005). Effects of inoculum density of pinewood nematode on the development of pine wilt disease in Japanese black pine seedlings pretreated with simulated acid rain. *Forest Pathology* 35: 135–144.

ASAI, E. AND FUTAI, K. (2006). The effect of acid rain on defense response of pines to pinewood nematodes. In *Proceedings: IUFRO Kanazawa 2003 "Forest Insect Population Dynamics and Host Influences"* (eds. N. Kamata, A.M. Liebhold, D.T. Quiring and K.M. Clancy), pp. 21–24. Kanazawa University, Kanazawa.

DE GUIRAN, G. and BRUGUIR, N. (1989). Hybridization and phylogeny of the pine wood nematode (*Bursaphelenchus* spp.). *Nematologica* 35: 321–330.

FORESTRY AGENCY, MINISTRY OF AGRICULTURE, FORESTRY AND FISHERIES (ed.) (2005). *Statistical handbook of forest and forestry version 2005*. Japan Forestry Foundation, Tokyo, 214 pp. (in Japanese).

FUTAI, K. AND FURUNO, T. (1979). The variety of resistances among pine species to pine wood nematode, *Bursaphelenchus lignicolus*. *Bulletin of the Kyoto university Forest* 51: 23–36.

ICHIHARA, Y., FUKUDA, K. AND K. SUZUKI (2000). Early symptom development and histological changes associated with migration of *Bursaphelenchus xylophilus* in seedling tissues of *Pinus thunbergii*. *Plant Disease* 84: 675–680.

IWAHORI, H., TSUDA, K., KANZAKI, N., KATSURA, I. AND FUTAI, K. (1998). PCR-RFLP and sequencing analysis of ribosomal DNA of *Bursaphelenchus* nematodes related to pine wilt disease. *Fundamental and Applied Nematology* 21: 655–666.

KIYOHARA, T. AND BOLLA, R. I. (1990). Pathogenic variability among populations of the pinewood nematode, *Bursaphelenchus xylophilus*. *Forest Science* 36: 1061–1076.

KIYOHARA, T. AND TOKUSHIGE, Y. (1971). Inoculation experiment of a nematode, *Bursaphelenchus* sp., onto pine trees. *Journal of Japanese Forestry Society* 53: 210–218 (in Japanese with English summary).

MOTA, M. M., BRAASCH, H., BRAVO, M. A., PENAS, A. C., BURGERMEISTER, W., METGE, K. AND SOUSA, E. (1999). First report of *Bursaphelenchus xylophilus* in Portugal and in Europe. *Nematology* 1: 727–734.

SNEDECOR, G. W. AND COCHRAN, W. G. (1989). *Statistical methods*, 8th ed. Iowa State University Press, Ames. 503 pp.

SOKAL, R. R. AND ROHLF, F. J. (1995). *Biometry: the principles and practice of statistics in biological research*, 3d ed. Freeman, New York. 887 pp.

TAKEMOTO, S., KANZAKI, N. AND FUTAI, K. (2005). PCR-RFLP image analysis – a practical method for estimating isolate-specific allele frequency in a population consisting of two different strains of the pinewood nematode, *Bursaphelenchus xylophilus* (Aphelenchida: Aphelencoididae). *Appl. Entomol. Zool.* 40: 529–535.

YANO, S. (1913) Investigation on pine death in Nagasaki prefecture. *Sanrin-Kouhou* 4: 1–14 (in Japanese, This English title is a tentative translation by authors).

ASAI, E. AND FUTAI, K. (2005). Effects of inoculum density of pinewood nematode on the development of pine wilt disease in Japanese black pine seedlings pretreated with simulated acid rain. *Forest Pathology* 35: 135–144.
ASAI, E. AND FUTAI, K. (2006). The effect of acid rain on defense response of pines to pinewood nematodes. In *Proceedings IUFRO Kanazawa 2003 "Forest Insect Population Dynamics and Host Influences"* (eds. N. Kamata, A.M. Liebhold, D.T. Quiring and K.M. Clancy), pp. [illegible]. Kanazawa University, Kanazawa.
DE GUIRAN, [illegible]
[illegible]

# Defense Systems of *Pinus densiflora* Cultivars Selected as Resistant to Pine Wilt Disease

**Keiko Kuroda**

**Abstract** In the stems of pine species resistant to pine wilt, such as *Pinus taeda* and *P. strobes* growing in North America, migration and propagation of the pinewood nematode (PWN) are suppressed and the nematodes disappear from pine tissue in contrast to the highly susceptible Japanese pine species, *P. thunbergii* and *P. densiflora*. Resistant cultivars of these susceptible species have been found in heavily damaged forests. Although they are potential saviors of pine forests in Japan, certain proportions of seedlings obtained from those cultivars are susceptible and are killed after infection. To obtain reliable seedlings with stable high resistance, it is important to find some criteria that can be used to select truly resistant trees for seed orchards. In the tissue of resistant cultivars, mechanisms that prevent nematode activities must be present even if the effect is weaker than those in *P. taeda*. The initial migration of PWN in the shoots was investigated on the cuttings of non-resistant and resistant cultivars of *P. densiflora* and compared with that in *P. taeda*. PWN was inoculated on the apices of 20 cm long cuttings. Every day or two, cuttings of each cultivar were sectioned into short segments (less than 5 cm). Nematodes were extracted from each segment and were counted. PWN in the cortex and xylem tissue was counted separately for the cuttings of *P. densiflora*. Then the anatomical characteristics were investigated on seedlings inoculated with PWN. In *P. taeda* cuttings, the PWN distribution was restricted to the inoculated area during 4 days from inoculation. On the other hand, suppression of nematode migration was not detected in resistant cultivars of *P. densiflora* judging from the PWN numbers in each stem segment. When PWN population in xylem tissue was compared, a tendency was detected: In resistant cultivars, PWN populations during 5 days from inoculation were smaller in the area more than 5 cm below from inoculated sites. In contrast, PWN population in cortex indicates no specific tendency in resistant cultivars. These results suggested that xylem tissue contributes to the defense system in the early period of infection although it is not yet clear whether the structural barrier is effective or toxic substances exist in xylem.

---

K. Kuroda
Forestry and Forest Products Research Institute, Kansai Research Center, Momoyama, Fushimi, Kyoto 612-0855, Japan
e-mail: keiko@affrc.go.jp

## Introduction

Two native Japanese pine species are susceptible to pine wilt caused by the pinewood nematode (PWN), *Bursaphelenchus xylophilus*; one is the Japanese black pine (*Pinus thunbergii*) growing along the sea coast, and the other is the red pine (*P. densiflora*), which is important for forestation and Japanese gardens. Breeding of resistant trees has been practiced by the selection of trees that survived in heavily damaged stands of the susceptible species (Kyushu Regional Association of Forest Research Institutions, 1999). At present, clones of more than 100 cultivars have been kept in seed orchards. Forestation with the seedlings of resistant cultivars has been recently promoted to decrease pine wilt, as it is very difficult to eradicate this disease with pesticides. Although these seedlings are expected to be the "saviors" of pine forests in Japan, there is a problem. Up to 50% of seedlings obtained from the open pollination of mother trees (cultivars) are susceptible to and die after inoculation with the PWN (Toda, 1999; Kuroda et al., 2007). This is because seed orchards contain cultivars with unstable resistance. To get truly resistant seedlings, seed orchards must be improved. For that purpose, breeders request scientific criteria that are useful to find truly highly resistant trees.

The susceptibility and resistance in pine wilt can be summarized as follows: in the stems of susceptible species, dehydration occurs in the sapwood, and the trees are killed by the stopping of the sap ascent within a short period after infection with PWN (Kuroda et al., 1988; Kuroda, 1991; Ikeda, 1996; Fukuda, 1997). In contrast, in the stems of the resistant species *Pinus taeda* growing in North America, migration and propagation of the PWN are strongly suppressed (Kuroda et al., 1991), and the blockage of sap ascent is very restricted. The resistant cultivars of susceptible species indicate intermediate resistance (Kuroda, 2004; Kuroda et al., 2007). The dispersal and propagation of nematodes in the pine tissue are retarded. Blockage of sap ascent due to embolism (Zimmermann, 1983) occurs in the xylem but does not become serious enough to cause wilt. Some resistant seedlings have many branches, about three times those in common trees. The complicated structure of resin canals actually prevents the PWN from going through the knots (Kawaguchi, 2006b). Due to the delay of PWN migration, the defense system might become effective in those seedlings. In the tissue of resistant cultivars of susceptible species, there must exist mechanisms that prevent nematode activities even if the effect is weaker than it is in *P. taeda*. The purpose of the present investigation is to determine the criteria that can be used to select trees that are truly resistant.

## Materials and Methods

### *Comparison Between Species*

The initial migration of the PWN in the shoots was investigated in 20-cm long cuttings of the resistant species, *P. taeda*, and the most susceptible species, *P. thunbergii*.

The cuttings were inoculated with 100 virulent PWN on an incision wound made on the apex of the cuttings (Fig. 1). The cut ends were soaked in water and kept at room temperature (25°C). Every day for 4 days, two or three cuttings were cut into short segments (Fig. 1, Exp. 1). Then, the PWN was extracted from each segment and counted (Thome, 1961).

## *Comparison Between Resistant Trees of* **P. Densiflora**

PWN migration was compared between resistant and non-resistant cultivars of *P. densiflora.* Eight cultivars of the most resistant ones (grades 5 and 4) (Kyushu Regional Association of Forest Research Institutions, 1999) and non-resistant ones were selected. Twenty cuttings (length: 20 cm) from each cultivar were inoculated with 160 PWN (Fig. 1). At 1- or 2-day intervals, three cuttings of each cultivar were cut into four segments (Fig. 1, Exp. 2: ca. 5 cm in length), and the cortex (including the phloem) was peeled from each segment. Nematodes were separately extracted from the cortex and xylem and counted (Fig. 1). Some of the specimens were sliced with a sledge microtome, stained with Nile blue, and prepared for microscopy. The anatomical characters were compared between cultivars and a non-resistant tree under the light microscope.

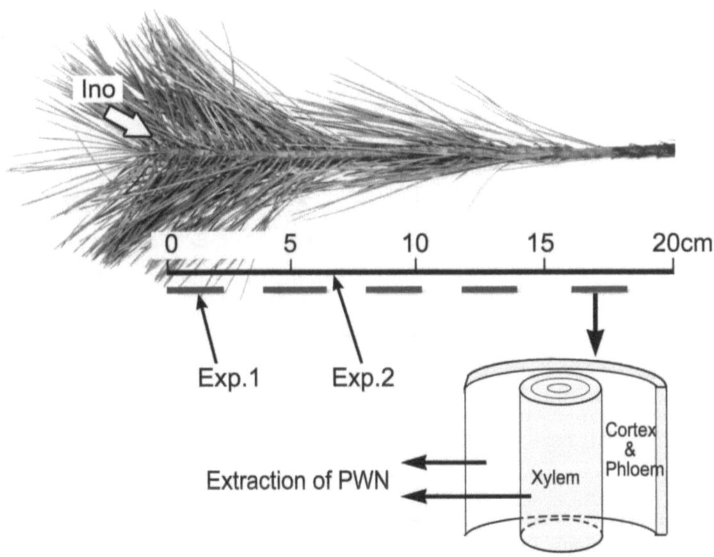


**Fig. 1** Diagram of inoculation on pine cuttings with *Bursaphelenchus xylophilus* and sampling procedure. Exp. 1: First experiment with cuttings of *Pinus taeda* and *P. thunbergii*. Exp. 2: Second experiment with resistant cultivars of *P. densiflora*. Ino: Inoculation site

## Results and Discussion

### *Comparison Between Species*

The nematode distribution comparison between species is shown in Fig. 2. In the cuttings of highly resistant *P. taeda*, the PWN distribution was restricted to the inoculated area at 4 days from inoculation. In the most susceptible species, *P. thunbergii*, PWN migrated from the inoculation point to the lower part of the cuttings. The prevention of nematode migration was clearly shown in resistant species.

### *Comparison Between Cultivars of* **P. Densiflora**

Figure 3 shows the PWN distribution in the cortex (including phloem) and xylem 5 days after PWN inoculation. In the cortex of resistant cultivars, the PWN population in the segments (specimens) 5 cm below the inoculation wound was higher than that in the non-resistant cultivar or almost the same. In contrast, there was a difference in the xylem: in the segments that were 5–10 cm from the inoculated sites, the PWN population was lower in all resistant cultivars than in non-resistant

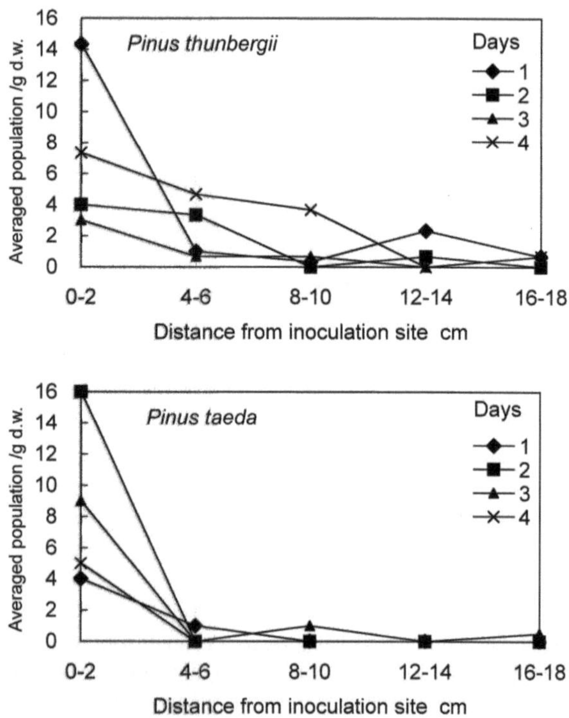


**Fig. 2** Distribution of *B. xylophilus* in the cuttings of resistant *P. taeda* and susceptible *P. thunbergii* during the 4 days from inoculation

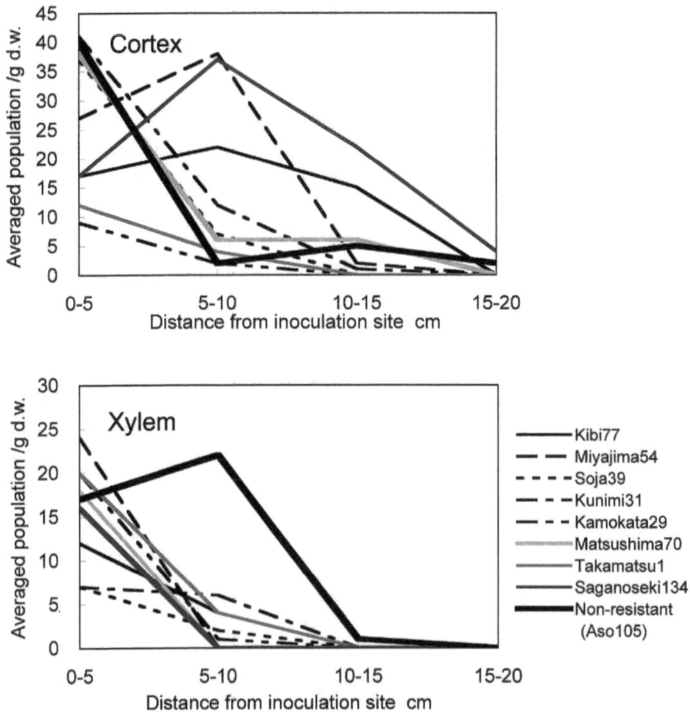


**Fig. 3** Distribution of *B. xylophilus* in the xylem and cortex of eight resistant cultivars of *P. densiflora* at 5 days after inoculation

ones, although it was high around the inoculated site in both cultivars. Nematode distribution seemed to be suppressed in the xylem of resistant cultivars.

These results suggested that the xylem tissue might be contributing to block the nematode migration, at least in the early period of infection. When the PWN distribution and population in each tissue segment were compared including the cortex and the xylem, there was no distinctive difference between cuttings of resistant and non-resistant cultivars at 5 days from inoculation. Therefore, to detect the characteristics of resistant trees, the PWN population in the cortex and xylem must be checked separately.

Among 100 resistant cultivars of *P. densiflora*, 1-year old seedlings of eight cultivars used in this experiment have been reported to indicate the highest survival rate of about 90% against PWN inoculation (Toda, 1997). The mortality rates (including partial dieback) of the 2-year old seedlings of three of the resistant cultivars used in the present experiment, Kibi77, Kunimi13, and Soja39, were 5–20%; in contrast, that of the non-resistant cultivar Aso103 was 46% (unpublished data). The survival rates of seedlings shown by these inoculation experiments are enough to guarantee protection of pine stands when those seedlings are used for reforestation. If the characteristics of these cultivars are specified, they will be applicable to the selection of highly resistant cultivars in the pine stands.

## *Anatomical Observations*

The anatomical structure of the non-resistant cultivar, Aso103, and one of the resistant cultivars, Soja39, was compared on a cross section, as shown in Fig. 4. PWN migrates in the pine tissue via resin canals. *Pinus* species have horizontal and vertical resin canals in the xylem and vertical canals in the cortex. The phloem does not contain resin canals. Epithelial cells that surround resin canals contain precursors of resin. Epithelial cells and ray parenchyma cells live many years and contribute to the defense system by the synthesis of secondary metabolites, such as monoterpenes and phenolic compounds (Hillis, 1987). In the shoot tissues of resistant and non-resistant cultivars that were harvested 5 days after inoculation, the epithelial cells of the vertical resin canals were slightly affected by the activities of PWN, judging from the appearance of nuclei and the stainability of cell contents with Nile blue. One of the anatomical incidences is the formation of tylosoids, which are ballooning epithelial cells formed by the stimuli of PWN. This is evidence that

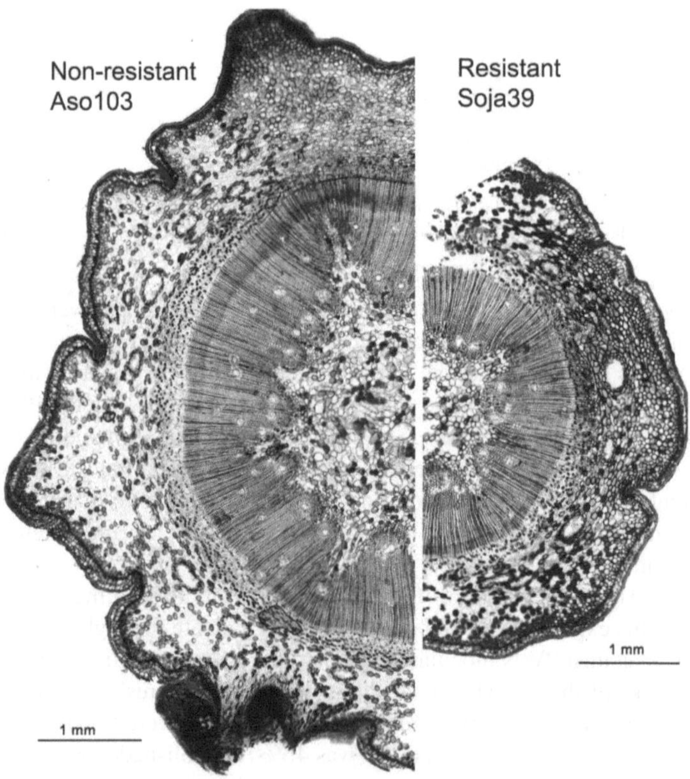


**Fig. 4** Cross sections of the non-resistant cultivar, Aso103, and the resistant cultivar, Soja39, stained with nile blue

the PWN goes through the resin canal. In some resin canals, epithelial cells were broken by the PWN activities. Oily droplets stained purple by Nile blue were observed in tracheids. These were released from the epithelium and ray parenchyma cells as a reaction to infection (Kuroda, 1989; Hara and Takeuchi, 2006). These incidences are important to determine the effect of nematode migration and feeding of cells.

It was revealed by anatomy that the diameter and distribution of the vertical resin canals in the xylem were different between the non-resistant cultivar, Aso103, and the resistant cultivar, Soja39. From the data shown in Fig. 3 and anatomical observation, it seems possible that the resin canals in the xylem may be playing a role to retard PWN migration in the xylem of some resistant cultivars. A detailed observation is necessary to determine whether anatomical characters, such as size and number of resin canals in the xylem, are related to the distribution of the PWN in the xylem just after the inoculation and whether it is available as criteria to identify resistant trees in damaged stands.

## Conclusions

The defense or protective reaction against infection is categorized into two types: one exists before the infection, and the other develops after the infection stimulated by the pathogen. The earlier population growth of PWN in the pine tissue segments that was boiled prior to the PWN inoculation to kill all living cells (Togashi and Matsunaga, 2003) supports the idea that the defense system is functional to a certain extent even in susceptible seedlings. As one of the mechanisms of the former type, the contribution of structural characteristics has been suggested by anatomical investigations (Kawaguchi, 2006a), in addition to chemical compounds being the most likely candidate (Yamada and Ito, 1993). The diameter and/or the distribution density of resin canals may contribute to the difficulty of PWN migration in pine tissue. The active synthesis of monoterpenes detected in infected trees is of the latter type, although this is not effective to prevent PWN migration in the stems of pine wilt-susceptible trees (Bolla et al., 1984; Kuroda, 1989).

A rough scenario of resistance in the resistant cultivars of pine wilt-susceptible species is as follows: inhibition of PWN migration and propagation in the xylem is important for the survival of infected trees, physically and chemically. The resistance level (strength) may be determined by the difficulty for PWN to go through the resin canals in the xylem. The contribution of several genes to resistance was suggested from the variety observed in the resistance level of cultivars. As a criterion to select resistant trees, the population density of PWN in the shoot xylem within a few days of inoculation seems to be useful. If the anatomical characters of various *Pinus* species, including those native to Europe, are examined, important information related to the resistance will be found.

## References


Bolla, R., Shaheen, F. and Winter, R.E.K. (1984). Effect of phytotoxin from nematode-induced pine wilt on *Bursaphelenchus xylophilus* and *Ceratocystis ips*. *Journal of Nematology* 16: 297–303.

Fukuda, K. (1997). Physiological process of the symptom development and resistance mechanism in pine wilt disease. *Journal of Forest Research* 2: 171–181.

Hara, N. and Takeuchi, Y. (2006). Histological analysis for mechanism of pine wilt disease. *Journal of the Japanese Forest Society* 88: 364–369.

Hillis, W.E. (1987). *Heartwood and tree exudates*. Springer-Verlag, Berlin, 268 pp.

Ikeda, T. (1996). Responses of water-stressed *Pinus thunbergii* to inoculation with avirulent pine wood nematode (*Bursaphelenchus xylophilus*): water relations and xylem histology. *Journal of Forest Research* 1: 223–226.

Kawaguchi, E. (2006a). Relationship between the anatomical characteristics of cortical resin canals and migration of *Bursaphelenchus xylophilus*. *Journal of the Japanese Forest Society* 88: 240–244.

Kawaguchi, E. (2006b). Relationship between the number of branches of *Pinus thunbergii* and the migration of *Bursaphelenchus xylophilus*. *Transactions of 117th Meeting of the Japanese Forest Society*, 15.

Kuroda, K. (1989). Terpenoids causing tracheid-cavitation in *Pinus thunbergii* infected by the pine wood nematode (*Bursaphelenchus xylophilus*). *Annals of the Phytopathological Society, Japan* 55: 170–178.

Kuroda, K. (1991). Mechanism of cavitation development in the pine wilt disease. *European Journal of Forest Pathology* 21: 82–89.

Kuroda, K. (2004). Inhibiting factors of symptom development in several Japanese red pine (*Pinus densiflora*) families selected as resistant to pine wilt. *Journal of Forest Research* 9: 217–224.

Kuroda, K., Ohira, M., Okamura, M. and Fujisawa, Y. (2007). Migration and population growth of the pine wood nematode (*Bursaphelenchus xylophilus*) related to the symptom development in the seedlings of Japanese black pine (*Pinus thunbergii*) families selected as resistant to pine wilt. *Journal of the Japanese Forest Society* 89: 241–248.

Kuroda, K., Yamada, T. and Ito, S. (1991). *Bursaphelenchus xylophilus* induced pine wilt: factors associated with resistance. *European Journal of Forest Pathology* 21: 430–438.

Kuroda, K., Yamada, T., Mineo, K. and Tamura, H. (1988). Effects of cavitation on the development of pine wilt disease caused by *Bursaphelenchus xylophilus*. *Annals of the Phytopathological Society of Japan* 54: 606–615.

Kyushu Regional Association of Forest Research Institutions (1999). Characteristics of Japanese cypress and resistant families of pine trees. *In* Publication of the Breeding Working Party, Kyushu Regional Association of Forest Research Institutions, 58 pp.

Toda, T. (1997). Breeding of pine trees resistant to the pine wilt. In *The pine wilt disease: The review of recent researches* (ed. T. Tamura), pp. 168–274. Zenkoku Shinrin Byochujugai Bojo Kyokai.

Toda, T. (1999). Studies on the improvement of resistance to pine wood nematode. *Forest Breeding* 192: 1–4.

Togashi, K. and Matsunaga, K. (2003). Between-isolate difference in dispersal ability of *Bursaphelenchus xylophilus* and vulnerability to inhibition by Pinus densiflora. *Nematology* 5: 559–564.

Yamada, T. and Ito, S. (1993). Chemical defense responses of wilt-resistant pine species, *Pinus strobus* and *P. taeda*, against *Bursaphelenchus xylophilus* infection. *Annals of the Phytopathological Society* 59 : 666–672

Zimmermann, M.H. (1983). *Xylem structure and the ascent of sap*. Springer-Verlag, Berlin, 143 pp.

# Histopathological Observations of *Bursaphelenchus xylophilus* in Symptomatic Tissues of Pinewood

**Yasuharu Mamiya**

**Abstract** Five-month old seedlings and three-year old *Pinus densiflora* were used to examine histopathological damage of wood tissues and symptomatic progress of seedlings at established intervals after the inoculation with the pinewood nematode. At the inoculation site, nematodes entered most axial resin canals of the cortex and xylem resulting in destruction of parenchyma cells. Just after inoculation, only a few nematodes were found in the resin canals in the cortex, and beyond the inoculation site. Cell death recognized by granulation of the cytoplasm and brown cell contents were sporadically observed among axial and radial xylem parenchyma cells through the stem as early as 3 days after inoculation. At the sixth day after inoculation, death of axial and ray parenchyma cells of seedlings which showed the typical disease symptom – marked reduction of oleoresin exudation – was widely distributed in the stem. At this stage, no population growth of nematodes was observed throughout the seedling and no destruction of parenchyma cells necessarily occurred in wood tissues. After complete stop of oleoresin exudation in the seedling, destruction of wood tissues, such as parenchyma cells of axial and radial resin canals, ray, cambium and phloem became more advanced as nematode populations grew rapidly in wood of the seedling. Cell death occurring at the initial stage of pathogenesis was indicated as one of the most remarkable pathological progresses of the nematode inoculated seedling. Destruction of wood tissues resulted from nematode feeding on parenchyma cells following disease progress.

## Introduction

Pine wilt disease caused by the pinewood nematode (PWN), *Bursaphelenchus xylophilus,* has been the most serious epidemic in Japanese forests since the disease was first found in Nagasaki, Kyusyu, early last century. At present, this devastating

Y. Mamiya
5-6-8 Kitanodai, Hachioji, 192-0913 Tokyo, Japan
Formerly, Tamagawa University, Japan
e-mail: cbl01545@nifty.com

M.M. Mota, P. Vieira (eds.), *Pine Wilt Disease: A Worldwide Threat to Forest Ecosystems*, 

epidemic spreads throughout Japan except Hokkaido, the northernmost island, and kills millions of pine trees every year.

Anatomical observations of dead pine trees naturally infected with the pinewood nematode and nematode inoculated pine seedlings demonstrated that nematodes inhabited mostly axial and radial resin canals of wood and destroyed parenchyma cells (Mamiya and Kiyohara, 1972). The evidence pointed to the resin canals and rays as the site of infestation and parenchyma cells as the most likely source of food for nematodes. It was also indicated that nematodes used the axial and radial resin canals, which formed a network in the wood, as passages for spreading throughout the wood. Histopathological studies on the nematode inoculated pine seedlings proved that death of parenchyma cells occurred at the initial stage of pathogenesis, and that a steady increase in number and extent of dead parenchyma cells reflected the disease development in pine seedlings (Mamiya, 1975; Mamiya, 1980; Mamiya, 1982; Mamiya, 1984; Fukuda et al., 1992; Hara et al., 2006).

The objective of the present study is to document the histopathological damage and its relationship to the progressive pathology in pine trees after inoculation with the pinewood nematode. Results of the inoculation experiments of pine seedlings (Mamiya, 1974; Mamiya, 1975; Mamiya, 1980; Mamiya, 1982; Mamiya, 1985) are reviewed and anatomical photographs of diseased wood tissues, some of which have not been previously published, are used to support the experimental results.

## Materials and Methods

### *Nematode Inoculation*

Cultures of PWN (isolate S6-1) on the fungus *Botrytis cinerea* grown on potato-dextrose agar or on barley grains were used as inoculum sources through the experiments. Three-year-old pine seedlings, *Pinus densiflora,* were grown in 30-cm pots with soil. The average seedling height was 48 cm. A two-year-old branch of each seedling was cut off at 5–7 cm from the stem. Inoculation was made by injecting a water suspension of nematodes, 5,000 per seedling, into a rubber tube firmly attached to the cut end of the branch. Distilled water was injected through the rubber tube of seedlings used as control. The seedlings were placed in a greenhouse kept at 27–30° C.

Five-month old seedlings, *P. densiflora*, were grown in polyethylene cups, 6.5 cm in diameter, filled with vermiculite, placed in a growth chamber and kept at 26°C. The seedlings were 7.4 cm in average height. The inoculation of seedlings with PWN was made by dropping a water suspension of nematodes, 500 per seedling, on a piece of filter paper inserted in the stem at 1 cm height.

### *Population Growth of PWN*

Three nematode inoculated three-year-old seedlings and ten five-month-old seedlings were collected at each time of collection, 24 hours after inoculation and then at intervals of 3 days. The nematode population of each seedling was estimated by extracting nematodes from the entire seedling (without leaves) by the Baermann funnel technique.

## *Histopathological Study*

**Three-year-old seedlings:** One-cm$^3$ wood blocks were collected from the inoculated branch, the stem just below the inoculated branch, the basal part of the stem and a middle part of the shoot, respectively, of seedlings used for nematode extraction. They were fixed in FPA (formalin-propionic acid-alcohol). Fixed blocks were dehydrated and embedded in 10% celoidine. Radial, tangential and cross sections of wood samples were cut 12 μm thick with a sliding microtome and double-stained with safranin and fast green

**Five-month-old seedlings:** Sections, 5 mm length, were collected from the stem, the epicotyl and the taproot of the nematode inoculated seedling at each sampling time. Collected samples were fixed in FPA and embedded in paraffin after dehydration. Paraffin blocks were cut 12 μm thick and double-stained with safranin and fast green.

**Pupal chambers of Japanese pine sawyer (JPS), *Monochamus alternatus*:** One-cm$^3$ wood blocks were cut off from wood around pupal chambers in logs of dead pine trees which were collected in Chiba Prefecture early spring. Wood blocks were treated the same way as the three-year-old seedlings were prepared for anatomical observations.

# Results

Population dynamics of PWN in wood of nematode inoculated seedlings are shown in Fig. 1 and Fig. 2.

*Three-year-old seedlings*: Almost all nematodes from a seedling were found in the inoculated branch until the sixth day after inoculation, even though a few nematodes were recognized to spread widely throughout the seedling beyond the inoculated branch. An increase in nematode population clearly initiated at the 9th day after inoculation and synchronized with the appearance of visible symptoms such as the cessation of oleoresin exudation from resin canals. The nematode population increased markedly 12 days after inoculation.

*Five-month-old seedlings*: The nematode population on a seedling showed a tendency to increase in the same way as the nematode population in the three-year-old seedlings. At the sixth day after inoculation, most nematodes were located at the inoculation site, even though only a few nematodes were distributed through the seedling beyond the inoculation site. On the ninth day after inoculation, oleoresin exudation came to a stop while a remarkable increase in the nematode population was recognizable.

Results of histopathological observations of nematode inoculated seedlings at fixed intervals after inoculation were as follows.

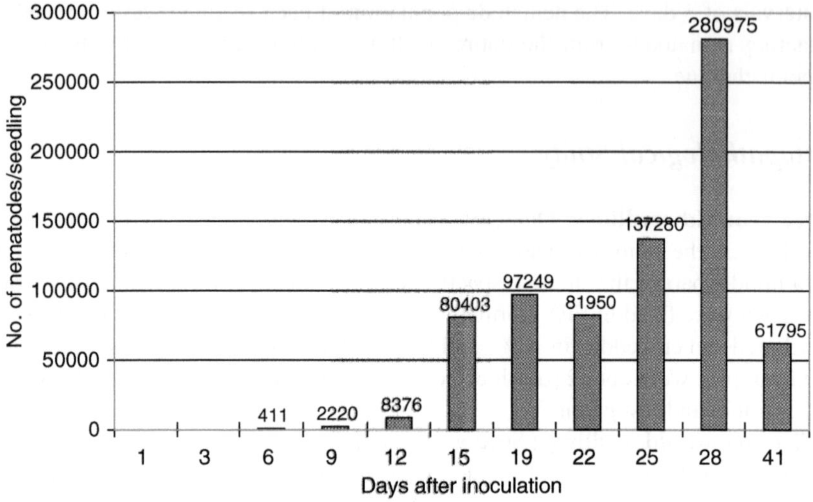


**Fig. 1** Population growth of *Bursaphelenchus xylophilus* in 3-year-old pine seedlings, *Pinus densiflora*. Mean number of nematodes per seedling. Three seedlings were used to estimate number of nematodes at each sampling date

## *24 Hours After Inoculation*

*Three-year-old seedlings* (Fig. 3: A–C): At the cut end of inoculated branches nematodes entered mostly axial and radial resin canals and destroyed their epithelial cells. At the part of a inoculated branch just a little away from the cut end, nematodes

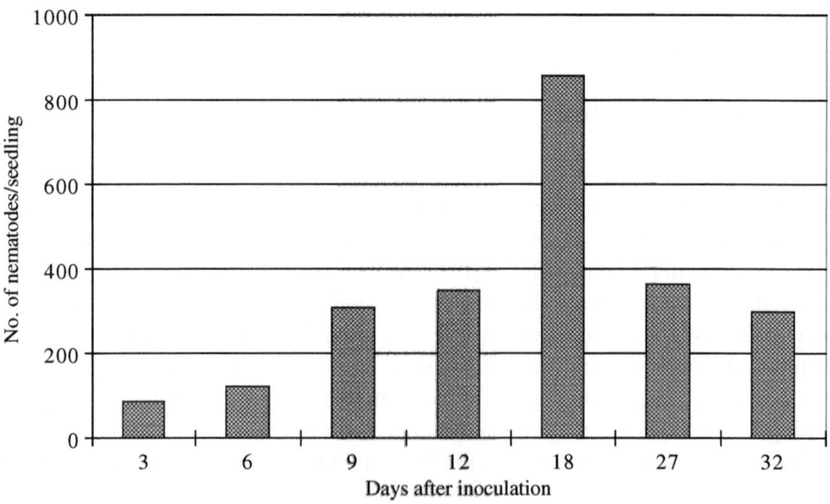


**Fig. 2** Population growth of *Bursaphelenchus xylophilus* in 5-month-old pine seedlings, *Pinus densiflora*. Mean number of nematodes per seedling. Ten seedlings were used to estimate number of nematodes at each sampling date

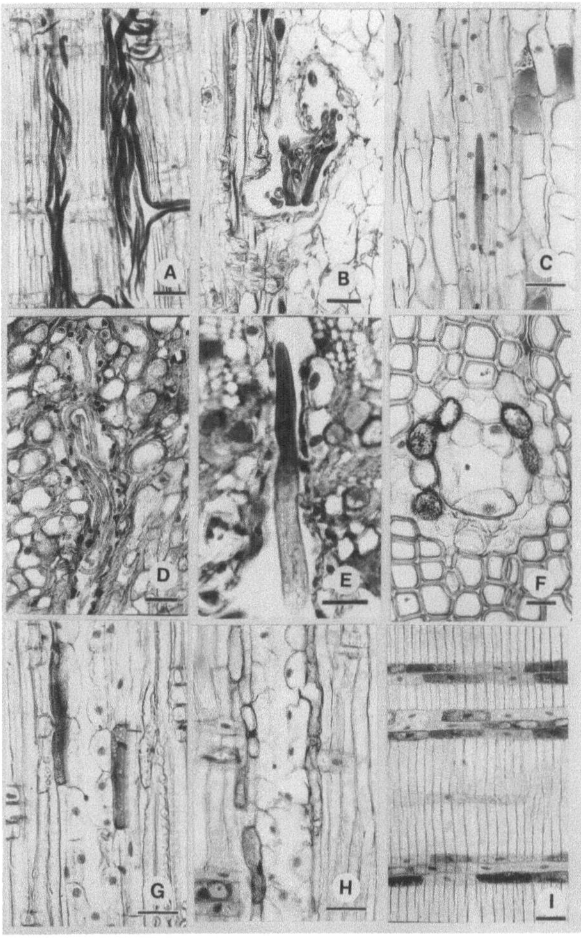


**Fig. 3** Histopathology of 3 years old pine seedlings inoculated with *Bursaphelenchus xylophilus*. **A–C** (RS): 24 hours after nematode inoculation. A: PWN in axial and radial resin canals at the inoculation site. B: PWN in cortical resin canals at the inoculation site. **C**: A nematode in cortical resin canal of the shoot. **D–F** (CS), **G, H** (RS): 3 days after nematode inoculation. D: PWN in cortical resin canals of the stem. E: PWN moving from cortex to resin canal in xylem of the inoculated branch. **F–H**: Denatured axial parenchyma cells in the stem. **I, K** (RS), **J** (CS): 6 days after nematode inoculation. I, J: Denatured ray parenchyma cells and axial parenchyma cells in the stem. **K**: A nematode in ray in the stem. **L, M and O–Q** (RS), **N** (TS): 9 days after nematode inoculation. L: Moving nematode from axial resin canal to radial resin canal. **M, N**: PWN in axial resin canals and destruction of epithelial cells. **O**: A mating pair of PWN in axial resin canal in the stem. P: Eggs in axial resin canal in the stem. Q: PWN in ray and tracheids in the stem. **R–T** (RS): 15 days after nematode inoculation. R, S: Many eggs and juveniles of PWN in axial resin canals destroyed by nematodes in the stem. **T**: Many eggs in cavities along cambial layer in the stem. **U** (CS), **V, W** (RS): 19 days after nematode inoculation. V, W: PWS in axial resin canals in the stem. Epithelial cells and axial parenchyma cells were completely destroyed. **X, Y** (RS): 22 days after nematode inoculation, A nematode in ray and destruction of ray parenchyma cells. RS: Radial sections; TS: Tangential sections; CS: Cross sections. (Scale bars: A-D, G-J, L, M, Q, R = 100 μm; E, F, K, N-P, S-V, X, Y = 50 μm; W = 200 μm)

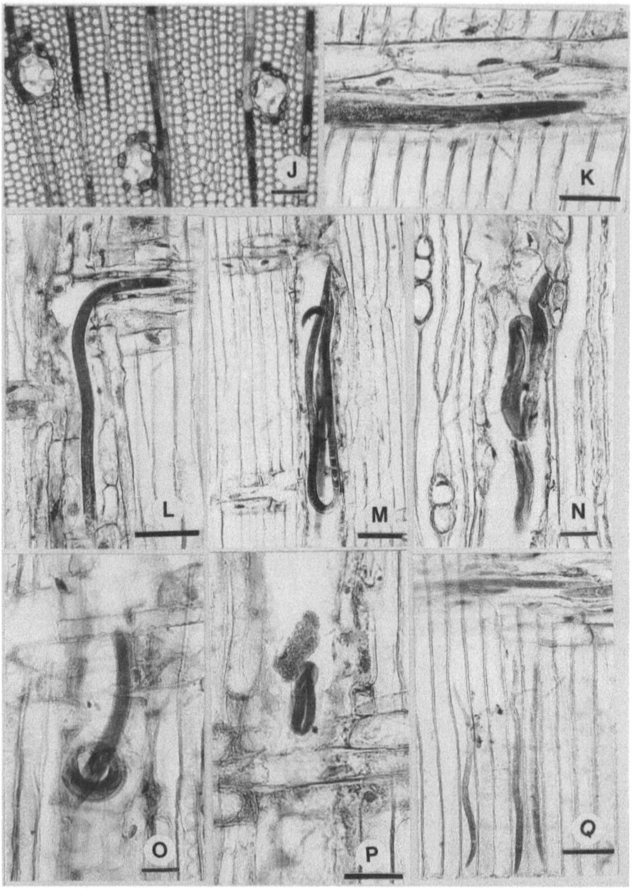


**Fig. 3** (continued)

were observed mostly in the cortical resin canals. A few nematodes were found in cortical resin canals of the shoot far from the inoculated branch.

*Five-month-old seedlings* (Fig. 5: A, B): At the inoculation site, nematodes invaded directly cortical resin canals and axial and radial resin canals in xylem resulting in the destruction of their epithelial cells.

## *Three Days After Inoculation*

*Three-year-old seedlings* (Fig. 3: D–H): Destruction of epithelial cells of axial and radial resin canals became more advanced in the inoculated branch. Cell death, indicated by a granulation of the cytoplasm, deformed nucleus and browned cell contents, was commonly observed among axial parenchyma cells, which surrounded epithelial cells of an axial resin canal, and ray parenchyma cells throughout the

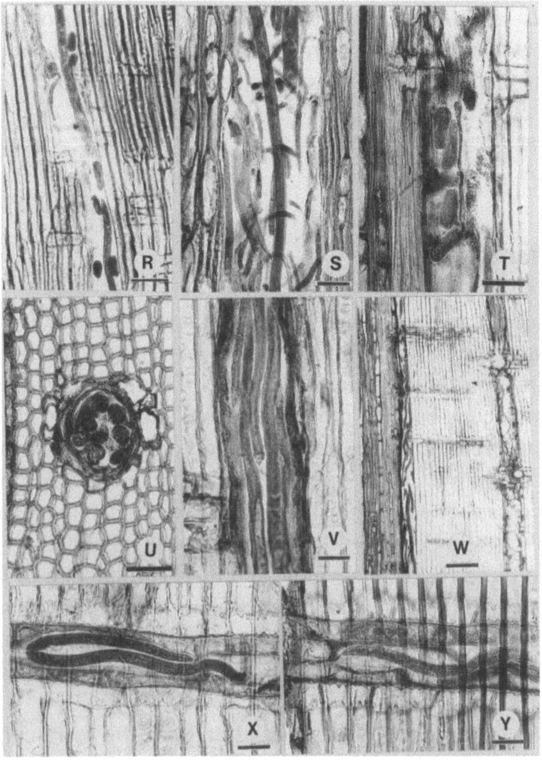


**Fig. 3** (continued)

inoculated branch. Though only a few nematodes were recognized in the resin canals of the stem, death of parenchyma cells spread widely to xylem of the stem.

*Five-month-old seedlings*: At the slit made on the stem for inoculation, nematodes directly invaded the resin canals of the cortex and xylem resulting in destruction of epithelial cells. A few nematodes were observed in resin canals of both xylem and cortex more than 1 cm away from the inoculation site. Denaturation or death of parenchyma cells was observed to a certain extent from the inoculation site.

## Six Days After Inoculation

*Three-year-old seedlings* (Fig. 3: I–K): Almost all resin canals of the inoculated branch were destroyed and occupied with nematodes. Death of parenchyma cells prevailed much more widely through xylem of the stem than at the third day. However, nematode distribution in the stem was uncommon and destruction of epithelial cells of the resin canals was not observed yet. Cytological changes of parenchyma cells had not occurred in the shoot.

*Five-month-old seedlings*: Destruction of parenchyma cells of cortex and xylem developed to a considerable extent from the inoculation site. Damage of the cambial layer caused by nematode activity was conspicuous by cavities along cambial area of the stem.

## Nine Days After Inoculation

*Three-year-old seedlings* (Fig. 3: L–Q): Seedlings did not exude oleoresin on the cross sections cut at the basal part of the stems. An increase in nematode population was clearly recognizable on those seedlings. Death of axial parenchyma cells and ray parenchyma cells became more noticeable than at the sixth day throughout xylem of the stem, even though destruction of epithelial cells was not common in the resin canals. In contrast, a group of axial resin canals, which occupied nearly a quarter of the cross section of the stem, was severely damaged and most epithelial cells were completely destroyed. Many nematodes inhabited those resin canals. Many eggs laid in those resin canals indicated that reproduction of nematodes took place there and became rapidly active. Nematodes were observed not only in resin canals and rays but also in tracheids. The result of the observation suggested that nematodes moved from ray to tracheid passing through window-like pit, and vice versa (Fig. 4: A–E).

*Five-month-old seedlings*: Most epithelial cells of the resin canals throughout the stem were destroyed and nematodes and their eggs were commonly observed in resin canals. Marked reproduction of nematodes must have progressed resulting in remarkable destruction of parenchyma cells. Cavities along the cambial zone were also formed as the results of nematode activities.

## Twelve Days After Inoculation

*Three-year-old seedlings*: The nematode population in seedlings increased in numbers more than those collected at the ninth day. The extent of damage to resin canals and death of parenchyma cells became more advanced than before throughout the stem. Cavities along the cambial zone came to prominence as the result of population increase of nematodes in the wood. There were only a few pathological responses in the shoot. A few nematodes were found in the cortical resin canals of the shoot, but not in xylem resin canals. Death of parenchyma cells was also uncommon in the shoot.

*Five-month-old seedlings*: Death of ray parenchyma cells progressed remarkably in the stem and the tap root. Almost all epithelial cells of resin canals in the epicotyl were destroyed and nematodes were commonly observed in those resin canals. The formation of cavities along the cambial zone developed markedly in the epicotyl.

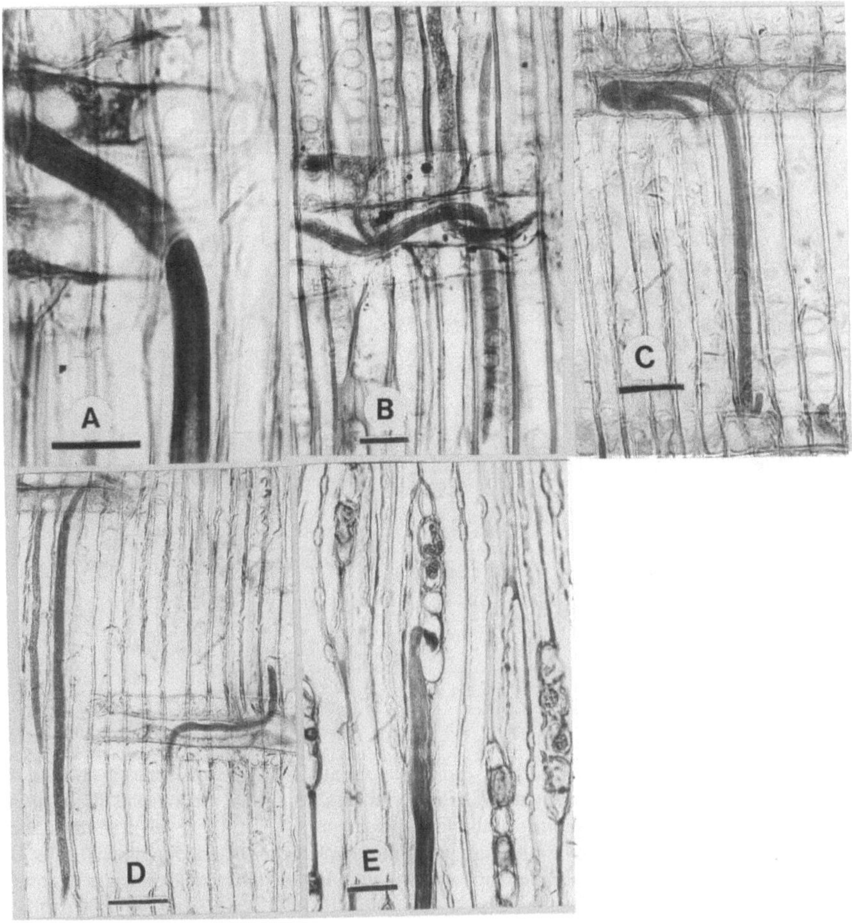


**Fig. 4** PWN in tracheids in wood of 3 years old pine seedlings. **A–D** (RS) **and E** (TS): PWN passing through the window-like pits from ray to tracheid, and vice versa. (Scale bars: A, B, E = 50 μm; C, D = 100 μm)

## *Fifteen to Twenty Two Days After Inoculation*

*Three-year-old seedlings* (Fig. 3: R–Y): Rapid growth of nematode populations was recognizable and epithelial cells of almost all resin canals were destroyed throughout the stem and the shoot. All parenchyma cells of a seedling seemed to be dead judging from their appearance such as shrinkage, discoloration and granulation of cytoplasm. Many cavities were formed along the cambial zone.

*Five-month-old seedlings* (Fig. 5: C–E): As the results of damage caused by nematodes, the cavity spread over almost the whole of cortex and developed into the hollow which surrounded xylem.

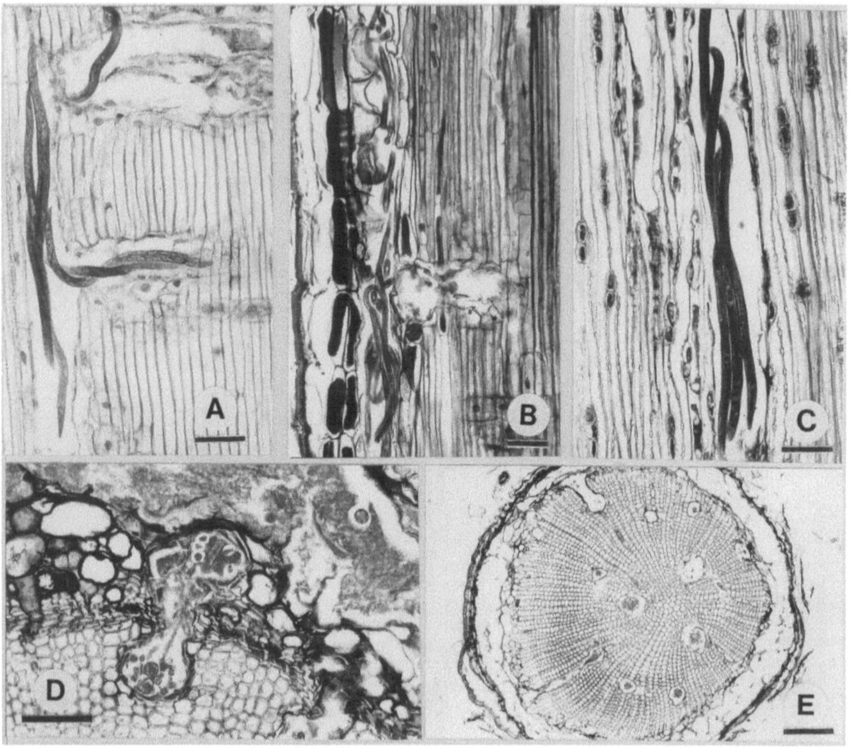


**Fig. 5** Histopathology of 5 months old pine seedlings inoculated with *Bursaphelenchus xylophilus*. **A, B** (RS): 24 hours nematode inoculation. **A**: PWN in axial and radial resin canals in xylem at the inoculation site. **B**: PWN in cortex at the inoculation site. **C** (TS), **D, E** (CS): 19 days after nematode inoculation. **C**: PWN in axial resin canals and complete destruction of epithelial cells. **D**: PWN in axial resin canals and destruction of cortical tissues. **E**: Circumferential damages of cortex and cambial zone caused by PWN. (Scale bars: A, C, D = 100 μm; B = 50 μm; E = 200 μm)

## *Wood Around the Pupal Chamber*

Anatomical observations revealed that almost all nematodes aggregating around the pupal chamber were the dispersal third stage juveniles ($J_{iii}$) and were located within 1–2 mm of the wood tissues surrounding the pupal chamber (Fig. 6: A–C). Most nematodes were located not only in the resin canals but also in the tracheids of spring wood. In the course of anatomical observations, it was recognizable that the dispersal third stage juveniles molted into the dispersal fourth stage juveniles ($J_{iv}$ = "dauer" juveniles) in wood. The beginning of this molt coincided with the pupation of JPS in early May.

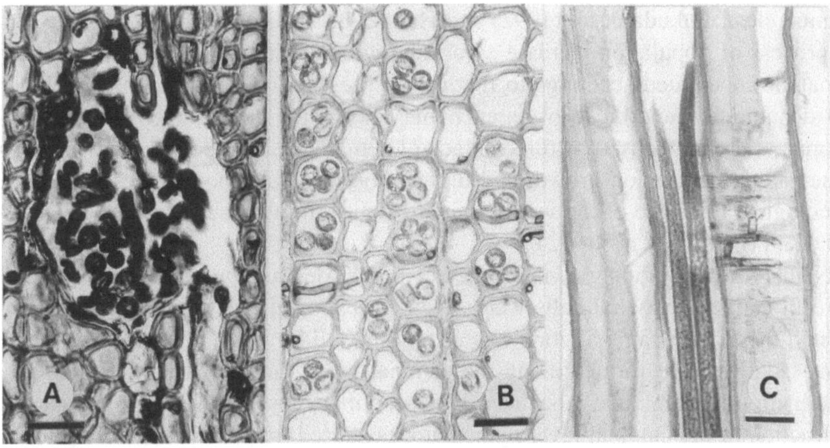


**Fig. 6** Histology of wood around the pupal chamber of JPS. **A** (CS): Dispersal third stage juveniles in axial resin canal. **B** (CS): Dispersal third stage juveniles in tracheids. **C** (RS): A dispersal fourth stage juvenile just after molting in tracheid. (Scale bars: A–C $= 50\,\mu\text{m}$)

## Discussion

Denaturation and death of ray and axial parenchyma cells occurred as the first visible pathological response of the pine seedlings infested with PWN at early stage of pathogenesis. Cytological changes of parenchyma cells developed, in relation to time, after nematode inoculation. It might be concluded that a steady increase in number and extent of denaturated or dead parenchyma cells reflected the disease development in the pine seedling. Histochemical studies proved that denaturation and death of xylem parenchyma cells were the first cytological symptom of the host tissue caused by the nematode infection (Fukuda et al., 1992; Fukuda, 1997; Hara and Futai, 2001; Hara et al., 2006). Death of parenchyma cells occurred and spread over xylem tissues before reduction and cessation of oleoresin exudation became clear through the seedling. When a nematode inoculated seedling ceased to exude oleoresin, dead axial parenchyma cells, which surrounded epithelial cells, had increased in numbers and spread over wood of the seedling. In addition, at that time no death or destruction of epithelial cells was observed. This fact provides an interesting point regarding the supposed relationship between death of axial and ray parenchyma cells and reduction of oleoresin exudation. Cytological changes in wood of the pine seedlings inoculated with each of an avirulent isolate of PWN and nonpathogenic *Bursaphelenchus mucronatus* were restricted spatially to the near inoculation site (Fukuda et al., 1992). Hara et al. (2006) showed that in wood of *Pinus taeda* seedlings recognized to be resistant tree species to PWN, cytological changes occurred only near the inoculation site.

After the occurrence of the cessation of oleoresin exudation in a seedling, the nematode population started to grow in wood and destruction of epithelial cells became clear. Fukuda et al. (1992) and Hara et al. (2006) demonstrated the similar process of population increase associated with cytological changes in wood of nematode inoculated seedlings to that of the present study. It may be concluded that denaturation and death of parenchyma cells at the early stage of disease development are not caused by direct attack of nematodes. Cell death occurring prior to nematode population growth and distribution through wood tissues suggests a pathological response of the wood tissue to some stimulus, such as hypersensitive cell death (Fukuda et al., 1992).

Nematode population growth resulted in marked destruction of parenchyma cells. In resin canals, epithelial cells were severely destroyed and many nematodes inhabited there. The cambial zone was also destroyed by nematodes and as a result many cavities were formed along the cambial layer. On the basis of histological examination of nematode inoculated pine seedlings, Myers (1986) postulated that destruction of the cambial layer by nematodes caused tree death. According to the present study, destruction of the cambial layer was considered a result of population growth of PWN. Disease symptoms such as cell death and oleoresin cessation developed progressively through a seedling prior to population growth. Sugawa (1979) demonstrated that axial traumatic resin canals appeared along the cambial layer in wood of nematode inoculated pine seedlings. In the present study, the axial traumatic resin canals were recognized as the results of circumferential damage to the cambial layer caused by nematodes in wood with elapsed time as the nematode population increased. Many nematodes were observed in these traumatic resin canals as well as in cavities along the cambial layer.

At the early stage of nematode infection, nematodes were commonly found in resin canals of the cortex around the infection site. The cortical resin canals were designated as the main route of nematode movement in wood, especially at the time just after nematode infection occurred (Ishida et al., 1993; Ichihara et al., 2000). Histological observations of the inoculation site indicated that many nematodes invaded and located not only in cortical resin canals but also in xylem resin canals. However, most nematodes located in the cortical resin canals in wood are a bit far from the inoculation site. It was clearly demonstrated that xylem resin canals and rays were used as main routes for movements throughout wood by nematodes after population increase simultaneous with disease development. Nematode movements from ray to tracheid, and vice-versa, passing through window-like pits were commonly observed. Because of the size of the tracheid, length of 1.5~6 mm and diameter of 40 ~ 60 μm, the passage through the tracheid and ray may not be so efficient as the passage through resin canals.

It is common view that the appearance of disease symptoms in the shoot is behind those in the other parts of a seedling. This was supported by histological observations of the shoot.

Histopathological observations revealed that the progressive pathology of three-year old seedlings was almost the same as 5-month old seedlings concerning cytological damages, development of symptoms and the population growth of PWN.

The accumulation of the dispersal third stage juveniles around pupal chambers was clearly demonstrated through histological observations. It was also confirmed that the dispersal third stage juveniles molted to the dispersal fourth stage juveniles in wood around the pupal chamber coinciding with the time of JPS pupation.

As concluding remarks, the following chart is presented.

| **Pinewood nematode** | **Seedling** | **Time after nematode Inoculation** |
|---|---|---|
| Invasion and establishment | Resin canals of cortex and xylem | 24 h |
| ↓ | ↓ | ↓ |
| Distribution | Denaturation of parenchyma cells | 3 days |
| ↓ | ↓ | ↓ |
| Population growth | Cessation of oleoresin exudation | 6 days ∼ 9 days |
| | Destruction of parenchyma cells | |
| ↓ | ↓ | ↓ |
| Marked increase in population | Discoloration of leaves and death of seedling | 15 days ∼ |

## References

FUKUDA, K. (1997). Physiological process of the symptom development and resistance mechanism in pine wilt disease. *Journal of Forest Research* 2: 171–181.

FUKUDA, K. HOGETSU, T. AND SUZUKI, K. (1992). Cavitation and cytological changes in xylem of pine seedlings inoculated with virulent and avirulent isolates of *Bursaphelenchus xylophilus* and *B. mucronatus*. *Journal of the Japanese Forestry Society* 74: 289–299.

HARA, N. AND FUTAI, K. (2001). [Histological changes in xylem parenchyma cells and the effects of tracheids of Japanese black pine inoculated with pine wood nematode, *Bursaphelenchus xylophilus*.] *Journal of the Japanese Forestry Society* 83: 285–289.

HARA, N., TAKEUCHI, Y. AND FUTAI, K. (2006). Cytological changes in ray parenchyma cells of seedlings of three pine species infected with the pine wilt disease. *Japanese Journal of Nematology* 36: 23–32.

ICHIHARA, Y., FUKUDA, K. AND SUZUKI, K. (2000). Early symptom development and histological changes associated with the migration of *Bursaphelenchus xylophilus* in seedling tissues of *Pinus thunbergii*. *Plant Disease* 84: 675–680.

ISHIDA, K., HOGETSU, T., FUKUDA, K. AND SUZUKI, K. (1993). Cortical response in Japanese black pine to attack by the pine wood nematode. *Canadian Journal of Botany* 71: 1399–1405.

MAMIYA, Y. (1974). [Population increase of *Bursaphelenchus lignicolus* in wood of pine seedlings in relation with the time after inoculation.] *Transactions of 85th Meeting of the Japanese Forestry Society*: 249–251.

MAMIYA, Y. (1975). [Behavior of *Bursaphelenchus lignicolus* in wood of pine seedlings and pathological responses of pine to nematode infection.] *Transactions of 86$^{th}$ Meeting of the Japanese Forestry Society*: 285–286.

MAMIYA, Y. (1980). [Inoculation of the first year pine (*Pinus densiflora*) seedlings with *Bursaphelenchus lignicolus* and the histopathology of diseased seedlings.] *Journal of the Japanese Forestry Society* 62: 176–183.

MAMIYA, Y. (1982). Pine wilt and the pine wood nematode: histopathological aspects of disease development. In *Resistance to diseases and pests in forest trees* (eds. H.M. Heybroek, B.R. Stephan and K. von Weissenberg), pp. 153–160. Center for Agricultural Publishing and Documentation, Wageningen, The Netherlands.

MAMIYA, Y. (1984). Behavior of the pine wood nematode, *Bursaphelenchus xylophilus*, associated with the disease development of pine wilt. In *Proceedings of the United States-Japan Seminar, The resistance mechanisms of pines against pine wilt disease* (ed. V. Dropkin), pp. 14–25. East-West Center, Honolulu, Hawaii.

MAMIYA, Y. (1985). Initial pathological changes and disease development in pine trees induced by the pine wood nematode, *Bursaphelenchus xylophilus*. *Annals of the Phytopathological Society of Japan* 51: 546–555.

MAMIYA, Y. AND KIYOHARA, T. (1972). Description of *Bursaphelenchus lignicolus* n. sp. (Nematoda: Aphelenchoididae) from pine wood and histopathology of nematode infested trees. *Nematologica* 18: 120–124.

MYERS, R.F. (1986) Cambium destruction in conifers caused by pinewood nematodes. *Journal of Nematology* 18: 398–402.

SUGAWA, T. (1979). [Occurrence of traumatic resin canals in the stem of Japanese black pine seedlings suffered from pine wood nematode (*Bursaphelenchus lignicolus*).] *Journal of the Japanese Forestry Society* 60: 460–463.

# Development of External and Internal Symptoms in Pine Seedlings (*Pinus sylvestris*) Due to Inoculation with *Bursaphelenchus vallesianus*

**Janina Polomski, Daniel Rigling and Fritz Schweingruber**

**Abstract** Development of needle discoloration and histological symptoms induced by *Bursaphelenchus vallesianus* were investigated in 3-year-old *Pinus sylvestris* plants. Seedlings were inoculated with *B. vallesianus* isolated from declining *P. sylvestris* in Valais, Switzerland. Plants were grown in the greenhouse at 28°C mean temperature and a low watering regime. The histological symptoms in the inoculated plants appeared as necrosis of parenchyma cells in the cortex, rays and cambium, about two weeks after inoculation. Tissue necrosis expanded to the xylem during the next days followed by the formation of cavitations. The external symptoms appeared as browning of needles on the stem and branches below the inoculation site, followed by wilting of the whole plant. Needle discoloration was delayed by few days to the internal symptoms. Our study with *B. vallesianus* demonstrated that external and internal symptoms followed a pattern similar to that caused by *B. xylophilus* and supported the hypothesis that the mechanism causing pine wilting is comparable among the pathogenic *Bursaphelenchus* species.

## Introduction

Symptoms of pine wilt disease (PWD) caused by *Bursaphelenchus xylophilus* in artificially inoculated plants are well documented (Mamiya, 1985; Kuroda, 1991; Riga et al., 1991; Ichihara et al., 2000; Utsuzawa et al., 2005). Disease symptoms typically develop into two distinct stages. The early stage is characterised by a slight destruction of cells in the cortex and xylem that appears about 7–14 days after inoculation (Kuroda, 1991; Ichihara et al., 2000). During the second, advanced stage of the disease, *B. xylophilus* multiplies intensively and migrates throughout the plant. A rapid expansion of cell necrosis from cortex to cambium and formation of tracheid cavitation follows. Finally, the plant wilts because of the blockage of water conduction (Mamiya, 1985; Kuroda, 1991; Ichihara et al., 2000; Utsuzawa et al., 2005). Disease development is accompanied by an increased production of

J. Polomski
Swiss Federal Research Institute WSL, Birmensdorf 8903, Switzerland
e-mail: Janina.polomski@wsl.ch

M.M. Mota, P. Vieira (eds.), *Pine Wilt Disease: A Worldwide Threat to Forest Ecosystems*,  

ethylene (Fukuda et al., 1994) and terpenes (Kuroda et al., 1991) by the tree and exudation of cellulase by the nematodes (Suzuki, 2004). External symptoms appear as discoloration of the needles to yellow and brown and lack of resin production. Disease development progresses very rapidly under favourable conditions, such as a high temperature (25–30°C) and water stress, leading to death of the plant within a few days (Kuroda et al., 1991; Utsuzawa et al., 2005). At temperatures below 20°C *B. xylophilus* multiplies poorly and seldom kills pine trees (Rutherford et al., 1992).

Several investigations indicate a potential pathogenicity of *B. mucronatus*, a species closely related to *B. xylophilus.* However, its virulence varies strongly depending on the isolate, experimental conditions or plant age (Bakke et al., 1991; Mamiya, 1999; Fukuda et al., 1994; Braasch et al., 1999). External symptoms or the histological changes caused by *B. mucronatus* have not been studied in detail. Other *Bursaphelenchus* species were rarely investigated. Some *Bursaphelenchus* species such as *B. hellenicus*, *B. leoni* or *B. tusciae* were described as harmless to pine seedlings, whereas *B. sexdentati* proved highly virulent (Braasch et al., 1999; Skarmoutsos and Michalopoulos-Skarmoutsos, 2000).

*B. vallesianus*, a newly described species in the *B. sexdentati*-Group (Braasch et al., 2004), is very common in declining pine forests in Switzerland (Polomski et al., 2006). However, its role in pine decline or its pathogenicity is not known. The objectives of this paper were to describe the temporal development of (i) external symptoms of young *Pinus sylvestris* inoculated with *B. vallesianus* and (ii) histological reactions of the host plants to *B. vallesianus* invasion.

## Material and Methods

Seedlings of 3-year-old *P. sylvestris* (provenance Tschlin, 1000 m a.s.l.) were grown in pots containing 3.7 l of organic substrate (pH 5.5). In July 2005, the plants were inoculated with *B. vallesianus* isolated in March 2005 from declining *Pinus sylvestris* in Valais, Switzerland. Nematodes were reared at 25°C on *Botrytis cinerea* grown on malt agar in 90 mm petri dishes and extracted using the Baermann funnel method. Inoculations were performed as described by Braasch et al. (1999), using 6000 nematodes per seedling suspended in 0.5 ml water. The nematodes were applied into 1 cm length slit cut at the stem below the new main shoot. Control plants were treated in the same way with nematode-free water suspension. The plants were placed in the greenhouse and subjected to a low watering treatment of 50 ml $H_2O$ per pot twice a week. The greenhouse was set to a day temperature of 25–32°C resulting in a mean day temperature of 28°C for the experimental period.

Wilt symptoms were assessed from 40 plants inoculated with *B. vallesianus* and 10 control plants. Discoloration of needles was recorded weekly for four plant segments separately: (1) main shoot above the inoculation site (MA), (2) branches above the inoculation site (BA), (3) stem below the inoculation site (SB), (4) branches below the inoculation site (BB).

The needle discoloration was grouped in the following classes: class 1 (0–10%), class 2 (11–40%), class 3 (41–80%) and class 4 (81–100%). Symptom development was assessed for a period of 50 days after inoculation and expressed as a percentage of seedlings with needle discoloration of classes 1–4, separately for each plant segment (mean of 40 inoculated and 10 control plants). For histological examinations the inoculated seedlings were harvested at intervals of 7, 11, 13 and 26 days post inoculation, one plant per harvest date. Stem segments of 1 cm in length were cut from the middle part of the stem below the inoculation point. Cross and tangential sections of 40 μm thickness were cut and embedded in glycerine and microscopically examined with a Zeiss MS16 microscope (Carl Zeiss, Germany).

The effect of nematode inoculation on water movement through the plants was studied by comparing controls with plants inoculated with *B. vallesianus* using a staining method (Ichihara et al., 2000). One control plant and three inoculated plants showing different degree of needle browning were harvested 13 days after inoculation. The root system was washed with tap water and immersed in a 0.05% solution of acid fuchsin for 14 hours. The stem was then cut at two locations: (i) 1 cm above the inoculation site, and (ii) 1 cm below the inoculation site. The cut surfaces were examined for reddish discoloration, indication of uptake of the stain solution.

## Results

### *Development of External Symptoms Caused by* B. vallesianus

#### Needle Discoloration

Symptoms induced by *B. vallesianus* developed first as browning of needles on the stem and branches, followed by wilting of the whole plant. The first browning of needles appeared on the stems below the inoculation site 7–14 days after inoculation and progressed rapidly along the stems in the following 12 days (Fig. 1c). At 22 days after inoculation, 60–80% of the plants showed complete needle browning (discoloration class 4) on the stem and branches below the inoculation site compared to 20% of the plants above the inoculation site. At 33 days after inoculation, all plants showed complete browning of the needles on the stem below the inoculation site (Fig. 1c). On the other plant segments, the browning of the needles was delayed by a few days during the early disease period (the first 21 days). At 36 days after inoculation, however, all plant segments showed severe needle discoloration and the mortality rate reached 100% (Fig. 1a, b, c, d). The control plants remained symptomless throughout the 50 days observation period. Figure 2 shows a typical pattern of the symptom development of a pine seedling inoculated with *B. vallesianus*.

#### Water Conduction

Figure 3 illustrates changes of water conduction in relation to the development of the external symptoms in one control and three inoculated plants. The control plant absorbed and transported the stain solution through the stem, resulting in a red stain

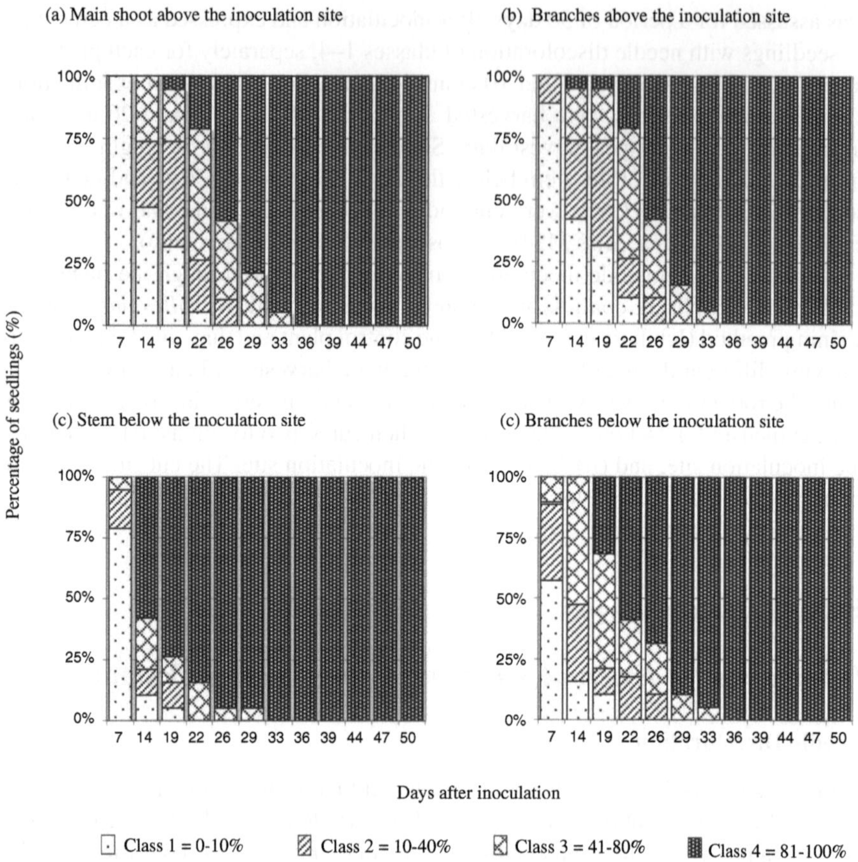


**Fig. 1** Progress of wilting symptoms on 3-year-old *P. sylvestris* plants inoculated with *B. vallesianus*. The symptoms were assessed as a needle discoloration and grouped in four classes, for the main shoot and the branches above the inoculation site (Fig. 1a, b) and the stem and branches below the inoculation site (Fig. 1c, d). Percentage of seedlings with needle discoloration of classes 1 to 4 were calculated for each plant segment, separately, as a mean of 40 inoculated plants

of the stem tissues. In plant 1, no needle browning was observed (class 1), as the first symptoms of water blockage appeared as white patches on the stem surface above and below the inoculation site (Fig. 3). The white patches (approx. 20%–50% of the stem surface) indicated that the conducting system was partially destroyed. Needle symptoms were more developed in plant 2 showing advanced discoloration on the branches and stem below the inoculation site. The white area on the stem surface of plant 2 reached the xylem and occupied almost the entire stem surface (Fig. 3). The stem surfaces of plant 3 exhibited a cavitation pattern similar to that of plant 2, indicating a complete blockage of water conduction.

**Fig. 2** A typical pattern of symptom development of 3-year-old *Pinus sylvestris* plant inoculated with *B. vallesianus* and grown under low water regime. Needle discoloration occurred first on the stem below the inoculation site and was now expanding to other parts of the plant

## Development of Histological Symptoms Caused by *B. vallesianus*

To study the histological response of the host plant to nematode invasion, plants inoculated with *B. vallesianus* were harvested at intervals of 7, 11, 13 and 26 days. The pathological changes were examined on the cross and tangential stem sections cut from the middle part of the stem, below the inoculation point. No visible changes of the tissue of the inoculated plants were observed in the early disease phase, 7 days after inoculation (not shown). However, remarkable pathological symptoms developed in the tissue 11 days after inoculation, i.e. before the external symptoms progressed, as necrosis of parenchyma cells in cortex, rays and cambium (Fig. 4a), which expanded rapidly during the next three days. Almost all parenchyma cells of the cortex, cambium, phloem and rays appeared brown 13 days after inoculation.

| Location of symptoms | Needle discoloration | | | |
|---|---|---|---|---|
| | Control | Plant 1 | Plant 2 | Plant 3 |
| Branches: above the inoculation site | no symptoms | 0–10% | 0–10% | 0–10% |
| Main shoot: above the inoculation site | no symptoms | 0–10% | 0–10% | 40–80% |
| Branches: below the inoculation site | no symptoms | 0–10% | 10–40% | 40–80% |
| Stem: below the inoculation site | no symptoms | 0–10% | 40–80% | 40–80% |
| | Control | Cross cut surfaces of stem with cavitation | | |
| Stem: 1 cm above the inoculation site | | | | |
| Stem: 1 cm below the inoculation site | | | | |

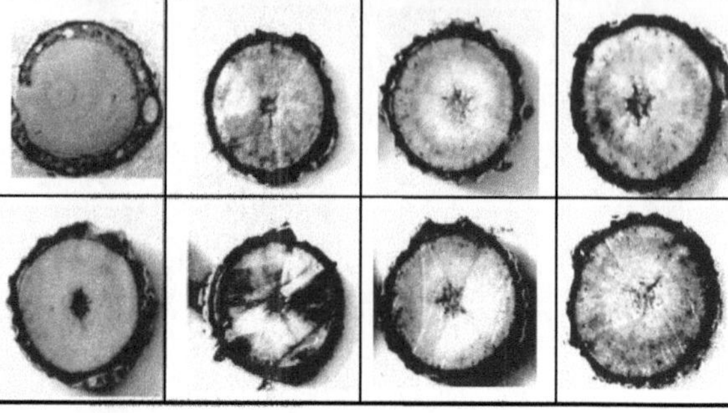

**Fig. 3** Water conduction in 3-years-old *Pinus sylvestris* inoculated with *B. vallesianus* in relation to the progress of needle discoloration. The plants were harvested 13 days after inoculation. Water conduction was visualised as a discoloration of the stem cross sections (1 cm above and 1 cm below the inoculation site) due to uptake of dye solution (0.05% fuchsin acid, 14 h) by the roots. Failure of water uptake is indicated by white areas on stem cross surfaces

Large part of the epithelial cells of the resin canals was also necrotic (Fig. 4b). At this time, the plant segments above the inoculation site showed small symptoms. Tissue necrosis of the inoculated plants seemed to develop slightly advanced to the external symptoms. Destruction of all cells of cortex, phloem, rays, cambium and resin ducts characterised the end stage of symptom development, 26 days after inoculation (Fig. 4c). The control plant was symptomless during the experimental observation (Fig. 4d). Many bordered pits of axial tracheids were filled by air, whereas the pits of control plant were air free (Fig. 5a, b).

## Discussion

Under the experimental conditions applied – high mean temperatures of 28°C and low watering – the external symptoms, induced by *B. vallesianus* on 3-year-old *P. sylvestris* could be divided into two phases: (i) the initial phase until 13–19 days after inoculation, characterised by a gradual appearance of the first wilting symptoms on all plant segments and (ii) the advanced phase between 19 and 33 days after inoculation, characterised by a rapid spread of needle wilting and death of the plants.

Characteristic for the symptom development in the initial stage was the earlier wilting of the needles on the stem and branches, below the inoculation site compared to other plant segments, where the browning of needles developed later and was variable in time and position (Fig. 1). These variations in the needle browning were probably due to different migration patterns of the nematodes. During the active

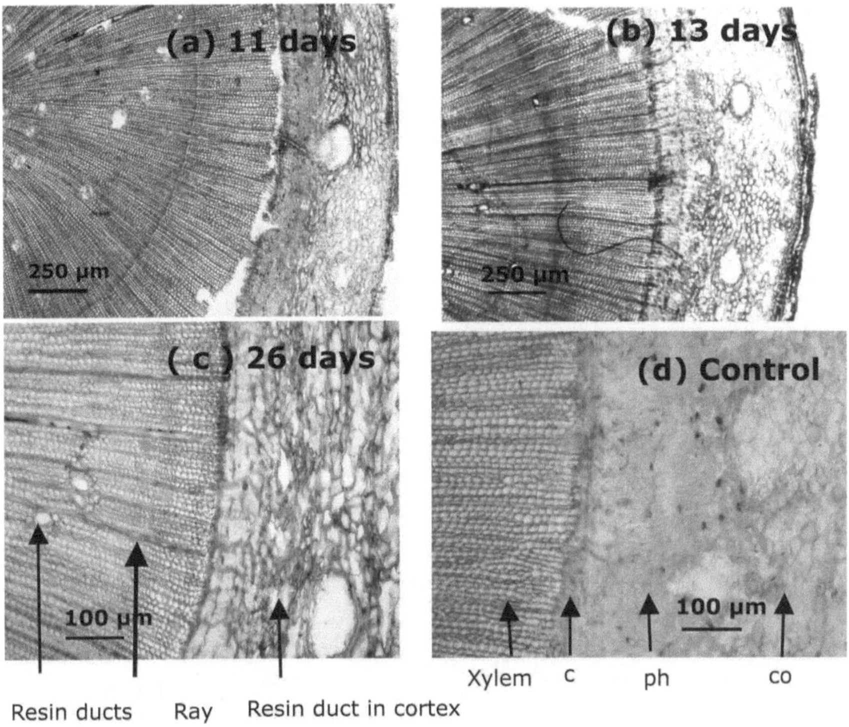


| External symptoms (class) | | | | | Internal symptoms: |
|---|---|---|---|---|---|
| Day | MS | BA | SB | BB | |
| 11 | 1 | 1 | 2 | 2 | Part of co and c: brown |
| 13 | 1 | 1 | 4 | 4 | Most of co, ph, c: brown |
| 26 | 4 | 4 | 4 | 4 | All cells of co, c, r, resin ducts: brown |
| Control | 1 | 1 | 1 | 1 | Co, ph, c, xylem: light |

BA=Branches above the inoculation site
MS = Main shoot
SB= Stem below the inoculation site
BB= Branches below the inoculation site

c = cambium
ph = phloem
co = cortex
r = ray

**Fig. 4** Cross sections of the middle part of stems from a control plant and seedlings inoculated with *B. vallesianus*, harvested at time intervals of 11, 13 and 26 days after inoculation. Slices (40 µm thickness) were embedded in glycerine before examination

phase, between 19 and 33 days after inoculation, the severity of needle browning increased from 10% to 100% for all plant segments.

Histological observation (Fig. 4) showed that damage of the tissues expanded drastically within a very short time span, between 11 and 13 days after inoculation. Such a rapid shift from the early to the advanced disease stage has also been described for *B. xylophilus* and considered a crucial moment in disease development (Utsuzawa et al., 2005). Disease progress from early to advanced stage depends on the population growth of the nematodes. Inoculation studies with

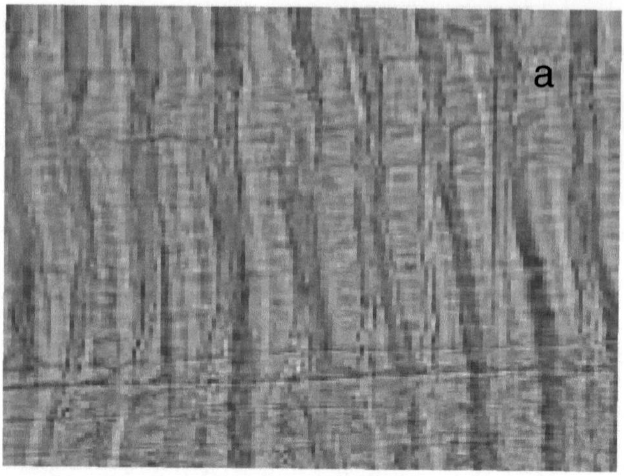


(**a**) Air free pits of control seedling (x400)

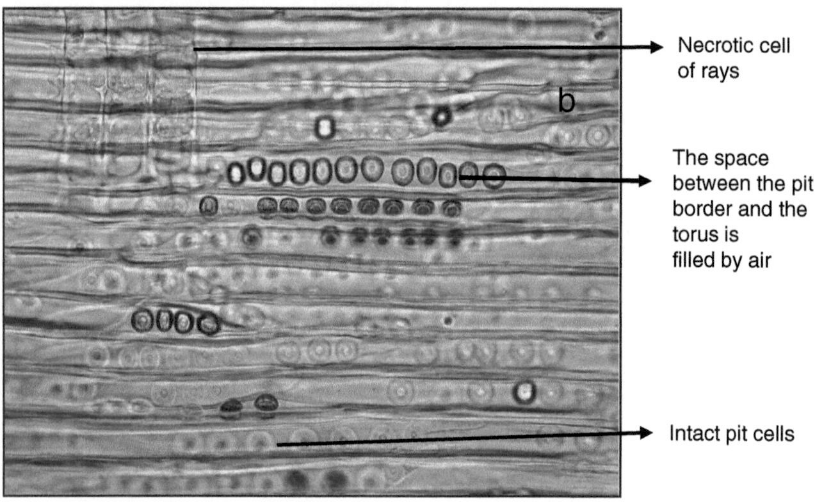


(**b**) Pits in an inoculated seedling 26 days after inoculation. The space between the pit border and the torus is filled by air. The ray cells appaer brown (x400).

**Fig. 5** Intact pits of the control plant; (**a**) compared to the gas filled pits in plants inoculated with *B. vallesianus*, 26 days after inoculation (**b**)

*B. xylophilus* indicated that nematode multiplication is affected by experimental conditions (Rutherford et al., 1992; Braasch et al., 1999). Under unfavourable conditions for the nematodes such as low temperatures or optimal plant water status, the nematode density did not increase or even decreased, resulting in stagnation of disease progress (latent stage). The rapid disease development observed in our study indicated that the experimental conditions applied were favourable for nematode multiplication. Both, high temperature and low watering might have directly or indirectly stimulated nematode growth. In accordance with this expectation, high nematode densities were found in the wilted plants (data not shown).

Destruction of epithelial cells of resin canals and embolism of many pits observed during the intensive phase of the disease (Figs. 4 and 5), are considered an important process in the wilting mechanism. Bolla and Wood (2004) suggested that any *Bursaphelenchus* species which destroyed the epithelial cells of the resin canals causing a leaking of resin into the plant tissues, is pathogenic.

External symptoms and disease development induced by *B. vallesianus* followed a pattern very similar to that caused by *B. xylophilus* (Braasch et al., 1999; Ichihara et al., 2000; Utsuzawa et al., 2005). Histological changes caused by *B. vallesianus* were also similar to that described for *B. xylophilus*. Both species caused destruction of the tissues progressing from the cortex to the cambium and xylem that results in cavitation and blockade of the water transport in the xylem (Kuroda et al., 1991; Ichihara et al., 2000; Utsuzawa et al., 2005). The detailed mode of action of the nematodes that leads to cell death and cavitation, however, requires further researches. Our results support the hypothesis that the mechanism causing pine wilting is very similar among the pathogenic *Bursaphelenchus* species.

**Acknowledgement** We thank Ursula Heiniger for helpful discussions and comments on the manuscript. This study was supported in part by the WSL research program "Forest dynamics" and by Canton Valais.

## References

Bakke, A., Anderson, R. and Kvamme, T. (1991). Pathogenicity of the nematodes *Bursaphelenchus xylophilus* and *Bursaphelenchus mucronatus* to *Pinus sylvestris* seedlings: A greenhouse test. *Scandinavian Journal of Forest Research* 6: 407–412.

Bolla, R. and Wood, R. (2004). Pine wood nematode: pathogenic or political? In *The pine wood nematode,* Bursaphelenchus xylophilus. *Nematology Monographs and Perspectives, vol. 1* (eds. M. Mota and P. Vieira), pp. 31–54. E. J. Brill, Leiden, The Netherlands.

Braasch, H., Caroppo, S., Ambrogioni, L., Michalopoulos, H., Skarmoutsos, G. and Tomiczek, Ch. (1999). Pathogenicity of various *Bursaphelenchus* species to pines and implications to European forests. In *Sustainability of Pine Forests in Relation to Pine Wilt and Decline. Proceedings of an International Symposium, Tokyo, Japan, 26–30 October 1998* (eds. K. Futai, K. Togashi and T. Ikeda), pp. 14–22. Shokado Shote, Kyoto, Japan.

Braasch, H., Schönfeld, U., Polomski, J. and Burgermeister, W. (2004). *Bursaphelenchus vallesianus* sp. n. – a new species of the *Bursaphelenchus sexdentati* group (Nematoda: Parasitaphelenchidae). *Nematologia Mediterranea* 32: 71–79.

Fukuda, K., Hogetsu, T. and Suzuki, K. (1994). Ethylene production during symptom development of pine-wilt disease caused by Bursaphelenchus xylopilus. *European Journal of Forest Pathology* 24: 193–202.

Ichihara, Y., Fukuda, K. and Suzuki, K. (2000). Early symptom development and histological changes associated with migration of *Bursaphelenchus xylophilus* in seedling tissues of *Pinus thunbergii. Plant Disease* 84: 675–680.

Kudora, K. (1991). Mechanism of cavitation development in the pine wilt disease. *European Journal of Forest Pathology* 21: 82–89.

Kuroda, K., Yamada, T. and Ito, S. (1991). *Bursaphelenchus xylophilus* induced pine wilt: factors associated with resistance. *European Journal of Forest Pathology* 21: 430–438.

Mamiya, Y. (1985). Initial pathological changes and disease development in pine trees inoculated by the pine wood nematode, *Bursaphelenchus xylophilus*. *Annals Pathology of the Society of Japan* 51: 546–555.

Mamiya, Y. (1999). Review on the pathogenicity of *Bursaphelenchus mucronatus*. In *Sustainability of Pine Forests in Relation to Pine Wilt and Decline. Proceedings of an International Symposium, Tokyo, Japan, 26–30 October 1998* (eds. K. Futai, K. Togashi and T. Ikeda), pp. 57–64. Shokado Shote, Kyoto, Japan.

Polomski, J., Schönfeld, U., Braasch, H., Dobbertin, M., Burgermeister, W. and Rigling, D. (2006). Occurrence of *Bursaphelenchus* species in declining *Pinus sylvestris* in a dry Alpine valley in Switzerland. *Forest Pathology* 36: 110–118.

Riga, E., Sutherland, J. and Webster, J. (1991). Pathogenicity of pine wood nematode isolates and hybrids to Scots pine seedlings. *Nematologica* 37: 285–292.

Rutherford, T., Riga, E. and Webster, J. (1992). Temperature-mediated behavioral relationships in *Bursaphelenchus xylophilus*, *B. mucronatus* and their hybrids. *Journal of Nematology* 24: 40–44.

Skarmoutsos, G. and Michalopoulos-Skarmoutsos, H. (2000). Pathogenicity of *Bursaphelenchus sexdentati, Bursaphelenchus leoni* and *Bursaphelenchus hellenicus* on European pine seedlings. *Forest Pathology* 30: 149–156.

Suzuki, K. (2004). Pine wilt disease – a threat to pine forests in Europe. In *The pine wood nematode,* Bursaphelenchus xylophilus. *Nematology Monographs and Perspectives, vol. 1* (eds. M. Mota and P. Vieira), pp. 25–30. E. J. Brill, Leiden, The Netherlands.

Utsuzawa, S., Fukuda, K. and Sakaue, D. (2005). Use of magnetic resonance microscopy for the non-destructive observation of xylem cavitation caused by pine wilt disease. *Phytopathology* 95: 737–743.

# Part VII
# Pinewood Nematode and Insect Vector Control Methods

**Katsunori Nakamura and Takefumi Ikeda**

## Summary

Since pine wilt disease (PWD) epidemic constitutes a serious threat to pine-dominated landscapes that carry out important functions in the local ecosystem and human life, control of PWD is of great concern not only to researchers but also to the general public in the damaged areas. When talking about control of PWD, we would refer to various aspects such as detection technique of the nematode, resistance of pine trees to the disease, ecology of the pathogenic nematode and its insect vector, prediction of the potential damage, floral study for re-vegetation of the damaged forest, and strategies for isolating the affected area. The scope of PWD control extends to such a wide area that we could say all research tasks relating to PWD ultimately aim for control of PWD.

The main subject of PWD control, however, is the methodology to prevent infection of the disease and spread of the affected area. Conventionally, *Monochamus* beetles, the insect vector of the pinewood nematode, has been regarded as the main target in PWD control, because the pathogenic nematode itself is difficult to control directly. Eradication of dead pine trees inhabited by immature *Monochamus* is the primary method for PWD control and has been employed to reduce (and hopefully exterminate) the vector beetle population. The dead trees are felled and treated, chemically or biologically, to kill the immature beetle in them. When there is a good chance of introduction of adult beetles by flying, the foliage of pine trees may be sprayed with insecticide so that the feeding adults are killed and the nematode infection minimized. In addition, enhancement of natural enemies of *Monochamus* beetles, such as entomogenous fungi, parasitic insects and bird predators, will be helpful to reduce PWD damage.

In this session, three papers dealt with the control or monitoring of *Monochamus* beetles. Two of them were presented by researchers from Japan, and their main topics were prevention spraying of insecticide and attraction trap for *Monochamus* beetles. Both methods have been developed and widely used in Japan, thus the points of the papers would contain beneficial ideas in practical use. Attempts to use parasitoid insect in controlling *M. alternatus*, in China, is challenging and may lead the way toward the use of natural enemies.

Recently, injection of nematicides into the trunk of live pine trees has become more and more popular. This method targets the pinewood nematode entering a live pine tree, thus the material must be introduced into the tree body prior to the nematode infection season. Several kinds of chemicals are already available, and many are still under development. Three reports relating to the nematicidal chemicals, were all of natural origin, natural compounds of low environmental toxicity and ready degradability which are desired worldwide. If one of those chemicals shows strong and stable nematicidal activity, it will serve as a substitute for conventional nematicides. The papers presented herein could contribute to the development of future trunk injection chemicals.

Implementation of control methods is more of a social activity than a scientific practice. Here, we need a strategy in which people and materials performing control operations are allocated in terms of time and space under a limited budget, and are required to evaluate the result. A presentation was made on the process of PWD control implementation and its result in a sea coast pine forest of a famous sightseeing spot in Japan (Amanohashidate). The example of successful control of PWD taught us the importance of local government efforts and cooperation of the community.

Wood materials contaminated with PWD can still be valid as a useful resource when they are properly treated. But if the treatment is not enough to eradicate the pathogenic nematode and its vectors, use of those materials may facilitate PWD spread. A report from China dealt with this important issue and proposed a heat treatment standard of PWD-damaged trees.

Severe damage of PWD had long been found only in East Asian countries (Japan, China, Korea) until the recent introduction of the nematode in Portugal. These countries have accumulated experiences to cope with PWD, including many failures, and so if information is shared, we can chose successful procedures as well as avoid the way to a failure. The experiences in East Asian countries could contain lessons for the practice of PWD control, especially in the area where PWD had been newly introduced.

# Screening and Isolation of Anti-Nematodal Metabolites Against *Bursaphelenchus xylophilus* Produced by Fungi and Plant

**Jinyan Dong, Guohong Li and Keqin Zhang**

**Abstract** Several plants and fungi were tested in random screening of anti-nematodal metabolites against *Bursaphelenchus xylophilus* using an immersion test. Among 30 species of the terrestrial plants tested, *Magnolia grandiflora*, *Michelia hedyosperma*, and *Nerium indicum* showed nematicidal activity. From leaf extracts of *M. grandiflora*, a new sesquiterpene, 4,5-epoxy-1(10)E, 11(13)-germacradien-12, 6-olide (**1**), was isolated as a nemticidal principle against *B. xylophilus* with an $ED_{50}$ value of 71.4 μg/mL. From 212 terrestrial fungi tested, 83 fungal strains were found to be pathogenic to the pine wood nematode, *B. xylophilus*. Surprisingly, 88 strains among 119 fresh water-derived fungi tested were found to have the ability to produce anti-nematodal substances against *B. xylophilus*. The degree of activity varied with the fungal isolate/species, extraction part, pH values of test solutions, length of exposure time, age of nematode, and media composition. From the nematicidal cultures of *Gliocladium roseum* YMF1.00133, *Paraniesslia* sp. YMF1.01400 and *Pseudohalonectria adversaria* YMF1.01019 and *Caryospora carllicarpa* YMF1.01026, several novel compounds, including gliocladines A-D (**2-6**), glioclatine (**7**), (2*S*, 2′*R*, 3*R*, 3′E, 4E, 8E) – 1 – O - (β - D - glucopyranosyl) – 3 – hydroxyl – 2 - [N - 2′ – hydroxyl - 3′ - eicosadecenoyl] amino – 9 – methyl - 4,8 - octadecadiene (**8**), 3, 5 – dihydroxyaldehyde – 4 – hydroxyl - acetophenone (**9**), pseudohalonectrin A-B (**10-11**), caryospomycins A-C (**12-14**), were isolated and identified along with known compounds such as verticillin A (**15**), 11′-deoxyverticillin A (**16**), sch52900 (**17**), sch52901 (**18**), (2*S*, 2′*R*, 3*R*, 3′E, 4E, 8E) – 1 – O - (β - D - glucopyranosyl) – 3 – hydroxyl – 2 - [N - 2′ – hydroxyl - 3′ - octadecenoyl] amino – 9 – methyl - 4,8 - octadecadiene (**19**), 1,3,6,8 - tetrahydroxy-anthraquinone (**20**), inosine (**21**), and adenosine (**22**). All the abovementioned compounds were examined for nematicidal activity against *B. xylophilus*.

K.Q. Zhang
Key Laboratory for Conservation and Utilization of Bioresources, Yunnan University, 650091 Kunming China
e-mial: kqzhang 111@yahoo.com.cn

M.M. Mota, P. Vieira (eds.), *Pine Wilt Disease: A Worldwide Threat to Forest Ecosystems*, 

## Introduction

*Bursaphelenchus xylophilus* is both a plant-parasitic and a fungal-feeding nematode that causes multi-million dollar loses to pine forests, especially in some Asian countries (Mamiya, 1984; Sutherland and Webster, 1993). Control of this disease depends primarily on fumigation of disease-infected trees, aerial application of synthetic pesticides against the vector of the pine wood nematode, *Monochamus alternatus*, or killing the pine wood nematode by injection of tree trunks with nematicides, such as mesulfenfos, morantel tartrate and levamisol hydrochloride, etc. (Takai et al., 2000). However, nematicides do not provide long-term suppression of nematodes, and environmental and human health concerns are resulting in increased restrictions on their use. Some safe procedures for nematode control have been developed based on biological control agents; however, there is still a need for alternative, environmentally-friendly measures or compounds for effective nematode control to be developed (Noling and Becker, 1994; Whitten et al., 1996). One way of searching for such nematicidal compounds is to screen naturally occurring compounds in plants and fungi.

Kawazu et al. (1980) first screened the nematicidal activities of extracts of 61 randomly selected terrestrial plants against *B. lignicolus* using the novel cotton ball-fungal mat bioassay and reported the nematicidal properties of 11 plants which included *Artemisia capillaries, Cirsium japonicum, Coreopsis lanceosata, Erigeron annus, Prunus verecunda, Sanguisorba officinalis, Dendropanax trifidius, Aleurites cordata, Aleurites fordii, Sapium japonicum and Triadica sebifera* at the screening dose of 20 μg/cotton ball. Later nematicidal properties towards the pine wood nematode were demonstrated in *Sophora flavescens* and active principles were identified as quinolizidine alkaloids: N-methylcytisine, sophocarpine, anagyrine, sophoramine and matrine (Matsuda et al., 1989, 1991). At present there are over 90 species of plants (39 families), which have shown to possess nematicidal properties towards the pine wood nematode. Apparently, there may be many plants, not yet tested, which could prove to be effective for the control of the pine wood nematode.

Fungi have already appeared to be a source of effective pesticidal compounds and may come to be regarded as an inexhaustible source of harmless pesticides having low plant and human toxicity and being easily biodegradable (Siddiqui and Mahmood, 1996). Many fungi have been reported to possess anti-nematodal activities against nematodes such as *Heterodera glycines, Meloidogyne incognita, Meloidogyne javanica, Caenorhabditis elegans, Panagrellus redivivus, Rotylenchulus reniformis*, etc. (Desai et al., 1972; Alam et al., 1973; Sing et al., 1983; Khan and Hussain, 1989; Khan and Kgan, 1992; Chattopadhyay and De, 1995; Pathak and Kumar, 1995; Sankaranarayanan et al., 1997). Many fungal metabolites with nematicidal activities have also been isolated and identified as alkaloids, peptides, terpenes, fatty acids etc. (Anke and Sterner, 1997; Chitwood, 2002). However, among these, only very few reports are related to anti-nematicidal fungi and metabolites against the pine wood nematode (Kawazu et al., 1993; Huang et al., 2004).

Therefore, a screening of anti-nematodal substances from fungi and plants against *B. xylophilus* was carried out by the authors in recent years. In the following, we present the results of this screening, including the isolation of several metabolites with nematicidal activities from extracts of *M. grandiflora* leaves and fungal cultures of *Gliocladium roseum* YMF1.00133, *Paraniesslia* sp. YMF1.01400 and *Pseudohalonectria adversaria* YMF1.01019 and *Caryospora carllicarpa* YMF1.01026.

## Materials and Methods

### *Cultivation of Bursaphlenchus xylophilus*

The fungus, *Bottyis cinerea*, was cultured on potato dextrose agar medium in Petri dishes (diam. 90 mm) at 26°C. Petri dishes with fully-grown fungus were inoculated with *B. xylophilus* and left until fungal mycelia were completely consumed. The cultured nematodes (mixed stages) were separated from the culture medium by the Baermann funnel technique and counted on a grid under a microscope (×20). An aqueous suspension of the nematode (ca.15 000 nematodes $ml^{-1}$) was prepared by appropriate dilution for use as a working stock.

### *Test Plant and Preparation of Aliphatic Extracts*

The plant materials were collected from the Kunming Institute of Botany, P. R. China, washed with running tap water and then dried in an oven at 50°C. Dry plant material (10 g) was chopped into small pieces, and extracted three times with 80% ethanol (100 ml) at room temperature (72 h each time). The ethanol extracts were filtered and then concentrated under vacuum, which were then dissolved in 80% ethanol. These samples were conserved at 4°C prior to using.

### *Test Fungi, Media, Culture Conditions, and Preparation of Test Solutions*

Terrestrial fungi were obtained from the fresh fruiting bodies or their basidiospores collected in mountains of Pu Er County of Yunnan Province in August 2000 and were identified by Prof. Zang Mu, senior taxonomist, Kunming Institute of Botany (Chinese Academy of Sciences, Kunming, Yunnan, China). Fresh water-derived fungi were isolated from the submerged woody substrates collected from various freshwater habitats (Cai et al., 2002, 2003). All cultures are deposited in the strain collection of the Laboratory for Conservation and Utilization of Bio-resources, Yunnan University, Yunnan Province, People's Republic of China. Cultures on PDA were kept at 4°C and transferred once every 6 months.

These fungi were inoculated in 250 mL Erlenmeyer flasks each containing 70 mL medium (**PDB**; **Czapek**; **NSK**: 0.5% $(NH_4)_2SO_4$, 0.1% $K_2HPO_4$, 0.05% KCl, 0.05% $MgSO_4{\cdot}7H_2O$, 0.001% $FeSO_4{\cdot}7H_2O$, 0.5% CaCl, 3% sucrose; 30, 2%

glucose, pH 7.0; **SGP**: 2% soybean power, 2% glucose, 0.2% peptone, 0.5% starch, 0.2% yeast cream, 0.4% NaCl, 0.05% $K_2HPO_4$, 0.05% $MgSO_4 \cdot 7H_2O$, 0.2% $CaCO_3$, pH 7.0 or **NNK**: 0.3% $NH_4NO_3$, 0.1% $K_2HPO_4$, 0.05% KCl, 0.05% $MgSO_4 \cdot 7H_2O$, 0.001% $FeSO_4 \cdot 7H_2O$, 0.5% CaCl, 1% molasses, 2% glucose, pH 7.0) and incubated at 28°C on a rotary shaker (200 rpm) until the carbon source was used up. The culture filtrates and the 10 mL water-soluble extracts of broken hyphae, respectively, served as test solutions for screening nematicidal fungi. When the pH value of the test solutions was measured at over 7.5 or under 5, the mobility of the nematodes was adversely affected. So, the influence of pH on nematicidal properties was pointed out by adjusting the test solutions to chemically neutral with NaOH and HCl water solution. Freshly prepared sterilized medium was clarified by centrifugation and designated as control.

### *Preparation of Test Solution of Aliphatic Extracts and Metabolites*

The extracts and metabolites to be tested were added in DMSO solution up to 3% of the final volume and diluted with Tween-20 water solution to prepare a stock solution with defined concentrations.

### *Immersion Test*

The nematode-toxin bioassay was performed in 5 cm Petri-dishes. Three hundred nematodes, in a volume of 20 uL of water, were transferred to Petri-dishes containing 2 mL stock solutions and mixed gently. All dishes were kept at 25°C. Numbers of live and paralyzed nematodes were counted under a binocular microscope after different incubation times, for 48 h. Nematodes were considered dead if they gave no response to physical stimuli such as mechanical stirring and picking with the point of a needle. Toxicity was estimated according to the mean percentage of dead nematodes. Each treatment was replicated four times and the data obtained were revising analyzed.

## Results and Discussion

### *Screening of Plants Antagonistic to B. xylophilus*

Ethanol extracts of 30 plants representing 24 families were assessed for anti-nematodal activity against *B. xylophilus* using the immersion test. The toxicity of plants was found to vary according to plant species and extraction part. Of these, the leaves extracts of *Magnolia grandiflora* (Magnoliaceae) and *Michelia hedyosperma* (Magnoliaceae), and the branch extract of *Nerium indicum* (Apocynaceae), respectively, showed 72.6, 66.4 and 45.3% nematicidal activity towards *B. xylophilus* after

48 h incubation at the concentration of 5 mg/mL (Hong et al., 2007). This is the first report of Magnoliaceae species with nematicidal activity. It is suggested that more plants belonging to this family should be screened to search for new sources of nematicidal substances. Results also indicate that nematicidal effect varies among different parts of plant, which may suggest different part of plant comprise of different chemical components.

## A New Nematicidal Sesquiterpene from Plant *M. grandiflora*

The activity-guided chromatographic purification of the methanol extract of *M. grandiflora* leaves led to a new sesquiterpene, 4,5-epoxy-1(10)E, 11(13)-germacradien - 12, 6-olide (**1**) (Fig. 1), as the sole compound responsible for killing the worms. The median lethal concentrations ($LC_{50}$) of **1** against *B. xylophilus* and *Panagrellus redivivus* were 71.4 and 46.2 μg/mL after 48 h incubation (Hong et al., 2007).

Besides 4,5-epoxy-1(10)E, 11(13)-germacradien -12, 6-olide, many other sesquiterpenoids have been discovered to be nematotoxic from various plants or microbes, which include aldehydes hemigossypol, 6-methoxyhemigossypol, $C_{30}$ dimers gossypol, 6-methoxygossypol, alantolactone, rishitin, 2β, 13-dihydroxyledol etc. (Datta and Saxena 2001; Chitwood, 2002; Huang et al., 2004). Therefore, the sesquiterpene class of compounds may have the potential to be developed as effective and alternative nematode-managing natural products may serve as a substitute for some of the synthetic nematode-managing agents on the market. Further studies are required to determine the effective dose and potential of the sesquiterpene class of compounds as nematode-magnging activity agents under field conditions.

## Screening of Fresh Water Derived Fungi for the Production of Nematicidal Metabolites

From 119 freshwater derived fungal strains grown on two different media (SGP and NNK) (22 species: *Aniptodera, Annulatascus, Camposporium, Caryospora, Diaporthe, Dictyosporium, Dyrithiopsis, Eutypa, Eutypella, Gliocladium, Helicomyces, Leptosphaeria, Massarina, Nectria, Ophioceras, Paraniesslia, Phoma, Phomatospora, Pseudohalonectria, Savoryella, Torula,* and *Xylaria*), the submerged cultures of 88 fungal strains were found to be pathogenic to the pine wood nematode, *B. xylophilus* following a 48-h exposure *in vitro* screening. Of these, over 90% of *B. xylophilus* were immobilized in 12 cell-free filtrates (CFs), 13 pH-adjusted cultural filtrates (AFs), 3 water-soluble extracts of broken hyphae (Hs), and 5 pH-adjusted water-soluble extracts of broken hyphae of 24 fungal strains (AHs) (see Table 1). As can be seen from Table 1, the degree of toxicity of the fungal cultures was found to vary according to fungal species/isolates, extraction parts, pH values of test solutions, and length of exposure time.

Pronounced anti-nematodal activity was often encountered in the 9 genera of *Aniptodera* (4 among 5 species), *Annulatascus* (6 among 6 species), *Caryospora*

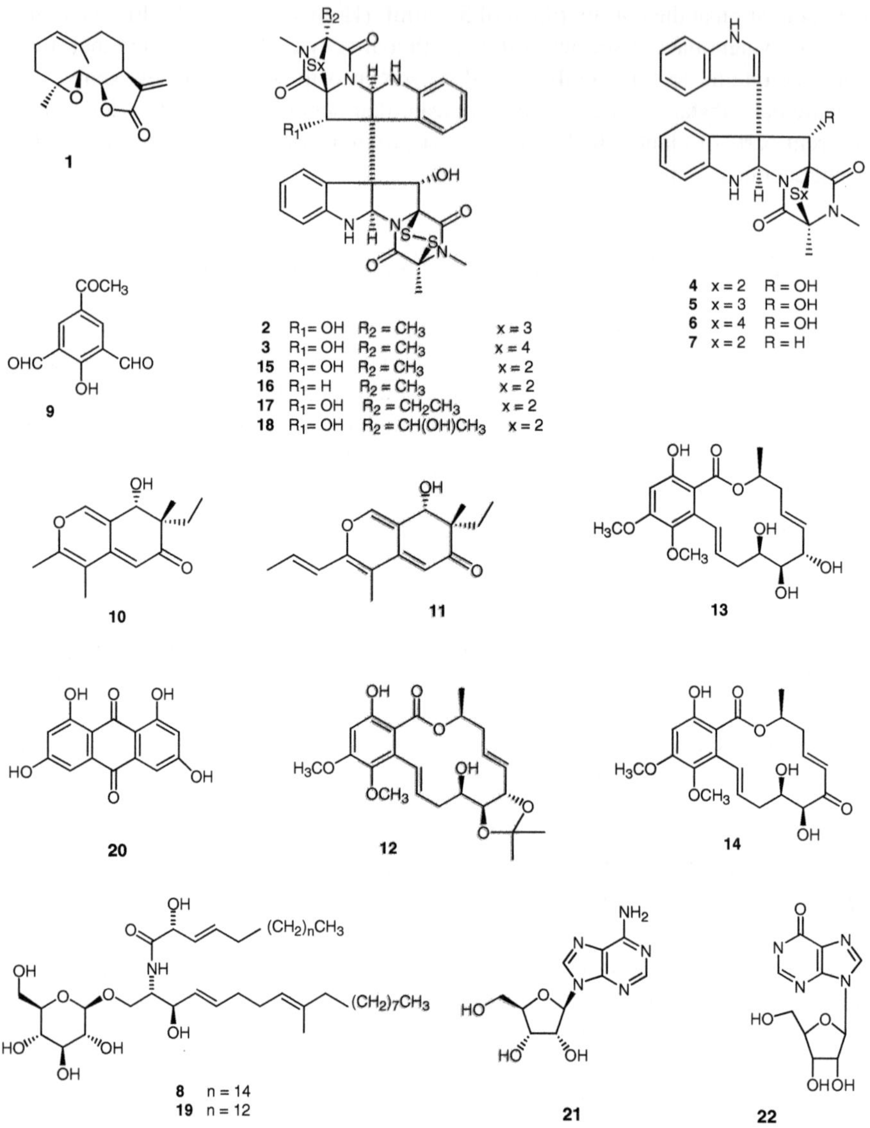


**Fig. 1** Chemical structures of compounds 1–22

(5 among 6 species), *Diaporthe* (4 among 4 species), *Massarina* (7 among 8 species), *Ophioceras* (8 among 10 species), *Pseudohalonectria* (7 among 8 species), *Savoryella* (3 among 3 species), *and Torula* (3 among 4 species) (Dong et al., 2004; Zhao, 2004). The wide occurrence of nematicidal substances in freshwater-derived fungi may be explained by the need to overcome the nutrient deficiencies (N limiting) of the natural habitat as did some wood-inhabiting Basidiomycetes like

**Table 1** List of fungi pathogenic to over 90% *B. xylophilus in vitro* test (Zhao,2004)

| Strain | Genus | Test solution | pH of test solution | Exposure time | | |
|---|---|---|---|---|---|---|
| | | | | 12 h | 24 h | 48 h |
| YMF1.01101 | *Aunulatascus* sp. | CF | 4.09 | 48.60 | 91.70 | 93.20 |
| | | AF | 7.06 | 29.68 | 87.48 | 97.21 |
| YMF1.01103 | Aunulatascus sp. | CF | 7.96 | 65.79 | 93.68 | 96.12 |
| | | AF | 6.02 | 96.34 | 98.14 | 98.02 |
| YMF1.01299 | *Caryospora callicarpa* | CF | 8.18 | 25.69 | 93.10 | 95.2 |
| | | AF | 7.41 | 56.61 | 91.36 | 92.06 |
| YMF1.01095 | *Diaporthe* sp. | AH | 6.29 | 89.65 | 79.54 | 97.22 |
| YMF1.01096 | *Diaporthe* sp. | AF | 7.36 | 72.87 | 83.43 | 96.46 |
| YMF1.01266 | *Dictyosporium heptasporum* | AH | 6.35 | 89.63 | 96.23 | 94.34 |
| YMF1.01286 | Dyrithiopsis lakefuxianensis | CF | 2.27 | 87.64 | 95.34 | 96.84 |
| YMF1.01082 | *Helicomyces roseus* | H | 6.63 | 89.73 | 73.51 | 98.95 |
| YMF1.0088 | *Leptosphaeria* sp. | AF | 6.28 | 1.51 | 45.25 | 98.85 |
| YMF1.01275 | *Nectria* sp. | CF | 3.74 | 78.65 | 85.32 | 96.49 |
| | | AF | 6.61 | 94.63 | 99.61 | 98.58 |
| YMF1.01276 | *Nectria* sp. | CF | 3.90 | 99.63 | 86.38 | 98.76 |
| YMF1.01074 | *Nectria* sp. | CF | 2.55 | 61.32 | 77.73 | 91.74 |
| YMF1.01273 | *Ophioceras cummune* | AF | 6.92 | 9.23 | 77.13 | 91.53 |
| YMF1.02124 | *Ophioceras cummune* | CF | 7.60 | 70.21 | 79.34 | 97.18 |
| | | AF | 7.31 | 74.26 | 98.99 | 97.17 |
| YMF1.02125 | *Ophioceras cummune* | CF | 3.17 | 85.96 | 98.32 | 97.59 |
| YMF1.01285 | *Phoma* sp. | AH | 6.48 | 87.65 | 90.23 | 90.00 |
| YMF1.01274 | *Phomatospora berkeley* | H | 4.81 | 17.01 | 95.38 | 94.96 |
| YMF1.01288 | *Pseudohalonectria adversaria* | AF | 6.45 | 68.46 | 95.73 | 95.86 |
| YMF1.01297 | *Pseudohalonectria adversaria* | AF | 7.41 | 78.57 | 97.73 | 90.00 |
| YMF1.00947 | *Pseudohalonectria lignicola* | CF | 7.94 | 88.28 | 95.44 | 96.15 |
| | | AF | 7.33 | 93.02 | 97.12 | 98.56 |
| | | H | 7.51 | 84.10 | 88.02 | 95.59 |
| | | AH | 6.95 | 90.38 | 97.95 | 97.94 |
| YMF1.01217 | *Savoryella lignicola* | CF | 2.84 | 21.05 | 79.34 | 93.24 |
| YMF1.01290 | *Torula herbarum* | AH | 6.06 | 55.87 | 91.54 | 96.59 |
| YMF1.01291 | *Torula herbarum* | AF | 7.43 | 3.72 | 51.92 | 94.54 |
| YMF1.01558 | *Xylaria* sp. | CF | 2.31 | 41.36 | 14.10 | 92.81 |
| | | AF | 6.64 | 92.66 | 97.89 | 98.28 |

CF: cell-free filtrate; AF: adjusted filtrate; H: water-soluble extract of broken hyphae; AH: adjusted water-soluble extract of broken hyphae.

*Pleurotus* (Thorn and Barron, 1984; Barron, 1992). In freshwater ecosystems, submerged woody substrata are the main energy input (Wong et al., 1998). Wood is, however, a substrate greatly deficient in nitrogen and therefore the nitrogen utilized by freshwater fungi may be obtained from other sources. Nematodes, are cosmopolitan organisms, adapted to living in soils and water. They have been shown to be an integral part of various ecosystems, serving as food for small invertebrates or fungi (Dropkin, 1980). With their high nitrogen content, nematodes are considered to play an important role in providing nitrogen to other organisms in freshwater ecosystems. Several nematophagous fungi have previously been reported from wood submerged in fresh water, e.g. *Dactylella ellipsospora* (Hyde and Goh, 1998) and *Dactylella aquatica* (Kane et al., 2002), and these species are normally found from the dead bodies of nematodes. It would also make sense if other wood inhabiting fungi occurring on wood in fresh water were able to supplement their diets by obtaining nitrogen via digesting nematodes. The ability for these fungi to produce nematicides that can kill nematodes, which they can subsequently consume, would be advantageous.

## *Screening of Terrestrial Fungi for the Production of Nematicidal Metabolites*

From 212 terrestrial fungi grown on four different media (NSK, SGP, PDB and Czapek), the submerged cultures of 83 fungal strains were found to be pathogenic to the pine wood nematode, *B. xylophilus* following 48 h exposure *in vitro* screening. Of these, over 80% of *B. xylophilus* were immobilized in 12 fungal strains of *Amanita fulva, Amauroderma macer, Boletus* sp. (3 strains), *collybia dryophila, Cyathus* sp., *Pezinza* sp., *Pluteus fulve, Xerocomus chrysenteron, Leatiporus sulphureus, Laccaria tortilis*. The pathogenic effect of both *Amauroderma macer* and *Peziza* sp. isolates grown on PDB on pine wood nematode were interesting as they had different biological activity on the adults and juveniles of *B. xylophilus*. When the tested nematodes were juveniles, 100% pathogenicity within 72 h of exposure was observed with both. On adult nematodes, nematicidal activities of the cultural filtrates of *Amauroderma macer* and *Peziza* sp. were only 71.4%, and 65.4% respectively (Dong et al., 2006c; Zhao, 2004).

It was also observed that the growth and nematicidal activity of fungi were dependent on the nutrient component of the culture medium tested. On the basis of the growth conditions and nematicidal effects of fungi grown on PDB or Czapek broth, all tested terrestrial fungal species could be grouped into four distinct categories. In the first category, most fungi grew well in both PDB and Czapek broth but their cultural filtrates were inactive. In the second category, growth and nematicidal activities were observed in cultural filtrates from fungi grown on PDB and Czapek broth as illustrated in the cultural filtrates of *Tylopilus scabrusus*. In the third category, fungi grew well in PDB medium and was found to be pathogenic to nematodes, whereas they were all inactive because no growth was observed in the Czapek broth medium. The fungi were *Amauroderma brunreopilus, Laccarie tortilis, Lentinula edodes, Oudemansielle longipeo, Oudermansielle mucida, Sinoterminomyces carnosus, Strobilomyces floccsus, Strobilomyces floccsus* and *Terminomyces albimusus*. In the fourth category, the nematicidal effect of the fungi was not proportional to their growth condition. For example, the growth of *Amanita junguullea* and *Amanita* sp. on PDB was better than on Czapek broth, but their nematicidal activities were significantly less toxic than those on Czapek broth (Dong et al., 2006c).

It has been already demonstrated that the cultures of many fungi possess potential nematicidal compound(s), the secretion of which is largely influenced by the nutrient component of the selected culture medium (Smith and Moss, 1985; Cayrol et al., 1989; Anke et al., 1995; Chen et al., 2000; Siddiqui and Shaukat, 2002). Our results further support this view. Hence it appears that the conditions necessary for mycotoxin production (media, culture conditions, age etc.) vary from one fungus to the other. Consequently, when the ability of fungi to produce toxins is studied, it is absolutely necessary to test several parameters. Also, to establish the identity of the toxic metabolite in the near future, it is absolutely necessary to define a well known synthetic media which gives optimal nematicidal results. The fact that modification of fermentation parameters is a valuable tool to increase the number of new nematicidal metabolites has been comfirmed by Anke et al. (1995). There is no doubt that the knowledge of the biosynthetic pathways, the understanding of their regulation, and the ecological functions of fungal secondary metabolism may help greatly in making use of these rich natural resources.

## *Nematicidal Metabolites from Fresh Water Derived Fungi*

The ability of several freshwater-derived fungi (*Gliocladium roseum* YMF1.00133, *Paraniesslia* sp. YMF1.01400 *Pseudohalonectria adversaria* YMF1.01019 and *Caryospora carllicarpa* YMF1.01026) to produce nematicidal metabolites was further investigated .

From cultures of *Gliocladium roseum* YMF1.00133 grown on wheat, six new epipolysulfanyldioxopiperazine alkaloids, gliocladine A-E (**2-6**) and glioclatine (**7**), have been isolated along with four known compounds, verticillin A (**15**), 11'-deoxyverticillin A (**16**), sch52900 (**17**) and sch52901 (**18**). *In vitro* motility tests showed that these epipolysulfanyldioxopiperazines possessed anti-nematodal activity against *Caenorhabditis elegans* and *Panagrellus redivivus*, but no distinct mortalities were observed on *B. xylophilus* at concertrations of up to 400 μg/mL (Dong et al., 2005a, 2006a; Dong, 2005).

Six metabolites from cultures of *Paraniesslia* sp. YMF1.01400 on Sabouraud's medium (Peptone 10 g, glucose 40 g and water 1000 mL) were isolated and identified as two sphingolipids, (2S, 2′R, 3R, 3′E, 4E, 8E) -1 - O - (β - D - glucopyranosyl) - 3 - hydroxyl - 2 - [N -2′ - hydroxyl - 3′ - eicosadecenoyl] amino - 9 - methyl - 4 -, 8 - octadecadiene (**8**) and (2S, 2′R, 3′E, 3R, 4E, 8E) - 1- O - (β - D- glucopyranosyl) -3 - hydroxyl - 2 - [N - 2′ - hydroxyl - 3′ -octadecenoyl] amino -9 - methyl - 4 -, 8 - octadecadiene (**19**), and four other metabolites, 3, 5-dihydroxyaldehyde-4-hydroxy-acetophenone (**9**), 1, 3, 6, 8-tetrahydroxy-anthraquinone (**20**), inosine (**21**), and adenosine (**22**). Among these, **8** and **9** were new compounds. In motility tests, compounds **8**, **9**, **19**, **20** showed nematicidal activity against *B. xylophilus* with $LC_{50}$ values of 110, 200, 110, 90 μg/mL, respectively whereas other compounds were not active at concentrations up to 500 μg/mL (Dong, 2005; Dong et al., 2005b).

Two new azaphilone metabolites from cultures of *Pseudohalonectria adversaria* YMF1.01019, named pseudohalonectrin A (**10**) and B (**11**) were isolated as the main responsibles for the nematicidal activity of the extracts of the fungus. They have approximately the same activity and respectively immobilized 53.8% and 52.7% pine wood nematodes at 100 ppm after 24 h (Dong et al., 2006b).

Three novel tetradecalactone metabolites from cultures of *Caryospora carllicarpa* YMF1.01026, caryospomycins A~C (**12**~**14**), were isolated. Their structures were elucidated by detailed NMR spectroscopic analysis to be 14-membered macrolides with a fused 1,2,4-trihydroxybenzene ring that is rare among the resorcylides. In *in vitro* tests, compounds **12-14** showed nematicidal activity against *B. xylophilus* with $LC_{50}$ values at 100, 80, 80 ppm respectively.

In the present screening, cultures of *Gliocladium roseum* YMF1.00133, *Paraniesslia* sp. YMF1.01400 *Pseudohalonectria adversaria* YMF1.01019 and *Caryospora carllicarpa* YMF1.01026 showed strong nematicidal activity towards the nematode, and the active compounds showed only moderated activity, which implied that some other active compound had still not been isolated, and maybe a synergistic action existed in the nematicidal cultures.

According to our knowledge, the compounds epipolysulfanyldioxopiperazine, azaphilone, resorcylide types are rare in nature. However, they have been reported to present a wide range of biological activities such as antifungal, antitumor, antiprotozoan, antimalarial, antiviral, antiparasitic etc. (Arai et al., 2003; Gardiner et al., 2005; Jayasuriya et al., 2005; Dong et al., 2006b.) Thus, it would be interesting to find that some common freshwater-derived fungi can produce these metabolites. Further studies are required to examine the capacities of this fungus to biosynthetic these metabolites and to obtain enough quantities of them and their analogues for evaluating nematicidal and other biological activities.

**Acknowledgement** This study was financially supported by the National Natural Science Foundation of China (NSFC30570059 and NSFC20562015), Yunnan Provincial Natural Science Foundation (2005C0005Q, 2005NG03 and 2005NG05) and State Key laboratory of phytochemistry and plant resources in west China, Kunming Institute of Botany, China.

## References

ALAM, M.M., KHAN, M.W. AND SAXENA, S.K. (1973). Inhibitory effect of culture filtrates of some rhizosphere fungi of okra on the mortality and larval hatch of certain plant parasitic nematodes. *Indian Journal of Nematology* 3: 94–98.

ANKE, H., STADLER, M., MAYER, A. AND STERNER, O. (1995). Secondary metabolites with nematicidal activity from nematophagous fungi and ascomycetes. *Canadian Journal of Botany* 72: 932–939.

ANKE, H. AND STERNER, O. (1997). Nematicidal metabolites from higher fungi. *Current Organic Chemistry* 1: 427–440.

ARAI, M., YAMAMOTO, K., NAMATAME, I., TOMODA, H. AND OMURA, S. (2003). New monordens produced by amidepsine-producing fungus *Humicola* sp. FO-2942. *Journal of Antibiotics* 56: 526–532.

BARRON, G.L. (1992). Lignolytic and cellulolytic fungi as predators and parasites. In *The Fungal Community* (eds. G.C. Carroll and D.T. Wicklow), pp. 311–326. Marcel Dekker, New York.

CAI, L., TSUI, C.K.M., ZHANG, K.Q. AND HYDE, K.D. (2002). Aquatic fungi from Lake Fuxian, Yunnan, China. *Fungal Diversity* 9: 57–70.

CAI, L., ZHANG, K.Q., MCKENZIE, E.H.C. AND HYDE, K.D. (2003). Fresh water fungi from bamboo and wood submerged in the Liput River in the Philippines. *Fungal Diversity* 13: 1–12.

CAYROL, J.C., DJIAN, C. AND PIJAROWSKI, L. (1989). Study of the nematicidal properties of the culture filtrate of the nematophagous fungus *Paecilomyces lilacinus*. *Revue de Nematologie* 12: 331–336.

CHATTOPADHYAY, P.R. AND DE, A.B. (1995). Inhibitory effects of certain fungi on the larval hatch of *Meloidogyne inconita*. *Indian Journal of Nematology* 25: 204–205.

CHEN, S.Y., DICKSON, D.W. AND MITCHELL, D.J. (2000). Viability of *Heterodera glycines* exposed to fungal filtrates. *Journal of Nematology* 32: 190–197.

CHITWOOD, D.J. (2002). Phytochemical based strategies for nematode control. *Annual Review of Phytopathology* 40: 221–249.

DATTA, S. AND SAXENA, D.B. (2001). Pesticidal properties of parthenin (from Parthenium hysterophorus) and related compounds. *Pest Management Science* 57: 95–101.

DESAI, M.V., SHAH, H.M. AND PILLAI, S.N. (1972). Effect of *Aspergillus niger* on root-knot nematode, *Meloidogyne incognita*. *Indian Journal of Nematology* 2: 210–214.

DONG, J.Y. (2005). Screening for nematicidal fungi and secondary metabolites from freshwater fungi against *Bursaphelechus xylophilus*. PhD thesis, Yunnan University, Yunnan, China.

DONG, J.Y., HE, H.P., SHEN, Y.M. AND ZHANG, K.Q. (2005a). Nematicidal epipolysulfanyldioxopiperazines from *Gliocladium roseum*. *Journal of Natural Products* 68: 1510–1513.

DONG, J.Y., LI, R., HE, H.P. AND ZHANG, K.Q. (2005b). Nematicidal sphingolipids from the fresh water fungus *Paraniesslia* sp. YMF 1.01400. *European Journal of Lipid Science Technology* 107: 779–785.

DONG, J.Y., LI, X.P., LI, L., LI, G.H., LIU, Y.J. AND ZHANG, K.Q. (2006c) Preliminary results on nematicidal activity from culture filtrates of Basidiomycetes against the pine wood nematode, *Bursaphelenchus xylophilus*. *Annals of Microbiology* 56: 163–166.

DONG, J.Y., ZHAO, Z.X., CAI, L., LIU, S.Q., ZHANG, H.R., DUAN, M. AND ZHANG, K.Q. (2004). Nematicidal effect of freshwater fungal cultures against the pine-wood nematode, *Bursaphelenchus xylophilus*. *Fungal Diversity* 15: 123–133.

DONG, J.Y., ZHOU, W., LI, L., LI, G.H., LIU, Y.J. AND ZHANG, K.Q. (2006a). A new epidithiodioxopiperazine metabolite isolated from *Gliocladium roseum* YMF1.00133. *Chinese Chemical Letters* 17: 922–924.

DONG, J.Y., ZHOU, W., LI, L., LI, G.H., LIU, Y.J. AND ZHANG, K.Q. (2006b). New nematicidal azaphilones from the aquatic fungus *Pseudohalonectria adversaria* YMF 1.01019. *FEMS Microbiology Letters* 264: 65–69.

DROPKIN, V.H. (1980). *Introduction to Plant Nematology*. John Wiley & Sons, New York.

GARDINER, D.M., WARING, P. AND HOWLETT, B.J. (2005). The epipolythiodioxopiperazine (ETP) class of fungal toxins: distribution, mode of action, functions and biosynthesis. *Microbiology* 151: 1021–1032.

HONG, L.J., LI, G.H., ZHOU, W., WANG, X.B. AND ZHANG K.Q. (2007). Screening and isolation of a nematicidal sesquiterpene from *Magnolia grandiflora* L. *Pest Management Science* 63: 301–305.

HUANG, Z.L., DAN, Y., HUANG, Y.C., LIN, L.D., LI, T.H., YE, W.H. AND WEI, X.Y. (2004). Sesquiterpenes from the mycelial cultures of *Dichomitus squalens*. *Journal of Natural Products* 67: 2121–2123.

HYDE, K.D. AND GOH, T.K. (1998). Fungi on submerged wood in the Riviere St Marie-Louis, The Seychelles. *South African Journal of Botany* 64: 330–336.

JAYASURIYA, H., ZINK, D.L., POLISHOOK, J.D., BILLS, G.F., DOMBROWSKI, A.W., GENILLOUD, O., PELAEZ, F.F., HERRANZ, L., QUAMINA, D., LINGHAM, R.B., DANZEIZEN, R., GRAHAM, P.L., TOMASSINI, J.E. AND SINGH S.B. (2005). Identification of diverse microbial metabolites as potent inhibitors of HIV-1 *tat* transactivation. *Chemistry & Biodiversity* 2: 112–122.

KANE, D.F., TAM, W.Y. AND JONES, E.B.G. (2002). Fungi colonising and sporulating on submerged wood in the River Severn, UK. *Fungal Diversity* 10: 45–55.

KAWAZU, K, MURAKAMI, T, ONO, Y, KANZAKI, H., KOBAYASHI, A., MIKAWA, T. AND YOSHIKAWA, N. (1993). Isolation and characterization of two novel nematicidal depsipeptides from an imperfect fungus, strain D1084. *Bioscience, Biotechnology and Biochemistry* 57: 98–101.

KAWAZU, K., NISHII, Y., ISHII, K. AND TADA, M. (1980). A convenient screening method for nematicidal activity. *Agricultural and Biological Chemistry* 44: 631–635.

KHAN, S.T. AND KGAN, T.A. (1992). Effect of culture filtrates of soil fungi on the hatching and mortality of root knot nematode (*Meloidogyne incognita*). *Current Nematology* 3: 53–60.

KHAN, T.A. AND HUSSAIN, S.I. (1989). Effect of culture filtrates of soil fungi on the mortality and hatching of reniform nematode (*Rotylenchulus reniformis*). *Indian Journal of Nematology* 19: 35–40.

MAMIYA, Y. (1984). The pine wood nematode. In *Plant and Insect Nematodes* (ed. W.R. Nickle), pp. 589–626. Marcel Dekker, Inc., New York.

MATSUDA, K., KIMURA, M., KOMAI, K. AND HAMADA, M. (1989). Nematicidal activities of (–)-N-Methylcytisine and (–)-anagyrine from *Sophora flavescens* against pine wood nematodes. *Agricultural and Biological Chemistry* 53: 2287–2288.

MATSUDA, K., YAMADA, K., KIMURA, M. AND HAMADA, M. (1991). Nematicidal activity of matrine and its derivatives against pine wood nematodes. *Journal of Agricultural Food Chemistry* 39: 189–191.

NOLING, J.W. AND BECKER, J.O. (1994). The challenge of research and extension to define and implement alternatives to methyl bromide. *Journal of Nematology* 26: 573–586.

PATHAK, K.N. AND KUMAR, B. (1995). Nematotoxic effects of *Trichoderma Harzianum* culture filtrate on second stage juveniles of rice root knot nematode. *Indian Journal of Nematology* 25: 223–224.

SANKARANARAYANAN, C., HUSSAINI, S., SREERAMAKUMAR, S. AND PRASAD, R.D. (1997). Nematicidal effect of fungal filtrates against root-knot nematodes. *Journal of Biological Control* 11: 37–41.

SIDDIQUI, I.A. AND SHAUKAT, S.S. (2002). Zinc and glycerol enhance the production of nematicidal compounds in vitro and improve the biocontrol of *Meloidogyne javanica* in tomato by fluorescent pseudomonads. *Letters in Applied Microbiology* 35: 212–217.

SIDDIQUI, Z.A. AND MAHMOOD, I. (1996). Biological control of plant-parasitic nematodes by fungi: a review. *Bioresource Technology* 58: 229–239.

SING, S.P., PANT, V., KHAN, A.M. AND SAXENA, S.K. (1983). Inhibitory effect of culture filtrate of some rhizosphere fungi of tomato as influenced by oil cakes on mortality and larval hatch of *Meloidogyne incognita*. *Nematologia Mediterranea* 11: 119–123.

SMITH, J.E. AND MOSS, M.O. (1985). *Mycotoxins, formation analysis and significance*. Wiley and Sons, Chichester, UK.

SUTHERLAND, J.R. AND WEBSTER, J.M. (1993). Nematode pests of forest trees. In *Plant-Parasitic Nematodes in Temperate Agriculture* (eds. K. Evans, D.L. Trudgill and J.M. Webster), pp. 351–380. CAB International, Wallingford, Oxon.

TAKAI, K., SOEJIMA, T., SUZUKI, T. AND KAWAZU, K. (2000). Emamectin benzoate as a candidate for a trunk infection agent against the pine wood nematode, *Bursaphelechus xylophilus*. *Pest Management and Science* 56: 937–941.

THORN, R.G. AND BARRON, G.L. (1984). Carnivorous mushrooms. *Science* 224: 76–78.

WHITTEN, M.J., JEFFERSON, R.A. AND DALL, D. (1996). Needs and opportunities. In *Biotechnology and Integrated Pest Management* (ed. G.J. Persley), pp. 1–36. CAB International, Wallingford, Oxon.

WONG, K.M.K., GOH, T.K., HODGKISS, I.J., HYDE, K.D., RANGHOO, V.M., TSUI, C.K.M., HO, W.H., WONG, W.S. AND YUEN, T.C. (1998). The role of freshwater fungi in freshwater ecosystems. *Biodiversity and Conservation* 7: 1187–1206.

ZHAO, Z.X. (2004). *Screening of Nematicidal Strains and Mutation Breeding*. MD thesis, Yunnan University, Yunnan, China.

# Microbial Control of *Bursaphelenchus xylophilus* by Fungi

**Noritoshi Maehara and Kazuyoshi Futai**

**Abstract** To reduce the number of pinewood nematodes (*Bursaphelenchus xylophilus*) carried by the Japanese pine sawyer (*Monochamus alternatus*), we attempted to change the mycoflora and also to prevent blue-stain fungi, the main food-source fungi of the nematodes, from spreading throughout pine wilt-killed *Pinus densiflora* wood by inoculating other fungi into the dead logs. *Trichoderma* sp. 3 and *Verticillium* sp. inoculation treatments tended to decrease the number of the nematodes carried by the beetles.

## Introduction

The pinewood nematode (PWN), *Bursaphelenchus xylophilus*, the causal agent of pine wilt disease (Kiyohara and Tokushige, 1971), is vectored from wilt-killed to healthy pines by the Japanese pine sawyer, *Monochamus alternatus*, in Japan (Mamiya and Enda, 1972; Morimoto and Iwasaki, 1972). The number of nematodes that enter a healthy tree is directly proportional to the number of nematodes carried by *Monochamus* beetles (Togashi, 1985). The rate of disease development is directly related to the number of nematodes in the inoculum (Kiyohara et al., 1973). Thus, to understand the dynamics of pine wilt disease development, it is important to identify the factors affecting the number of nematodes carried by an individual beetle, which ranges from zero to over 200 000.

Previous studies showed that *M. alternatus* emerging from extremely dry or wet pupal chambers (Morimoto and Iwasaki, 1973; Maehara et al., 2005), or logs (Terashita, 1975; Kobayashi et al., 1976; Togashi, 1989; Fukushige, 1990) carried relatively few nematodes. There were differences in the numbers of nematodes carried by *Monochamus* beetles among individual pine trees from which the beetles emerged (Maehara et al., 2005). Another important factor, which affects the number of nematodes carried by beetles, is the fungal flora in the wood. In vitro studies

N. Maehara
Tohoku Research Center, Forestry and Forest Products Research Institute, 92-25 Nabeyashiki, Shimo-Kuriyagawa, Morioka, Iwate 020-0123, Japan
e-mail: maehara@ffpri.affrc.go.jp

showed that beetles carried a greater number of nematodes when the blue-stain fungus *Ophiostoma minus* was dominant in wood around artificial pupal chambers of the beetles, while the numbers of nematodes decreased when *Trichoderma* spp. or *Verticillium* sp. was prevalent (Maehara and Futai, 1996, 1997). The former fungus is suitable and the latter two fungi are unsuitable for nematode propagation (Fukushige, 1991; Maehara and Futai, 2000). In field surveys, intense blue-stain on the pupal chamber walls of *M. alternatus* increased the number of nematodes aggregating around such chambers and the number carried by the beetles which emerged from the chambers (Maehara et al., 2005). In the present study, to reduce the number of nematodes carried by the beetles, we attempted to change the mycoflora and also to prevent blue-stain fungi from spreading throughout pine wilt-killed *Pinus densiflora* wood, by inoculating other fungi into the dead logs.

## Materials and Methods

### *Inoculation of Fungi into Pine Trees Killed by Pine Wilt in 1990*

Four 30 to 40-year-old *P. densiflora* trees, killed by the PWN in 1990, were felled at Kamigamo Experimental Station, Field Science Education and Research Center, Kyoto University, on November 14, 1990. Eighty, 25 cm long, ca. 10 cm in mid-diameter logs were cut from the trees. The following day, each log was inoculated with 7 mm diameter mycelial disks of one of the test fungi growing on potato dextrose agar (PDA). Inoculation was made via two drill holes, 11 mm diam. × ca. 20 mm deep; one hole at each cut end and opposite side of each log, then each hole was plugged with a rubber stopper. Ten species of fungi were inoculated: *Arthrobotrys* sp., *Coriolus hirsutus*, *Cryptoporus volvatus*, *Cystidiophorus castaneus*, *Gloeophyllum striatum*, *Pycnoporus coccineus*, *Trichaptum abietinum*, *Trichoderma* sp. 2, *Trichoderma* sp. 3, and an unidentified species of a basidiomycete. These fungi could be isolated from dead pine trees. The controls were logs without any fungus inoculation, with the holes filled with rubber stoppers. After the inoculation, each log was placed lengthwise in a plastic container, 25 cm diam. × 28 cm deep, then covered with a steel mesh. The containers were placed in a shed at Kitashirakawa Experimental Station, Field Science Education and Research Center, Kyoto University. The logs were watered weekly during the experiment, and observed daily from June to July 1991.

### *Inoculation of Fungi into Pine Trees Killed by Pine Wilt in 1992*

Six 30 to 40-year-old *P. densiflora* trees, killed by wilt in 1992, were felled at Kamigamo Experimental Station on October 22, 1992. One hundred and four, 25 cm long, 7.4–14.8 cm in mid-diam. logs were cut from the trees. Five hundred beech wood chips, 9 mm diam. × 20 mm long, were autoclaved with 100 ml distilled water for 25 min. The chips were placed on one of the test fungi growing on PDA and cultured at 20°C for 25 days. On October 23, 1992, each log was inoculated with the chips of one of the test fungi. The reason why we did not use PDA disks but wood chips with fungi in this year was that PDA dries faster than the wood chips. Inoculation was made through two drill holes, 9 mm diam. × ca. 20 mm deep; one hole at each cut end and opposite side of each log. Fungi used for inoculation were *Trichoderma* sp. 2, *Trichoderma* sp. 3, *P. coccineus*, *Pleurotus ostreatus*, and *Verticillium* sp. *P. ostreatus* and *Verticillium* sp. could be also isolated from dead pine trees. Further inoculations with *Trichoderma* sp. 3 were also carried out, by boring another two or four holes in the side of each log (four or six points in total). Logs serving as controls were inoculated with two axenic wood chips. After the inoculation with fungi, each log was placed lengthwise in a plastic container. The containers were placed in the same shed as in 1990, and each was covered with a steel mesh on May 21, 1993. The logs were watered at least once a week during the experiment, and observed daily from June to August 1993.

## *Inoculation of* Trichoderma *spp. into Pine Trees Killed by Pine Wilt in 2001*

Seven 20 to 30-year-old *P. densiflora* trees, killed by pine wilt in 2001, were felled and cut into 1 m segments at Chiyoda Experimental Station of Forestry and Forest Products Research Institute, Kasumigaura City, Ibaraki Prefecture, Japan on October 30, 2001. The 1 m long logs were placed in a screen cage at Forestry and Forest Products Research Institute, Tsukuba City, Ibaraki Prefecture, Japan. One hundred and sixty-eight, 25 cm long, 7.8–20.8 cm in mid-diam. logs were cut from the 1 m long logs on November 19–22, 2001. Two thousand one hundred and sixty beechwood chips, 9 mm diam. × 18 mm long, were prepared. Every thirty chips were autoclaved with 20 ml potato dextrose broth (PDB) in a 70 ml wide-mouthed bottle for 30 min. The chips were placed on one of the test fungi growing on PDA and cultured at 20°C in the dark for 22–24 days. On December 10–12, fourteen logs each were inoculated with the chips of each test fungus by the randomized block design. Inoculation was made through six drill holes, 9 mm diam. × ca. 20 mm deep; two holes at each cut end and another two holes in the side of each log. Fungi inoculated were *Trichoderma* sp. 2, *Trichoderma* sp. 3, *Trichoderma harzianum*, *Trichoderma* A, B, E, G, H, I, J, and K. *Trichoderma harzianum* was isolated from a bed log, *Trichoderma* A, B, E, H, and I from hybrids (F1) of *Pinus thunbergii* × *P. massoniana* killed by wilt, G and J from wilt-killed *P. thunbergii*, and K from soil in *Pinus* hybrid (*P. thunbergii* × *P. massoniana*) stand. Fourteen logs serving as controls were inoculated with six axenic wood chips. After the inoculation with fungi, each log was placed lengthwise in a plastic container. The containers were placed in the screen cage and each was covered with a steel mesh on December 17. The logs were watered two to three times a week during the experiment, and observed daily from June to August 2002.

## *Inoculation of* Trichoderma *spp. into Pine Trees Killed by Pine Wilt in 2002*

Seven 20 to 30-year-old *P. densiflora* trees, killed by pine wilt in 2002, were felled and cut into 1 m segments at Chiyoda Experimental Station on October 22, 2002. The 1 m long logs were placed in the same screen cage as in 2001. Two hundred and nine, 25 cm long, 5.7–15.2 cm in mid-diam. logs were cut from the 1 m long logs on December 13–17, 2002. Two thousand three hundred and forty beech wood chips, 9 mm diam. × 18 mm long, were prepared. Every thirty chips were autoclaved with 30 ml PDB in a 70 ml wide-mouthed bottle for 30 min. The chips were placed on one of the test fungi growing on PDA and cultured at 20°C in the dark for 28–30 days. On December 25–27, 2002, nineteen logs each were inoculated with the chips of each test fungus by randomized block design. Fungi inoculated were *Trichoderma* sp. 3, *T. harzianum*, *Trichoderma* K, L, M, and N. *Trichoderma* L, M, and N were isolated from wilt-killed *P. densiflora*. Inoculation was made through six drill holes, 9 mm diam. × ca. 20 mm deep: two holes at each cut end and another two holes in the side of each log. Further inoculation with *Trichoderma* sp. 3, *T. harzianum*, and *Trichoderma* K were also carried out, by boring another six holes (twelve points in total). Thirty-eight logs serving as controls were inoculated with six axenic wood chips. After the inoculation with fungi, each log was placed lengthwise in a plastic container. The containers were placed in the screen cage and each was covered with a steel mesh on June 4, 2003. The logs were watered two to three times a week during the experiment, and observed daily from June to August 2003.

## *Number of Nematodes Carried by* Monochamus alternatus

Upon emergence of *M. alternatus* from the logs, each beetle was ground for 10 s in 40 ml tap water and the suspension was placed in a Baermann funnel overnight to extract the nematodes in 1991, 1993, 2002 and 2003. These were then counted for each sample using a stereomicroscope. When the nematodes were too abundant for counting, the suspension was diluted and the number of nematodes was estimated. In 2002 and 2003, after emergence of *M. alternatus* each pupal chamber was located.

## *Statistical Analysis*

One-way analysis of variance (ANOVA) was used to analyse the differences in the numbers of nematodes among individual trees or fungus treatments into logs. Student's *t*-test was used to analyse the differences in the numbers of nematodes between locations of pupal chambers in logs. For ANOVA and Student's *t*-test, the numbers of nematodes were $\log_{10}$-transformed. Chi-square test was used to determine if the percentage of *M. alternatus* carrying less than 100 or 1000 nematodes was dependent on fungus treatments.

## Results

From 20 June–11 July in 1991, 101 *Monochamus* beetles emerged. The numbers of nematodes (fourth-stage dispersal juveniles: $J_{IV}$) carried by the individual beetle ranged from zero to 85 500, the mean being 4585.0. The beetles that emerged from Tree 1 (n = 7) carried small numbers of $J_{IV}$, although there was no significant difference ($P = 0.289$). Therefore, we analysed the effect of fungus inoculation on the numbers of $J_{IV}$ carried by a beetle with the exception of the data from Tree 1 (Table 1) (Maehara et al., 2006). One beetle only emerged from the logs inoculated with *C. hirsutus* and the number of $J_{IV}$ carried by the beetle was 1. The datum was not included in Table 1 because the beetle emerged from Tree 1. *Trichoderma* sp. 2, *P. coccineus*, and *Trichoderma* sp. 3 treatments tended to decrease the number of $J_{IV}$ carried by a beetle, however, the differences were not significant ($P = 0.219$) because of great variance in the numbers of $J_{IV}$ among individual beetles. The percentage of beetles carrying less than 100 nematodes also tended to be high in the logs inoculated with such fungi ($P = 0.180$). On the contrary, the number of $J_{IV}$ carried by a beetle from the logs inoculated with *Arthrobotrys* sp. and the basidiomycetes except *P. coccineus* was similar to or higher than that in the controls.

From 21 June–5 August, 1993, 135 *Monochamus* beetles emerged. The numbers of $J_{IV}$ carried by the individual beetle ranged from zero to 22 600, the mean being 1037.3. The beetles that emerged from Tree 5 (n = 23) carried smaller numbers of $J_{IV}$ than those from the other trees ($P < 0.0001$). Therefore, we analysed the effect of fungus inoculation on the numbers of $J_{IV}$ carried by a beetle with the exception of the data from Tree 5 (Table 2) (Maehara et al., 2006). As in the 1991 result, the number of $J_{IV}$ carried by a beetle varied even when beetles emerged from the logs receiving the same fungus. Therefore, there was no significant difference among fungus treatments ($P = 0.649$). The *Trichoderma* sp. 3 (four- and six-point

**Table 1** Effect of inoculating pine wilt-killed *Pinus densiflora* logs with fungi on the numbers of pinewood nematodes carried by emerging *Monochamus alternatus* in 1991

| Fungus treatment | Number of beetles | Number of $J_{IV}$ carried by a beetle[1] | Percentage of the beetles carrying less than 100 nematodes |
|---|---|---|---|
| *Trichoderma* sp. 2 | 5 | 299.6 ± 648.9 | 80.0 |
| *Pycnoporus coccineus* | 10 | 1205.9 ± 2784.9 | 60.0 |
| *Trichoderma* sp. 3 | 10 | 2204.5 ± 5609.4 | 80.0 |
| *Cystidiophorus castaneus* | 6 | 3255.2 ± 2860.6 | 16.7 |
| Unidentified species of basidiomycetes | 15 | 4353.3 ± 12 519.9 | 60.0 |
| *Trichaptum abietinum* | 12 | 5044.8 ± 6591.9 | 41.7 |
| *Arthrobotrys* sp. | 9 | 7473.8 ± 15 709.0 | 44.4 |
| *Gloeophyllum striatum* | 12 | 10 224.7 ± 24 144.1 | 33.3 |
| *Cryptoporus volvatus* | 4 | 12 063.5 ± 24 091.0 | 75.0 |
| Control (no fungus) | 11 | 3965.8 ± 5167.4 | 36.4 |

[1]Values are means ± SD (modified from Maehara et al., 2006).

**Table 2** Effect of inoculating pine wilt-killed *Pinus densiflora* logs with fungi on the numbers of pinewood nematodes carried by emerging *Monochamus alternatus* in 1993

| Fungus treatment | Number of beetles | Number of $J_{IV}$ carried by a beetle[1] | Percentage of the beetles carrying less than 100 nematodes |
|---|---|---|---|
| *Trichoderma* sp. 3 (four inoculation points/log) | 11 | 333.5 ± 408.0 | 45.5 |
| *Verticillium* sp. | 15 | 442.8 ± 644.1 | 46.7 |
| *Trichoderma* sp. 2 | 16 | 687.2 ± 1232.4 | 56.3 |
| *Trichoderma* sp. 3 (six inoculation points/log) | 16 | 700.6 ± 1841.8 | 68.8 |
| *Pleurotus ostreatus* | 11 | 736.4 ± 1489.0 | 45.5 |
| *Pycnoporus coccineus* | 20 | 1085.0 ± 1911.9 | 35.0 |
| *Trichoderma* sp. 3 | 12 | 2723.6 ± 6201.9 | 33.3 |
| Control (no fungus) | 11 | 3917.1 ± 7695.8 | 36.4 |

[1]Values are means ± SD (modified from Maehara et al., 2006).

inoculation), *Verticillium* sp., and *Trichoderma* sp. 2 treatments, however, seemed to decrease the number of $J_{IV}$ carried by a beetle. The percentage of beetles carrying less than 100 nematodes also seemed to be high in the logs inoculated with *Trichoderma* sp. 3 (six-point inoculation), although there was no significant difference ($P = 0.521$).

From 14 June–26 July 2002, 116 *Monochamus* beetles emerged. The numbers of $J_{IV}$ carried by the individual beetle ranged from zero to 53 000, the mean being 5595.8. The beetles that emerged from Tree 6 (n = 24) carried smaller numbers of $J_{IV}$ than those from the other trees ($P < 0.0001$). The number of $J_{IV}$ carried by a beetle from the chamber beneath the bark (n = 3) was significantly smaller than the one from the sapwood ($P = 0.029$). Therefore, we analysed the effect of fungus inoculation on the numbers of $J_{IV}$ carried by a beetle with the exception of the data from Tree 6 and the chambers beneath the bark (Table 3) (Maehara, 2008). *Trichoderma* K and *Trichoderma* sp. 3 treatments decreased the number of $J_{IV}$ carried by a beetle ($P = 0.035$). The percentage of beetles carrying less than 1000 nematodes was also high in the logs inoculated with such fungi ($P = 0.030$).

From 21 June–28 July 2003, 219 *Monochamus* beetles emerged. The numbers of $J_{IV}$ carried by the individual beetle ranged from zero to 47 800, the mean being 3766.3. The number of $J_{IV}$ carried by a beetle from the chamber beneath the bark (n = 1) was smaller than the one from the sapwood. Therefore, we analysed the effect of fungus inoculation on the numbers of $J_{IV}$ carried by a beetle with the exception of the data from the chamber beneath the bark (Table 4) (Maehara, 2008). *Trichoderma* sp. 3 (twelve-point inoculation) treatment tended to decrease the number of $J_{IV}$ carried by a beetle, although the differences were not significant ($P = 0.209$). The percentage of beetles carrying less than 1000 nematodes was also high in the same treatment logs ($P = 0.006$).

**Table 3** Effect of inoculating pine wilt-killed *Pinus densiflora* logs with fungi on the numbers of pinewood nematodes carried by emerging *Monochamus alternatus* in 2002

| Fungus treatment | Number of beetles | Number of $J_{IV}$ carried by a beetle[1] | Percentage of the beetles carrying less than 1000 nematodes |
|---|---|---|---|
| *Trichoderma* K | 5 | 1061.0 ± 2314.1 | 80.0 |
| *Trichoderma* sp. 3 | 5 | 1259.2 ± 1122.9 | 60.0 |
| *Trichoderma harzianum* | 8 | 2709.9 ± 1960.1 | 25.0 |
| *Trichoderma* H | 7 | 4868.4 ± 4798.6 | 28.6 |
| *Trichoderma* B | 6 | 4957.2 ± 9433.8 | 66.7 |
| *Trichoderma* A | 9 | 5845.7 ± 8460.5 | 22.2 |
| *Trichoderma* I | 11 | 6890.0 ± 9562.0 | 9.1 |
| *Trichoderma* G | 6 | 8708.5 ± 9566.8 | 16.7 |
| *Trichoderma* J | 8 | 9398.5 ± 13 130.8 | 12.5 |
| *Trichoderma* sp. 2 | 9 | 11 475.9 ± 15 904.3 | 11.1 |
| *Trichoderma* E | 6 | 18 281.7 ± 19 102.2 | 0 |
| Control (no fungus) | 9 | 8799.6 ± 9751.7 | 33.3 |

[1]Values are means ± SD (modified from Maehara, 2008).

**Table 4** Effect of inoculating pine wilt-killed *Pinus densiflora* logs with fungi on the numbers of pinewood nematodes carried by emerging *Monochamus alternatus* in 2003

| Fungus treatment | Number of beetles | Number of $J_{IV}$ carried by a beetle[1] | Percentage of the beetles carrying less than 1000 nematodes |
|---|---|---|---|
| *Trichoderma* sp. 3 (twelve inoculation points/log) | 22 | 1143.5 ± 1634.8 | 81.8 |
| *Trichoderma* K (twelve inoculation points/log) | 16 | 2440.0 ± 2169.6 | 37.5 |
| *Trichoderma* sp. 3 | 11 | 2646.2 ± 2673.6 | 45.5 |
| *Trichoderma* K | 28 | 3405.9 ± 6707.1 | 35.7 |
| *Trichoderma harzianum* | 28 | 4059.3 ± 5401.4 | 28.6 |
| *T. harzianum* (twelve inoculation points/log) | 15 | 4162.7 ± 4760.3 | 26.7 |
| *Trichoderma* M | 22 | 4190.7 ± 10 228.2 | 45.5 |
| *Trichoderma* N | 22 | 4866.8 ± 6243.2 | 22.7 |
| *Trichoderma* L | 18 | 5645.8 ± 7881.3 | 38.9 |
| Control (no fungus) | 36 | 4421.1 ± 8513.1 | 25.0 |

[1]Values are means ± SD (modified from Maehara, 2008).

## Discussion

Several authors isolated blue-stain fungi at higher frequency from nematode-infested or nematode-inoculated pine trees (Kobayashi et al., 1975; Fukushige and Futai, 1987; Wingfield, 1987; Maehara et al., 2005). Kobayashi et al. (1974, 1975), Fukushige (1991), and Maehara and Futai (2000) compared nematode propagation on various fungi isolated from healthy and wilt-killed pines and showed that, while the nematodes fed and multiplied on some of these fungi, others were unsuitable for nematode propagation. Blue-stain fungi were extremely suitable for the nematode propagation. Fukushige (1990) found a positive relationship between the degree of blue-stain and the density of the nematodes in dead wood of *Pinus koraiensis*. Fukushige (1991) and Maehara and Futai (2000) also reported that blue-stain fungi were highly suitable for the occurrence of third-stage dispersal juveniles ($J_{III}$) of the nematodes because a high population density is a prerequisite for the occurrence of $J_{III}$. $J_{III}$ moults to $J_{IV}$ in the presence of insect vectors of the genus *Monochamus* (Morimoto and Iwasaki, 1973; Warren and Linit, 1993; Maehara and Futai, 1996, 2001; Necibi and Linit, 1998), and then $J_{IV}$ transfers from infested wood to vectors.

In vitro studies showed that when the blue-stain fungus *O. minus* dominated wood around artificial pupal chambers of *M. alternatus*, the beetles emerging from the chambers carried many nematodes (Maehara and Futai, 1996, 1997). In field surveys, intense blue-stain of the pupal chamber walls in wilt-killed pine trees also increased the numbers of $J_{IV}$ carried by the beetles which emerged from the chambers, i.e., blue-stain fungi greatly affected the number of $J_{IV}$ carried by the beetles (Maehara et al., 2005). Blue-stain fungi also affected the differences in the number of $J_{IV}$ carried by several other species of beetles which emerged from wilt-killed pine trees (Maehara and Futai, 2002).

In the present study, to reduce the number of $J_{IV}$ carried by *M. alternatus*, we changed the mycoflora composition and also prevented blue-stain fungi from spreading throughout killed pine wood by inoculating other fungi into the dead logs. Eradicating the main food-source fungi of the nematodes in the wood should reduce the density of the nematodes, and in turn reduce the number of $J_{IV}$ carried by a beetle. *Trichoderma* sp. 3 and *Verticillium* sp. treatments tended to decrease the number of $J_{IV}$ carried by a beetle. In vitro studies also showed that the numbers of nematodes carried by a beetle decreased when these fungi dominated wood around artificial pupal chambers of the beetles (Maehara and Futai, 1996, 1997). Fukushige (1991) and Maehara and Futai (2000) reported that the nematode population on *Verticillium* sp. decreased drastically because this fungus is a nematode parasite, and that *Trichoderma* sp. 3 was not suitable for the nematode propagation. This resulted in the decrease of the number of $J_{IV}$ carried by a beetle by the inoculation of these fungi into the logs.

*Pleurotus ostreatus* immobilizes and digests nematodes (Thorn and Barron, 1984; Barron and Thorn, 1987), and also preys on PWN (Mamiya et al., 2005), and so is unsuitable for the propagation of the nematodes on PDA (Dozono, 1974). Mamiya et al. (2005) also reported that *T. abietinum* and *C. volvatus* had the low ability to prey on the PWN. Saiki et al. (1984) tried to control the nematodes in *P. densiflora* and *P. thunbergii* seedlings by spraying spores of *Arthrobotrys* sp., a nematode-trapping fungus, and confirmed the effect of the fungus on the nematodes to a certain extent. In the present study, there were, however, no clear effects of treatments of these fungi on the number of $J_{IV}$ carried by a beetle. Therefore, it is suggested that these fungi may not spread well in the logs. The microbial control of the PWN by fungi would be successful when interactions of the nematodes, wood-inhabiting fungi, and the beetles are better understood.

## References


BARRON, G.L. and THORN, R.G. (1987). Destruction of nematodes by species of *Pleurotus*. *Canadian Journal of Botany* 65: 774–778.

DOZONO, Y. (1974). [Reproduction of *Bursaphelenchus lignicolus* on the colony of various kinds of fungi.] *Transactions of the 27th Annual Meeting of Kyushu Branch of the Japanese Forestry Society*: 161.

FUKUSHIGE, H. (1990). The number of *Bursaphelenchus xylophilus* carried by *Monochamus alternatus* and some possible factors regulating the number. *Japanese Journal of Nematology* 20: 18–24.

FUKUSHIGE, H. (1991). Propagation of *Bursaphelenchus xylophilus* (Nematoda: Aphelenchoididae) on fungi growing in pine-shoot segments. *Applied Entomology and Zoology* 26: 371–376.

FUKUSHIGE, H. and FUTAI, K. (1987). Seasonal changes in *Bursaphelenchus xylophilus* populations and occurrence of fungi in *Pinus thunbergii* trees inoculated with the nematode. *Japanese Journal of Nematology* 17: 8–16.

KIYOHARA, T., DOZONO, Y., HASHIMOTO, H. and ONO, K. (1973). [Correlation between number of inoculated nematodes and disease occurrence in pine wilt disease.] *Transactions of the 26th annual Meeting of Kyushu Branch of the Japanese Forestry Society*: 191–192.

KIYOHARA, T. and TOKUSHIGE, Y. (1971). [Inoculation experiments of a nematode, *Bursaphelenchus* sp., onto pine trees.] *Journal of Japanese Forestry Society* 53: 210–218.

KOBAYASHI, K., OKUDA, M. and HOSODA, R. (1976). [Influence of wood size and humidity on the emergence of Japanese pine sawyer and the number of pine wood nematodes per insect.] *Transactions of the 87th Annual Meeting of the Japanese Forestry Society*: 239–240.

KOBAYASHI, T., SASAKI, K. and MAMIYA, Y. (1974). [Fungi associated with *Bursaphelenchus lignicolus*, the pine wood nematode (I)]. *Journal of Japanese Forestry Society* 56: 136–145.

KOBAYASHI, T., SASAKI, K. and MAMIYA, Y. (1975). [Fungi associated with *Bursaphelenchus lignicolus*, the pine wood nematode (II)]. *Journal of Japanese Forestry Society* 57: 184–193.

MAEHARA, N. (2008). Reduction of *Bursaphelenchus xylophilus* (Nematoda: Parasitaphelenchidae) population by inoculating *Trichoderma* spp. into pine wilt-killed trees. *Biological Control* 44: 61–66.

MAEHARA, N. and FUTAI, K. (1996). Factors affecting both the numbers of the pinewood nematode, *Bursaphelenchus xylophilus* (Nematoda: Aphelenchoididae), carried by the Japanese pine sawyer, *Monochamus alternatus* (Coleoptera: Cerambycidae), and the nematode's life history. *Applied Entomology and Zoology* 31: 443–452.

MAEHARA, N. and FUTAI, K. (1997). Effect of fungal interactions on the numbers of the pinewood nematode, *Bursaphelenchus xylophilus* (Nematoda: Aphelenchoididae), carried by the Japanese pine sawyer, *Monochamus alternatus* (Coleoptera: Cerambycidae). *Fundamental and Applied Nematology* 20: 611–617.

MAEHARA, N. and FUTAI, K. (2000). Population changes of the pinewood nematode, *Bursaphelenchus xylophilus* (Nematoda: Aphelenchoididae), on fungi growing in pine-branch segments. *Applied Entomology and Zoology* 35: 413–417.

MAEHARA, N. and FUTAI, K. (2001). Presence of the cerambycid beetles *Psacothea hilaris* and *Monochamus alternatus* affecting the life cycle strategy of *Bursaphelenchus xylophilus*. *Nematology* 3: 455–461.

MAEHARA, N. and FUTAI, K. (2002). Factors affecting the number of *Bursaphelenchus xylophilus* (Nematoda: Aphelenchoididae) carried by several species of beetles. *Nematology* 4: 653–658.

MAEHARA, N., HATA, K. and FUTAI, K. (2005). Effect of blue-stain fungi on the number of *Bursaphelenchus xylophilus* (Nematoda: Aphelenchoididae) carried by *Monochamus alternatus* (Coleoptera: Cerambycidae). *Nematology* 7: 161–167.

MAEHARA, N., TSUDA, K., YAMASAKI, M., SHIRAKIKAWA, S. and FUTAI, K. (2006). Effect of fungus inoculation on the number of *Bursaphelenchus xylophilus* (Nematoda: Aphelenchoididae) carried by *Monochamus alternatus* (Coleoptera: Cerambycidae). *Nematology* 8: 59–67.

MAMIYA, Y. and ENDA, N. (1972). Transmission of *Bursaphelenchus lignicolus* (Nematoda: Aphelenchoididae) by *Monochamus alternatus* (Coleoptera: Cerambycidae). *Nematologica* 18: 159–162.

MAMIYA, Y., HIRATSUKA, M. and MURATA M. (2005). Ability of wood-decay fungi to prey on the pinewood nematode, *Bursaphelenchus xylophilus* (Steiner and Buhrer) Nickle. *Japanese Journal of Nematology* 35: 21–30.

MORIMOTO, K. and IWASAKI, A. (1972). [Role of *Monochamus alternatus* (Coleoptera: Cerambycidae) as a vector of *Bursaphelenchus lignicolus* (Nematoda: Aphelenchoididae).] *Journal of Japanese Forestry Society* 54: 177–183.

MORIMOTO, K. and IWASAKI, A. (1973). [Studies on the pine sawyer (IV). Biology of the pine sawyer and the pine wood nematode in the pupal cell.] *Transactions of the 26th Annual Meeting of Kyushu Branch of the Japanese Forestry Society*: 199–200.

NECIBI, S. and LINIT, M.J. (1998). Effect of *Monochamus carolinensis* on *Bursaphelenchus xylophilus* dispersal stage formation. *Journal of Nematology* 30: 246–254.

SAIKI, H., SAITO, T., YONEDA, K., KAIJO, M., UCHIDA, K. and YAMANAKA, K. (1984). Biological control of the pine-wood nematode by spraying a nematode-trapping fungus. *Journal of Japanese Forestry Society* 66: 30–32.

TERASHITA, T. (1975). [Relations between water content of wood, *Bursaphelenchus lignicolus* density in wood and the numbers of the nematode carried by beetles emerged from diseased trees.] *Transactions of the 26th Annual Meeting of Kansai Branch of the Japanese Forestry Society*: 279–281.

THORN, R.G. and BARRON G.L. (1984). Carnivorous mushrooms. *Science* 224: 76–78.

TOGASHI, K. (1985). Transmission curves of *Bursaphelenchus xylophilus* (Nematoda: Aphelenchoididae) from its vector, *Monochamus alternatus* (Coleoptera: Cerambycidae), to pine trees with reference to population performance. *Applied Entomology and Zoology* 20: 246–251.

TOGASHI, K. (1989). Factors affecting the number of *Bursaphelenchus xylophilus* (Nematoda: Aphelenchoididae) carried by newly emerged adults of *Monochamus alternatus* (Coleoptera: Cerambycidae). *Applied Entomology and Zoology* 24: 379–386.

WARREN, J. and LINIT, M.J. (1993). Effect of *Monochamus carolinensis* on the life history of the pinewood nematode, *Bursaphelenchus xylophilus*. *Journal of Nematology* 25: 703–709.

WINGFIELD, M.J. (1987). Fungi associated with the pine wood nematode, *Bursaphelenchus xylophilus*, and cerambycid beetles in Wisconsin. *Mycologia* 79: 325–328.

# Attraction Trap for Monitoring *Monochamus alternatus* Adults – Its Usefulness and Limitations

**Katsunori Nakamura**

**Abstract** Monitoring of the prevalence of *Monochamus alternatus* adults and its load of *Bursaphelenchus xylophilus* is useful for understanding the local epidemic pattern of pine wilt disease (PWD). In Japan, an attraction trap for capturing *M. alternatus* adults has been commercially available and widely used to monitor the flying adults. It can be converted for live capture trap with a small modification, so that we may examine nematode load of the flying adults. The attraction trap is an effective tool for monitoring *Monochamus* beetles as the vector of *B. xylophilus*, but has some restrictions or limitations in field use: (1) The estimates of the adult density and nematode load provided by the trap data would be tentative because the power of attraction is susceptible to some environmental conditions and the number of nematodes carried by an adult continuously decrease after its emergence. (2) Use of attraction trap should be refrained in the protected pine forest from PWD unless it could induce an incidence of PWD brought by the attracted *Monochamus* beetles. (3) Attraction trap is ineffectual in control use because the attractants only lures the matured adults that already transmit a considerable number of the carried nematodes and is considered not to collect enough number of the flying adults to decrease PWD damage.

## Introduction

The Japanese pine sawyer, *Monochamus alternatus*, is the main vector of the pathogenic pinewood nematode, *Bursaphelenchus xylophilus*, which causes pine wilt disease (PWD), in East Asia (Mamiya and Enda, 1972; Morimoto and Iwasaki, 1972; Kobayashi et al., 1984; Lee et al., 1990; Yang, 2004). As the occurrence of PWD is strongly affected by the adult's density and its load of nematode, monitoring of the adults and its nematode load is helpful to predict or explain the pattern of PWD epidemic, and to make plans to cope with the disease.

K. Nakamura
Tohoku Research Center, Forestry and Forest Products Research Institute, Morioka 020-0123, Japan
e-mail: knakam@affrc.go.jp

M.M. Mota, P. Vieira (eds.), *Pine Wilt Disease: A Worldwide Threat to Forest Ecosystems*, 

In Japan, flight interception traps with chemical attractants have been developed and widely used for monitoring or mass trapping of the forest insect pests (Miyazaki, 1980; Makino et al., 1993; Nakai et al., 1994; Kobayashi and Hagita, 2000; Kobayashi and Ida, 2004; Sato and Maeto, 2006), and have also been adopted to faunal investigations of the forest insects (Shibata et al., 1996; Sakakibara et al., 1998; Ueda, 1998; Makihara et al., 2001; Maeto et al., 2002;). Among them, the Sankei trap was originally devised for capturing flying adults of *M. alternatus* (Ikeda, 1981). It has since become one of the most popular insect traps in Japan widening its use for other insects or for other purposes like faunal investigations. In this paper, I will introduce the trap and its live-capture modification (Nakamura et al., 1999; Nakamura and Sone, 2004) for monitoring flying adults of *M. alternatus*, and discuss the use and restrictions of attraction trap in field.

## Attraction Traps Developed for Capturing *Monochamus alternatus*

Chemical attractants for *M. alternatus* had originally been explored for use in controlling PWD, e.g. mass trapping of the adults to reduce their population for lessening transmission of *B. xylophilus*. It had long been known that wood boring insects in pine trees, including *M. alternatus,* were attracted to newly dead trees. Those insects were also lured to artificially weakened pine trees by injecting chemicals such as insecticide (Inoue and Yamaguchi, 1959), herbicide (Yamasaki et al., 1980; Yamasaki and Suzuki, 1982), ethanol and acetone (Ikeda et al., 1981). Volatiles released from the cut logs or weakened trees inoculated with *B. xylophilus* and behavioral reaction of *M. alternatus* adults to the volatiles were studied (Oda, 1974; Yamane et al., 1975; Yamane and Asada, 1977; Ikeda and Oda, 1980; Ikeda et al., 1980b) to show that an appropriate mixture of terpenoids (mainly alfa-pinene) and ethanol could strongly attract them (Ikeda and Oda, 1980; Ikeda et al., 1980a; Ikeda, 1981; Ikeda et al., 1986). After those findings, a flight interception trap (the Sankei trap) and attractants contained in special releasers (Madara-call®) were developed (Ikeda, 1981) and commercialized by Sankei Chemical Co., Kagoshima, Japan.

The Sankei trap mainly consists of a roof, two collision plates (cross vane) and an insect-catching bucket, all made of vinyl chloride (Fig. 1). The original trap for capturing *M. alternatus* was colored black, but the manufacturer provides yellow and white traps so that a suitable color for the target insects may be selected. The baited insects were supposed to hit the collision plate and drop inside the bucket. Usually the bucket is used as a water pan for collecting the downed insects.

The attractant was to be set on the top of the cross vane. The releaser of the attractant is disposable and separated into two packs; one for ethanol and the other for alfa-pinene and the other minor components. The releaser have projections or nipples and when used we make 3 mm holes for the pinene releaser and 1 mm holes for the ethanol releaser, then set the ethanol releaser under the pinene releaser.

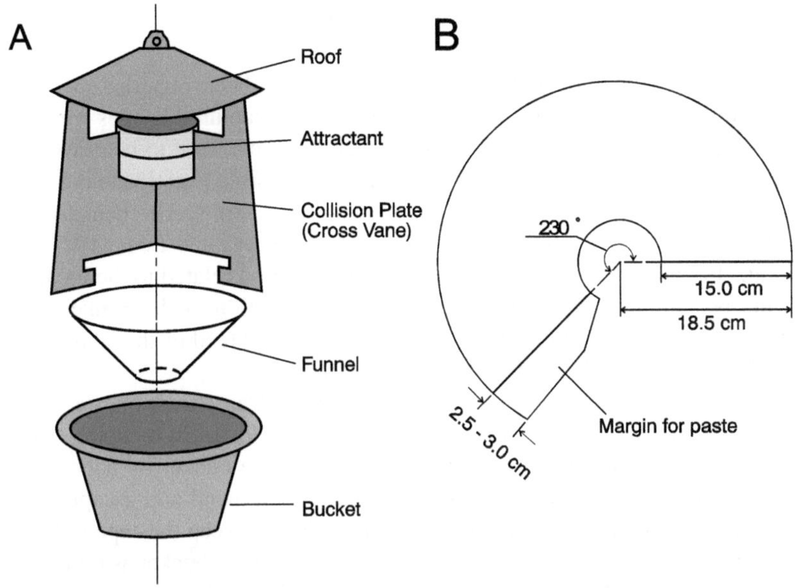


**Fig. 1** Outline of the trap. A: The Sankei trap and live-capture modification. Original trap is not furnished the funnel and supposed to use the bucket as water pan for catching the baited insect. B: Plane plan of the funnel for live capture trap. It could be easily handcrafted from transparent plastic plates. The radius of the sector may need to be adjusted to fit the upper edge of the bucket.

These procedures were developed for the proper diffusion and mixing ratio of the volatiles.

Attempts for mass trapping of *M. alternatus* adults using the Sankei trap or other kinds of traps were made by some researchers, but turned to be not successful (Makino et al., 1978; Ikeda, 1981). Moreover the adult sawyers were proven not to be lured to the attractant until they were sexually matured (Yamane, 1975), which requires 2–4 weeks after emergence (Enda and Nobuchi, 1970, Nobuchi, 1976). Departure of *B. xylophilus* from the adult's body starts soon after emergence and reaches its peak in 1–5 days (Kishi, 1978; Togashi, 1985) or after 10 days (Shibata and Okuda, 1979; Togashi, 1985). Thus the adults are supposed to have already transmitted their carrying nematodes to some extent before they are capture by the trap. Indeed, the number of nematodes extracted from the trap-captured adults was reported to be less than those carried by the newly emerged ones (Mamiya and Enda, 1972; Hagihara et al., 1975). Finally, the attraction traps for capturing *M. alternatus* adults came to be considered ineffective for controlling PWD epidemic, and have served for research or monitoring use exclusively.

The existence of pheromone, or other types of semi-chemicals, that can lure the newly emerged adults of *M. alternatus*, has been inferred by some researchers (Fauziah et al., 1987; Ito, 1987; Kishi and Ohtsu, 1987; Sakai and Yamasaki, 1991; Kishi, 1995; Yamasaki et al., 1997), but their availability has not been yet thoroughly examined.

## Live-Capture Modification of the Trap

The Sankei trap with the Madara-call® attractant has been regarded as a useful tool for monitoring flying adults of *M. alternatus*. The captured beetles were, however, dead and water soaked because the collected insects were to be drawn in the bucket containing water, and this made it almost impossible to estimate the number or frequency of *B. xylophilus* carried by the captured adults. The nematode load of the flying adults have already been decreasing (Mamiya and Enda, 1972; Hagihara et al., 1975) but still provides significant information related to the local PWD epidemic. Nakamura et al. (1999) proposed a modification of the Sankei trap for capturing *M. alternatus* adults alive so that the nematode load of the adults as well as their flight season and/or density could be examined.

The modification itself was as simple as attaching a transparent plastic funnel in the insect-catching bucket to prevent the captured beetles from escaping (Fig. 1). The funnel was roughly fixed on the upper ridge of the bucket by adhesive tape to avoid getting out of the position when shaken by the wind and escaping of the beetles at the contacting end of the funnel with bucket. To keep the captured beetles alive, cut pine twigs (needles detached) were supplied in the bucket as food.

Use of the original bucket originated a saving in labor and cost, but caused some serious problems. One was the frequent escape of the captured beetles. The original bucket was not so deep, thus the lower part of the funnel was as close as 2.5 cm to the bottom of the bucket (Fig. 2). This could allow a walking adult in the bucket reach the lower part of the funnel and go outside. The escape rate was estimated ca. 30% per day, per trap (Nakamura et al., 1999). High mortality of the captured beetles was the other problem. The black colored bucket was easily heated by the sunlight. Narrow and poorly ventilated space for the captured beetles would worsen the situation. Consequently, the temperature inside the bucket reached 45–49°C in the afternoon of a sunny summer day (Nakamura et al., 1999), which could be lethal

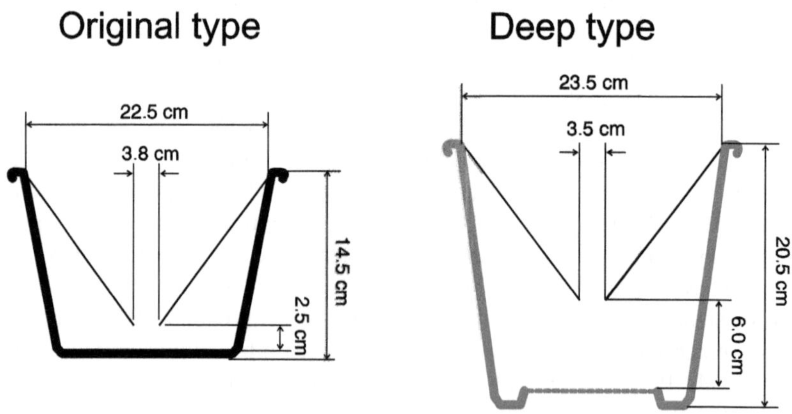


**Fig. 2** Vertical sections of the insect-catching containers

to the captured beetles. These problems seemed to be solved by using a deep container with whitish color.

Nakamura and Sone, (2004) adopted a commercial plant pot as the insect-catching container of the trap instead for the original black bucket. The pot was made of biodegradable plastic (detailed components unknown) and the color was somber white. With this alteration, the distance from the lower part of the funnel to the bottom of the container became 6 cm, which was long enough for the walking adults of *M. alternatus* to reach the funnel (Fig. 2). The upper ridge of the plant pot was almost the same size as the original bucket, so that the pot was easily set on the lower part of the cross vane without special modification.

Escape of *M. alternatus* adults contained in the insect catching container with original bucket (original type) and deep plant pot (deep type) was estimated under laboratory conditions (Table 1). In the original type container, escaped beetles accounts for 25% of the total counts, corresponding to the estimation in the previous study (Nakamura et al., 1999). On the other hand, the occurrence of escape was 5% per a day in the deep type container. These results indicate that the escape of the captured beetles was strongly restrained by using the deep type container, although this alteration could not prevent all of the beetles from escaping. The occurrences of escape of the trap-captured beetles for 3–4-day intervals were estimated at 4–6%, which was less than that in the laboratory experiment (Nakamura et al., unpublished data). This may be due to the lower density of the captured beetles in the field settings. Consequently, we can conclude that the escape rate of the captured beetles in the deep type container was low enough for monitoring use of *M. alternatus* in field. Mortality of the captured beetles was also less in the deep type container, probably due to the whitish color and wide space inside.

We have used the live capture trap to monitor *M. alternatus* adults and its nematode load in some different locations and the relationships among the number of captured beetles, their nematode load and occurrence of PWD is being pursued (Nakamura, 2003).

**Table 1** Escape of *Monochamus alternatus* adults contained in the insect-catching containers for live capture trap. Eight adults (4 males and 4 females) were released in each container and funnel was set on it, and placed in a separated small cage. The numbers of adults inside and outside of the containers were recoded once a day for two weeks. Dead adults found in the observation time were discarded

| Container[1] | Sex | Total count of the adults (A) | Total count of the adults escaped (B) | %escaped (B/A)[2] |
|---|---|---|---|---|
| Original type | male | 59 | 16 | 27.1 |
| | female | 60 | 14 | 23.3 |
| | combined | 119 | 30 | 25.2 |
| Deep type | male | 59 | 1 | 1.7 |
| | female | 60 | 5 | 8.3 |
| | combined | 119 | 6 | 5.0 |

[1]See Fig. 2.
[2]The difference between the treatments was highly significant in Fisher's exact test ($P < 0.001$). The differences between male and female in both treatments were not significantly ($P = 0.68$ and 0.21 for Original and Deep types, respectively).

## Limitations in the Effecturalness of Attraction Traps

Attraction trap like the Sankei trap would be a useful tool for monitoring *Monochamus* beetles, because it will provide rough estimates of the local density of the flying adults and the adult flight season without much labor. If one adopts the live capture modification to the traps, one can access the information about the nematode load of the captured adults, which will be valuable to understand local PWD epidemic situation. But this methodology has obvious limitations or restrictions as follows, thus we must be careful with the introduction of attraction traps in field and interpretation of the obtained data.

1) Tentative Estimates of Adult Density and Nematode Load

The number of baited and collected beetles would roughly correspond to the population size of the flying adults around the trap site, but hardly give the exact estimate. The attraction efficiency would be swayed by wind, temperature, consumption of the volatile attractants, site conditions such as topology, etc. Moreover, existence of newly dead pine trees that have superior attraction activity should strongly affect the captures in the traps nearby. The number of *B. xylophilus* carried by an adult *M. alternatus* continuously decrease along the time after emergence (Kishi, 1978; Shibata and Okuda, 1979; Togashi, 1985; Kishi, 1995), and the flying adults need 2–3 weeks before they can respond to the attractants (Yamane, 1975; Enda and Nobuchi, 1970, Nobuchi, 1976). Therefore, nematode load of the captured adults is always less than that at the time of emergence.

2) Potential Risk of Inducing PWD

Attraction trap attracts *Monochamus* beetles. Not all of the lured beetles, however, would be captured by the trap, but wandering around the area. In the Sankei trap with the Madara-call® attractant, the field density of *M. alternatus* adults tended to be high in the 5 m radius of the trap site (Iwasaki and Morimoto, 1975), and we often experience the incidence of PWD in the pine tree on which the trap have been hanged or nearby trees. Thus we should expect that the use of attraction traps can induce PWD affection by luring vector insects carrying *B. xylophilus* to the trap site. Care must be exercised when using attraction traps in protected pine forests adjacent to the PWD affected area, or we might promote the spread of PWD.

3) Ineffectiveness in Control Use

The attraction trap for *Monochamus* beetles was originally developed as a control method against PWD. Mass trapping (Ikeda, 1981; Makino et al., 1978) or pesticide application in the area where the adult density was supposed to be raised by the effect of the attractants (Iwasaki et al., 1976) were attempted but not very successful.

Now we know the alfa-pinene baited traps only lures the matured adults of *Monochamus* beetles that already transmit a considerable number of *B. xylophilus* carried by them. Moreover, it is likely that attraction traps can not collect enough number of *Monochamus* beetles to decrease PWD damage, which is estimated ca. 90% or more of the local population (Togashi, 1991). So far the attractants for *Monochamus* beetles are considered to be imitations of dying or insect damaged pine trees, even when we add some effective semiochemicals like bark beetle pheromones (Groot and Nott, 2004; Pajares et al., 2004). Thus if there are actual dead trees in the surroundings, the majority of the beetles would not be collected by the trap but baited to the dead trees.

Consequently, the attraction traps for *Monochamus* beetles are not recommendable for control use, whereas they serve effectively in monitoring use.

**Acknowledgement** I thank Prof. K. Sone of Kagoshima University and his students, Prof. N. Kameyama of University of the Ryukyus and his students, Dr. Y. Sato, Mr. T. Tashiro, Ms. E. Kawaguchi and Mr. Y. Usui of Kagoshima Pref. Forestry Exp. Stn., Mr. H. Irei, Mr. C. Kyuna, Ms. Y. Nakahira and Mr. T. Miyagi of Okinawa Pref. Forestry Exp. Stn. for their arrangements and a great contribution in the field work. Thanks are also due to Ms. M. Nakamura for her help in making the apparatus.

## References

Enda, N. and Nobuchi, A. (1970). [Studies on the pine bark and wood boring beetles: maturation of the ovary and its parasitic nematodes.] *Transactions of the 81st Annual Meeting of the Japanese Forestry Society*: 274–276.

Fauziah, B.A., Hidaka, T. and Tabata, K. (1987). The reproductive behavior of *Monochamus alternatus* Hope (Coleoptera: Cerambycidae). *Applied Entomology and Zoology* 22: 272–285.

Groot de, P. and Nott, R.W. (2004). Response of the whitespotted sawyer beetle, *Monochamus scutellatus*, and associated woodborers to pheromones of some *Ips* and *Dendroctonus* bark beetles. *Journal of Applied Entomology* 128: 483–487.

Hagihara, Y., Ogawa, S. and Nakashima, Y. (1975). [Temporal changes of the number of trapped beetles and their nematode load using attractants for *M. alternatus*.] *Transactions of the 28th Annual Meeting of the Japanese Forestry Society Kyushu Branch*: 147–148.

Ikeda, T. (1981). [Chemical attractants for *Monochamus alternatus* and its use]. *Shinrin-boeki (Forest Pests)* 35: 95–100.

Ikeda, T. and Oda, K. (1980). The occurrence of attractiveness for *Monochamus alternatus* Hope (Coleoptera: Cerambycidae) in nematode-inoculated pine trees. *Journal of Japanese Forestry Society* 62: 432–434.

Ikeda, T., Enda, N., Yamane, A., Oda, K. and Toyoda, T. (1980a). Attractants for the Japanese pine sawyer, *Monochamus alternatus* Hope (Coleoptera: Cerambycidae). *Applied Entomology and Zoology* 15: 358–361.

Ikeda, T., Oda, K., Yamane, A. and Enda, N. (1980b). Volatiles from pine logs as the attractant for the Japanese pine sawyer *Monochamus alternatus* Hope (Coleoptera: Cerambycidae). *Journal of Japanese Forestry Society* 62: 150–152.

Ikeda, T., Yamane, A., Enda, N., Matsuura, K. and Oda, K. (1981). Attractiveness of chemical-treated pine trees for *Monochamus alternatus* Hope (Coleoptera: Cerambycidae). *Journal of Japanese Forestry Society* 63: 201–207.

Ikeda, T., Yamane, A., Enda, N., Oda, K., Makihara, H., Ito, K. and Okochi, I. (1986). Attractiveness of volatile components of felled pine trees for *Monochamus alternatus* Hope (Coleoptera: Cerambycidae). *Journal of Japanese Forestry Society* 68: 15–19.

Inoue, M. and Yamaguchi, H. (1959). [Mass trapping of scolytid beetles using artificially weakened trees injected with pesticide.] *Bulletin of Forestry Experiment Station* 111: 1–34.

Ito, K. (1987). [Field experiments of the attractiveness of *Monochamus alternatus* adults to the other individuals.] *Transactions of the 33rd Annual Meeting of the Japanese Forestry Society Kansai Branch*: 319–321.

IWASAKI, A. AND MORIMOTO, K. (1975). [Studies on the Japanese pine sawyer (XIII) The range where adults feeding on pine twigs increase around the attraction traps.] *Transactions of the 28th Annual Meeting of the Japanese Forestry Society Kyushu Branch*: 197–198.

IWASAKI, A., MORIMOTO, K. AND YOSHIDA, Y. (1976). [Studies on the Japanese pine sawyer (XVIII) Effect of pesticide application combined with luring of the adults using attractant.] *Transactions of the 29th Annual Meeting of the Japanese Forestry Society Kyushu Branch*: 195–196.

KISHI, Y. (1978). [Invasion of pine trees by *Bursaphelenchus lignicolus* (Nematoda: Aphelenchoidae) from *Monochamus alternatus* Hope (Coleoptera: Cerambycidae).] *Journal of Japanese Forestry Society* 60: 179–182.

KISHI, Y. (1995). *The pinewood nematode and the Japanese pine sawyer*. Thomas Company Ltd., Tokyo, Japan, 302 pp.

KISHI, Y. AND OHTSU, S. (1987). [Preliminary study on pheromone of *Monochamus alternatus* (Coleoptera: Cerambycidae).] *Transactions of the 98th Annual Meeting of the Japanese Forestry Society*: 537–538.

KOBAYASHI, M. AND HAGITA, M. (2000). Process of mass mortality of oak trees and capture of *Platypus quercivorus* Murayama (Coleoptera: Platypodidae). *Applied Forest Science* 9: 133–140.

KOBAYASHI, M. AND IDA, A. (2004). [Trapping scarabaeid beetles in a tree nursery using several kinds of insect traps.] *Shinrin-boeki (Forest Pest)* 53: 230–238.

KOBAYASHI, F., YAMANE, A. AND IKEDA, T. (1984). The Japanese pine sawyer beetle as the vector of pine wilt disease. *Annual Review of Entomology* 29: 115–135.

LEE, S.L., CHOO, H.Y., PARK, N.C., MOON, Y.S. AND KIM, J.B. (1990). Nematode and insects associated with dead trees, and pine wood nematode detection from the part of *Monochamus alternatus*. *Korean Journal of Applied Entomology* 29: 14–19.

MAETO, K., SATO, S. AND MIYATA, H. (2002). Species diversity of longicorn beetles in humid warm-temperate forests: the impact of forest management practice on old-growth forest species in southwestern Japan. *Biodiversity and Conservation* 11: 1919–1937.

MAKIHARA, H., GOTO, H., MAETO, K. AND KITAJIMA, H. (2001). Investigation of longicorn beetles (Coleoptera, Cerambycidae) to search for environmental indicator in Satoyama (1) longicorn beetles living in the natural forests of a low mountain area in Ibaraki Prefecture and surveying effective traps for catching longicorns. *Bulletin of Hoshizaki Green Foundation* 5: 1–16.

MAKINO, S., MAKIHARA, H., FUJITA, K., KITAJIMA, H. AND ISONO, M. (1993). Seasonal changes in the number and ovarian conditions of the Cryptomeria twig borer, *Anaglyptus subfasciatus* Pic (Coleoptera: Cerambycidae), collected with an odor attractant. *Applied Entomology and Zoology* 28: 581–584.

MAKINO, Y., MORIMOTO, K., KURANAGA, Z., IWASAKI, A. AND YOSHIDA, S. (1978). [Field experiments for effective way of trapping *Monochamus alternatus* by attractants] *Transactions of the 31st Annual Meeting of the Japanese Forestry Society Kyushu Branch*: 247–248.

MAMIYA, Y. AND ENDA, N. (1972). Transmission of *Bursaphelenchus lignicolus* (Nematoda: Aphelenchoididae) by *Monochamus alternatus* (Coleoptera: Cerambycidae). *Nematologica* 18: 159–162.

MIYAZAKI, Y. (1980). [Studies on consevation of seaside pine forest in Sawara Experiment Forest of Kyushu University (1) Experiment for edadicating *Monochamus alternatus* adults baited to attractants.] *Bulletin of Kyushu University Experiment Fores*t 27: 59–98.

MORIMOTO, K. AND IWASAKI, A. (1972). Rôle of *Monochamus alternatus* (Coleoptera: Cerambycidae) as a vector of *Bursaphelenchus lignicolus* (Nematoda: Aphelenchoididae). *Journal of the Japanese Forestry Society* 53: 210–218.

NAKAI, I., NAKANE, I., FURUNO, T. AND FUTAI, K. (1994). [The catching of the Japanese sawyer, *Monochamus alternatus* Hope, by attractant at Kamigamo Experiment Station of Kyoto University Forest, Kyoto City.] *Bulletin of Kyoto University Experiment Forest* 26: 9–19.

NAKAMURA, K. (2003). Monitoring of the Japanese pine sawyer, *Monochamus alternatus*, and its load of the pinewood nematode, *Bursaphelenchus xylophilus*, with a commercial attraction trap modified for live capture. *Korea-Japan Joint Conference on Applied Entomology and Zoology*: 180.

NAKAMURA, K. AND SONE, K. (2004). Modification of capturing bucket of the attraction trap for live trapping of *Monochamus alternatus* adults to prevent escaping of the captured beetles. *Kyushu Journal of Forest Research* 57: 110–112.

NAKAMURA, K., SONE, K. AND OOKUMA, H. (1999). Modification of a commercial attraction trap for live trapping of *Monochamus alternatus* (Hope) adults (Coleoptera: Cerambycidae). *Japanese Journal of Applied Entomology and Zoology* 43: 55–59.

NOBUCHI, A. (1976). [Fertilization and oviposition of *Monochamus alternatus* Hope.] *Transactions of the 87th Annual Meeting of the Japanese Forestry Society*: 247–248.

ODA, K. (1974). [Volatile components released from cut pine log.] *Bulletin of Forestry Experiment Station* 266: 1–11.

PAJARES, J.A., IBEAS, F., DIEZ, J.J. AND GALLEGO, D. (2004). Attractive responses by *Monochamus galloprovincialis* (Col., Cerambycidae) to host and bark beetle semiochemicals. *Journal of Applied Entomology* 128: 633–638.

SAKAI, M. AND YAMASAKI, T. (1991). (+)-*cis*-Pinen-2-ol: Attractant for male cerambycid beetle, *Monochamus alternatus* Hope. *Journal of Chemical Ecology* 17: 757–765.

SAKAKIBARA, Y., KIKUMA, A., IWATA, R. AND YAMANE, A. (1998). Performance of four chemicals with floral scents as attractants for longicorn beetles (Coleoptera: Cerambycidae) in a broadleaved forest. *Journal of Forest Research* 3: 221–224.

SATO, S. AND MAETO, K. (2006). Attraction of female Japanese horntail *Urocerus japonicus* (Hymenoptera: Siricidae) to alpha-pinene. *Applied Entomology and Zoology* 41: 317–323.

SHIBATA, E. AND OKUDA K. (1979). Transmission of pine wood nematode, *Bursaphelenchus xylophilus* (Steiner et Buhrer) Nickle (Nematode: Aphelenchoididae), by the Japanese pine sawyer, *Monochamus alternatus* Hope (Coleoptera: Cerambycidae) to pine twigs under laboratory conditions. *Japanese Journal of Nematology* 18: 6–14.

SHIBATA, E., SATO, S., SAKURATANI, Y., SUGIMOTO, T., KIMURA, F. AND ITO, F. (1996). Cerambycid beetles (Coleoptera) lured to chemicals in forests of Nara Prefecture, central Japan. *Annals of the Entomological Society of America* 89: 835–842.

TOGASHI, K. (1985). Transmission curves of *Bursaphelenchus xylophilus* (Nematoda: Aphelenchoididae) from its vector, *Monochamus alternatus* (Coleoptera: Cerambycidae), to pine trees with reference to population performance. *Applied Entomology and Zoology* 20: 246–251.

TOGASHI, K. (1991). [Estimation of the control measures against pine wilt disease by computer simulations.] *Ringyou-to-Yakuzai (Forestry Chemicals)* 116: 1–10.

UEDA, A. (1998). Capture of beetles at different heights by using alpha-pinene and ethanol as attractants. *Applied Forest Science* 7: 109–112.

YAMANE, A. (1975). [Behavioral characteristics of *Monochamus alternatus*.] *Shinrin-boeki (Forest Pest)* 24: 211–213.

YAMANE, A. AND ASADA, T. (1977). [Change in the production if the odor from fresh cut pine bolts after cutting in relation to its oviposition attractiveness to *Monochamus alternatus*.] *Transactions of the 88th Annual Meeting of the Japanese Forestry Society*: 283–284.

YAMANE, A. MIYAZAKI, M., ODA, K. AND ASADA, T. (1975). [Response of *Monochamus alternatus* adults to the odor from cut pine bolts and the other materials in their oviposition season]. *Transactions of the 86th Annual Meeting of the Japanese Forestry Society*: 329–330.

YAMASAKI, T. AND SUZUKI, N. (1982). Luring of Japanese pine sawyer *Monochamus alternatus* Hope by Paraquat-treated pine trees (II) The attractiveness of volatile oil from lightwood. *Journal of Japanese Forestry Society* 64: 340–345.

YAMASAKI, T., HATA, K. AND OKAMOTO, H. (1980). [Luring of Japanese pine sawyer *Monochamus alternatus* Hope by Paraquat-treated pine trees (I) The effect of Paraquat and Ethephon o pine trees.] *Journal of Japanese Forestry Society* 62: 99–102.

YAMASAKI, T. SATO, M. AND SAKOGUCHI, H. (1997). (−)-Germacrene D: Masking substance of attractants for the cerambycid beetle, *Monochamus alternatus* (Hope). *Applied Entomology and Zoology* 32: 423–429.

YANG, B.J. (2004). The history, dispersal and potential threat of pine wood nematode in China. In *The pine wood nematode*, Bursaphelenchus xylophilus. *Nematology Monographs and Perspectives, vol. 1* (eds. M. Mota and P. Vieira), pp. 9–20. E. J. Brill, Leiden, The Netherlands.

# Studies on *Scleroderma guani* to Control the Pine Sawyer Beetle, *Monochamus alternatus*

**Fuyuan Xu, Keqin Xu, Chunxia Xie, Pei Zhang, Sangchul Shin and Youngjin Cheong**

**Abstract** After clear-cutting dead pine trees caused by the pinewood nematode (PWN), *Bursaphelenchus xylophilus*, some remained as dead branches and stumps which were bored by larvae of pine sawyer beetle (PSB), *Monochamus alternatus*, which could not be removed from the forest, constitute a potential source of inoculum. Several natural enemies for controlling PSB were found, in Dongshanqiao district near Nanjing. *Scleroderma guani* is one of the most successful natural enemies that parasitizes the larvae of PSB and is easy to reproduce. Three hundred thousand larvae of PSB were collected in the forest annually for the purpose of massive rearing of *S. guani* in the lab. The studies of the biology, reproductive methods, release techniques and control effect of the larvae of PSB by *S. guani* were investigated. Mass rearing of *S. guani* in the lab resulted in 13 million female adults of *S. guani* for 0.3 million larvae of PSB annually from 2004–2006. *S. guani* was released to control the larvae of PSB in the same and other affected areas by PWN. The dead pine trees were significantly reduced in the test area. Both collecting the larvae of PSB and releasing the adults of *S. guani* were a good and ecological way to control PSB and PWN in a large area because of its low risk to humans and to the environment.

## Introduction

Pine wilt disease (PWD), caused by *Bursaphelenchus xylophilus*, is one of the most important diseases of conifer trees around the world. The pinewood nematode, a pathogen of pines, is distributed in Japan, United States of America, Canada, Mexico, South Korea, Portugal and China. In Japan and China, it has caused extensive mortality to Japanese red pine (*Pinus densiflora*), Japanese black pine (*P. thunbergii*) and Masson pine (*P. massoniana*). It has also caused an increase in dead pine trees in South Korea in recent years. Recently, the main damage to pine trees caused by PWN changed from *P. thunbergii* to *P. massoniana* in southeast China. The total number of dead pine trees (*P. massoniana*) in the forest increased from 10% to

F. Xu
Forestry Academy of Jiangsu Province, Nanjing, 211153, China

50%, from 1985 till now. The attention has been now focused on the methods of PWN controlling techniques and its vector.

At present, PWN is mainly controlled by intensified quarantine service, clear-cutting and methyl bromide fumigation of dead pine trees. Both aerial and ground spray insecticides can significantly reduce the death of pine trees. Other treatments can effectively control the pine sawyer beetle (PBS), *M. alternatus*, such as spraying MPP oil emulsion on stumps, hot water treatment, chipping dead wood into pieces, high temperature treatment, submerging in water, burying wood remains in the soil and attractants to lure and kill the pine sawyer beetle. All of them are an effective way to control the PWN. However, in large areas, the effectiveness of those methods was not satisfactory (Yang, 1995; Dwinell, 1997; Xu, 1998).

After clear-cutting the dead pine trees, some dead branches and stumps have remained which are then bored by the larva of the PSB. Biological control is one of the useful methods because of low risk to humans and because it is a specific interaction. Several natural enemies for controlling PSB have been found. *Scleroderma guani* is one of the most successful natural enemies that parasitizes the larvae of PSB and is easy to reproduce through the larvae of pine sawyer beetles. The study of the biology of *S. guani*, its release techniques and control effects to PSB is being carried out systematically. The research results are hereby presented.

## Material and Methods

## Biology of *S. guani*

For the purpose of mass rearing, the biological study was carried out in the laboratory and in a test forest respectively.

### *Collected Larvae of PSB for Massive Rearing Natural Enemies and Their Control Effects*

1. The number of larvae of PSB collected in the forest annually in the remained the dead branches and stumps in the forest area after the clear-cutting.
2. The use of larvae of PSB to mass rear *S. guani* in the lab. A tube (1 cm in the diameter and 5 cm in length) was used. One larva of PSB was placed inside one tube and inoculated with 3–4 female adults of *S. guani*. One side of the tube was sealed and the other was sealed with cotton ball (Xu K., Xu F. et al., 2002).

### *The Release Techniques of* **S. guani**

1. Two different release methods were conducted: (1) One, the point release method and (2) the other, the model of point and single trees release method.
2. Different release time conducted in different months respectively.

3. Different release density conducted releasing 5 000 and 10 000 wasps/ha respectively.
4. Different release sites were carried out in different places of Nanjing, Zhenjiang and Wuxi.

### *The Control Effect of* S. guani *to PSB and PWN in the Large Area*

1. The experiments of the control effect on the PSB were carried out in the test field. All tests set up five standard plants of pine trees and in the repetition of three times in one test field (Xu K., Xu F. et al., 2002).
2. The control effect of *S. guani* to PWN in the large area: In the following year of releasing *S. guani*, investigations were carried out by comparing the number of dead pine trees caused by the PWN and living pine trees in the test field to calculate the control effect to PWN, both in the test field and CK field.

### *The Exam Methods of the Control Effect of* S. guani

1. The experiments of the control effect on the PSB were carried out in the laboratory and the test field respectively and observed daily in the lab.
2. In the test field all of the tests set up five standard plants of pine trees and in the repetition of three times in one test field.
3. All pine trees were inspected for the parasitized effect of wasp by peeling off the bark of trees in a certain period of time and area of bark and split wood of dead pine trees which was caused by PWN.
4. In the next year of releasing *S. guani* the investigations were carried out by the comparison of the number of dead pine trees which was caused by PWN and living pine trees in the test field to calculate the control effect to PWN both in the test field and CK field.

## Results and Discussion

### *The Biology of* S. guani

The activities of the adults of *S. guani*: In open field, the adults of *S. guani* are diffused by creeping and being blown away by wind after crawling to the top of the branches and searching for new host on pine trees. By observation in July 19, in the field, the crawling speed of *S. guani* could reach 50–150 cm/m. After found the boring hole of the larvae of PSB the adults of *S. guani* bored into immediately. One bored hole of the larva of PSB could get into 1–3 of the adults of *S. guani* and after that they began parasitizing the larvae of PSB immediately.

Table 1 The activity of oviposition of the adults of *S. guani*

| No. | Date of release | Sting time to the larva body of PSB (hours) | Replenishing feeding time (d) | Oviposit time (d) | Oviposit number (eggs) |
|---|---|---|---|---|---|
| 1 | 07-19 | 6 | 2.5 | 2.5 | 60 |
| 2 | 07-19 | 24 | 3.0 | 3.0 | 55 |
| 3 | 07-19 | 6 | 2.0 | 2.0 | 31 |
| 4 | 07-19 | 15 | 3.0 | 2.5 | 53 |
| 5 | 07-19 | 7 | 3.5 | 3.0 | 88 |

ADULTS OVIPOSITION: After the adults were released in July 19, the female wasp sought the larvae of PSB and crawled into its body. The female wasp stung the body of the larvae of PSB repeatedly. It could persist 6–15 hours until paralysis of the body of larvae of PSB after anesthetization of the larva of PSB. The replenishing feeding lasted 2.0–3.5 days. After this stage, the female wasp began to lay eggs. The oviposition stage lasted 2.0–3.0 days. Each female wasp could lay 31–88 eggs (Table 1).

Hatch and Growth of the Young Larvae of *S. guani*: After oviposition, the egg stage lasted 3.5–5.0 days. The young larvae of *S. guani* hatched quickly. The rate of hatch was 60.0%–87.1%. Until August 1, the larvae of *S. guani* began to produce a cocoon. The larvae stage lasted 4.0–5.5 days. The rate of survival of the larvae of *S. guani* was 76.0%–92.5%. Until August 10, the adults of *S. guani* began to eclosion (Table 2).

## *Improved Techniques of Mass Rearing Natural Enemies and Their Control Effects*

We have collected the larvae of PSB in the remainder dead branches and stumps in the forest area following clear-cutting. Three hundred thousand larvae of PSB were collected in the forest annually for the purpose of mass rearing in the lab.

Usage of larvae of PSB for massive rearing *S. guani* in the lab. In 2000 we studied the techniques of massive rearing *S. guani* in the lab (Xu K., Xu F. et al., 2002), and in 2001 we built a natural enemy massive rearing lab. At the beginning of 2000 the lab could not reproduce enough *S. guani* for the need in controlling PWN in Jiangsu. After many tests the success of massive rearing *S. guani* was significantly increased from 5.0 times in 2000 to 43.3 times in 2004. The mass rearing of *S. guani* in the lab raised 13 million of female adults of *S. guani* by use the same number of 0.3 million of the larvae of PSB annually from 2004–2006 (Table 3).

## *The Releasing Techniques of* S. guani *and Its Control Effect*

Comparison of the Control Effect of the Different Releasing Models: After releasing the wasp in early July, we examined the parasitic effect of *S. guani* to the larva of

**Table 2** The hatch and growth of the young larva of *S. guani*

| No. | Oviposit number (eggs) | The date of hatch | The end date of hatch | The rate of hatch (%) | Egg stage (d) | The started date cocoon stage | The end date cocoon stage | The young larva wasp stage (d) | The started date of eclosion stage |
|---|---|---|---|---|---|---|---|---|---|
| 1 | 60 | 07.27 | 07.28 | 71.67 | 4.0 | 08.01 | 08.05 | 4.5 | 08.10 |
| 2 | 55 | 07.28 | 07.29 | 60.00 | 3.5 | 08.02 | 08.05 | 4.5 | 08.11 |
| 3 | 31 | 07.26 | 07.28 | 87.10 | 3.0 | 08.01 | 08.04 | 5.5 | 08.10 |
| 4 | 53 | 07.27 | 07.28 | 70.00 | 3.5 | 08.01 | 08.03 | 4.0 | 08.12 |
| 5 | 88 | 07.27 | 07.28 | 75.00 | 4.0 | 08.02 | 08.06 | 5.5 | 08.12 |

**Table 3** The usage of the larvae of PSB for massive raising *S. guani* in the lab

| year | Collected larvae (million) | massive raising *S. guani* (million) | Reproduce *S.guani* (times) | Increase times |
|---|---|---|---|---|
| 2000 | 0.2 | 1.0 | 5.0 | 1.0 |
| 2001 | 0.3 | 5.0 | 16.7 | 3.34 |
| 2002 | 0.3 | 6.0 | 20.0 | 4.0 |
| 2003 | 0.3 | 10.0 | 33.3 | 6.66 |
| 2004 | 0.3 | 13.0 | 43.3 | 8.67 |
| 2005 | 0.3 | 13.0 | 43.3 | 8.67 |
| 2006 | 0.3 | 13.0 | 43.3 | 8.67 |

PSB in August 31-September. The rate of parasitism of the model of point & single trees release method was 48.68%, and the rate of parasitism of the model of point release method was 37.50%. The model of point & single trees release method was much higher than that of the model of point release method (Table 4).

Comparison of the Control Effect of the Different Release Times: After testing in different years and different months, the test results showed that the rate of parasitism was 37.50% and 66.82% respectively in the release time of July. However, the rate of parasitism was 28.75% in the release time of August. Regarding the rate of parasitism, the release time in July was much higher than that of the release time in August (Table 5).

Comparison of the Control Effect of the Different Release Density: After releasing 5 000 wasps/ha and 10 000 wasps/ha respectively in the middle of July, the control effects were examined. The rate of parasitism of 5 000 wasps/ha and of 10000 wasps/ha were 48.68% and 52.86% respectively in late September. The control effects of two releasing densities were not significantly different. The release density of 5 000 wasps/ha was one of the economical and effective releasing densities to control the larvae of PSB (Table 6).

Comparison of the Control Effect of the Different Release Sites: After releasing 5 000 wasps/ha at three sites in Nanjing, Zhenjiang and Wuxi, the control effects were examined. The test results showed that the control effects at different places

**Table 4** The comparison of the control effect of the different release models

| Release models | Release places | Release number (wasps/ha) | The parasite effect to the larva of PSB | | |
|---|---|---|---|---|---|
| | | | The number of larva | Parasitized number of larva | The corrected rate of parasite (%) |
| the model of point & single tree release | Jurong Changlong Shan | 5000 | 38 | 25 | 48.68 |
| the model of point release | Dongshanqiao linchang | 5000 | 120 | 70 | 37.50 |
| CK | Dongshanqiao linchang | 0 | 90 | 30 | – |

**Table 5** The comparison of the control effect of the different release times

| Release places | Release number (wasps/ha) | The parasite effect to the larva of PSB | | |
|---|---|---|---|---|
| | | The number of larva | Parasitized number of larva | The corrected rate of parasite (%) |
| Lisheilinchang | 5000 | 120 | 70 | 37.50 |
| Lisheilinchang | 5000 | 400 | 210 | 28.75 |
| Dongshanqiao linchang | 5000 | 140 | 109 | 66.82 |
| CK | 0 | 360 | 120 | – |

**Table 6** The comparison of the control effect of the different releasing density

| Release places | Release number (wasps/ha) | The parasite effect to the larva of PSB | | |
|---|---|---|---|---|
| | | The number of larva | Parasitized number of larva | The corrected rate of parasite (%) |
| Jurong Changlong shan | 5000 | 38 | 25 | 48.68 |
| Jurong linyuan | 10000 | 35 | 24 | 52.86 |
| CK | 0 | 90 | 30 | – |

**Table 7** The comparison of the control effect of the different release places

| Release places | Release number (wasps/ha) | The parasite effect to the larva of PSB | | |
|---|---|---|---|---|
| | | The number of larva | Parasitized number of larva | The corrected rate of parasite (%) |
| Jurong Changlongshan | 5000 | 38 | 25 | 48.68 |
| Lishelinchang | 5000 | 120 | 70 | 37.50 |
| Dongshangqiaolinchang | 5000 | 120 | 70 | 37.50 |
| Wuxicuishan | 5000 | 27 | 15 | 33.33 |
| CK | 0 | 90 | 30 | – |

were not significantly different (Table 7). *S. guani* was able to control the larvae of PSB effectively in Jiangsu.

## *Release Techniques of* S. guani *and Its Control Effect to PSB and PWN in the Large Area*

Control Effect of *S. guani* to PSB in the Large Area: Using the model of point & single trees releasing method, 5 000 wasps/ha were released in the test area of 540 ha, the rate of parasites to the larvae of PSB were 66.82%–84.21% in late September *S. guani* could effectively control the larvae of PSB (Table 8). By the observation both in the laboratory and the test field, the reasons for this effectiveness were that there were two generations of *S. guani* parasitizing the larvae of PSB. After release, first generation *S. guani* wasps parasitised the young larvae of PSB under the bark

**Table 8** The control effect of *S. guani* to PSB in the large area

| Release places | Release area (ha) | The parasite effect to the larva of PSB | | |
|---|---|---|---|---|
| | | The number of larvae | Parasitized number of larvae | The corrected rate of parasite (%) |
| Dongjinlinchang-1 | 40.0 | 140 | 109 | 66.82 |
| Dongjinlinchang-2 | 100.0 | 38 | 33 | 84.21 |
| Dongjinlinchang-3 | 150.0 | 23 | 17 | 68.69 |
| Dongjinlinchang-4 | 150.0 | 36 | 30 | 80.00 |
| CK | 0 | 6 | 1 | – |
| Total area | 440.0 | – | – | – |

of dead pine trees in July and the second generation of *S. guani* parasitized the third instar larvae of PSB which of them bored into the timber in September.

The Control Effect of *S. guani* to PWN in the Large Area: After clear-cutting the pine forest area which was damaged by PWN, 0.3 million of the larvae of PSB were collected from the test forest annually. At the same time by the use of the model of point & single trees method, 5 000 wasps/ha were released in the test area. We found that: (1) The rate of parasites to the larvae of PSB was 66.82–84.21% in late September. In the test area not only could *S. guani* control the larvae of PSB effectively, but also the rate of dead pine trees which of them were caused by PWN could be reduced by more than 97.0%. The PWN could be controlled effectively and significantly. (2) The effects of control dead pine trees which was caused by pine wood nematode were in fact significant. At the beginning of the year 2000 it was easy to collect 0.3 million of the larvae of PSB in the two forest farms in order to raise *S. guani*. Since 2001 we went to another forest farm in order to collect enough larvae of PSB for mass raising *S. guani*. It was very difficult to collect the larvae of PSB in the two forest farms in the year 2002. (3) On the other hand, pine death rate which was caused by PWN were significantly reduced from 18.51%, 26.26% in the year 2000 to 0.29%, 0.38% in the year of 2003 respectively in the two test forest farms (Table 9). (4) We found that chopping and collecting the larvae of PSB in the

**Table 9** The effect of collecting the larvae of PSB in the after clear-cutting area*

| Address | Test area (ha) | Collected larvae (million) | PWN death rate (%) | 2002 PWN death rate (%) | 2003 PWN death rate (%) |
|---|---|---|---|---|---|
| Dongshan forest farm | 4020 | 0.09*<br>0.12**<br>0.006*** | 18.51**<br>3.26*** | 4.59 | 0.29 |
| Tiexin forest farm | 3150 | 0.11*<br>0.18**<br>0.02*** | 26.26**<br>4.18*** | 5.48 | 0.38 |

*1.*2000 collected larvae number, **2001 collected larvae number, ***2002 collected larvae number;

2. ** 2000 PWN death rate, ***2001 PWN death rate;

3. Since 2001 we've went to other forest farm in order to collect enough larvae of PSB for the massive raising *S. guani*.

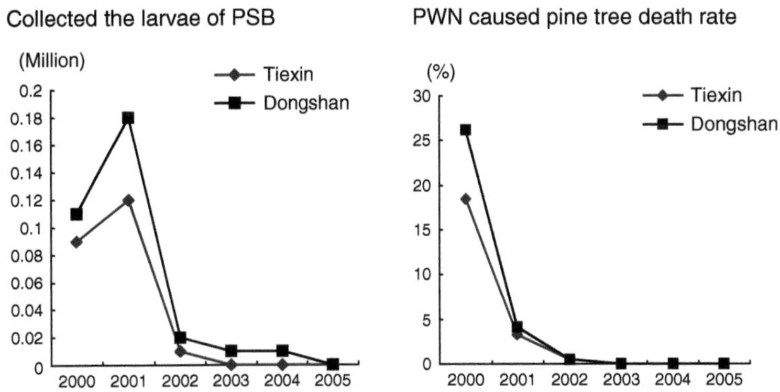


**Fig. 1** Collecting effect larvae of PSB, after the clear-cutting area. N°. 1, 2, 3, 4, 5 and 6 represent the year of 2000, 2001, 2002, 2003, 2004 and 2005, respectively

after clear-cutting area was a good way to control PWN as it is an ecological way of using the larvae of PSB for the mass raising *S. guani*. Following that, we released *S. guani* to control the larvae of PSB in the same and other infected area was caused by PWN in Jiangsu. So both collecting the larvae of PSB in the dead pine forest area after clear-cutting and releasing *S. guani* were an ecological way and good way to control PWN.

## References

DWINELL, L.D. (1997). The pinewood nematode: regulation and mitigation. *Annual Review of Phytopathology* 35: 153–166.

XU, F. (1998). Advances in the research in the natural enemy of *Monochamus alternatus* in the world. *World Forestry Research* 11: 41–45.

XU, F. and GE, M. (1999). Studies on Provenance of Masson Pine Resistance to Pine Wood Nematode (PWN) and Resistance Mechanisms. In *Symposium on Sustainability of Pine Forests in Relation to Pine Wilt and Decline*, pp. 213–216.

XU, K., XU, F., WANG, M., ZHAO, J., JIANG, Q., ZHANG, P., XU, D., HE, R. and JIANG, X. (2002). The techniques of *Scleroderma guani* Xiao et Wu to control pine sawyer beetles. *Journal of Nanjing Forestry University* 26: 48–52.

YANG, B. (1995). The pine wood nematode in China. In *International Symposium on Pine Wilt Disease Caused by Pine Wood Nematode, Beijing, China*, pp. 207–209.

Collected the larvae of PSB
PWN caused pine tree death rate
(Million)
Texan
Dongshan
(%)
10

# Effect of Aerial Spraying of Insecticide as a Control Measure for Pine Wilt Disease

**Shin Ugawa and Kenji Fukuda**

## Introduction

In Japan, two pine species (*Pinus densiflora* and *P. thunbergii*) have been heavily damaged by pine wilt disease (PWD). The pathogen is the pinewood nematode *Bursaphelenchus xylophilus* (Kiyohara and Tokushige, 1971). The nematode is pathogenic not only to Asian pine species but also to European species (Bedker et al., 1987, Riga et al., 1991). Unfortunately, the nematode was detected in Portugal in 1999 (Mota et al., 1999).

The nematode is primarily vectored by a cerambycid beetle *Monochamus alternatus* (Mamiya and Enda, 1972; Morimoto and Iwasaki, 1972) in Japan. To prevent PWN transmission by insect vectors, the government and local public authorities have conducted aerial sprayings with insecticide, since 1973. However, this control measure has been somewhat questionable, because the effect of the aerial spraying is not clarified enough in scale of a stand. This results from the difficulty to evaluate the damage from pine wilt disease.

A critical problem on the evaluation is that the damage from pine wilt disease changes along the years. This change is thought to be mainly caused by the synergistic effect of the number of emerging beetles together with the amount of damaged pine trees from the previous year (Togashi et al., 1992; Inada, 1993; Yoshimura et al., 1999; Takasu et al., 2000). To solve this problem, the application of the logistic model is needed (Inada, 1993).

In this study, to clarify the effect of aerial spraying of insecticide to PWD, we evaluated the damage from the disease by the approximation of the logistic formula, and compared the damage between the stands with and without the aerial spraying of insecticide. In addition, we surveyed the change of total basal area with the cumulative mortality.

S. Ugawa
Institute of Environmental Studies, Graduate School of Frontier Sciences, University of Tokyo, Kashiwanoha 5-1-5, Kashiwa-shi, Chiba 277-8563, Japan
e-mail: shin-u@nenv.k.u-tokyo.ac.jp

M.M. Mota, P. Vieira (eds.), *Pine Wilt Disease: A Worldwide Threat to Forest Ecosystems*,  

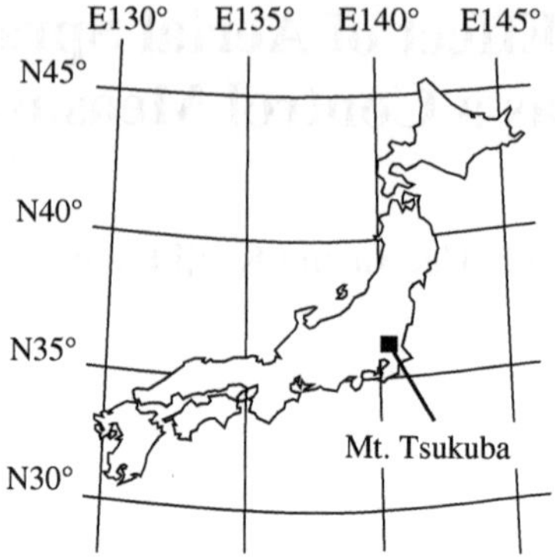


**Fig. 1** Location of study site. Mt. Tsukuba (36°13'31" N, 140°06'24" E; altitude, 877 m) is 60 km northeast of Tokyo

## Study Site

The study site was situated in artificial pine stands around Mt. Tsukuba (36° 13'31"N, 140° 06'24"E, altitude, 877 m) in Ibaraki prefecture, central Japan (Fig. 1). The stands are composed of *P. densiflora* and *P. thunbergii* below an altitude of 450 m.

## Methods

At the study site, eleven squares were established in the pine stands with different level of the damage in 1999 (Table 1). The size of each area was 30 m by 20 m. The squares were divided into three groups (P, C and M) by the years that the aerial spraying of insecticide (fenitrothion: $C_9H_{12}NO_5PS$, an organic phosphorus compound) was conducted. In group P, the aerial spraying of insecticide was continuous from 1996–2004, and named P1–5. In group C, the aerial spraying has never been conducted, and named C1–5. In group M, which was composed of only one square, the aerial spraying was conducted only in 1996 and 1997, and named M1. Other control methods have not been applied in these squares. In a preliminary survey in 1999, the wood of three dead pine individuals was sampled from each square, and the pine wood nematode was found in at least one dead individual of each square.

A tree census was conducted in all squares from September to February in each year between 1999–2004. For each tree >3.0 m tall, we recorded the species, and measured the diameter at breast height (DBH) to the nearest 0.5 cm. In addition, for each pine tree, we checked whether individual pines were live or dead through discoloration of tree crown.

The progress of pine wilt disease in a fixed area can be approximated by the logistic model (Togashi et al., 1992; Inada, 1993). Therefore, to evaluate the progress of the disease, we transformed the logistic formula into a formula where the parameters y, n, N, $n_0$ and r are the year, the number of dead pine individuals, the number of all pine individuals, the number of dead pine individuals in y=0, the potential progress speed of pine wilt disease, respectively.

**Table 1** Community characterisitcs and structure in each quadrat in 1999

| Quadrat | Group P | | | | | Group C | | | | | Group M |
|---|---|---|---|---|---|---|---|---|---|---|---|
| | P1 | P2 | P3 | P4 | P5 | C1 | C2 | C3 | C4 | C5 | M1 |
| Altitude (m) | 150 | 410 | 390 | 230 | 250 | 370 | 420 | 340 | 300 | 370 | 370 |
| Aspect | NE | N | NE | SE | N | SE | E | NW | E | W | NW |
| Inclination ($^o$) | 5 | 23 | 12 | 26 | 11 | 14 | 17 | 19 | 21 | 18 | 25 |
| Number of species | 23 | 24 | 22 | 13 | 15 | 11 | 13 | 20 | 20 | 14 | 19 |
| Number of stems per hectare | 2767 | 3400 | 2650 | 1717 | 1933 | 1800 | 1800 | 2967 | 3217 | 1850 | 3033 |
| BA per hectare ($m^2$) | 49.4 | 40.5 | 32.2 | 39.3 | 42.8 | 54.6 | 21.3 | 32.8 | 37.0 | 36.9 | 33.4 |
| Maximum DBH | 36.0 | 34.0 | 30.5 | 35.5 | 37.5 | 50.0 | 36.0 | 34.0 | 35.0 | 37.0 | 32.0 |
| Dominant species[1] | Pd | Pd | Pd<br>Qs<br>Pj | Pd | Pd | Pd<br>Co | Pd | Pd | Pd | Pd | Pd<br>Qs<br>Pt |
| Number of pine individuals[2] | 70 | 65 | 51 | 57 | 59 | 39 | 35 | 47 | 94 | 79 | 59 |
| Accumulative mortality[3] (%) | 21.4 | 36.9 | 37.3 | 21.1 | 23.7 | 25.6 | 40.0 | 19.1 | 40.4 | 43.0 | 32.2 |
| RBA of live trees | | | | | | | | | | | |
| *Pinus* spp.[3] | 0.89 | 0.69 | 0.52 | 0.82 | 0.86 | 0.78 | 0.76 | 0.83 | 0.75 | 0.78 | 0.60 |
| Other species | 0.11 | 0.31 | 0.48 | 0.18 | 0.14 | 0.22 | 0.24 | 0.17 | 0.25 | 0.22 | 0.40 |
| RBA of dead trees | | | | | | | | | | | |
| *Pinus* spp.[3] | 0.99 | 0.93 | 0.94 | 0.86 | 1.00 | 0.99 | 1.00 | 0.97 | 1.00 | 0.99 | 0.98 |
| Other species | 0.01 | 0.07 | 0.06 | 0.14 | – | 0.01 | 0.00 | 0.03 | – | 0.01 | 0.02 |

[1]Pd; *Pinus densiflora*, Qs; *Quercus serrata*, Pj; *Prunus jamasakura*, Co; *Chamaecyparis obtuse*, Pt; *Pinus thunbergii*.
[2]For both live and dead individuals in each quadrat (30 m × 20 m).
[3]For *Pinus densiflora* and *Pinus thubergii*.

$$[\text{Formula}]\frac{1}{n} - \frac{1}{N} = \left(\frac{1}{n0} - \frac{1}{N}\right) \exp\left(-r \cdot t\right)$$

The transformed logistic formula was approximated to the sequential cumulative mortalities in 1999–2004. The parameter r in the approximated formula represents the potential progress speed of pine wilt disease, and was considered as the "disease progress index (DPI)".

## Statistical Analysis

For the damage from pine wilt disease, annual mortalities in 1999–2004 and the DPI were calculated in each square. To clarify the effect of the aerial spraying of insecticide, annual mortalities were compared among all squares using ANOVA. Moreover, the average of the annual mortalities and the DPI were compared between square group P and C using t-tests.

For the change of pine population, total basal area (BA) was calculated in each square. The changes of the BA with the cumulative mortality were analyzed using Pearson's product-moment correlation.

## Results

In all quadrats, the cumulative mortality was 19.1–43.0% in 1999, and reached 33.3–94.3% in 2004 (Fig. 2). The annual mortalities in 1999–2004 were not significantly different among all quadrats (p=0.220, Fig. 3). The average value of the annual mortality in group P was significantly lower than that in group C (Fig. 4).

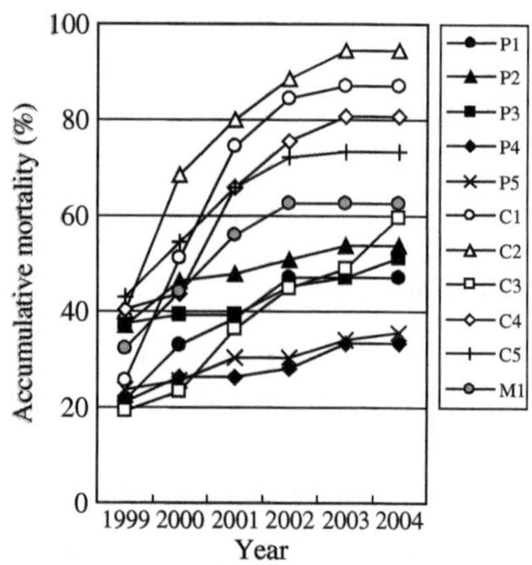


**Fig. 2** Cumulative mortality of *Pinus* spp. in 1999–2004 in each quadrat

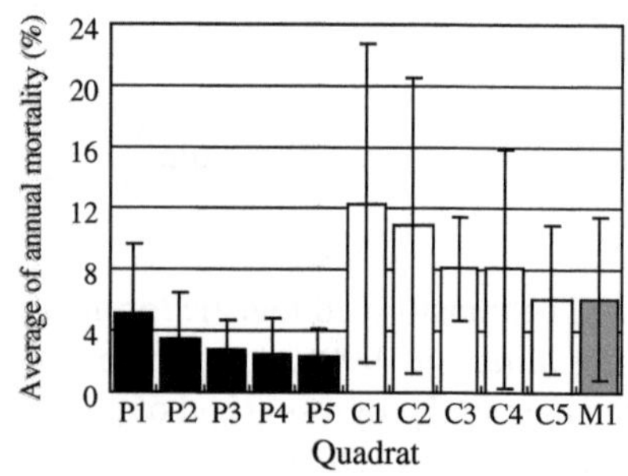


**Fig. 3** Average annual mortality of *Pinus* spp. in 1999–2004 in each quadrat. Bars indicate standard deviation. There is no difference among all quadrats with ANOVA (p = 0.220)

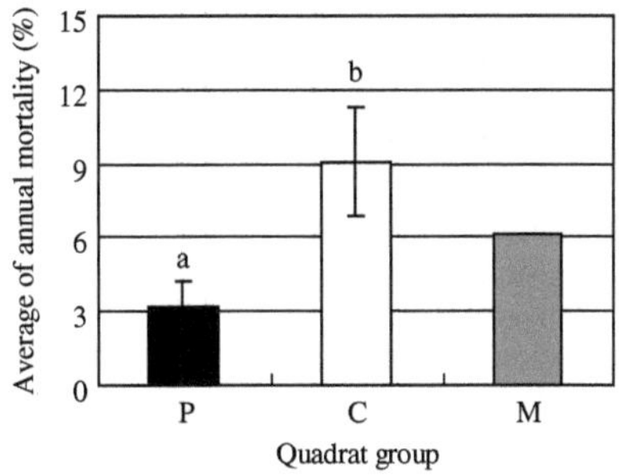


**Fig. 4** Average annual mortality of *Pinus* spp. in 1999–2004 in each quadrat group. Bars indicate standard deviation. Different letters indicate significant differences between quadral group P and C (t-test: $p < 0.001$)

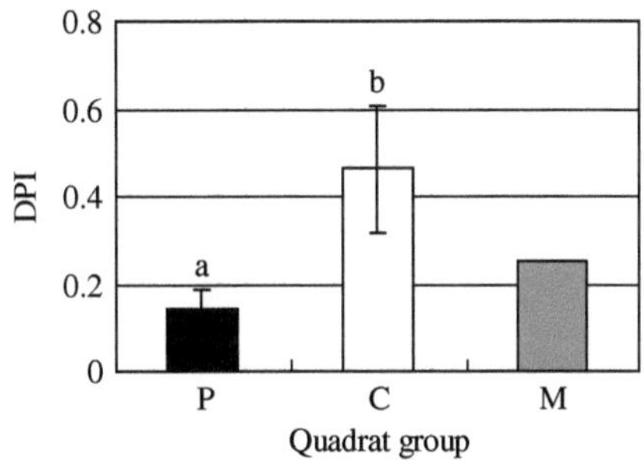


**Fig. 5** Disease progress index (DPI) in each quadrat group. Bars indicate standard deviation. Different letters indicate significant differences between quadrat group P and C (t-test: $p = 0.005$)

The average annual mortality in group M was higher than the maximum value in group P, and almost the same as the minimum value in group C.

The coefficient of determination varied from 0.82 to 0.98 in each quadrat for the approximation by the logistic formula. The DPI was 0.11–0.23, 0.27–0.65 and 0.25 in quadrat P1–5, C1–5, M1, respectively. The DPI in group P was significantly lower than that in group C (Fig. 5, $p=0.005$). The DPI in group M was between the maximum value in group P and the minimum value in group C.

In quadrats of group P, the BA sometimes increased, and the decrease with the cumulative mortality was not significant except the most damaged quadrat P1 (Fig. 6, $p=0.009$). However, in all quadrats of group C, the BA decreased significantly with the cumulative mortality ($p<0.001$). The correlation coefficient between the cumulative mortality and the BA decreased with the annual mortality (Fig. 7).

## Discussion

Even in the quadrat P1–5 with the aerial spraying of insecticide, the cumulative mortality increased with the year (Fig. 2). Thus, aerial spraying alone cannot stop damage from the disease. However, the average value of the annual mortalities in the stands with the aerial spraying of insecticide was significantly lower than that in the stands without the aerial spraying (Fig. 4). Thus, aerial spraying of insecticide seems to influence the damage from pine wilt disease, however, the annual mortality was not significantly different among the stands because of the high variance of the annual mortality in quadrat C1–5 (Fig. 3). This seems to result from the sequential

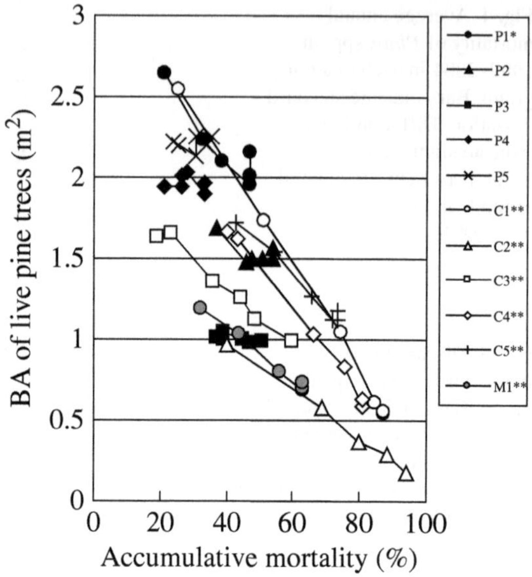


**Fig. 6** Total basal area (BA) of live *Pinus* spp. with cumulative mortality in each quadrat. Asterisks on quadrat's names indicate significant correlation between the BA and the cumulative mortality (Pearson's product-moment correlation: $^{*}p < 0.05$, $^{**}p < 0.01$)

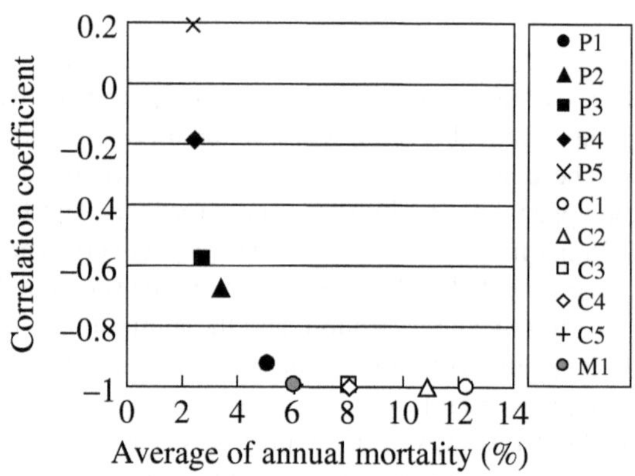


**Fig. 7** Correlation coefficient between total basal area (BA) of live *Pinus* spp. and cumulative mortality with average of annual mortality in each quadrat

change of the annual mortality like the logistic model (Fig. 2). Therefore, the concept of the approximation by the logistic formula is needed to evaluate the damage from pine wilt disease.

The coefficient of determination means that the value of DPI represents the sequential change of damage from pine wilt disease in all stands. The DPI in stands with the aerial spraying of insecticide was significantly lower than that in the stands without the aerial spraying (Fig. 5). Therefore, aerial spraying slows the progress of pine wilt disease.

In group M, the average annual mortalities and the DPI was between that in group P and C (Figs. 4 and 5). In southern Japan, the high annual mortality was reported one year after the stop of aerial spraying of insecticide (Kawabata and Kojo 1976). Thus, stopping aerial spraying may result in heavy damage.

The biomass of live pine trees decreases surely with the cumulative mortality in the stands with high annual mortality without aerial spraying of insecticide (Figs. 6 and 7). However, the biomass of live pine trees sometimes increases regardless of the cumulative mortality in the stands of the low annual mortality with the aerial spraying. Therefore, it is thought that the growth of pine individuals can compensate for the damage if the average of annual mortality was $<5\%$.

In this study, it is concluded that the aerial spraying of insecticide does not stop the damage from pine wilt disease perfectly, however, the control measure delays the progress of the disease, and allow the pine population to recover its biomass.

**Acknowledgement** We are indebted to the members of the Department of Natural Environmental Studies, Graduate School of Frontier Sciences, The University of Tokyo for help with this research. We thank Dr. K. Suzuki (Graduate School of Bioresource Sciences, Nihon University) for supporting this research. Permission to study this National Forest site was provided by the Forestry Technology Center, Kanto Regional Forest Office, Forestry Agency. This research was supported by the JSPS Grants-in-Aid No. 18380088 for Scientific Research and the ESPEC Foundation for Global Environment Research and Technology (Charitable Trust).

## References

BEDKER, P.J., WINGFIELD, M.J. AND BLANCHETTE, R.A. (1987). Pathogenicity of *Bursaphelenchus xylophilus* on three species of pine. *Canadian Journal of Forest Research* 17: 51–57.

INADA, M. (1993). The simulation model for the dynamics of wood volume damaged by the pine wilt disease. *Bulletin of the Faculty of Agriculture, Shimane University.* 27: 1–6.

KAWABATA, K. AND KOJO, M. (1976). The amount of damaged pine one year after the stop of aerial spraying of insecticide. *Transactions of the Annual Meeting of Kyushu Branch, the Japanese Forest Society.* 29: 181–182.

KISHI, Y. (1995). *The pinewood nematode and the Japanese pine sawyer.* Thomas Company Ltd., Tokyo, Japan, 302 pp.

KIYOHARA, T. AND TOKUSHIGE, Y. (1971). Inoculation experiments of a nematode, *Bursaphelenchus* sp. onto pine trees. *Journal of Japanese Forestry Society* 53: 210–218.

MAMIYA, Y. AND ENDA, N. (1972). Transmission of *Bursaphelenchus lignicolus* (Nematode: Aphelenchoididae) by *Monochamus alternatus* (Coleoptera: Cerambycidae). *Nematologica* 18: 159–162.

MORIMOTO K. AND IWASAKI A. (1972). Role of *Monochamus alternatus* (Coleoptera: Cerambycidae) as a vector of *Bursaphelenchus lignicolus* (Nematode: Aphelenchoididae). *Journal of Japanese Forestry Society* 54: 177–183.

MOTA, M.M., BRAASCH, H., BRAVO, M.A., PENAS, A.C., BURGERMEISTER, W., METGE, K. AND SOUSA, E. (1999). First report of *Bursaphelenchus xylophilus* in Portugal and in Europe. *Nematology* 1: 727–734.

RIGA, E., SUTHERLAND, J.R. AND WEBSTER, J.M. (1991). Pathogenicity of pinewood nematode isolates and hybrids to scots pine seedlings. *Nematologica* 37: 285–292.

TAKASU, F., YAMAMOTO, N., KAWASAKI, K., TOGASHI, K., KISHI, Y. AND SHIGESADA, N. (2000). Modeling the expansion of an introduced tree disease. *Biological Invasions* 2: 141–150.

TOGASHI, K., NAKAMURA, K. AND TAKAHASHI, F. (1992). An index of susceptibility of pine stands to pine wilt disease. *Applied Entomology and Zoology* 27: 341–347.

YOSHIMURA, A., KAWASAKI, K., TAKASU, F., TOGASHI, K., FUTAI, K. AND SHIGESADA, N. (1999). Modeling the spread of pine wilt disease caused by nematodes with pine sawyers as vector. *Ecology* 80: 1691–1702.

# Control Program of Pine Wilt Disease for Landscape Conservation – The Case of Amanohashidate, Kyoto, Japan

**Takefumi Ikeda**

## Introduction

Amanohashidate in Kyoto, Japan, has been given many titles, and is well known as one of the three most beautiful scenic spots in Japan (Fig. 1), *The Three Great Views of Japan*. The others are Matsushima and Miyajima. For a long time, Amanohashidate has been best known for its coastline of white sands and green pines (Fig. 2) on a sandbar spreading 3.2 km long by 40–170 m wide (Fig. 1). The splendid cluster of pine trees growing there has been the subject matter of books and theater over time. Amanohashidate is located in the innermost part of Miyadu-bay and is separated from land. This means that the pine community in Amanohashidate does not connect with the pine forests of surrounding mountainous areas that play an important role as background landscape (Fig. 1). Therefore, it is very important to preserve the entire area, including the surrounding mountainous area, from any damage to the pine forest. Japanese people highly appreciate such coastal sceneries since ancient times. Such a landscape is one of the typical Japanese sceneries. It is important that we leave the legacy of Amanohashidate in good condition for the next generation.

In 2001, increase in severe pine death occurred in Amanohashidate, and when compared to 2000. Consequently, various measures have been implemented to prevent the spreading of pine wilt disease. These measures have been successful, and are highlighted in this paper.

## Occurrence of Pine Death and Control Programs

There are about 5 000 pine trees of over 10 cm DBH at Amanohashidate (Table 1). The most common pines species are *Pinus thunbergii* and a few *Pinus densiflora*. *P. thunbergii* is growing along the coast. *P. densiflora* remains behind *P. thunbergii*

T. Ikeda
Department of Forest Science, Kyoto Prefectural University, Shimogamo-hangicho, Sakyo, Kyoto 606-8522, Japan
e-mail: tikeda@kpu.ac.jp

M.M. Mota, P. Vieira (eds.), *Pine Wilt Disease: A Worldwide Threat to Forest Ecosystems*, 

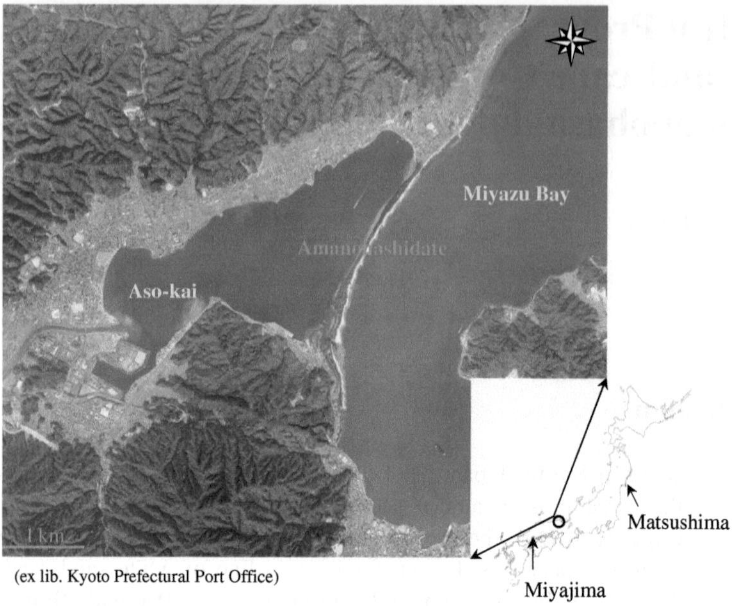


**Fig. 1** Aerial photo of Amanohashidate and its surrounding area

because the latter is more resistant to salt than *P. densiflora*. Many old pine trees are still presently growing. The oldest pine is about 600 years old and is described in the literature of the time. Twenty-one pine trees have pet names, which are appreciated by the people.

Recently, the number of dead pines caused by pine wilt disease has ranged from 10 to 30 per year (Fig. 3). The disease doubled in frequency after 2000, and 178 pine

**Fig. 2** Coastline of white sands and green pines

**Table 1** Number of trees at Amanohashidate

| Year | Pine[1] | Broad-leaved tree |
|---|---|---|
| 1934 | 3 954 | |
| 1954 | 4 551 | |
| 1974 | 4 720 | |
| 1979 | 4 715 | |
| 1988 | 5 144 | |
| 1997 | 5 208 | 1 169 |
| 2001 | 4 937 | 1 269 |

[1] >10 cm of dbh.

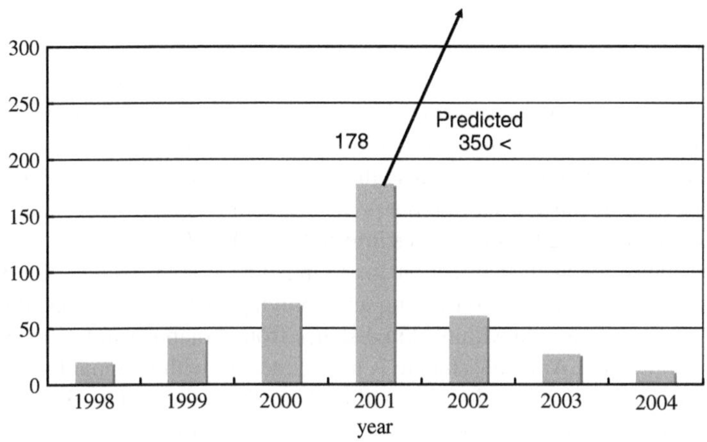


**Fig. 3** Number of pine death at Amanohashidate. Pine death more than 350 was predicted on 2002 if any control measures would not be conducted

trees died in 2001. We predicted that pine death in 2002 would exceed 350 trees. Nevertheless, the number of dead pines in 2001 is not very serious when compared with that occurring in other pine forests. However, as Amanohashidate is ranked as one of the most scenic places in Japan it is very important economically for tourism. Even one dead pine is not allowed.

In autumn of 2001, some experts conducted a field survey and identified that pine death occurring in Amanohashidate was caused by pine wilt disease. Consequently, several measures have been instigated to prevent pine wilt death from spreading (Kishi, 1995). Cooperation between administration, experimental and research organizations, as well as local residents, carried out a cooperative protection action of the pine forest at this area.

Before 2001, the control program has been conducted as follows: (1) felling dead pine trees, removing and spraying to clear sources of infection; (2) ground spraying (twice) of insecticides to prevent the insect vector, *Monochamus alternatus*, maturation feeding from the surrounding mountainous area during June. Judging from the beneficial effect of insecticidal spray, two times preventive ground spraying was considered to be incomplete to exterminate *M. alternatus*. There seems to be a blank period between sprayings. *M. alternatus* appears to come flying from the surrounding mountainous area in late May and July.

After 2002, the control measures were as follows: firstly, as precautionary measures (1) felling, removing and spraying of dead pine trees during autumn to winter, (2) three times preventive ground spraying with insecticides such as fenitrothion or thiacloprid during late May to late June. To compensate ground spraying, aerial spraying with insecticides using a sprinkler and remote-controlled small-scale helicopter (Fig. 4) was conducted. In July, ground spraying was not carried out although *M. alternatus* was flying in the area, because Amanohashidate is used as bathing resort during July and August, (3) 4 291 pine trees have been injected with nematicides such as morantel-tartarate, levamisol or emamectin to kill the pinewood nematode introduced by maturation feeding of *M. alternatus*. Approximately 700 pines trees were not injected with nematicide because the circulation of chemicals throughout the whole tree was not expected to yield results.

Secondarily, pine forests are distributed at the surrounding mountainous area and pine wilt damage has become serious in those areas. However, measures have not been taken at all, and it was thought that it functioned as an outbreak source of pine death at Amanohashidate. To suppress the damage in an affected pine forest, the following measures must be carried out precisely: removing the surrounding source of infection, eradicating the dead pines infested with *M. alternatus*, and conducting preventive measures against infection for living pine trees (Yoshida, 2006). This control program was focused on felling and spraying of the dead pines at the surrounding mountainous area within 2 km distance from Amanohashidate to eradicate a source of infection. Amanohashidate is not connected with the pine forests of the surrounding land area but is located within 2 km from there. The flight distance of *M. alternatus* is known to be at least 2 km per year (Yamamoto et al., 2000). Before 2001, several dead pines had been left in the location. With compliance of the forest owner, felling and spraying of the area has been almost completed until 2005.

**Fig. 4** Spraying with insecticide by remote-controlled small-scale helicopter

Thirdly, (1) educational activity on pine death caused by pine wilt disease to the local residents, because it is important to gain their understanding and support to put the control program into effect. Therefore we held some meetings to explain what pine wilt disease is, why pine trees die, what control measures are available, (2) current situation surveys such as vegetation and soil, fungi, and landscape of Amanohashidate.

Meanwhile, nematicide injection was conducted once because there is concern that it has negative effects on pine physiology resulting in debility of pine trees.

As a result of the control program, we were able to dramatically decrease the number of dead trees (Fig. 3), and prevent the damage from spreading. These measures have been successful. However, the damage of pine death by pine wilt disease has not been halted. Felling and spraying of dead pines in the future in surrounding forests will be continued, and changes of tree species composition will have to be considered. Furthermore, to predict a future major outbreak of pine death and to establish effective measures, the damage situation must be monitored at the most outer area within 2 km from Amanohashidate.

Has this mission been completed? We believe not. More importantly, problems still remain.

## Current Situation and Assignments of Pine Forests

To draw up integrated measures to protect the landscape of Amanohashidate, surveys of vegetation and soil, fungi, and landscape have also been carried out along with pine wilt disease control measures. The following situations have emerged: (1) increase in broad-leaved trees in pine forests (Table 1, Figs. 5 and 6), exuberance of herbaceous plants (Fig. 6) and fertilization of forest soil. There is concern that pine forests are gradually substituted for a broad-leaved forest with time because the climax vegetation of this area is the evergreen broad-leaved forest, (2) the root

**Fig. 5** Vegetations at Amanohashidate

**Fig. 6** Pine forest floors at Amanohashidate. A: typical pine forest floor on the beach. Very few herbaceous plants are there. B: pine forest floor where herbaceous plants grow thickly

**Fig. 7** Wind-fallen pine trees by typhoon on 2004

system of big trees have declined because of higher ground water level of forest land and soil dressing covering the original forest floor which has taken place in the past. Soil improvement conducted in the past encouraged the growth of the terrestrial part of pine trees inducing the increase in T/R ratio. This has increased the risk of windfall. In 1998 about 100 pines fell and broke as a result of snow damage and in 2004 about 200 pine trees fell down by a typhoon (Fig. 7), (3) the stand density is too high and is not so great for pine forest, (4) montage photo test showed that up to 30% decrease in pine trees did not have an influence on landscape.

## Future Works for Conservation of Pine Forests

Most parts of current Amanohashidate are in an unfavorable situation for growth of pine. Vegetation succession to conserve the landscape of Amanohashidate must be considered. Amanohashidate is located in a temperate zone and its vegetation is characterized by evergreen broad-leaved forest. Pine forest is not the climax vegetation but an intermediate one. Not all pines change to broad-leaved trees because pine trees at Amanohashidate are growing on the sandbar crossing into the sea and only *P. thunbergii* can grow along the coast. The landscape of Amanohashidate is characterized by pine tree. Therefore we have to regain the pure pine forest. In rich (fertile) soil conditions broad-leaved trees grow well in comparison with pine, which is a pioneer species and can grow even in poor soil condition. We must consider the vegetation succession besides control of pine wilt disease, that is, pine forest change to mixed forest of pine and broad-leaved trees and then broad-leaved forests with time.

To keep the pine forest healthy we must perform the following ecological measures to establish progress of succession to an artificial stand: (1) thinning of pine trees to improve the light condition in the forest, thus encouraging balanced growth of pine trees in part, (2) improving cutting of broad-leaved trees, (3) removing litter such as leaves and twigs, and humus layer of soil to make a poor soil, (4) replanting gaps of forest occurred by various reasons, (5) maintenance of root system. In addition, it is necessary to take care of old pines with pet name by a tree doctor.

## Proposals to Make Above Mentioned Measures Done in Time

(1) Sensitizing people to the present situation of Amanohashidate, (2) zoning of vegetation at Amanohashidate considering its history and culture, (3) making precise conservation measures on each vegetation zone and validation of its results, (4) making a selection old pine trees to preserve and apply the necessary measures, (5) development of management system for conservation of pine forest in cooperation with local residents.

## Conclusions

We can judge that the control measure of pine death by pine wilt disease is evaluated as one of several success examples in collaboration with various relevant organizations such as Kyoto Prefectural Government, Miyadu City Government, Forestry Agency of Japanese Government, Kyoto Prefecture Tree Doctor Association and local residents of Miyadu City. To preserve Amanohashidate in the future we must continue the various actions herein described, and it is also important to follow up the action in the next generation.

## References

KISHI, Y. (1995). *The pinewood nematode and the Japanese pine sawyer*. Thomas Company Ltd., Tokyo, Japan, 302 pp.

YAMAMOTO, N., TAKASU, H., KAWASAKI, H., TOGASHI, K., KISHI, Y. AND SHIGESADA, N. (2000). Local dynamics and global spread of pine wilt disease. *Japanese Journal of Ecology* 50: 269–276.

YOSHIDA, N. (2006). A strategy for controlling pine wilt disease and its application on-site. *Journal of the Japanese Forestry Society* 88: 422–428.

# Index

MIX
Papier aus verantwortungsvollen Quellen
Paper from responsible sources
FSC® C105338

If you have any concerns about our products,
you can contact us on
**ProductSafety@springernature.com**

In case Publisher is established outside the EU,
the EU authorized representative is:
**Springer Nature Customer Service Center GmbH**
**Europaplatz 3, 69115 Heidelberg, Germany**

Printed by Libri Plureos GmbH
in Hamburg, Germany